T0242047

Communications
in Computer and Information Science 540

More information about this series at http://www.springer.com/series/7899

Leon Wang · Shiro Uesugi
I-Hsien Ting · Koji Okuhara
Kai Wang (Eds.)

Multidisciplinary Social Networks Research

Second International Conference, MISNC 2015
Matsuyama, Japan, September 1–3, 2015
Proceedings

 Springer

Editors

Leon Wang
National University of Kaohsiung
Kaohsiung City
Taiwan

Koji Okuhara
Osaka University
Osaka
Japan

Shiro Uesugi
Matsuyama University
Matsuyama
Japan

Kai Wang
National University of Kaohsiung
Kaohsiung City
Taiwan

I-Hsien Ting
National University of Kaohsiung
Kaohsiung City
Taiwan

ISSN 1865-0929   ISSN 1865-0937 (electronic)
Communications in Computer and Information Science
ISBN 978-3-662-48318-3   ISBN 978-3-662-48319-0 (eBook)
DOI 10.1007/978-3-662-48319-0

Library of Congress Control Number: 2015947121

Springer Heidelberg New York Dordrecht London

Printed on acid-free paper

Springer-Verlag GmbH Berlin Heidelberg is part of Springer Science+Business Media
(www.springer.com)

Preface

Welcome to MISNC 2015 in Matsuyama, Japan. The conference, in its second year, is going global. The Organizing Committee envisions a world of synergy to combine cyber and physical networks, and starts visiting hometowns of MISNC participants every year, worldwide. This year, we have the pleasure to socialize under authentic Japanese culture and network with international researchers, professionals, experts, students from all over the world.

We are proud to report that there are participants from 15 different countries/areas including Japan, Taiwan, China, UK, USA, Greece, Denmark, Canada, Guatemala, India, South Korea, New Zealand, Saudi Arabia, Thailand, and Vietnam. We have also invited five keynote speakers from Korea, Taiwan, the UK, and Vietnam. There are also two joint workshops and one special session. The venue of the conference is the traditional setting of Yamatoya Honten, in *Do-Go Onsen*, one of the oldest *Onsen* Spa resorts in Shikoku, Japan.

The success of MISNC 2015 depends on the people and organizations that support it. We must thank all the volunteers who helped organize this conference. In particular, we thank the Honorary Chair, Professor Jiawei Han of the University of Illinois at Urbana-Champaign, USA, for his kind support. We thank Program Chairs Professor Shiro Uesugi of Matsuyama Junior College, Japan; Professor I-Hsien Ting of the National University of Kaohsiung, Taiwan; Professor Koji Okuhara of Osaka University, Japan; and Professor Hiroki Idota, Kindai University, Japan, who, together with the Program Committee, created a great technical program. In addition, we would like to thank Publicity Chairs Yu-Hui Tao, Chung-Hong Lee, Takashi Okamoto, and Been-Chian Chien; Financial Chairs Kai Wang, and Shiro Uesugi; Publications Chairs Hsin-Chang Yang, and Hidenobu Sai; Steering Committee Chairs Christine Largeon, Johann Stan, Tzung-Pei Hong, Vincent Tseng, Kai Wang, and Yuua Dan; Track Chairs Jerry Chun-Wei Lin, Yoko Orito, Takashi Okamoto, Hidenobu Sai, Eizen Kimura, and Yuya Dan; and all the Program Committee members.

The conference would not be possible without partners and sponsors. They include IEEE Technical Committee of Granular Computing; IEEE Computational Intelligence Society; IEEE Society of Social Implications of Technology Japan Chapter; Matsuyama University's International Academic Research Collaboration Funds; Hokyo Inc. Japan; KDDI Research Institute; Yamatoya Honten, Japan The Matsuyama Convention and Visitors Bureau; Ehime Prefectural International Center; Matsuyama City and Ehime Prefecture, Japan; National University of Kaohsiung and its Social Network Innovation Center (SNIC); Ministry of Science and Technology of ROC; Taiwanese Association for Social Networks (TASN); and Harbin Institute of Technology, China.

Last and most importantly, we thank all of you, the authors and attendees for participating in MISNC 2015, sharing the knowledge and experience, and contributing to the advancement of science and technology for the improvement of the quality of our lives. We wish each and every one a most pleasant experience at MISNC 2015. We also look forward to seeing you in MISNC 2016 in New York City, USA, and MISNC 2017 in Bangkok, Thailand.

July 2015   Leon Wang
                                          On Behalf of the Organizing Team

Organization of MISNC 2015

Honorary Chair

Jiawei Han   University of Illinois at Urbana-Champaign, USA

General Chair

Leon Shyue-Liang Wang   National University of Kaohsiung, President of TASN,
                                    Taiwan

Program Chairs

Shiro Uesugi   Matsuyama University, Japan
I-Hsien Ting   National University of Kaohsiung, Taiwan
Koji Okuhara   Osaka University, Japan
Hiroki Idota   Kindai University, Japan

Publicity Chairs

Yu-Hui Tao   National University of Kaohsiung, Taiwan
Chung-Hong Lee   National Kaohsiung University of Applied Sciences,
                                    Taiwan
Takashi Okamoto   Ehime University, Japan
Been-Chian Chien   National University of Tainan, Taiwan

Financial Chairs

Kai Wang   National University of Kaohsiung, Taiwan
Shiro Uesugi   Matsuyama University, Japan

Publication Chairs

Hsin-Chang Yang   National University of Kaohsiung, Taiwan
Hidenobu Sai   Ehime University, Japan

Steering Committee Chairs

Christine Largeon   University of Jean-Monnet at St. Etienne, France
Johann Stan   NIH, USA
Tzung-Pei Hong   National University of Kaohsiung, Taiwan
Vincent Tseng   National Chiao-Tung University, Taiwan

Kai Wang	National University of Kaohsiung, Taiwan
Yuya Dan	Matsuyama University, Japan

Track Chairs

Track of Information Technology and Social Networks Mining

Jerry Chun-Wei Lin	Harbin Institute of Technology Shenzhen Graduate School, China

Track of Ethics Issues in Social Networks

Yoko Orito	Ehime University, Japan

Track of Electronic Commerce and Social Networks

Takashi Okamoto	Ehime University, Japan

Track of Business Strategy and Social Networks

Hidenobu Sai	Ehime University, Japan

Track of Medical Applications in Social Networks

Eizen Kimura	Ehime University, Japan

Track of Mathematical Approach in Social Networks

Yuya Dan	Matsuyama University, Japan

Program Committee Members

Ajith Abraham	Machine Intelligence Research Labs (MIR Labs), USA
Nada Marie Anid	New York Institute of Technology, USA
Guido Barbian	Leuphana University Lueneburg, Germany
Marenglen Biba	University of New York in Tirana, USA
Babiga Birregah	University of Technology of Troyes, France
Dan Braha	NECSI, USA
Piotr Bródka	Wroclaw University of Technology, Poland
Chien-Chung Chan	University of Akron, USA
Bao-Rong Chang	National University of Kaohsiung, Taiwan
George Chang	Kean University, USA
Steven Chang	Long Island University, USA
Wei-Lun Chang	Tamkang University, Taiwan
Goutam Chakraborty	Iwate Prefectural University, Japan
Richard Chbeir	CNRS, France
Frank Hsu	Fordham University, USA
Chian-Hsueng Chao	National University of Kaohsiung, Taiwan
Shin-Horng Chen	Chung-Hua Institution for Economic Research, Taiwan
Yunwei Chen	Chengdu Library of Chinese Academy Science, China

Been-Chian Chien	National University of Tainan, Taiwan
João Cordeiro	University of Saint-Joseph, Macau
Shing-Huang Doong	Shu-Te University, Taiwan
Schahram Dustdar	TU Wien, Austria
Michael Farrugia	Planitas, Ireland
Mathilde Forestier	IMS-Bordeaux, France
Terrill Frantz	Peking University, China
Paolo Garza	Politecnico di Torino, Italy
Tyrone Grandison	Proficiency Labs, USA
William Grosky	University of Michigan, USA
Wu He	Old Dominion University, USA
David Hicks	Aalborg University, Denmark
Huu Hanh Hoang	Hue University, Vietnam
Tzung-Pei Hong	National University of Kaohsiung, Taiwan
Han-Fen Hu	University of Nevada, Las Vegas, USA
Pin-Rui Hwang	National United University, Taiwan
Eizen Kimura	Ehime University, Japan
Edward Roy Krishnan	Bangkok School of Management, Thailand
Adam Krzyzak	Concordia University, Canada
Christine Largeon	University of Jean-Monnet at St. Etienne, France
Gang Li	Deakin University, Australia
Paoling Liao	National Kaohsiung Marine University, Taiwan
Jerry Chun-Wei Lin	Harbin Institute of Technology, China
Wen-Yang Lin	National University of Kaohsiung, Taiwan
Lin Liu	Kent State University, USA
Wei-Chung Liu	Academia Sinica, Taiwan
Xumin Liu	Rochester Institute of Technology, USA
Chung-Hong Lee	National Kaohsiung University of Applied Sciences, Taiwan
Frank Lee	New York Institute of Technology, USA
SanKeun Lee	Korea University, South Korea
Jason Jung	Chung-Ang University, South Korea
Georgios Lappas	Technological Educational Institute of Western Macedonia, Greece
Massimo Marchiori	UNIPD and UTILABS, Italy
Nasrullah Memon	Southern Denmark University, Denmark
Kazuaki Naruse	Matsuyama University, Japan
Federico Neri	Synthema, Italy
Hitoshi Okada	National Institute of Informatics, Japan
Takashi Okamoto	Ehime University, Japan
Koji Okuhara	Osaka University, Japan
Kok-Leong Ong	Deakin University, Australia
Yohko Orito	Ehime University, Japan
Fatih Ozgul	National Police Academy, Turkey
Javier Bajo Perez	Salamanca University, Spain
Laura Pullum	Oak Ridge National Laboratory, USA

Contents

International Workshop on Ethical Issues Related to SNS

Special Session on Information Technology and Social Networks Mining

MISNC 2015 Papers

Switching Motivations on Instant Messaging: A Study Based on Two Factor Theory

Avus C.Y. Hou[(✉)]

Oriental Institute of Technology, Taipei 22061, Taiwan
avushou@gmail.com

Abstract. Instant messaging, like Skype, is facing a challenge from mobile instant messaging in recent year and some instant messaging might to close their service in the near future. Therefore, understanding user's motives to switch to mobile instant messaging has great influence on operators' business performance, as they could adopt appropriate strategies to retain the users. The study attempts to explore the switching factors and to examine the relationships between those and user's intentions to switching to mobile instant messaging, based on the two-factor theoretical perspective. Structural equation modeling was applied to analyse data collected from 186 users by empirical survey. According to our finding, the study suggests that operators should satisfy users' pursuit of freshness, and carefully examine whether their services could address users' needs of sociality and entertainment.

Keywords: Instant messaging · Mobile instant messaging · Service switching · Two-factor theory · Structural equation modeling

1   Introduction

Mobile instant messaging (mIM) is a way of real-time communication and social interaction. It costs less, is less disturbing to other receivers, and is favoured by modern consumers who attach great importance on privacy. Therefore, mIM services (e.g. Line) are gradually taking the place of some forms of voice communication. Line, a mIM App that attained to overnight success in 2013, is a telling example of this trend.

There are many mIM products in the market, such as mobile Skype, WeChart, and Line. These mIM products have similar functions such as text chat and emotion icons, and there exists intense competition among them. Thus a challenge facing mobile service providers is to retain their customers. On the other hand, the switching cost is low. Users can easily switch from a mIM platform to another. As the traditional instant messaging social is rife with heavy competition, it is important for service operators to retain their current customers as well as to attract new ones. However, existing studies have focused on the user's intention to participate [1,2,3]. Few studies have shed light on users' switching behavior toward mIM. User turnover affects the success of a messaging service and is therefore crucial in understanding the factors affecting users' intention to switch messaging services. For service operators,

© Springer-Verlag Berlin Heidelberg 2015
L. Wang et al. (Eds): MISNC 2015, CCIS 540, pp. 3–15, 2015.
DOI: 10.1007/978-3-662-48319-0_1

understanding the behavioral intention of mIM users can help them design features for a target groups of particular users and sustain their business.

The key objective of this study is to examine the factors that drive users to switch instant messaging services, especially from traditional IM to mIM. In this study, we propose a switching framework which is adapted from the two-factor theory to demonstrate users' intention. The theory has also been applied to online service migration. Park & Ryoo [4] did the research claiming that customers who switch clouding computing service was not only positively influenced by motivators, but also negatively influenced by hygiene factors.

2   Theoretical Background

2.1   Service Switching

Customer switching is one of the central concepts in relationship marketing field, and it refers to the people migration between service providers or firms. There are three major stream of research on understanding customers' switching behavioral intentions: 1. using process models for customer service switching [5]; 2. the heterogeneous characteristics between stayers and switchers [6]; 3. the factors that drive customers to switch. The last stream has been attracting the most attention among researchers [4],[7,8,9,10,11]. Prior research on customer switching covers various areas, such as traditional service. Keaveney [9] has examined reasons for customer service switching by critical incidents method and categorized as eight major causes of service switching, some could be associated with feelings of dissatisfaction with the service (e.g., core service failure, failed service encounters), but others were extrinsic or situational factors (e.g., price, inconvenience, ethics) [6].

Compared with relationship marketing in offline environment, research in online environment has just recently been attracting increasing attention. For instance, Kim et al. [10] attempted to understand the association between customer satisfactions for email service switching. Zhang et al. [12] investigated the role of gender in bloggers' switching behaviors. Park and Ryoo [4] attempted to explore the switching factors in cloud computing service. However, due to distinction between online and off-line contexts, customer satisfaction with an online service was measured in multiple dimensions including: information quality; system quality; and service quality of information technology department [13].

In addition the element of service satisfaction, of customers' switching intention also are influenced by psychological and non-psychological barriers. Such barriers may be largely associated with the availability of advantage of alternative. Several studies indicated that advantage of alternative have direct influence on customers' service switching decision but also play as an critical role [8],[12],[14].

The IM service offered by operators is different from other paid services because IM service offer with free. Operators do not generate revenue directly from IM services but to expand network externality and to boost customer loyalty, which in turn translate into larger users and generate revenue by other business activities such as advertisement. Besides, there are also elements that distinguish the IM service from other online services, which may significantly affect customer satisfaction for an IM service, including privacy policy, the way that the providers handle subscriber information, and free sticker cartoons. With its uniqueness as an online service, better understanding on the IM service switching is necessary.

2.2   Two Factor Theory

Two-factor theory of technology usage which is adapted from Herzberg's two-factor theory of motivation [15,16] has emerged recently in IS research [17,18]. One of the earliest applications of two-factor theory is to explain several factors that lead to employee satisfaction and dissatisfaction [16]. After several decades, two-factor theory has been applied to other research context, such as cloud service switching [4].

Two-factor theory characterizes factors affecting user's intentions as both enablers (motivators) and inhibitors (hygiene factors). In the online service switching context, enablers refer to the factors that motivate users to adopt a service, and inhibitors refer to the factors that inhibit users from using it [8]. Each of the two factors is not opposite to the other; independently, each of them has effects on individual intents as dual-factor constructs; that is, an individual can hold perceptions of both inhibitors and enablers simultaneously [4]. Many objects or behaviors can be simultaneously evaluated both favorably and unfavorably [19]. Furthermore, inhibitors can hinder intentions despite the presence of enablers facilitating those same intentions. Cenfetelli and Schwarz [17] posited inhibitors are not the polar opposites of enablers. They proposed that inhibitors are factors that discourage adoption but whose absence does not necessarily encourage adoption.

Switching behavior to mIM services is a decision to reject old technology (IM) and/or service and adopt new technology (mIM) and/or service simultaneously. Although extant theory, for instance, TAM helps us make solid progress in understanding adoption, their focus has been primarily on adoption enablers [4]. Thus, to better understand the switching behavior to mIM services, the two factor theory explaining the dual roles of enablers and inhibitors should be relevant to the present study.

3   Research Hypotheses

Figure 1 shows the research model that the study adopts. Hygiene factor consists of three: sociality, entertainment, and mIM system quality. Motivator consists of three: advantage of alternative, peer influence, and critical mass. This section will explain our chose of these factors and their positions in the switching model.

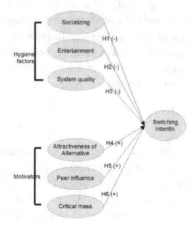

Fig. 1. The research model

3.1   Hygiene Factors

Due to the novelty nature of the mIM context, only a few prior studies have identified which predictors might be effective in predicting users' "defection". Thus the study adopts the "hygiene factors" from the marketing discipline. mIM could be seen as an information system that helps users to contact with others and maintain social relationship. In analysing users' satisfaction towards information system, Delone and McLean [20] found that it is resulted from two major antecedents: information quality and system quality. Based on this analogy, this study suggests that IM that fail to satisfy users' expectation in these three aspects will push users away.

Information quality is assessed by evaluating the output of an information system has to be relevant and correct when it is delivered to the user [20]. Validity and relevance of the output should be guarantee. People's main purposes for using IM are fulfilling their needs of sociality and entertainment [21]. System quality, which refers to the evaluation of the system itself, is assessed based on its reliability and perceived ease of use [20]. The abovementioned ideas about information and system quality of IMs will be discussed in an IM context below.

Information quality (sociality and entertainment)
Sociality
Communication with peers and sociality is considered to be one of the most important concepts in IM usage [21,22]. In Barker's [23] study, he indicated the four aspects of sociality: to get support from others, to meet interesting people, to converse with others, and to stay in touch with friends. Therefore, the "communication with peers" should be including in sociality. Base on such studies, the IM must at least provide the function for users to keep and maintain their social ties. Users have better tendency to continue a specific IM service that is able to maintain or enhance users' current social ties. Conversely, users tend to discontinue an IM service that is not capable of satisfying their sociality urges:

H1: Users' perceived sociality with the current IM service is negatively associated with the intention to switch to another IM service.

Entertainment

In order to attract and retain users in an IM for longer online and active duration, IM operators offer many add-on applications. Many interesting add-on applications are provided by an IM service, such as social games and emotional stickers. The most attractive of all is the emotional stickers, for example, the "LINE" stickers. Joining the LINE, users can not only text but also deliver their feeling with emotional stickers as an interesting interaction. The emotion stickers created a new age in the IM history; users now utilize the IM platform for usual communication and for entertainment as well. Zhou & Lu [22] observed this phenomenon and indicated that entertainment was one of significant factors for users to participate in IM. Hence, the factor of entertainment is considered to be a strong element for users to continue a specific IM service:

H2: Users' perceived entertainment with the current IM service is negatively associated with the intention to switch to another IM service.

System quality

System quality refers to the accountability, flexibility and ease of use of the system [24]. When it comes to system quality of service used for IM, access speed, ease of use, and security are valued the most by users [21]. System quality of mIMs, rooted in Apps, is evaluated by its reaction speed: how fast the system could access photos or data, run additional applications, or maintain the flow of instant text. If the system failed to provide a smooth flow of information on these aspects, then users would consider leaving the system due to poor efficiency:

H3: Users' perceived system quality of the current IM is negatively associated with the intention to switch to another IM.

3.2 Motivators

Advantage of alternative

Competitor that provides quality customer service can induce consumers to switch to the new provider [7],[9],[25]. IM, which provide communication services mediated by Internet technology, are also within the scope of the service industry. Therefore, the aforementioned strategies for improving service quality can also be applied to online services when enticing new customers to switch services. The core service of IM lie in the sociality and entertainment it brings [26,27]. A new IM that is more exciting than an existing IM and provides better customer service will attract users to switch to it. Accordingly, this study presents the motivator that creates switching behavior in terms of advantage of alternative.

H4: Users' perceived advantage of alternative IM service is positively associated with the intention to switch to another IM.

Peer influence

Peer influence has been considered as a crucial predictor to affect behaviors adopting a new technology [28]. Recently, study has paid attentions to users' peer influence regarding switching behavior between online services [8],[29]. As most people use IMs to maintain or search new relationships with friends, an alternative IM seems more attractive because of their friends' preferences. The influence of peers can work through creating "opportunities for new activities, environment, or people" for an IM user. When one receives invitations by a large number of friends, he/she would be likely to switch to the alternative.

H5: Users' perceived peer influence resulted from alternative IM service is positively associated with the intention to switch to another IM.

Critical Mass

In many Internet-based services, network externality have a significant impact on the evolution of the market. This means that the technology's value for a user increases as the number of people using this specific technology grows [30]. In particular, critical mass is a fundamental element due to its relevance to the growth process of these kinds of services [31]. Since a IM can be classified as an Internet-based service, it thus can be considered to have network externality. Prior studies have shown that having a sufficient number of participants in a IM positively influences the entertainment value perceived by the users [26]. The larger the growth in the number of users, the greater entertainment they would experience with the service. Hence, we hypothesized as below:

H6: Users' perceived critical mass of alternative IM service is positively associated with the intention to switch to another IM.

4   Research Method

4.1   Instrument Development

According to the hypotheses presented in last section, a survey instrument was developed based on prior research. Questionnaire items were modified slightly to fit our specific research context. There are 7 concepts with a total of 24 items. All the questionnaire items are listed in Appendix A. Except for the two sections of the questionnaire used to collect users' experience and demographic backgrounds, all items were measured using either 7-point Likert scale or 7-point semantic differential scale.

The scale used for sociality was referring to Park's [32] study which considers users immersed in IM usage. Entertainment was adopted from Ghani and Deshpande's enjoyment scale [33]. System quality was measured by Delone & McLean's scale [24]. The motivators were measured with advantage of alternative, peer influence, and critical mass. The advantage of alternative was measured using the scale described by Bansal et al. [7]. Peer influence was adopted from Hung et al. [28] scales with 4 items. The critical mass was measured using the three-item scale proposed by Hsu and Lu [34]. Finally, switching intentions was measured using Park & Ryoo's [4] method.

4.2   Data Collection

Given that our study focused on examining why users tend to switch IM service, it was important to be able to probe perceptions of those who had prior experience on using IM (Yahoo messaging). Thus we adopted a survey approach in order to test our research model. We placed messages on the most popular online survey sites in Taiwan, Myservey.com (www.mysurvey.com) to recruit volunteers who have had experience using Yahoo messaging to join this survey. The data was collected over a span of two months between November 2014 and December 2014.

4.3   Survey Respondents

The survey yielded 217 responses. After removing those with unanswered items, we ended up with 186 usable responses. Approximately 54% of the respondents were female. The majority of respondents were between 21 and 35 years old (61%), and 19% were 20 years old or less. Approximately 33% of the respondents were students. This sample represents a population that averagely spends 5 minutes every time when checking IM, and checks it eight times a day.

5   Results

The measurement model was first examined and then the structural models were assessed [35]. First, test for Conbach' alpha, convergent, and discriminant validity were made to examine the reliability and construct validity of instrument. Second, the theoretical model was tested by using Partial Least Squares (PLS), a latent structural equations modeling technique with a component-based approach to estimation. Because of this, minimal demands were made on sample size and residual distributions [36].

5.1   Measurement Model

All scales were assessed in terms of reliability, convergent validity, and discriminant validity. We used Cronbach's alpha to verify the reliability of the survey instrument. According to Bagozzi & Yi's suggestions [37], all of the constructs obtained a Cronbach's alpha value greater than 0.6 and composite reliability score greater than 0.7, indicating the internal consistency of the measurement. Descriptive statistics for the constructs are given in Table 1.

Table 1. Descriptive statistics and scales reliability

Construct	Mean	S.D.*	Alpha	Items factor loading	C.R.	AVE
Switching Intentions	4.44	1.44	0.90	0.85; 0.87; 0.88; 0.83; 0.86	0.93	0.73
Sociality	5.65	1.35	0.74	0.86; 0.13; -0.08	0.27	0.25
Enjoyment	4.25	1.92	0.85	0.71; 0.89; 0.79; 0.90	0.89	0.68
System Quality	4.95	1.70	0.85	0.85; 0.79; 0.85; 0.79	0.89	0.67
Advantage of alternative	2.70	1.37	0.85	0.90; 0.89; 0.86	0.91	0.78
Peer Influence	4.08	1.30	0.72	0.91; 0.86	0.88	0.78
Critical Mass	4.21	1.62	0.71	0.83; 0.87; 0.68	0.84	0.63

*S.D.: Standard division; Alpha: Cronbach's Alpha * C.R.: Composite Reliability; AVE: Average variance extracted

Convergent validity can be assessed against two standards: (1) the items coefficient should be greater than 0.7; and (2) the average variance extracted (AVE) should be greater than 0.5 [38]. As seen from the results in Table 1, all items exhibit high path coefficient (>.70) on their respective constructs. All item coefficients were greater than 0.7 and all AVE were larger than 0.5, confirming the convergent validity is sufficient.

Discriminant validity provides in highlights Table 2. According to the square root of AVE assessment of discriminant validity, all constructs share more variance with their indicators than with the other constructs, which reinforces the discriminant validity of our model [39].

Table 2. Correlation Matrix of Constructs

	SI	SO	EN	SQ	AA	PI	CM
SI	**0.85***						
SO	-0.14	**0.89**					
EN	-0.23	-0.12	**0.82**				
SQ	-0.06	0.03	-0.33	**0.82**			
AA	-0.15	-0.13	0.46	-0.15	**0.81**		
PI	0.31	0.04	0.05	-0.09	-0.04	**0.88**	
CM	0.26	0.12	-0.01	0.17	-0.04	-0.36	**0.86**

*Diagonals represent the square root of AVE, while the other matrix entries indicate the correlations.
SI: Switching Intention; SO: SOcializing; EN: ENtertainment; SQ: System Quality; AA: Advantage of alternative; PI: Peer Influence; CM: Critical Mass

5.2   Structural Model

The SmartPLS 3.0 package was used to test the hypotheses of our research model [40]. Figure 2 shows the result of the PLS analysis. The predictive validity was evaluated by examining the R square and the path coefficient. The results indicate that our research model explains a 41 percent variance, which is satisfactory [35]. It can be seen that all hypotheses, except H4, are supported. Hygiene factors show a significant negative influence on switching intentions while motivators do the opposite.

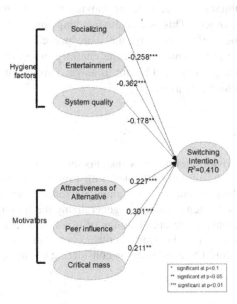

Fig. 2. Results

6   Conclusion and Discussion

The study utilizes two-factor theory to investigate instant messaging users' switching intentions. Empirical results show all factors, except consumer service satisfaction, were all able to trigger switching intentions. The discovery sides with the view of Park & Ryoo [4] that two-factor theory could serve as the theory basis of online service switching research, a field that is critical in recently [14].

The study hypothesized sociality, entertainment, and system quality of IM, as factors constituting hygiene factors. The finding shows that sociality, entertainment, and system quality are negatively influential to switching intentions, meaning that whether users could be entertained by IM function or could keep in touch with friends through IM is crucial in deciding users' retention. Therefore fulfilling the needs in entertaining and sociality becomes two important issues for IM or mIM designers. Entertainment can be generated by interaction with other users in communications, between users and applications, and through social games and more free emotion stickers. When this need is not fully addressed, naturally users would like to switch to another service provider.

The measure of "advantage of alternative" is conducted by assuming messaging service that claims higher sociality and higher entertainment. These measures are originated from existing literature on service switching [41]. Based on these perspectives, it is suggested that operators could formulate stronger forces to attract users to leave their incumbent by novelty of messaging service. Therefore it is suggested that existing IM operators can fulfil users' need for newness by providing new content and more functions.

When some inhabitants in a service are moving out, naturally it triggers the tendency to leave from others. This is a mutual driving force. Similarly, in the virtual world, because of network externality, the more participants a messaging has, the more users it will be able to attract. On the contrary, when there are fewer participants, even the existing users would want to leave. This factor is also significant in influencing the switching intentions.

7   Limitation

Though this study provides several new insights into the service process of IM users on switching, our results should be treated with caution for several reasons. We focuses on the switching intentions of Yahoo messenger users, therefore a second thought may be needed when considering generalizing the finding. Another limitation is this study sample only represents feelings of users of a specific region, therefore attempts to apply the result of this study in other contexts should be extra cautious. In sum, future research could be conducted by considering all the above discussions.

References

1. Premkumar, G., Rammurthy, K., Liu, H.: Internet Mes- saging: An Examination of the Impact of Attitudinal, Normative and Control Belief Systems. Information & Management **45**, 451–457 (2008)
2. Li, H., Gupta, A., Luo, X., Warkentin, M.: Exploring the impact of instant messaging on subjective task complexity and user satisfaction. European Journal of Information Systems **20**, 139–155 (2011)
3. Lien, C.H., Cao, Y.: Examining WeChat users' motivations, trust, attitudes, and positive-word-of-mouth: Evidence from China **41**, 104–111 (2014)
4. Park, S.C., Ryoo, S.Y.: An empirical investigation of end-users' switching toward cloud computing: A two factor theory perspective. Computers in Human Behavior **29**(1), 160–170 (2013)
5. Roos, I.: Switching processes in customer relationships. Journal of Service Research **2**(1), 68–85 (1999)
6. Keaveney, S.M., Parthasarathy, M.: Customer switching behavior in online services: An exploratory study of the role of selected attitudinal, behavioral, and demographic factors. Journal of the Academy of Marketing Science **29**(4), 374–390 (2001)
7. Bansal, H.S., Taylor, S.F., James, Y.S.: 'Migrating' to new service providers: toward a unifying framework of consumers' switching behaviors. Journal of the Academy of Marketing Science **33**(1), 96–115 (2005)
8. Hou, A., Chern, C.C., Chen, H.G., Chen, Y.C.: 'Migrating to new virtual world': exploring MMORPG switching through demographic migration theory. Computers in Human Behavior **25**(2), 578–586 (2011)
9. Keaveney, S.M.: Customer switching behavior in service industries: an exploratory study. Journal of Marketing **59**(2), 71–82 (1995)
10. Kim, G., Shin, B., Lee, H.G.: A study of factors that affect user intentions toward email service switching. Information & Management **43**(7), 884–893 (2006)

11. Oliver, R.L., Swan, J.E.: Consumer perceptions of interpersonal equity and satisfaction in transactions: a field survey approach. Journal of Marketing **53**(2), 21–35 (1989)
12. Zhang, Z.K., Lee, K.O., Cheung, M.K., Chen, H.: Understanding the role of gender in bloggers' switching behaviour. Decision Support System **47**(4), 540–546 (2009)
13. Hou, A., Wu, K-L., Huang, C.C.: The effect of push-pull-mooring on the switching model for social network sites migration. In: Proceedings of The 18th Pacific Asia Conference in Information System (PACIS 2014), Sichuan, China, June 24–28, 2014
14. Chiu, H., Hsieh, Y., Roan, J., Tseng, K.-J., Hsieh, J.: The challenge for multichannel services: Cross-channel free-riding behavior. Electronic Commerce Research and Applications **10**(2), 268–277 (2011)
15. Herzberg, F.: One more time: How do you motivate employees? Harvard Business Review **65**(5), 109–120 (1987)
16. Herzberg, et al.: The motivation to work. Wiley, New York (1959)
17. Cenfetelli, R.T., Schwarz, A.: Identifying and Testing the Inhibitors of Technology Usage Intentions. Information Systems Research **22**(4), 808–823 (2008)
18. Liu, C.-T., Guo, Y.M., Lee, C.-H.: The effects of relationship quality and switching barriers on customer loyalty. International Journal of Information Management **31**(1), 71–79 (2011)
19. Petty, R.E., Cacioppo, J.T., Goldman, R.: Personal involvement as a determinant of argument-based persuasion. Journal of Personality and Social Psychology **41**(5), 847–855 (1981)
20. DeLone, W., McLean, E.: The DeLone and McLean Model of Information Systems Success: A Ten-Year Update. Journal of Management Information Systems **19**(4), 9–30 (2003)
21. Ogara, S., Koh, C., Prybutok, V.: Investigating factors affecting social presence and user satisfaction with Mobile Instant Messaging. Computers in Human Behavior **36**(2), 453–459 (2014)
22. Zhou, T., Lu, Y.: Examining mobile instant messaging user loyalty from the perspectives of network externalities and flow experience. Computers in Human Behavior **27**(2), 883–889 (2011)
23. Barker, V.: Older Adolescents' Motivations for Social Network Site Use: The Influence of Gender, Group Identity, and Collective Self-Esteem. CyberPsychology & Behavior **12**, 209–213 (2009)
24. DeLone, W., McLean, E.: The DeLone and McLean Model of Information Systems Success: A Ten-Year Update. Journal of Management Information Systems **19**(4), 9–30 (2003)
25. Jones, M.A., Mothersbaugh, D.L., Beatty, S.E.: Why customers stay: Measuring the underlying dimensions of services switching costs and managing their differential strategic outcomes. Journal of Business Research **55**(6), 441–450 (2002)
26. Cheung, C., Chiu, P., Lee, M.: Online social networks: Why do students use facebook? Computers in Human Behavior **27**(4), 1337–1343 (2011)
27. Chang, Y.-P., Zhu, D.-H.: Understanding social networking sites adoption in China: A comparison of pre-adoption and post-adoption. Computers in Human Behavior **27**(5), 1840–1848 (2011)
28. Hung, S., Ku, C., Chang, C.: Critical factors of WAP services adoption and empirical study. Electronic Commerce Research and Applications **2**(1), 42–60 (2003)
29. Chang et al. (2009)
30. Shapiro, C., Varian, H.: Information rules: A strategic guide to the network economy. HBS Press, Cambridge (1998)

31. Arroyo-Barrigüete, J., Ernst, R., López-Sánchez, J., Orero-GiménezOn, A.: On the identification of critical mass in Internet-based services subject to network effects. The Service Industries Journal **30**(5), 643–654 (2010)
32. Park, N., Kee, K.F., Valenzuela, S.: Being immersed in social networking environment: Facebook groups, uses and gratifications, and social outcomes. CyberPsychology, Behavior **12**(6), 729–733 (2009)
33. Ghani, J.A., Deshpande, S.P.: Task characteristics and the experience of optimal flow in human-computer interaction. Journal of Psychology **128**(4), 381–391 (1994)
34. Hsu, C.L., Lu, H.P.: Why do people play on-line games? An extended TAM with social influences and flow experience. Information, Management **41**(7), 853–868 (2004)
35. Hair, J.F., Hult, G.T.M., Ringle, C.M., Sarstedt, M.: A Primer on Partial Least Squares Structural Equation Modeling (PLS-SEM) Sage. Thousand Oaks (2013)
36. Chin, W.W.: Commentary: issues and option on structural equation modeling. MIS Quarterly **22**(1), vii–xvi (1998)
37. Bagozzi, R.P., Yi, Y.: On the evaluation of structural equation models. Journal of the Academy of Marketing Science **16**(1), 74–94 (1988)
38. Fornell, C.A., Larcker, D.: Evaluating structural equations models with unobservable variables and measurement error. Journal of Marketing Research **18**(1), 39–50 (1981)
39. Williams, Clippinger (2002)
40. Ringle, C.M., Wende, S., Will, A.: SmartPLS 3.0 (2014). www.pls-sem.com
41. Ping, R.A.: The effects of satisfaction and structural constraints on retailer exiting, voice, loyalty, opportunism, and neglect. Journal of Retailing **69**(3), 320–352 (1993)

Appendix A Survey Instrument

Switching Intentions (SI)
SI1. I am considering switching from my current IM service (SD…SA)
SI2. The likelihood of my switch to another IM service is high (SD…SA)
SI3. I am determined to switch to another IM service (SD…SA)*
*: SD: Strongly Disagree; SA: Strongly Agree

Sociality (SO)
SO1. I can get peer support from others while using current IM (SD…SA)
SO2. I would like to share something with others while using current IM (SD…SA)
SO3. I can stay in touch with people whom I know while using current IM (SD…SA)
SO4. I can feel belonging to a specific community while using current IM (SD…SA)

Entertainment (EN)
Please describe your feelings of entertainment while using your current IM
EN1. Uninterested…Interested
EN2. Not fun…Fun
EN3. Dull…Exciting
EN4. Not enjoyable…Enjoyable

System quality (SQ)
SQ1. When using current IM, I value the ease of operation and use of the IM (SD…SA)

SQ2. When using current IM, I value the reaction time and speed of execution of the IM (SD...SA)

SQ3. When using current IM, I value the accountability of operation of the IM (SD...SA)

SQ4. When using current IM, I value the stability of operation of the IM (SD...SA)

Advantage of Alternative (AA)

AA1. I believe that this alternative IM offers much better connect functions than my current IM (SD...SA)

AA2. I believe that this alternative IM offers much better entertainment value than my current IM (SD...SA)

AA3. I believe that this alternative IM offers much better social function than my current IM (SD...SA)

Peer Influence (PI)

PI1. My friends have moved to the alternative IM from current IM (SD...SA)

PI2. My friends have sent me invitations from the alternative IM (SD...SA)

PI3. My friends recommend the alternative IM to me (SD...SA)

Critical Mass (CM)

CM1. Most of my relatives frequently use the alternative IM that current is popular (SD...SA)

CM2. Most people in my social circle frequently use the alternative IM that current is popular (SD...SA)

CM3. Most people in my class/office frequently use the alternative IM that current is popular (SD...SA)

The Role of Consumption Value and Product Types in Repurchase Intention of Printed and Online Music Products: The Taiwan's Case

Yu-Min Lin[1], Shu-Chen Kao[2,1], Yu-Hui Tao[1,2], and ChienHsing Wu[1(✉)]

[1] National University of Kaohsiung, 700, Kaohsiung University Rd., Nanzih District, Kaohsiung 811, Taiwan, R.O.C.
yuminhlin@gmail.com, kaosc@mail.ksu.edu.tw,
{ytao,chwu}@nuk.edu.tw
[2] Kun Shan University, No.195, Kunda Rd., Yongkang Dist., Tainan City 710, Taiwan, R.O.C.

Abstract. The paper presents the role of consumption value in repurchase behavior for printed and online music products. Based on the consumption value theory and expectation-confirmation theory, it conducts an empirical study that statistically examines the effect of consumption values on the satisfaction toward repurchase intention. Data analysis from 728 valid samples reveal findings as follows: (1) the functional, emotional, and epistemic value are the significant driving factors that make people still paying for both the printed and online product types, though it shows a limited distinction for the two types, (2) in comparing of the concepts of designed functional value, printed music consumers tend to have a stronger enthusiasm in music consumption, (3) social value and conditional value are not the predicators of satisfaction, implying that the music consumption is self-oriented for any normal amusement, and (4) for the factors that are significant, the epistemic value is significant at all levels of significance for the printed products, while insignificant if $\alpha=0.01$ for the online products. Implications and suggestions are also addressed.

Keywords: Music consumption · Consumption value · Printed/online music

1 Background

As the printed record market shrank, most of the recent studies with various topics in Taiwan market focus mainly on the online music domain, but many of which are focusing on the free online music file downloading, while limited literature concentrated on legal and paid music [1, 2, 3, 5, 20]. However, not many research findings in literature have addressed the intention of repurchasing digital music after Apple Computer's virtual online store, iTunes Store, officially launched Taiwan market on June 27, 2012. As reported to account for three fourths of the global digital music market [6], iTunes Store is also expected to make influences in Taiwan's music market.

Despite paying for music, there are so many ways for people to enjoy free music by computer these days, no matter it is legal or not. An Online media website, such as YouTube, could be a common one, or go to an online streaming website such as

© Springer-Verlag Berlin Heidelberg 2015
L. Wang et al. (Eds): MISNC 2015, CCIS 540, pp. 16–29, 2015.
DOI: 10.1007/978-3-662-48319-0_2

SoundCloud, and now even the Sweden based Spotify has already launched Taiwan market, which also provides free account for streaming. Needless to say the notorious illegal mp3 download, according to Record Industry Foundation in Taiwan, digital piracy rate accounts for an enormous 82%. Around only 1.5 million users are using the legal digital music out of 8 million music audience. Moreover, though the advances of information technology almost killing the industry, the popularization of 3G, Wi-Fi and mobile devices like smart phone and tablet give the industry another gleam to survive, and even to grow again. Importantly, purchasing physical compact disc is no longer the only way of buying music. When the music can be sold digitally and virtually, it can replace the basic function of tradition CDs. Nonetheless it raises a question at the same time, if the digital music file could be easily acquired, like 24/7 and everywhere by downloading, why would people still pay for a virtual item that they can be just freely obtained? In contrast, if those digital files could easily work with devices, why do people still buy the printed CDs, which still accounts for over 57% of the total sales of the music market in Taiwan.

To address these issues, the present research studies the value of people who still pay for music in the age and environment that free music are just so easy to obtain, and also try to contrast the different value provided by two product types in diverse intrinsic qualities at the same time. By specifying on the context of music industry, the current research proposes an integrated theory by incorporating the consumption value theory by Sheth et al. [7] into the expectation-confirmation theory by Bhattacherjee [8] to describe the repurchase intention of music products. Moreover, the research also conducts a comparison analysis of different product types, tradition music products and online music products. The rest of this paper is organized as follows; Ssection Two presents a review of literature in which research arguments are highlighted and hypotheses defined. The research method is described in Section Three where research model, sampling, measure, and statistics techniques used are included. Finally, Section Four provides research findings and implications, followed by the Conclusion section.

2   Music Industry

Taiwan was ranked No. 2 in Asia market, No. 13 in the world back in the year 1997, the highest peak in Taiwan's recorded music industry. In fact, if taking a closer look at the Taiwan market reports [9], there are two factors that might make the printed market looks better; the first one is the sales of vinyl records, the other one is parallel imports. According to RIT's market report, the annual unit sale of vinyl is only 1.11thousand in 2009, and then started to grow constantly to 12.66 thousand units in 2012, which account for an incredible growth rate for more 11 times within 4 years, especially in year 2012. The other factor in RIT's report is that all the sales figures exclud the parallel imports, which means the overall consumption for music products of Taiwan customer could be higher than those numbers. Just like digital music, parallel imports would cut down the sales of records released officially by the local music labels, but some of the local importers can benefit from it.

In this research, the tradition music product is defined mainly on the focus of its physical characteristics comparative to the virtual digital item. Compact disc is the most dominant type of tradition music product, including the normal CDs, high definition Super Audio CDs (SACD), DVDs, VCDs, Blu-Ray Discs, and so on, which accounts for 99.7% of the total printed market, while the vinyl record takes the remaining 0.3% (RIT market report, 2012). To the contrary, the online music product is defined as the virtual product or service transaction in the world of Internet, while this category of product type basically combines online streaming and digital download. The concept of online streaming is like to rent a legal account, a monthly certification to unlimitedly access the database of the service providers, and charge the subscriber the subscription fees every month. Digital downloading is more likely leading to item purchaseing as purchasing printed products, in the virtual way. Although somewhat different from each other, this research attemps to combine these two business models as a category of online music product.

3   Related Concepts

3.1   Satisfaction and Repurchase Intention

To predict repurchase intention, among theories depicted in literature is the Expectation Confirmation Theory (ECT) developed by Oliver [10] and Bhattacherjee [8] in marketing and information system domain. Basically, the fundamental idea of the ECT is that the assessment of satisfaction can be used as a reference for the assessment of repurchase intention [8, 10]. However, according to the concept of post-purchase expectation, Bhattacherjee [8] argued that the factors of expectation could also induce latent changes, such as usefulness and ease of use from technology acceptance theory by Davis [11]. On the one hand, most ECT-related research reported that satisfaction can be used to predict repurchase intention [4, 12, 13, 14]. On the other hand, the research findings for antecedents that contribute to the confirmation toward satisfaction vary. For example, perceived playfulness was significantly related to consumer satisfaction in the report of Ahn et al. [12] and Nambisan and Watt [13] among others, but insignificantly in Kim [15]. The current research argues that the factors that affect technology acceptance are not the equivalent of those that affect use continuance of similar or dissimilar types of product; that is, the independent nature and distinctiveness of products should be also considered concurrently to explain the continuance behavior of a product. For example, music products (e.g., music CDs) are naturally different from hardware products (e.g., cell phone) with respect to the perception of post-acceptance toward confirmation, and satisfaction as well. The first hypothesis is then defined as follow.

H1: Satisfaction significantly influences repurchase intention of music products.

3.2   Consumption Value

The consumption value theory was introduced by Sheth et al. [7] explaining why consumers choose to buy a certain product or not, why consumers prefer one product type over another, and why consumers make their choice of one brand over another in their work. It contains mainly five values, including functional, social, emotional, epistemic, and conditional. Importantly, music products do not have negative effect on human body, and it does not provide positive values if someone does not pay for it (while cigarette does in Sheth et al.'s research). But it is still possibly applicable to explain why people pay for music products because of they can still enjoy music without paying for it. On the other hand, to choose not buying music products can provide positive value instead, for example, saving money. Therefore, although this study only focusing on those who pay for music, this study believes Sheth et al.'s model is suitable to be applied to explain why people still pay for music in this research. After that, the choice of product type is another key element in this study, just like the original work. This paper tries to understand the reason of why those who still buy music products make their decision between printed and digital type of music products. Sheth et al. [7] identified five values in their theory, including functional value, social value, emotional value, epistemic value, and conditional value that lead to affect consumer choice behavior, and also suggested three fundamental propositions to the theory: (1) consumer choice is a function of multiple consumption values, (2) consumption values make differential contributions in any given choice situation, and (3) consumption values are independent. Based on these arguments, related research reports depicted in literature have demonstrated its use to explain repurchase behavior [1, 16, 17, 18]. Particularly, the current research defines the consumption values as the values from the consumers or users who actually paid for the music, in expect that this study can provide more insight and suggestion to the correlated businesses in the industry.

3.2.1　Functional Value

If a product or one product type can provide the functional, utilitarian or physical performance to the consumers, and which can satisfy the need of function and utility, then the product or the product type acquire the functional value [7]. In addition, attribute, need, quality, reliability and durability are seem as part of the element contained in it, and notice that price (economic utility) is included in functional value as well. However, because Chen's study [1] only focused on the digital type of music, this research develops the functional value into four sub-factors in order to do a further analysis of each feature of both the printed and digital types of music product, but the four sub-factors are combined into one as "Functional Value" as the factor influencing consumer's behavior. Based on the definition of functional value and the use of previous research [1, 16], the four sub-factors udrf in this study are worth-price, quality, attribute and need. Also based on Sweeney & Soutar's point of view [19], this paper only uses "worth-price" instead of economy utility to classify the factor from the overall functional value, because the research target of this paper is those who really spend their money on music. However, comparing to tradition music product, one of the advantages and/or characteristic of digital music product is that the prices are usually lower as an album, and especially as a single track. So this study extends "worth-price" as one of the sub-factors in the functional value, and expects it will make influence on the various product types.

This study defines the quality wider than Chen's point of view [1]; for example, not only try to compare the quality of sound performance, but also take "content" into comparison. Because of, first, although most of the audience cannot tell, printed music products usually provide higher sound quality than digital music. Secondly, sometimes the contents and the quality of music products, such as the tracks, video, gifts, lyrics or booklet, would have differentiation between printed and digital types of products. Therefore this study combines these elements into one "quality" sub-factor to examine whether there are disparities between the two or not. The "attribute" defined in the present study is related to marketing channel and management. One of the most advantages of online music is its convenience. Not only the online music store is always operating twenty-four seven, but it is also accessible at any place and on any device. The attributes of the products types would also influence the convenience when users are trying to use or to manage their purchased item. And for digital music, it only occupies the virtual storage space in a computer device, but printed products have to take up the space of the shelf, nonetheless, the preference of each might varied from person to person as well. Though this sub-factor is also based on the physical and digital attributes of the product types, this study takes these utilities as a different level from convenience of the "attribute", in order to inspect whether the product types would affect consumer's perceive value or not. The second hypothesis is then defined as follow.

H2: Functional value significantly influences satisfaction in printed/online product type

H2a: Worth-price significantly influences satisfaction in printed/online product type

H2b: Quality significantly influences satisfaction in printed/online product type

H2c: Need significantly influences satisfaction in printed/online product type

H2d: Attribute significantly influences satisfaction in printed/online product type

3.2.2   Social Value

When consumer can perceive the utility of a product by getting association with some specific social group, and also by getting association with some stereotyped demographic, socioeconomic and cultural-ethnic group, then they can acquire social value from the product. Sometimes consumers want to be accepted by a social group, to conform to the social norms, or to express some images of their own [1, 16]. This paper inspects the social value mainly based on the elements symbolic value, the normative component of attitudes, because the research believes that although free music is just so easy to get nowadays, many of them are still illegal, so paying for legal music must contain some correct value in it. And, paying, collecting, or supporting some kind of music genres or artists could also pertain to an image or value. Therefore, this study believes that customers would have perceived much more values comparing to those who are just downloading the music for free, because the consumers are actually spending their money on it.

H3: Social value significantly influences satisfaction in printed/online product type

3.2.3   Emotional Value

If a consumer can perceive the utility of a product by arouse feelings and emotions, then the consumer can perceive the emotional value from the product. Emotional value is also seen as one important factors affecting consumer behavior, usually some purchase on impulse, or some non-planned purchase are seen driving by emotional value. In Chen's study [1], the emotional value is measured mainly focusing on the element of "motivation research", like happiness, joyfulness emotions arouse by the music or the melody itself, which also affects consumer behavior significantly. However, this research develops the emotional value of music products into the elements of "personality", "nonverbal processing and hemispheral brain lateralization", focusing on consumer's personal emotion to the music, the song, and the artist, that would really make them like to spend money on it, to own the music themselves, and to collect those music products they like instead of just downloading the virtual files for free. And at the same time, focusing on this consumer behavior can express their style, taste, and personality or not. These emotional values are expected to influence quite well to the purchase of music products for music customers.

H4: Social value significantly influences satisfaction in printed/online product type

3.2.4   Epistemic Value

The epistemic value is define as a product or service that can arouse consumer's curiosity, provide novelty, or satisfy a desire for knowledge. This epistemic value drives consumer to purchase something that is non functional or no other needs required, like a new item or an entertaining item, with measuring elements coverng exploratory, variety-seeking and novelty-seeking behavior, optimal arousal and stimulation, and innovativeness [1, 7, 16]. Trying a new type of coffee, visiting a new night club, and the novelty of purchasing a new house are all examples of epistemic value. Although Chen's study [1] had already showed that the epistemic value has a positive relation with user's intention of music download, this research here still tends to retain the epistemic value in inspecting consumer's intention of music purchase. Because this paper thinks that people can still acquire curiosity, novelty, or knowledge when they purchase music products, such as the tracks they did not hear in advance of an album, the bonus gift and bonus video in the music products, which might not be contained in those downloaded mp3 file. And maybe the consumer is supporting the artist they like, and then make the purchase without hearing all the tracks beforehand, or the consumer is just attracted by an album's cover, package, or the promoting written descriptions, and so on. In addition, beside the music itself, the design, like the lyrics, booklets, and pictures, inside an album is also fresh to the music buyer. Therefore, this study still keeps the epistemic value as one of the test components.

H5: Epistemic value significantly influences satisfaction of in printed/online product type

3.2.5   Conditional Value

If the consumer can acquire perceived utility from a product or service under a specific situation or circumstance, then the product or service can provide conditional value. The conditional value can also be seen as derived from temporary functional value or social value, and the measuring elements include effects of situational contingencies, classifications of situational characteristics, antecedent states, physical surroundings, social surroundings, task definition, and temporal perspective [1, 7, 16]. This research keeps the conditional value to inspect consumer's purchase behavior on music products. Because this study think that some music purchases are made under certain situations and circumstances such as for doing some jobs or tasks, holding some events or parties, or attending some artists' promotion activities, and so. Therefore, the reason to inspect conditional value is to find out that if the purchase of music products has happened randomly under certain circumstances, by which of the product types will the consumers choose more directly and more often. Hence, this paper also retains the last conditional value as one of the test components.

H6: Conditional value significantly influences satisfaction in printed/online product
type

4   Method

4.1   Research Model

With all the derived research hypotheses, H1 to H5, developed, Figure 1 summarizes the research model of this study. It contains five independent variables, functional value, social value, emotional value, epistemic value, and conditional value, and their direct dependent variable, satisfaction that affects repurchase intention. In addition, in the purpose of comparing the affecting factors of product type, the two targeted product types are both pertained but examined separately in the developed research model.

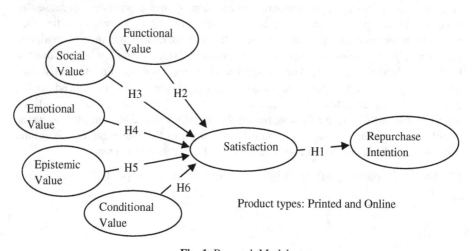

Fig. 1. Research Model

4.2   Sampling and Measure

The subjects of the research are the music fans who retaining their purchase convention rather than downloading music files online for free, within one year. Moreover, in order to make an exact comparison of the values between traditional printed music and digital online music, the research questionnaires were separated in two while designed exactly the same (only changed the object printed/online in sentences) for research subjects who had consumed either printed music or online music within one year. If both were consumed, the subject will be asked to fill the questionnaire twice by each experience they had. While in the early stage, this questionnaire was only distributed among friends via social network, and then turned to Taiwan's local BBS site (Ptt) to ask music fans for help. Also, in hoping to collect the more balanced data in music preference and gender, this study published the online research questionnaire on board "WesternMuisc", "J-PopStation"; boy groups "Mayday", "EXO", and "INFINITE"; girl group "APINK", "T-ara", and "SNSD", which are some popular boards that would gather many music fans on Ptt BBS site. The online questionnaire was officially released for a month. The questionnaire was designed and developed via a review of the related literature and the research arguments (e.g., functional value and their sub-variables, worth-price). For example, the function value was defined to be that subjects perceive a positive attitude to the worth-price, quality, attribute, and utility of music products. One of the items of functional value is "I perceive that (printed/ online) music are worth the price". The reliability and validity of the collected data were examined to ensure internal consistency, and linear regression model to test the research hypotheses proposed.

5   Data Analysis Results and Discussion

After deleting 97 returned questionnaire, there were 728 valid questionnaires in total, showing a validity rate of 88%. The descriptive statistics was presented in Table 1 for both types of music products.

The Alpha coefficients of all variables for the reliability confirmation were present, showing that the variables and sub-factors are all reliable. For example, the Functional Value of sub-factor, Worth-price, is 0.862 in printed type, and the corrected item-total correlation values range from 0.696 to 0.827. Moreover, the overall reliability for each variable in both printed and online type is acceptable. For the factor analysis, there are 8 factors extracted from the designed items of printed music product type. The cumulative extraction sums of squared loadings percentage is 71.082%, and the loadings of each item are from 0.591 to 0.889, which show that the items are extracted into appropriate factors. The originally designed 4 sub-factors of Functional Value are combined into 3 factors, Cost-performance, Need, and Attribute, and the 6-item Emotional Value is divided into 2 factors, Style and Affection, while the rest designed items are all extracted into the expected factors, i.e., Social value, Conditional value, and Epistemic value, and two dependent variables are extracted as Satisfaction and Repurchase Intention. For the online type, unlike printed music product type, the idea of the 3-item of worth-price is combined with the idea of the

Table 1. Consumption behavior statistics of product types

Total N=728 (M:198; Female:530)	Printed		Online	
Consumption experience	Frequency	Percentage	Frequency	Percentage
Less than 5 years:	200	28.8%	194	78.5%
5 to 10 years:	197	28.3%	44	17.8%
Over 10 years:	298	42.9%	9	3.6%
Consumption frequency (a year)				
1 to 5 times:	423	60.9%	117	47.4%
6 to 10 times:	187	26.9%	40	16.2%
11 to 15 times:	54	7.8%	49	19.8%
Over 16 times:	31	4.5%	41	16.6%
Money spent each time (TWD)				
Below 150:	4	0.6%	130	52.6%
151 to 300:	29	4.2%	74	30.0%
301 to 450:	382	55.0%	17	6.9%
Over 451:	280	40.3%	26	10.5%
Music language preference (choose at least 2)				
Mandarin	459	32.6%	186	36.3%
Korean	373	26.5%	116	22.6%
Japanese	274	19.4%	68	13.3%
English	206	14.6%	112	21.9%
Taiwanese	35	2.5%	22	4.3%
Other	62	4.4%	8	1.6%
Music genre preference (choose at least 3)				
Pop	684	32.6%	241	32.0%
Rock	420	20.1%	171	22.7%
Soundtrack	349	16.7%	125	16.6%
R&B	174	8.3%	68	9.0%
Electronic	121	5.8%	49	6.5%
Instrument	77	3.7%	23	3.1%
Rap	61	2.9%	25	3.3%
Classical	55	2.6%	14	1.9%
Jazz	46	2.2%	15	2.0%
Crossover	32	1.5%	8	1.0%
Religion	8	0.3%	2	0.3%
Other	69	3.3%	12	1.6%

fundamental need as the factor to consume online music, though this component can be seen as that consumer can obtain the basic requirement when consuming online music with a satisfactory cost. To denominate the factors, the cost achieving such needs is still considered the more critical factor that provides the positive value for consumers. Therefore, the extracted factor is denominated as Worth-price as one of the sub-factors in Functional Value, also, two other factors are extracted as sub-factor Quality and Attribute. Besides, the other extracted factors are all converged into the designed factors and sub-factors: Emotional Value, Conditional Value, Social Value, Epistemic Value, and Quality and Attribute as the sub-factors of Functional Value. All of the correlation coefficients for the 9 factors in online music product type are shown significant at the 0.01 level. All of the correlation coefficients for the 10 factors in printed music product type are shown significant at the 0.01 level.

5.1   Test Results

In the first level as seen in Table 2, Functional Value (FV), Emotional Value (EmV), and Epistemic Value (EpV) among the five consumption value in both printed and online music product are shown positive and significant at the 0.01 level, except Epistemic Value in online type. Therefore, the factors leading to customers' satisfaction toward printed music consumption support hypotheses 2, 4, and 5, while Social Value (SV) and Conditional Value (CV) are not statistically significant in relation with consumer's satisfaction, thus reject hypotheses 3 and 6. It also shows that satisfaction significantly influenced repurchasing intention for both product types. In the second level, the three sub-factor of Functional Value and two of Emotional Value are taken into exam, respectively. As seen in Table 3, all sub-factors of Functional Value and those of Emotional Value are all having a significant positive relation with the satisfaction.

Table 2. Hypothesis test results

Factors	Dependent: Satisfaction						Dependent: RepI	
	Beta		t		Sig.		Sig.	
	Pr	On	Pr	On	Pr	On	Pr	On
(Constant)			.000	.000	1.000	1.000		
FV (H2)	.603	.611	21.449***	12.627***	.000	.000		
SV (H3)	.044	-.027	1.464	-.506	.144	.613		
EmV (H4)	.164	.163	5.095***	2.877***	.000	.004		
EpV (H5)	.132	.114	4.415***	2.169**	.000	.031		
CV (H6)	-.039	.027	-1.488	.551	.137	.582		
	R-square		.608	.572	-		-	-
(Constant)			.000	.000			1.000	1.000
Sat. (H1)	.739		28.866***	16.258***			.000	.000
	R-square		.546	.519	-		-	

** $p<0.05$; *** $p<0.01$; FV: Functional Value; SV: Social Value; EmV: Emotional Value; EpV: Epistemic Value; Sat.: Satisfaction; RepI: Repurchase intention

Table 3. Second level hypotheses test results for functional value

| | Dependent: Satisfaction | | | | | |
| | Beta | | t | | Sig. | |
Factors	Pr	On	Pr	On	Pr	On
(Constant) -		-	.000	.000	1.000	1.000
FVCp	.445	-	13.483***		.000	
FVNe	.299	-	8.947***		.000	
FVAt	.128 -		4.237***		.000	
FVPr	-	.359		6.061***		.000
FVQu	-	.241		4.519***		.000
FVAt	-	.280		5.250***		.000
		R-square	.564	.549	-	

*** $p < 0.01$; FVCp: Cost and performance; FVNe: Need; FVAt: Attribute; FVPr: Worth price; FV Qu: Quality; FVAt: Attribute

5.2   Discussion and Implications

The test result of this study shows similar validated models of the two music product types. Nonetheless, it does not necessarily imply that the values, the factors, or the reasons that music consumer make their consumption are totally the same. There are implications. First, although the result is consistent with Chen's study [1], the measuring ideas are diverse. Both music consumers of the two product types perceive Emotional Value from their music consumption, which is more straightforward to understand the implications behind. As discussed, this value is seen as the most driving factor that makes the music fans willing to pay for music instead of listening to it free online. No matter it is printed or online music consumer, spending money on the track they like or supporting their favorite artist is the most obvious and direct way to express their emotion to the artist and the music, and to express their own personality and style as well. This is particularly important as the driving factor for the idolism and the younger generations.

Second, in both music product types, Social Value and Conditional Value do not significantly affect consumer's satisfaction, which suggests that although the society has the popularity and atmosphere of listening to free or illegal music online, the idea of Social Value is not the driving factor that makes the music fans still willing to pay for legal music. This finding is consistent with Chen's study [1], that users do not perceive Social Value in downloading music on Internet. On the other hand, the results can suggest that for subjects in both product types, typical music entertainment is the key issue in their consumption behavior instead of the conditions that happened rarely. These results are implying that the motivation of music consumption is simply self-oriented under normal entertainment purpose. Except of these rejected hypotheses factors, there are still some consumption values that drive the consumer willing to pay instead.

Third, consisting with Chen's study [1] surprisingly, the Epistemic Value is affecting consumer's satisfaction significantly. This indicates that even though under the popularity of Internet, it is quite possible for music consumers to listen to the music first before they make the purchase and music customers still can perceive novelty about music not only for online music but also printed music. For printed music product, besides the novelty perceived by listening to new music, the physical characteristic could be another factor provideing epistemic value for consumer. For example, like the novelty of opening an elaborate gift and discovering every subtle design of the artworks inside a printed album, or, the random lucky draw to get a star post card could also provide the epistemic value exclusively for printed music product type. While for online music product type, Epistemic Value is significant yet at the 0.05 level, this result might generate from the function of online streaming to stream unlimitedly on the online music database. At the same time, however, this is one of the research limitations of this study, the combination of digital download and online streaming as one online music product type, which means more comprehensive definition can be focused in the related future study.

Forth, the Functional Value is also significantly affecting customer's satisfaction thoroughly in both product types, even though some questionnaire items were deleted and the factor extraction is not exactly in the same way. For online music product type, all the three extracted factors, Worth-price, Quality, and Attribute all show significant in the result, which is pretty much consistent to the ideas of how the questionnaire is designed. Comparing to printed music, the average cost of online music consumption is undoubtedly lower, although it is only the digital item that the consumer can get, subjects can still perceive the price of online music satisfactory to fulfill their basic need of music. Moreover, different from the result obtained by Chen [1], it shows that subjects are also satisfied about many aspects of the quality that online music provides, which could be explained that to attract and satisfy customers, online music providers definitely need to provide their products in great quality, and fortunately this effort seems to be accepted by the online music consumers as well. In the end, for the result of the third sub-factor Attribute, it also agrees with the characteristics of cloud computing as the convenience and accessibility as online music is anytime and anywhere for any devices.

Finally, all of the sub-factors in Functional Value for printed music product type show significantly affecting consumer's satisfaction as well as online music type does. For example, even though the average cost of a copy of printed music album is certainly higher than the one in digital version, with the satisfactory quality, printed music consumers still perceived the price being worthy, affordable and reasonable. Secondly, the sub-factor Need is just a basic requirement that drives the consumer to buy a printed album, and notice that the requirement of great sound quality is seem as one of the basic needs, which indicates that most of the consumers are contented with the sound quality of the music product they purchase. And for the last sub-factor Attribute, in comparing to online music, the physical characteristic of printed music product certainly makes it not as convenient as online music either to buy, to play, or to manage. However, just like price, printed music consumers are still satisfied with the physical attributes of printed music product. To sum up, the Functional Value of printed music product type, the result could also be indicated that even though the higher price yet lower convenience comparing to online music, printed music consumers have a pretty strong enthusiasm in purchasing and collecting those printed music products.

Moreover, most of the online music consumers are also purchasing printed music at the same time, and printed music is still the dominant product type that most of the music consumer would like the buy in Taiwan market especially for the younger generation. On the other hand, consumers who aged from 26 to 45 and who with higher dispensable income level tend to consume more online music. For most of the music consumer, either printed or online product type, seldom happened circumstance and condition are not the driving factors, nor the factors related to the Social Value, which means the music consumption is generally occurred under a self-oriented motivation. Therefore, other driving factors like Emotional Value and/or Epistemic Value might be more relevant in affecting customer's behavior. Furthermore, both printed and online music provide the similar consumption values, beside the basic functional related consumption value, Emotional Value and Epistemic Value are the most factors that drive the music lovers still pay for both the two comparative product types. All the sub-factors extracted in the Functional Value are positively affecting customer's satisfaction in both music product types, however, it seems that the consumer of printed music tend to have a stronger enthusiasm in music consumption.

6   Concluding Remarks

This study develops and examines the research model that is based on the consumption value theory of Sheth et al. [7] to investigate the factors that will affect music lover's post-purchased consumer behavior in satisfaction and repurchase intention. Furthermore, two research models are involved for the two contrariety product types, tradition printed music and digital online music, and to make a comparison, trying to understand the reason why music fans are still willing to spend money on music products, and what factors make them choose between these two product types. After all of these fluctuation and alternation of the music industry within this decade, this paper provides a reference and correlated results for the future studies for the related paper especially for the prosperous printed parallel imports items and online music. Moreover, marketing strategy that promoting customers' perceived emotional value would be critical to attract people to spend money on legal music for both printed and online music, just as Soundtrack is among the most popular music genre, the visual video or film would be the most supportive element in stimulating people's perceived emotions to the artists or tracks they like. Though online music has its unique characteristics and many advantages different from printed music, upstream record companies should also keep publishing records in printed type instead of releasing the digital version only that let the printed music enthusiasts can still collect. Conversely, in Taiwan market, online music companies might need to have more exclusive advantages in competing with printed product type especially in attracting those loyal printed music fans, for example, the margin cost to increase the content within an album should be lower than a printed one, thus to contain more exclusive materials then build up its uniqueness different from printed products.

References

1. Chen, H.Y.: The effects of consumption values on behavioral intention to MP3 download. Unpublished Master Thesis, Department of Business Administration, Soochow University. (In Chinese) (2006)
2. Wu, W.-L.: Research of investigating digital music platform and users' needs through the aspect of knowledge innovation. Unpublished Master Thesis, Department of Information and Communication, Southern Taiwan University of Science and Technology. (In Chinese) (2007)
3. Chen, Y.C., Shang, R.A., Lin, A.K.: The intention to download music files in a P2P environment: Consumption value, fashion, and ethical decision perspectives. Electronic Commerce Research and Applications 7, 411–422 (2008)
4. Thong, J.Y.L., Hong, S.-J., Tam, K.-Y.: The effects of post-adoption beliefs on the expectation-confirmation model for information technology continuance. International Journal of Human-Computer Studies 64(9), 799–810 (2006)
5. Chai, T.Y.: The effects of virtual community services on Taiwan online music websites. Unpublished Master Thesis, Department of Graphic Arts and Communications, National Taiwan Normal University. (In Chinese) (2011)
6. AppleInsider, Apple's iTunes accounts for 75% of global digital music market, worth $6.9B a year. From: http://appleinsider.com/articles/13/06/20/apples-itunes-accounts-for-75-of-global-digital-music-market-worth-69b-a-year Access date: 2013/October (2013)
7. Sheth, J.N., Newman, B.I., Gross, B.L.: Why we buy what we buy: A theory of consumption value. Journal of Business Research 22, 159–170 (1991)
8. Bhattacherjee, A.: Understanding information systems continuance: An expectation-confirmation model. MIS Quarterly 25(3), 351–370 (2001)
9. Record Industry Foundation in Taiwan (RIT), Taiwan Music Market. From: http://www.ifpi.org.tw/ Access date: 2013/October (2013)
10. Oliver, R.L.: A cognitive model of the antecedents and consequences of satisfaction decisions. Journal of Marketing Research 17(4), 60–469 (1980)
11. Davis, F.D.: Perceived usefulness, perceived ease of use, and user acceptance of information technology. MIS Quarterly 13(3), 319–340 (1989)
12. Ahn, T., Ryu, S., Han, I.: The impact of Web quality and playfulness on user acceptance of online retailing. Information & Management 44, 263–275 (2007)
13. Nambisan, P., Watt, J.H.: Managing customer experience in online product communities. Journal of Business Research 64, 889–895 (2011)
14. Hossain, M.A., Quaddus, M.: Expectation–confirmation theory in information system research: a review and analysis. In: Dwivedi, Y.K. (eds.). Information Systems Theory: Explaining and Predicting Our Digital Society, vol. 1 (2012)
15. Kim, B.: An empirical investigation of mobile data service continuance: Incorporating the theory of planned behavior into the expectation–confirmation model. Expert Systems with Applications 37(10), 7033–7039 (2010)
16. Kim, H.W., Koh, J., Lee, H.L.: Investigating the intention of purchasing digital items in virtual communities. In: Proceedings, PACIS 2009, Hyderabad, India, July 10–12, 2009
17. Ho, C.H., Wu, T.Y.: Factors affecting internet to purchase virtual goods in online games. International Journal of Electronic Business Management 10(3), 204–212 (2012)
18. Pan, Y.T.: Applying consumption value theory to analysis consumer behavior of the netbook. National Chengchi University Advanced MBA Program Master's Thesis. (In Chinese) (2009)
19. Sweeney, J.C., Soutar, G.: Consumer perceived value: the development of multiple item scale. Journal of Retailing 77(2), 203–220 (2001)
20. Bureau of Audiovisual and Music Industry Development. Taiwan Pop Music Industry Survey 2011. Ministry of Culture, Taiwan (2012)

A Novel Multi-objective Optimization Algorithm for Social Network

I-HO Lee, Jiann-Horng Lin, and Chuan-Chun Wu[✉]

Department of Information Management, I-SHOU University, Kaohsiung, Taiwan
loki77m@gmail.com, {jhlin,miswucc}@isu.edu.tw

Abstract. In real-life social networks, the decisions of individual actors are often influenced by multiple sources of information, whose relative influence depends on several factors, in much the same way as many real world networks, such as the spread of viruses, or the spreading of a new product's reputation through a given human population. Previous attempts to model social networks have focused on single-diffusion processes. However, real social networks are usually more complicated, and attempting to model them with single-diffusion processes often fails to capture higher-order effects seen in the real world. Multiple-diffusion processes have to take into account not only multiple sources of information, but also multiple types of information source, where each of the information sources may potentially contradict one another. Complex calculations involving conflicting information sources must rely on heuristics to reduce the execution time. This study provides a multi-objective optimization algorithm for solving performance problems using a creative and heuristic algorithm. The results provide a motivation to utilize the algorithm in multiple diffusion and conflicting-information problems of social networking.

Keywords: Multi-objective optimization · Evolutionary algorithms · Information diffusion · Social networks

1   Introduction

Traditional single diffusion processes lack the power to model the complex interactions that often appear in real social networks. Linear threshold[2] and independent cascade[3][4] models are typically used for analysing the performance of single diffusion processes. In today's complicated social networking environment, an individual person typically engages with a variety of information from different sources, as well as information from different types of source. Therefore, using multi-objective optimization is becoming an important way to improve the execution performance of systems that analyze social networks. Young and Ramin have previously created multi-objective diffusion techniques that attempt to model social networks in which information is shared on a peer-to-peer basis. In situations where multiple types of information are exchanged, or where non-linear phenomena are present, the calculation cost can rise significantly. Multi-objective diffusion models (MODM) can apply differential evaluation techniques to each individual piece of information, calculating

© Springer-Verlag Berlin Heidelberg 2015
L. Wang et al. (Eds): MISNC 2015, CCIS 540, pp. 30–44, 2015.
DOI: 10.1007/978-3-662-48319-0_3

such properties as its influence, relevance or diversity, and maybe even assigning it an overall score. Previous attempts by researchers to apply heuristic algorithms to the problem of social network analysis have already opened the door to a new era of large-scale data analysis.

The growing maturity of heuristic algorithms has spawned a growing interest in applying these algorithms to new problems in different, and sometimes completely unrelated fields. Optimization problems and the diffusion of information through social networks are commonly modeled with well-defined algorithms that place an emphasis on reduction of execution time and avoidance of complex conditions. The design of multi-objective optimization algorithms is usually based on modifying single-objective optimization algorithms, with the addition of new mechanisms to achieve reasonable execution performance. Most multi-objective optimization methods consider the distance property of each Pareto solution or use some criteria to filter the evaluation of certain properties. Multi-objective optimization problems take many forms: including linear problems, nonlinear problems and constrained problems. The adoption of Particle Swarm Optimization (PSO) to solve different types of multi-objective optimization problems has resulted in the publication of many articles [5][6][7][8]. Algorithms based on a single heuristic have attempted to model many naturally-occurring phenomena, such as the cuckoo search algorithm [9], firefly algorithm[10], harmony search algorithm[11] and the bat algorithm[12]. This study adopts the bat algorithm in particular, and uses it as the basis for a multi-objective optimization algorithm.

The bat algorithm is a popular, well-known algorithm, used in fields such as engineering and financial markets. It is a recently-discovered meta-heuristic algorithm that incorporates intensification and diversification mechanisms in an attempt to find the point of optimal utility. Several powerful constructions allow us to design a multi-objective bat algorithm (MOBA), which we will discuss later in the paper. We are inspired by nature, which features several natural processes that are suitable for modeling this type of problem. Bats are the only mammals with wings and possess the remarkable ability of echolocation, but not all bats use echolocation in the same way. Bats typically use echolocation to detect their prey, foods and also to locate their roosts, all in complete darkness. They emit a series of very loud sound pulses, and listen for the echoes that reflect back from nearby objects. Different species of bats use different frequency-modulated signals for echolocation. In this way they even can distinguish between their families and other, non-related bats.

Multi-objective optimization problems can be broadly categorized into two kinds: assembling and linear. The ant colony optimization algorithm [13] [14] and the simulated annealing algorithm [15] can handle assembling problems of multi-objective optimization that are normally considered to fall into the category of shop-scheduling problems. The bat, PSO, cuckoo search, firefly algorithm (and so on) are suitable for handling linear problems, but can also be adapted for dealing with multi-objective optimization problems. Generally, this can be achieved by augmenting a single-objective optimization algorithm with some additional mechanism. Such mechanisms include neighborhood, dominated tree[16], criteria standard, and the weight allocation mechanism. However, these designs commonly exhibit several shortcomings, such as: getting stuck in extreme Pareto space during the search phase, the use of huge para-meters, and last but not least, poor performance[17][18]. This study attempts to reduce these shortcomings by constructing a multi-objective optimization algorithm with rea-sonable rules that are inspired by nature.

Many articles use the neighborhood mechanism to construct algorithms for dealing with multi-objective optimization problems. Xiaohui and Eberhart created a dynamic neighborhood PSO algorithm that compares pairs of closest particles[19]. It separates the search space into four regions in order to preserve potential Pareto optimality solutions in region II and IV. Additionally, it maintains potential Pareto optimality solutions in a candidate pool, storing this pool in extended memory. Ahmad and Ra-banimotlagh use two neighborhood structures in their local search algorithm to ex-plore a wider solution space. It deals with the flow-shop scheduling problem, which is NP-complete, and this unfortunately leads to slow execution performance when the input is large. However, those algorithms that utilize neighborhood mechanisms achieve outstanding levels of computation in a reasonable time.

This study develops a multi-objective optimization bat algorithm from the sin-gle-objective optimization bat algorithm, by adding Pareto optimality and density mechanisms. In multi-objective optimization, while searching within a huge space it is hard to find solutions that have the same value as an existing solution. Pareto opti-mality can reduce the search space, because the undominated space of Pareto optimal-ity is stable and limited. If we can discover enough solutions to dominate the maxi-mum Pareto space, then those solutions can be used for decision or resource control. We implement the Pareto mechanism and the density mechanism to achieve two main goals:

1. Finding a reference solution for other particles to explore more undominated Pareto space.
2. The creation of an automatic searching ability.

Multi-objective optimization algorithms have to efficiently tackle many conditions of high uncertainty, carefully managing the search space so as to avoid redundant calculations, or else performance will suffer. If the performance is too poor, this will greatly limit the application of such algorithms to new problems.

2   Multi-objective Bat Algorithm Searching Method and Pareto Solutions

The design of the density mechanism is based on weight allocation within a Pareto space, which has intensification and diversification features that are similar to those of Particle Swarm Optimization (PSO) algorithms. The first tier of the algorithm searches through physical space, and utilizes existing single-objective algorithms to search for solutions. The second tier searches through Pareto space for solutions, and the searching method is based on the probability of finding new Pareto solutions. When we discuss intensification and diversification, this is equivalent to talking about the local and global searching abilities of algorithms inspired by nature.

The evaluation criteria of MOBA are different from that of single objective optimization algorithms, in that intensification refers to searching within areas that have a higher density of existing Pareto solutions, and diversification refers to searching within areas that have higher probability of finding new Pareto solutions. The balance between intensification and diversification is dynamic, such that MOBA is able to prioritize searching within relatively undominated space, and de-prioritize searching within space near extreme boundaries. When designing MOBA, careful consideration must be given to the relative priorities of diversification and intensification. If the design of MOBA were to place too high an emphasis on diversification, and if it were to only take as reference values those solutions that have the highest probability of generating further Pareto solutions, this would cause the algorithm to spend too many resources searching within extreme boundary Pareto space, where the density of Pareto solutions is always lower.

Therefore, the density mechanism has to balance these priorities when adjusting the weights of each reference Pareto solution. It has to dynamically find a balance between intensification and diversification, so that other particles are able to change their search direction, leading to a higher probability of finding a new Pareto solution. This design fits the characteristics of particle swarm optimization under Pareto space, such that when more Pareto solutions have been found, there will consequently be less undominated space, and furthermore a smaller undominated space has a higher density.

2.1   Efficient Identification

We use the density mechanism to evaluate the set of existing solutions, and from among them find a single solution that is highly likely to yield further solutions (which we designate as a reference solution), basing our assessment on a calculation of the area of undominated space. Furthermore, to calculate the area of undominated space we make use of the Riemann Integral. For a multi-objective optimization problem, the area of the Pareto space is fixed. As we find more Pareto solutions, the area of undominated space will decrease accordingly. When MOBA compares two partitions, if a given partition contains relatively many solutions (said to be a high-density partition), that partition will have a correspondingly lower undominated space. Conversely, if a given partition contains relatively few solutions (said to be a low-density partition), that partition will have a correspondingly greater undominated space. In this study, we use the Riemann Integral to calculate the undominated area associated with each Pareto solution, and in conjunction with our density mechanism we can identify at any given stage which existing Pareto solution has the greatest probability of generating further Pareto solutions.

The Riemann Integral is used to calculate dominated space, which can then be used to assess each Pareto solution for its probability of generating new Pareto solutions. A closed interval [a, b] in Pareto space, n partitions are tagged in [a, b]. The sum of a function f is defined as follows:

$$\sum_{i=0}^{n} f(ti)\Delta ti$$

The ti is height and Δti i is width. So the total partitions are n. And integral is sum of 0 to n. Assuming weak law makes that n=3 and 9, line lengths of their own partitions are equal to their average. When n=3, that tree lines are $\overline{AX0} = \overline{X0X1} = \overline{X1B}$. And n=9, that lines are $\overline{AX0} = \overline{X0X1} = ... = \overline{X7B}$. We can compare the integrals of different n as the following figure. 1

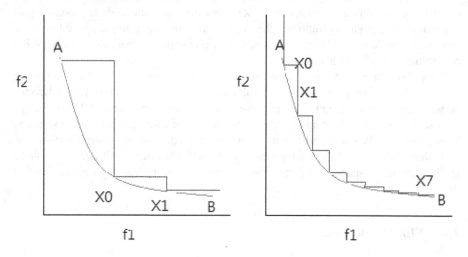

Fig. 1. Compared undominated space by two differential partitions in the same length line distance $\overline{AB}$

The dominated space of the left figure is less than that of the right, so by using point X0 from the left figure to be the reference, we have a higher probability of discovering new Pareto solutions.

2.2   Pareto Dominance

Pareto dominance is based on a rule of improvement that does not permit any individual criterion to be sacrificed in pursuit of increasing the total value. For engineering purposes, Pareto optimality considers a state to dominate other states when all conditions of that state are considered to be better than or equal to those of other states. In MOBA, assuming that functions $f_1 \sim f_n$ and vectors $v_1 \sim v_n$, If $f_i(v_i)$ have value fv_i which i=1 ~ n and i $\in$ N, if we can find new ($v_1 \sim v_n$) which is better than old set vectors to these functions $f_1 \sim f_n$ which has novel $fv_i >=$ original fv_i that can be assumed as follows:

$$\forall \ i=1 \ to \ n, \ i \in N \ ; \ f_i(v_i) = fv_i \ ; \ novel \ fv_i >= original \ fv_i$$

Pareto dominance is a well known theory used within life sciences, engineering and economics. The basic idea is to increase the total value of a function without lowering the values of that function's individual components.

2.3   Multi-objective Optimization Problems

Multi-objective optimization problems are normally constructed from several functions. Those functions often have mutual trade-offs between each other, or are otherwise in conflict with one another. These relationships can often cause optimization algorithms to produce solutions with attributes that differ from one another: these different solutions are considered important, since they can provide different levels of information for decision making. Multi-objective optimization problems are generally defined as follows:

$$\text{Min } f(\vec{X}) = [f_1(\vec{X}), f_2(\vec{X}), \dots \dots, f_n(\vec{X})]$$
$$\vec{X} = (X_1, \dots \dots, X_D) \epsilon R^N$$

And there p is quantity of inequality constraint functions and m is quantity of equality constraint functions. Therefore, subject to $g_j(\vec{X}) \leq 0$ for $j = 1, \dots., p$ and $h_j(\vec{X}) = 0$ for $j = p + 1, \dots \dots m$ The group of solutions of multi-objective optimization is based on the rules of improving way which cannot sacrifice any others to increase the total value of solutions

3   Modified Multi-objective Optimization Bat Algorithm

We follow the original BAT algorithm and append Pareto front and density mechanisms to the end of the code.It is clear that multi-objective bat algorithm is to serve as a model of Pareto front and Density mechanism because bat algorithm has intensification properties of particle swarm intelligence and diversification properties of simulated annealing properties. Their powerful features make it possible to design dynamic expansion to discover all the legitimate solutions, if possible, and avoid searching in extreme Pareto front solution. An interesting aspect of Pareto front is that contradictions of lacking intensification or diversification cause different problems during iterative times, so we need to add our density mechanism to improve the performance. Basically, bat algorithm is better than other algorithms in its construction. When we handle a problem which need to find global minimum or maximum that simulated annealing property is well known to achieve better performance. And the properties of particle swarm intelligences are swift to search the local minimum or maximum solutions.

The pseudo code of multi-objective bat is below, and we append Pareto front and density mechanism code in final evaluation in the end.

Pseudo Code of Multi-Objective Bat

> *Objective function* $f(x)$, $X = (x_1 \ldots \ldots, x_d)^T$
> *Initialize the bat population* $X_i (i = 1,2 \ldots \ldots, n)$ *and* V_i
> *Define pulse frequency* f_i *at* x_i
> *Initialize pulse rates* r_i *and the loudness* A_i
> *While(t< Max number of iterations)*
> *Generate new solutions by adjusting frequency*
> *And updating velocities and locations/ solutions*
>   *If   (rand >r_i)*
>   *Select a solution among the best solutions*
>   *Generate a local solution around the selected best solution*
>   *End if*
>   *Generate a new solution by flying randomly*
>   *If   (rand <A_i & $f(x_i) < f(x_*)$)*
>   *Accept the new solutions*
>   *Increase r_i and reduce A_i*
>   *End if*
> *Evaluating Pareto Front which be dominated or not*
> *and keep the un-dominated solutions($P_1 \ldots \ldots P_j$, j<=n, j>=0)*
> *Finding the shortest distance for each point of*
> *Pareto Front solutions SSD_j [(1)]*
> *Calculating the average distance of Pareto Front solutions*
> *μ [(2)]*
> *According the average distance to estimate the Density*
> *D_* [(3)]*
> *Setting the priorities by Density state (D_*-D_j)*
> *Selecting one to be the current best by with random number*
>
> *End while*
> *Postprocess results and visualization*

Those steps are developed for selecting better solutions and avoid waste times in searching the same area caused extreme solutions. The entire ideas is considering the distance of each Pareto front solutions and according the distance to set a priority of each Pareto front solutions. Then, we select a solution to be a current best solution and a direction of particles which selected solution guides other particles convergent to it. In the pseudo code, we have written 1 to 3 equations. The entire ideas can be demonstrated as follow steps.

First, finding shortest distance for each of Pareto solutions can be formula as follow:

$$SSD_j = \left(\sum_{L=1}^{K} |X_L - Y_L|^D \right)^{\frac{1}{D}} \tag{1}$$

The idea can be demonstrated as follow Fig 2. Assuming, we have 3 solutions which qualify the Pareto front and will be the shortest distance which connects all Pareto solutions by two lines $\overline{AB}$ and $\overline{BC}$.

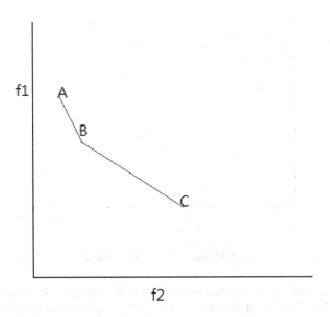

Fig. 2. Assuming 3 solutions, finding 2 shortest Distances of Pareto solutions

In equation (1), we evaluate the distance of Pareto front solutions, j, which is less or equal of q, q is total of Pareto solutions. D is norm distance and K is length of vectors. Next, we calculate the shortest distance in each point of Pareto front solutions.

Second, diversification is considered in density for each solution. For measuring the region to count the points of circle, we have to find the shortest distance SSD_j from one to the nearest solution and count the average μ which is going to be used to measure distances in Pareto space for counting one solution to other solutions.

$$\mu = \frac{\sum_{j=1}^{q} SSD_j}{q} \tag{2}$$

According to equation (1) that we can calculate the shortest distances of each Pareto solution, and equation (2) calculates the average distance for evaluating the density. When we have all shortest distances of current Pareto front solutions, we count the average of all shortest distances of current Pareto front solutions to be a metric standard. The metric standard is depended on estimating the density which will be the μ in the equation (2).But the highest values of density means that is not suitable to continuous be the best of particles. We have to inverse the value, which means we want to find a loose point to be the best of particles that can avoid straggling or crowding solutions.

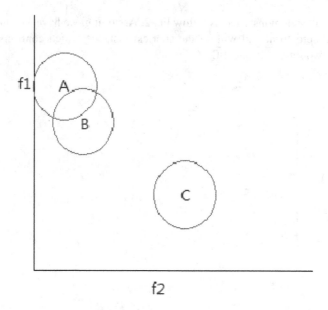

Fig. 3. Density of Pareto front solutions

We can calculate the density that A solution and B solution are included each and C solution is along. The result shows in the Fig.3 and we accord that to locate the priorities to decide the next best solution for redirecting and fixing positions.

$$\sigma_i = (x_i, y_i)$$
$$\text{For } t=1 \text{ to } q, \text{If } (\sigma_i - \sigma_t)^\wedge 2 < \mu \text{ then } \alpha_i = \alpha_i + 1 \quad \text{and } t \neq i \tag{3}$$

Final, according the μ scope measures distances between others Pareto solution that will be the density weights (3). Using the region for each solution calculates the included solutions of inside region which will be the density weight α. And the results are preserved in α, each α_m has its single measurement of density weight. However, the fundamental mechanism can handle problems of searching extreme solution. But the Pareto front problems are more tricking in variety of spaces which can be hardly to find or cause fault. Therefore, we need the other weight evaluation to adjust intensification and diversion abilities.

3.1   Weight Allocations

Before the step of selecting a particle to be the current reference particle from which other particles can be chosen, we must consider the distinctive features of multi-objective optimization. The density mechanism is one of the dynamic weight aggregations, and is based on diversification and intensification. The density mechanism mixes two weights, density and survival time, and can evaluate and adjust the probability of selecting a suitable reference Pareto solution. Those two weights are in conflict each other.

We use the survival time of a particle to adjust its weight β. If the algorithm discovers that a given particle has remained within the same position between successive iterations, the weight β will be increased for the following iteration.

The density and survival time mechanisms are designed to be in conflict with one another. The weights within the density mechanism are calculated in two steps:

1. Calculate the distance from each Pareto solution to its nearest neighbor, and then calculate the arithmetic mean of these distances, which we denote as the average μ.
2. For each Pareto solution (ρ) we calculate the density of solutions in its immediate space, which is defined by a circle with (ρ) at its center, and whose radius is equal to μ.

Ultimately, the algorithm mixes these two different weights in a ratio that is decided according to the relative need for intensification or diversification. The algorithm will generally prioritize its search within areas of Pareto space with relatively lower densities of existing solutions. As new solutions are found within these areas, and as the relative density of solutions in such areas rises (when compared with other areas), the algorithm will switch to searching within other areas accordingly. However, in order to increase the algorithm's propensity for diversification, the algorithm can be configured such that the density mechanism uses a higher weight ratio. Conversely, increasing the survival time weight will reduce the algorithm's propensity for diversification. Accordingly, the density mechanism can be used to transform any single-objective optimization algorithm into a multi-objective optimization algorithm. The Pareto density mechanism is specified in this way:

1. The density mechanism is based on a dynamic relationship between intensification and diversification. This design endows Pareto solutions with the ability to act as particles, similar to the way particles swarm within PSO algorithms. We have researched other algorithms whose design is based on simply strengthening the degree of intensification or diversification, but usually these designs are deficient in their ability to search either globally or locally. This deficiency led us to design a new algorithm with the ability to dynamically change its behaviour, tending toward more intensification or more diversification as and when necessary.
2. The survival time mechanism is based on iteration time. The mechanism in particular deals with those particles that remain in the same position across multiple iterations. The longer a given particle remains in the same position, the more redundant that solution is considered to be. This measure of redundancy is represented by a weight β that gradually decreases as iteration time increases. Pareto solutions considered to be more redundant are less likely to be chosen as reference Pareto solutions. We have designed a mechanism to adjust this weight according to the following equation, where q is the total number of Pareto solutions in the current working set:

$$\beta_i = 1 + \text{prev}\beta_i \quad, i=1,2,3...q$$

3. Using a mixture of two weights as the criterion for selection is advantageous when searching. We set the ratio for iteration weight β and density weight α^2. Using the squared value in this way predisposes the algorithm to more swiftly explore comparatively lower density areas, where the probability of finding new solutions is higher.

This mechanism is similar to survival analysis when we evaluate the iteration weights. Before combining the two weights α and β for each solution, we first calculate the maximum values of $Max(\alpha)$ and $Max(\beta)$ over all solutions, and then subtract the weights α and β from their respective maxima:

$$\beta_i = Max(\beta) - \beta_i + 1$$
$$\alpha_i = Max(\alpha) - \alpha_i + 1$$
$$priority_i = \beta_i + \alpha_i^2$$

In the end, the two weights β and α^2 are blended together into a single priority weight. Each Pareto solution has exactly one priority weight.

We first calculate the sum of all priority weights, and then on the basis of each Pareto solution's individual $priority_i$ weight, set that solution's region to fall within the range between 0 and the sum of all priority weights. We choose a random number between 0 and this sum, and use this number to select a region, and then we use the Pareto solution that lies within this region as a reference.

4   Conclusion

Multi-objective optimization problems commonly have a high complexity, requiring significant execution time. When compared with such algorithms, MOBA offers both a reduced computation time and improved searching ability. This study provides a new multi-objective optimization algorithm that makes use of a efficient density mechanism, and offers criteria to evaluate searching results. There are a few testing functions that measure the ability of multi-objective optimization algorithms to find solutions: ZDT1, ZDT2 and ZDT3. The results are shown in figures 4-6.

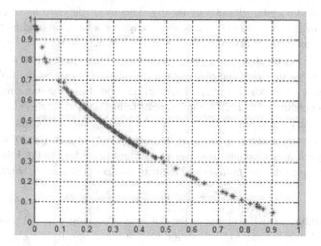

Fig. 4. ZDT1 result

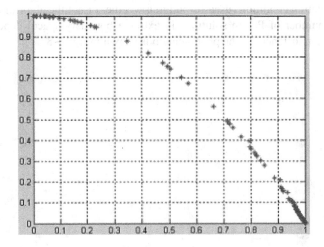

Fig. 5. ZDT2 result

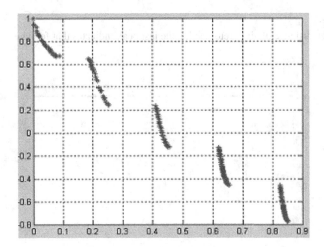

Fig. 6. ZDT3 result

This study also provides a new concept, whereby the search termination condition is obvious to reason about. MOBA improves the searching ability of multi-objective optimization algorithms and provides a way to evaluate a set of Pareto solutions to determine whether or not all the Pareto solutions in extreme boundary space have been found. After MOBA has already located all of the Pareto solutions in extreme boundary space, the arithmetic mean will gradually decrease, so we can use this phenomenon to detect when a sufficient number of Pareto solutions have been found.

The termination condition (used for judging whether or not the algorithm has found a sufficient number of Pareto solutions in extreme boundary space) depends on the arithmetic mean distance μ. The phenomenon is shown in fig. 7~8

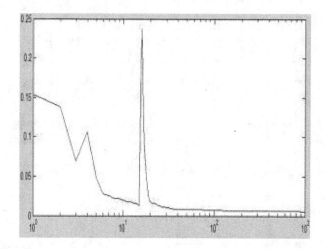

Fig. 7. Average distance of all Pareto results, the last peak indicates that found extreme position in ZDT1

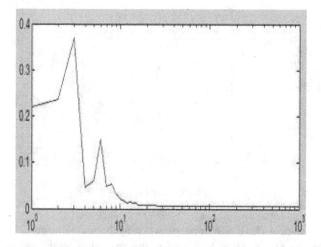

Fig. 8. Average distance of all Pareto results, the last and lower peak indicates that found extreme position in ZDT3

The last peak of μ in ZDT1 appears near the 40th iteration and in ZDT3 it appears near the 10th iteration. From the positions of these final peaks we can judge when the search has already completed.

In addition to solving two- or three-dimensional problems, multi-objective optimization algorithms can also be applied to higher dimensional problem. When tackling such problems, we use a similar criterion to judge when the search has completed, i.e. sufficient numbers of Pareto solutions in extreme boundary space have been located.

In the future, we will focus on three types of information provided by Young-Koo and Sungyoung[1]:

1. Independent information: this type of information is similar to that modeled by traditional single diffusion processes: the diffusion process only considers one source of information.
2. Mutually-exclusive information: this type of information comes from multiple sources or different types of information source. This type of information may involve conflicts or trade-offs. The relationship between separate pieces of information is static.
3. Competing information: this type of information can also come from multiple sources or different types of information source, and also may involve conflicts or trade-offs. The relationship between separate pieces of information is dynamic.

The complex relationships that exist between multiple sources of information are worthwhile target for MOBA. In the future we will determine the ability of MOBA to tackle multi-diffusion processes, and also measure its efficiency in doing so.

Pareto solutions can help firms and decision makers to allocate their resources more appropriately for a given level of efficiency. In the future, we will use MOBA to evaluate social networks that have multiple sources of information. Using MOBA to model a multi-diffusion environment, we should be able to handle multiple sources of information efficiently, even those sources that contain conflicts or involve trade-offs with each other. And by adding our density mechanism to different single-objective optimization algorithms we can estimate whether this approach leads to improved efficiency over those existing algorithms.

References

1. Fatima, I., Fahim, M., Lee, Y.-K., Lee, S.: MODM: multi-objective diffusion model for dynamic social networks using evolutionary algorithm. The Journal of Supercomputing 66(2), 738–759 (2013). Springer US, (2013-11-01)
2. Kempe, D., Kleinberg, J., Tardos, E.: Maximizing the spread of influence in a social network. In: Proceedings of the 9th International Conference On Knowledge Discovery And Data Mining, pp. 137–146 (2003)
3. Kempe, D., Kleinberg, J.M., Tardos, É.: Influential nodes in a diffusion model for social networks. In: Caires, L., Italiano, G.F., Monteiro, L., Palamidessi, C., Yung, M. (eds.) ICALP 2005. LNCS, vol. 3580, pp. 1127–1138. Springer, Heidelberg (2005)
4. Saito, K., Kimura, M., Ohara, K., Motoda, H.: Learning continuous-time information diffusion model for social behavioral data analysis. In: Zhou, Z.-H., Washio, T. (eds.) ACML 2009. LNCS, vol. 5828, pp. 322–337. Springer, Heidelberg (2009)
5. Kennedy, J., Eberhart, R.C., Shi, Y.: Swarm Intelligence. Morgan Kaufmann Publishers, San Francisco (2001)

6. Clerk, M., Kennedy, J.: The particle swarm-explosion, stability, and convergence in a multidimensional complex space. IEEE trans. Evol. Comput. **6**, 58–73 (2002)
7. Zitzler, E., Deb, K., Thiele, L.: Comparison of multiobjective evolutionay algorithms: empirical results. Evolutionary Computation **8**(2), 173–195 (2000)
8. Kennedy, J., Eberhart, R.: Particle swarm optimization. In: In: Proceedings of IEEE International Conference on Neural Networks, pp. 1942–1948 (1995)
9. Yang, X.-S., Deb, S.: Cuckoo Search via Lévy Flights. In: Nature & Biologically Inspired Computing, pp. 210–214. IEEE Publications, December 9–11, 2009
10. Yang, X.S.: Firefly Algorithm, Stochastic Test Functions and Design Optimisation. Papercore summary, pp. 1–11 (2008)
11. Mahdavia, M., Fesangharyb, M., Damangirb, E.: An improved harmony search algorithm for solving optimization problems. Applied Mathematics and Computation **188**(2), 1567–1579 (2007)
12. Yang, X.-S.: A new metaheuristic bat-inspired algorithm. In: González, J.R., Pelta, D.A., Cruz, C., Terrazas, G., Krasnogor, N. (eds.) NICSO 2010. SCI, vol. 284, pp. 65–74. Springer, Heidelberg (2010)
13. Rabanimotlagh, A.: An Efficient Ant Colony Optimization Algorithm for Multiobjective Flow Shop Scheduling Problem. World Academy of Science, Engineering and Technology **51**, 127–133 (2011)
14. Salmasi, N., Logendran, R., Skandari, M.R.: Total flow time minimization in a flowshop sequence-dependent group scheduling problem. Computers & Operations Research **37**(1), 199–212 (2010)
15. Kirkpatrick, S., Gelatt, C.D., Vecchi, M.P.: Optimization by Simulated Annealing. Science **220**(4598), 671–680 (1983)
16. Fieldsend, J.E., Singh, S.: A multi-objective algorithm based upon particle swarm optimization, an efficient data structure and turbulence. In: Proceedings of the 2002 U.K. workshop on computational intelligence, Birmingham, UK, pp. 37–44, September 2–4, 2002
17. Ling, H., Xiao, Y., Zhou, X., Jiang, X.: An improved PSO Algorithm for Constrained Multiobjective Optimization Problems. Computer Science and Service System (CSSS), pp. 3859–3863 (2009)
18. Worasucheep, C.: Solving constrained engineering optimization problems by the constrained PSO-DD. In: 5th International Conference on Electrical Engineering/Electronics, Computer, Telecommunications and Information Technology, vol. 1, pp. 5–8. IEEE, May 14–17, 2008
19. Hu, X., Eberhart, R.C., Shi, Y.: Particle swarm with extended memory for multiobjective optimization. In: Proceedings of the Swarm Intelligence Symposium, SIS 2003, pp. 193–197. IEEE, April 24–26, 2003

Individual Attachment Style, Self-disclosure, and How People Use Social Network

Rang-An Shang[1(✉)], Yu-Chen Chen[1], and Je-Wei Chang[2]

[1] Soochow University, Taipei 10048, Taiwan
rashang@scu.edu.tw
[2] Pegatron Corporation, Taipei 11261, Taiwan

Abstract. Social Network Site has been one of the fastest growing and popular applications in the Internet. People use SNS to maintain and enhance relationships with others. How they use SNS should be affected by how they evaluate social relationships with others, which is affected by individual attachment style. This study examined the impacts of individual attachment style on self-disclosure and user behavior in Facebook. The results of an online survey showed individual attachment style affects their posting and responding behaviors, willingness for self-disclosure also affected users posting, responding, and openness of their behavior.

Keywords: Social network service · Self-disclosure · Attachment style

1 Introduction

Social networking has been the most rapidly growth network applications for years. The social network sites (SNS) is a composite network application that helps people keep connected with their friends or with the others. It includes functions such as showing or sharing something on one's own space, signaling to others one's status, discussing in a circle of your friends or in public, and messaging to others in a one to one or many to many ways. Unlike traditional virtual communities, which primarily bring together a lot of strangers with a common interest to share their experiences, SNSs facilitate people building, maintaining, and showing their individual social network. Checking and showing on the social network such Facebook, Twitter, LinkedIn, have been an important activity in many people's everyday life [55].

The wildly spread SNS also attracts many studies to explore the secrets behind its successes. Users primarily use SNS to maintain pre-existing and create new friendships [22,35,55]; maintaining pre-exist friendships has been found to be more important than creating the new ones [11,41]. Park et al. [50] found socializing, entertainment, self-status seeking, and information as the primary needs for participating in groups within Facebook. The SNS provides users a personal space to show their profile, status, experiences, ideas, and social connections; therefore, it is also been perceived as a place for self-presentation that users can use to build and manage their personal image on the Internet [23,37,44].

© Springer-Verlag Berlin Heidelberg 2015
L. Wang et al. (Eds): MISNC 2015, CCIS 540, pp. 45–59, 2015.
DOI: 10.1007/978-3-662-48319-0_4

Many studies also showed the impacts of personality to explain individual differences in using SNS. In the big-five model of personality, previous studies have shown that people's online behavior is related to high neuroticism [12,27,72], high openness to experience [27], low extraversion [3,5] and low agreeableness [40,52]. Anyway, the results in the studies in SNSs are different that Krämer & Winter [37] found that extroversive posted more messages on the SNS. Ross [58] did not find significant connection between Facebook behavior and both agreeableness and openness, but found a partial link between behavior on Facebook and the traits of extroversion, neuroticism, and conscientiousness. Amichai-Hamburger and Vinitzky [3] using user-information upload on Facebook as objective measurements, showed strong connection between personality and Facebook behavior. Previous studies also found that the associations between personality and using SNS differ between male and female users [49,56].

Although there have been many studies on SNS, except for a few of them [49,56], most of these studies focused on an unidimensional usage and neglected that SNS is a composite service and people uses it in a diversity of ways. Users can actively post, share, passively respond, showing in public, contact other in private, or even just lurk and hide behind the screen. User showing up on the SNS to share his/her ideas, feelings, status, and experiences, is a self-disclosure behavior that one reveals something about him/herself to maintain or build friendships with others [2,17,18]. Mutual disclosure during social interactions on SNS helps people develop the feelings of familiar and trust and further enhance an intimate relationship [2,7,32].

People primarily use SNS to keep contact with others; how they expect from and respond to the relationship with others should be a main determinant of how they use SNS. Psychologists proposed attachment theory to describe the particular way in which one relates to other people. Attachment is a deep and enduring emotional bond that connects one person to another across time and space [1,10]. A person's attachment style is established from one's experiences with his/her main caregivers in one's early childhood and it is a central component that creates individual personality [28]. Previous studies also suggested individual attachment style as a major factor affecting one's self-disclosure behavior [15,33,45]. This study proposed and tested a model by a survey on the users of Facebook to show how personal attachment style, through the mediation of willingness for self-disclosure, affects how one use Facebook, including the time spend on it, posting, responding, liking, and openness of the posts and static profiles. The results of this study can help us understand better the psychological causes and value of using SNS, and explore the multi-dimensional nature of using SNS.

2   Theoretical Model

2.1   Self-disclosure

Previous studies on virtual communities showed that there are posters and lurkers in a virtual community and their behaviors differ a lot that the people who contribute information to the community may not be the people who browse for information [29,54,57,62]. In the same way, people also show a lot of differences in how they use SNS. Someone may spend a lot of time but just hide behind the screen to inspect what happened to their friends and feel they are connected. On the other hand, others may actively post or share something with others or passively response when they see something happens in their friends' wall.

Many network applications in the past were anonymous. Because no one knows who a user in the applications really is, these anonymous applications are often used as a tool for self-presentation, which users can create and maintain an image they would like to have [21,59]. On the contrary, many SNSs, such Facebook, use a real name system that users' identities are linked with their true identities; therefore, what they do in the application is directly tied up with the person they are in the real world. Deriega and Grzelak [18] defined self-disclosure as including "any information exchange that refers to the self, including personal states, dispositions, events in the past, and plans for the future." While people actively post something on the Facebook, no matter it's in the form of text, photo, picture, video, or just showing his/her status by liking or checking in, it's revealing something about oneself. The ability to reveal one's feelings and thoughts to another is a basic skill for developing close relationships [2,26]. Research on computer mediated communication suggested that the pattern of people's interaction in the computer mediated environment is similar with that in a face to face environment [67]. Revealing something about oneself to create a closer relationship with friends may be a major reason for people to actively participate in the Facebook. Therefore, we proposed the hypothesis that:

H1: User's willingness for self-disclosure positively affects his/her active posting behavior on Facebook.

Although revealing something about oneself can keep friends being connected and closer the distance to them, social penetration theory argues that intimate relationships can only be developed if both partners reciprocate disclosures. Reciprocity occurs when the person receiving disclosure from others self-discloses something in return. This process continues and gradually goes into a deeper and wider disclosure, and then mutual understanding and trust between the two also emerged [2,19,32]. Dietz-Uhler [20] showed that the norm of mutual disclosure exists in a synchronous computer-mediated discussion environment and the extent of disclosure will be expended under this norm. Disclosure reciprocity is a dimension of self-disclosure which refers to the tendency of recipients of disclosure to respond by disclosing about themselves at a comparable level of intimacy [45]. Except for active participant, users on Facebook often respond to others by leaving comments or likes. Responses on the Facebook, however, usually are not just between the two on conversation but open to the circle of all their friends. Therefore, responses on the Facebook may not be only the result of disclosure reciprocity, as in the offline mutual interaction, but also the result of the general willingness for self-disclosure. So we proposed the following hypothesis:

H2: User's willingness for self-disclosure positively affects his/her passive respondent behavior on Facebook.

Although disclosing personal information may help people have closer relationships with their friends, users still concern about the possibility of being hurt because some others, includes the SNS providers, strangers, or even one's peers, use the information in an inappropriate way. Privacy concern is a general concerns that reflect individuals' inherent worries about possible loss of information privacy [73]. Many studies have shown that privacy concern is a major factor that affects self-disclosure in SNSs [7,37,64,73]. In order to relieve users' concerns about privacy, Facebook provides many mechanisms for users to control the targets to whom their information is exposed. Protecting or spreading personal information is usually a trade-off [7,64]. As a personal characteristic, people with higher willingness for disclosure will be opener to others and have less restriction on how their information is shown on Facebook. So we proposed the hypothesis that:

H3: Users with higher willingness for self-disclosure will be opener in the privacy setting on Facebook.

2.2   Attachment Style

Attachment is an emotional bond to another person. Attachment theory attempts to describe the dynamics of long-term interpersonal relationships between humans. It assumes that the earliest bonds formed by a child with his or her major caregivers are strongly predictive of other close relationships later in life [6,9,10]. Attachment depends on the person's ability to develop basic trust in their caregivers and self. The early experiences of infants form their beliefs about themselves and others, then gradually create an inner working model and a static attachment pattern that influences how they reacts to their needs and go about getting them met [10]. Studies have shown that one's attachment style formed in the childhood still persists when one grows up [6,24,25,37]; it affects adult's behaviors such as how one interacts with his or her friends [51,75,76], one's romantic relationship [28], and even the leader-follower relationship at work [33,35,62].

Ainsworth et al. [1] developed the strange situation procedure to discover the nature of attachment behaviors. They proposed secure, ambivalent/anxious, and avoidant as three major styles of attachment. Children with secure attachment feel secure and are able to depend on their caregivers. They may feel upset when their parent leaves but will enjoy when the parent return. As adults, they are more likely to have trusting and long-term relationships with others. Both anxious and avoidant attachment are insecure styles. Anxiously attached children tend to be extremely suspicious of strangers. They usually become very distressed when a parent leaves and do not show comforted when their parent return. Adults with anxious style like to seek high levels of intimacy, approval, and responsiveness from their partners, but often feel reluctant about becoming close to others because they worry that others would not like to get as close as they do. Avoidant attached children show avoid their parents after they leave for a period of time. When they grow up, they tend to have difficulty with intimacy and close relationships.

Previous studies showed that secure attached people will be more cooperative [60], have more intimate friendship [6], and higher skills for conflict resolution [42], therefore, the self-disclosure intention of secure attached people will be higher than that of insecure attached people [33]. So we proposed the hypothesis that:

H4: Individual attachment style affects one's willingness for self-disclosure; the willingness of secure attached people will be higher than the willingness of insecure people.

Following the hypotheses H1 to H3, which about the effects of willingness for self-disclosure, we also proposed the following mediation hypotheses about the effects of secure attachment style on how people use Facebook:

H5a: Secure attached users will actively post in Facebook more than insecure users, through the mediation effect of willingness for self-disclosure.

H5b: Secure attached users will passively respond to others on Facebook more than insecure users, through the mediation effect of willingness for self-disclosure.

H5c: Secure attached users will be opener in privacy setting on Facebook than insecure users, through the mediation effect of willingness for self-disclosure.

Interactions in the computer mediated environment can increase the sense of security and reduce the social anxiety in the face to face interactions [52,60,69]. Anxious attached people would like to have a closer relationship with their partners, but are worry about other's responses. Online environment encourages them to be more active in seeking for more intimate relationships by relieving the pressure of facing instant responses. In the two types of insecure style, we proposed that:

H6a: Anxious attached users will actively post in Facebook more than avoidant attached users.

The study of Mikulincer and Nachshon [45] showed that in order to gain trust from others, anxious people tends to disclosed more information when they facing a high discloser partner. In contrast, avoidant people's behaviors were not affected by their partners. Therefore, anxious users may be more likely to response to others on Facebook than avoidant attached users. So we proposed the hypothesis that:

H6b: Anxious attached users will passively respond to others on Facebook more than avoidant attached users.

Anxious attached people would like to have intimate and close friendships with others, but they hesitate to actually build such relationships, and in the meantime, they usually worry about them. Anxious attached people also have higher level of separation anxiety than people with other types of attachment. Although they may not show up in front of others in the SNS, the SNS provides anxious style people a secure space to quietly browse the status of others and feel being connected with them. Previous studies in Internet addition found that higher levels of boredom proneness, loneliness, social anxiety, private self-consciousness and depression predict Internet addition [42,74]. Due to the special value of SNS for the anxious attached people, we proposed the following hypothesis:

H7: Anxious attached users will spend more time on Facebook than secure and avoidant attached users.

3   Research Method

We conducted a survey research to test our hypotheses. Hazan and Shaver [28] developed the Measure of Attachment Style (MAS) to measure the three styles of Ainsworth et al. [1]. Collins and Read [15] revised MAS and proposed the Adult Attachment Scale, which uses depend, anxiety, and close as dimensions to measure individual attachment style. Collins [14] further revised the scale into the Revised Adult Attachment Scale (RAAS). Huang and Chen [31] translated RASS into a Chinese version that showed a good reliability and construct validity. We used this Chinese version to measure individual attachment style and also followed the items used in Hu [29] to revise some wording problems found in the pretest. Willingness for self-disclosure was measured by the ten-item Self-Disclosure Index (SDI) [47]. This instrument has been widely used in previous studies in self-disclosure [45].

Previous studies usually designed their own instrument to measure how people used Facebook to satisfy the special needs of the study [4,16,48,49,58]. Active post was measured by the five items that asking the respondents about how often they shared links, posted text, checked in, posted photos and videos. Each item was measured in a 7-point scale of real frequencies, which had been revised in the pretest. Users on Facebook can respond to others by leaving messages and liking. Passive response was measured by asking the frequencies of responding to others' shared links, text messages, shared photos and videos, and check in status, by messages and by likings. These 8 items were all in a 5-point scale, from never to always.

Using time was measured by asking both the average log in and browsing time each day, in an 8-point scale of the actual time. Facebook provides a privacy protecting mechanism that the targets of personal information can be set as only me, custom, friends, friends of friend and public. This scale reflects the degree of openness of the privacy setting. We also added the responses of "not changed" or "by default" to the choice of public because they show that the respondent might not care about privacy. We divided personal information into the dynamic status and posts and the static profile. Status and posts include shared links, text messages, photos, videos, check in status, and tagged messages; Profile includes basic information, work and education, interests and loves, and contact information. We measured openness of privacy setting by 6 items about posting setting and 4 items about profile setting, both in the 5-poing scale mentioned above.

Because we measured actual usage as the dependent variables, we added facilitating conditions in the study as a control variable [65]. We revised the instrument of Venkatesh et al. [65] to ask respondents in four items about if they had convenient network environment and equipment to satisfy their needs. We also added a question that "I have never used computer and network equipment before" as the nonsensical item to check for the careless responses [45].

We posted invitations in several discussion boards in PTT (http://www.ptt.cc) to invite members to take part in the pretest. PTT is a non-commercial and the biggest electronic bulletin board system in Taiwan, with more than 300,000 concurrent users during the peak time. We offered respondents 250 Pcoin as incentive for participation; Pcoin is a currency used in the PTT for some privileges and requiring for special services. Finally, 70 valid responses were obtained from a total of 100 responses. The Cronbach's α of all variables were larger than or near 0.7. After the pretest, we revised the scale and wording of some items to improve the quality of the measurement.

Respondents in the formal study were obtained in the same way as in the pretest and we totally obtained 380 respondents. We dropped 101 invalid ones that showed logical error in the two questions of using time and in the nonsensical question, and that showed a large distance between the scores of a pair of reversed question when one of them was reversed. We finally got 279 valid responses. 55.6% of them are male. 45.2% are in the age between 19 and 22 and 36.9% are between 23 and 25. 82.4% of the respondents are student. Furthermore, 24% of them posted 1 to 2 text messages per week. 21.5% of them logged in Facebook about 1 to 2 hours each day, 21.5% about 3 to 4 hours, and 14.7% more than 6 hours. This results show that SNS has heavily penetrated into many users' everyday life.

4   Results

4.1   Measurement Model

We followed the procedure in previous studies using RASS to determine individual attachment style [13,15,31]. First, an exploratory factor analysis was conducted on the 18 RASS items. Three factors were extracted according to the result of scree test and these factors correspond to Collins and Read [15]'s dimensions of close, depend, and anxiety. 5 items were dropped because they were not loaded in the expected factor. The Cronbach's α of depend and anxiety were larger than 0.7 and the Cronbach's α of close were 0.68. Secondly, we used the standardized factor score of the three dimensions to conduct a cluster analysis using K-means to classify the subjects into three clusters. Finally, a multivariate analysis of variance and multiple comparisons using Scheffe's method were conducted and showed that there were significant differences between the three clusters. The result showed a similar structure with Collins and Read [15].

The cluster I, which comprises 91 subjects (32.6%), shows the highest score in close and depend dimension and lowest in anxiety, was classified as the secure attachment. Cluster II, which shows the lowest score in close and middle score in depend and anxiety, was avoidant attachment. There are 104 (37.3%) subjects in this cluster, which is the biggest among the three. This result is similar with the study of Hu [29]. Clusters III gets the highest score in anxiety, middle score in close, and lowest score in depend. We named cluster III as anxious attachment and there are 84 (30.1%) subjects in this cluster. The ratio of the three styles significantly differs between the two genders ($\chi^2(2)=6.28$, p=0.043). Anxious is the biggest and secure is the smallest cluster in the male sample, while in the female sample, avoidant is the biggest and anxious is the smallest. This result is also similar with the results in previous studies [13,70].

We also conducted an exploratory factor analysis using principal axis method and orthogonal rotation to test the measurement model of other variables. Response by leaving message and by liking, and openness of posts and openness of profile were all divided into different factors. So we used two variables to measure passive response and two variables for openness in the privacy setting in further analyses. In the items of using behavior, upload and post video, response to other's check in by liking and leaving message were not loaded in the predicted factor. We dropped these three items from the study. Furthermore, one item in willingness for self-disclosure was also dropped because the factor loading was too low. The Cronbach's α of active post is 0.652; except for that, the Cronbach's α of all other variables are larger than 0.7.

4.2 Hypotheses Testing

We first conducted regression analyses to test the impacts of willingness for self-disclosure on active post, responses by messaging, liking, using time, openness of posts, and openness of profile. Facilitating conditions were added in the model as a control variable and the results are shown in Table 1. The results show willingness for self-disclosure significantly increases active post, the two variables of passive responses, and the two variables of openness of privacy setting. The hypothesis 1, 2 and 3 are all supported. We also tested the effect of willingness for self-disclosure on using time for the exploratory purpose, and the result shows an insignificant effects.

Then we conducted an ANOVA to test hypothesis 4. The result shows there are differences in the willingness for self-disclosure among groups of different attachment style $(F(2/276)=4.60, p<0.05)$. H4 is supported that the results of multiple comparisons using Scheffe's method show that the mean in secure attached group significantly larger than that in avoidant and anxious attached groups, but the difference between the two insecure groups is not significant.

Finally, we conducted ANCOVA to test the effects of attachment styles on the six dependent variables. We tested in the model I the effects of attachment styles on the dependent variables, and then added willingness for self-disclosure in model II to test the mediation effects. Because the effect of willingness for self-disclosure on using time is insignificant, the model II for the effect on using time was not tested. Facilitating conditions was added in all of the models as a control variable. Table 2 shows the summary of these results.

Table 1. Regressions of willingness for self-disclosure on using behaviors

	Active post	Response	Liking	Using time	Openness of posts	Openness of profile
FC[a]	0.141*	0.122*	0.143*	0.134*	-0.101	-0.054
WSD	0.179**	0.215**	0.135*	0.088	0.193**	0.209**
R^2	.058	.067	.043	.028	.043	.044
Adjusted R^2	.051	.060	.036	.021	.036	.037
F(2/276)	8.469**	9.940**	6.234**	4.032*	6.193**	6.357**

FC: facilitating conditions; WSD: willingness for self-disclosure
[a] : standardized regression coefficient
*: $p<0.05$;   **: $p<0.01$

Table 2. ANCOVA of attachment style, willingness for self-disclosure on using behaviors

	Active post	Response	Liking	Using time	Openness of posts	Openness of profile
Model I						
Corrected model[a]	6.544**	6.117**	5.698**	2.713*	1.750	0.602
Intercept	32.299**	124.735**	183.324**	42.291**	187.254**	113.093**
AS	5.972**	6.046**	4.879**	1.128	1.741	0.776
FC	5.906*	5.086*	5.885*	5.856*	2.033	0.248
R^2	.067	.063	.059	.029	.019	.007
Adjusted R^2	.056	.052	.048	.018	.008	.004
Estimated mean						
Secure	2.643	3.402	3.936	4.745	3.560	3.295
Avoidant	2.304	3.040	3.645	4.474	3.331	3.156
Anxious	2.194	3.163	3.652	4.769	3.444	3.318
Pairwise Comparison[b]	Secure> [linefeed] Avoidant, [linefeed] Anxious	Secure> Anxious, Avoidant	Secure> Anxious, Avoidant	n.s.	n.s.	n.s.
Model II						
Corrected model	6.618**	7.325**	5.118**	--	3.628**	3.516**
Intercept	5.100*	33.070**	69.435**	--	58.909**	26.997**
AS	4.549*	4.461*	3.871*	--	1.060	0.689
FC	4.809*	3.867*	5.050*	--	3.014	0.719
WSD	6.450*	10.326**	3.238	--	9.108**	12.182**
R^2	.088	.097	.070	--	.050	.049
Adjusted R^2	.075	.083	.056	--	.036	.035
ΔR^2	.021	.034	.011	--	.031	.042
Estimated mean						
Secure	2.609	3.367	3.916	--	3.522	3.245
Avoidant	2.320	3.055	3.654	--	3.348	3.179
Anxious	2.212	3.181	3.662	--	3.463	3.344
Pairwise Comparison[b]	Secure> [linefeed] Avoidant, [linefeed] Anxious	Secure> Avoidant	Secure> Anxious, Avoidant	--	n.s.	n.s.

[a]: F value
[b]: using LSD method in 0.05 significant level
AS: attachment style; FC: facilitating conditions; WSD: willingness for self-disclosure
*: $p<0.05$; **: $p<0.01$

For the effects on active post, secure attached users show significantly higher score than anxious and avoidant users, but this effect is not changed when willingness for self-disclosure was added into the model. Hypothesis 5a is partially supported that secure attached users post more than anxious and avoidant users, but the effect is not mediated by willingness for self-disclosure. Similar results are found in the models on liking. On the other hand, adding willingness for self-disclosure decreased the impact of attachment style on response with messages. The difference between secure and anxious attached users is insignificant when the impact of willingness for self-disclosure is controlled. Therefore, the mediation hypothesis 5b is supported when the response is measured by text response. The effects of attachment style on the two variables of privacy setting are all insignificant. We failed to support hypothesis 5c. Although secure attached users show higher willingness for self-disclosure, there may be other mechanism that neutralizes this effect so the effect of attachment style on privacy setting was insignificant. Anyway, the pairwise comparisons in the model I on the openness of posts show that the p value of the difference between secure and avoidant style is 0.063. This effect can be checked again in future study.

Both hypothesis 6a and 6b are not supported because the differences between anxious and avoidant users in the models on active post, response, and liking are all insignificant. In the model II on response, however, the result shows a similar trend with our hypothesis 6b that the difference between secure and anxious users is insignificant when the impact of willingness for self-control is controlled. Finally, hypothesis 7 is not supported. Although anxious users show longer using time than secure and avoidant users, the difference is insignificant. For the exploratory purpose, we added active post into the model to exclude the impact of active participation. The result show that the p value of the difference between anxious and avoidant attached users is 0.068. This result can also be checked further with a different sample.

5   Conclusions

This study investigated the impacts of individual attachment style and willingness for self-disclosure on how people use Facebook, including active post, passive response, liking, using time, and openness of post and profile. The study found that, first, secure attached users more actively and passively participate in Facebook than insecure users. Except the effect on text response, these effects are not mediated by its effect on willingness for self-disclosure. Instead of willingness for self-disclosure, we need to find other mechanism which can explain the effect of attachment style on participating in Facebook.

Secondly, we tried but failed to show the different effects of anxious and avoidant attached users; however, although there is significant difference between secure and anxious users in active posting, the difference turned to be insignificant when it went to passive responses. Being consist with our conjecture, this result suggests that SNS may provide an easy way for anxious people to improve friendships with others by passively respond to them. Besides, although we did not find the effects of attachment style on using time, a further analysis shows a weak positive effect of anxious style when the impact of active post is controlled. Future studies can try to find more evidence for the differences between anxious and avoidant style users.

Finally, openness of privacy setting was found to be positively affected by individual willingness for self-disclosure. Secure style shows a weak effect that it could be opener in posting than the avoidant style and this effect is totally mediated by willingness for self-disclosure. Anyway, this effect also needs to be verified in future studies.

This study suggests that satisfying one's needs for attachment may be a major reason for one to use SNS. With a variety of functions, SNS provides a space for users with different attachment style to fulfill their needs for bond with others. Some of our hypotheses, however, only show a weak effect. Because we used object indicators to measure actual using behavior, it's possible that we need more control variables to show the effects on actual behaviors. Furthermore, we suggested willingness for self-disclosure as the mediation variable that explains the effects of attachment style on using behaviors. Self-disclosure is a multi-dimensional construct. Except for the willingness, it can also described by the dimensions such as breadth, depth, amount, honesty, accuracy, positiveness [2,19], flexibility, reciprocity [46], emotional/affective [67], or intimate/descriptive [65]. Future studies can add these dimensions into current model to form a more complete explanation.

References

1. Ainsworth, M.D.S., Blehar, M.C., Waters, E., Wall, S.: Patterns of Attachment: A Psychological Study of the Strange Situation. Lawrence Erlbaum Associates, Pub., Hillsdale (1978)
2. Altman, I., Taylor, D.A.: Social Penetration: The Development of Interpersonal Relationships. Holt, Rinehart & Winston, New York (1973)
3. Amichai-Hamburger, Y., Ben-Artzi, E.: Loneliness and Internet Use. Computers in Human Behavior 19(1), 71–80 (2003)
4. Amichai-Hamburger, Y., Vinitzky, G.: Social Network Use and Personality. Computers in Human Behavior 26, 1289–1295 (2010)
5. Bargh, J.A., McKenna, K.Y.A., Fitzsimons, G.M.: Can You See the Real Me? Activation and Expression of the "True Self" on the Internet. Journal of Social Issues 58(1), 33–48 (2002)
6. Bartholomew, K., Horowitz, L.M.: Attachment Styles among Young Adults: A Test of a Four–category Model. Journal of Personality and Social Psychology 61(2), 226–244 (1991)
7. Bateman, P., Pike, J.C., Butler, B.S.: To Disclose or Not: Publicness in Social Networking Sites. Information, Technology & People 24(1), 78–100 (2011)
8. Berg, J.H.: Responsiveness and self-disclosure. In: Derlega, Y.J., Berg, J.H. (eds.) Self Disclosure, pp. 101–130. Plenum Press, New York (1987)
9. Bowlby, J.: The Nature of the Child's Tie to His Mother. International Journal of Psycho-Analysis 39, 350–373 (1958)
10. Bowlby, J.: Attachment and Loss, vol. 1. Attachment. Basic Books, New York (1969)
11. Boyd, D.M., Ellison, N.B.: Social Network Sites: Definition, History, and Scholarship. Journal of Computer Mediated Communication 13(1), article 11 (2007)
12. Butt, S., Phillips, J.G.: Personality and Self Reported Mobile Phone Use. Computers in Human Behavior 24(2), 346–360 (2008)
13. Chiu, C.-Y.: An Exploratory Study on Personality and Experience of the Cyber Lovers. Unpublished Master's Thesis, National Chiao Tung University, Hsinchu, Taiwan (2003)

14. Collins, N.L.: Working Models of Attachment: Implications for Explanation, Emotion, and Behavior. Journal of Personality and Social Psychology **71**, 810–832 (1996)
15. Collins, N.L., Read, S.J.: Adult Attachment, Working Models, and Relationship Quality in Dating Couples. Journal of Personality and Social Psychology **58**(4), 644–663 (1990)
16. Correa, T., Hinsley, A., de Zuniga, H.: Who Interacts on the Web? The Intersection of Users' Personality and Social Media Use. Computers in Human Behavior **26**(2), 247–253 (2010)
17. Crosby, L.A., Bitner, M.J., Gill, J.D.: Organizational Structure of Values. Journal of Business Research **20**(2), 123–134 (1990)
18. Derlega, V.X., Grzelak, J.: Appropriateness of self-disclosure. In: Chelune, G.J. (ed.) Self-disclosure, pp. 151–176. Jossey-Bass, San Francisco (1979)
19. Derlega, V.J., Metts, S., Petronio, S., Margulis, S.T.: Self-Disclosure. Sage Pub., Newbury Park (1993)
20. Dietz-Uhler, B., Bishop-Clark, C., Howard, E.: Formation of and Adherence to a Self-disclosure Norm in an Online Chat. CyberPsychology & Behavior **8**, 114–120 (2005)
21. Dominick, J.R.: Who do You Think You Are? Personal Home Pages and Self-presentation on the World Wide Web. Journalism and Mass Communication Quarterly **76**(4), 646–658 (1999)
22. Ellison, N., Steinfield, C., Lampe, C.: The Benefits of Facebook "Friends": Exploring the Relationship between College Students' Use of Online Social Networks and Social Capital. Journal of Computer Mediated Communication **12**(3), article 1 (2007)
23. Fogel, J., Nehmad, E.: Internet Social Network Communities: Risk Taking. Trust, and Privacy Concerns Computers in Human Behavior **25**(1), 153–160 (2009)
24. Fraley, R.C., Brumbaugh, C.: A dynamical systems approach to understanding stability and change in attachment security. In: Rholes, W.S., Simpson, J.A. (eds.) Adult Attachment: Theory, Research, and Clinical Implications, pp. 86–132. Guilford Press, New York (2004)
25. Fraley, R.C., Shaver, P.R.: Attachment theory and its place in contemporary personality theory and research. In: John, O.P., Robins, R.W., Pervin, L.A. (eds.) Handbook of Personality: Theory and Research, pp. 518–541. Guilford Press, New York (2008)
26. Greene, K., Derlega, V.J., Mathews, A.: Self-disclosure in personal relationships. In: Vangelisti, A.L., Perlman, D. (eds.) The Cambridge Handbook of Personal Relationships, pp. 409–427. Cambridge University Press, New York (2006)
27. Guadagno, R.E., Okdie, B.M., Eno, C.A.: Who Blogs? Personality Predictors of Blogging. Computers in Human Behavior **24**(5), 1993–2004 (2008)
28. Hazan, C., Shaver, P.: Romantic Love Conceptualized as an Attachment Process. Journal of Personality and Social Psychology **52**(3), 511–524 (1987)
29. Hu, W.-K.: Studying the Influence of Attachment Style on Blogger's Self-Disclosure. Unpublished Master's Thesis, National Sun Yat-sen University, Kaohsiung, Taiwan (2008)
30. Huang, L.T., Farn, C.K.: Effects of virtual communities on purchasing decision-making: the moderating role of information activities. In: Proceedings of the 13th Pacific Asia Conference on Information Systems, Hyderabad, India (2009)
31. Huang, Y.-L., Chen, S.-H.: Psychometric Properties of the Chinese Version of Revised Adult Attachment Scale and Its Prediction to Psychological Adjustment. Chinese Journal of Psychology **53**(2), 209–227 (2011) (in Chinese)
32. Jourard, S.M.: Self-disclosure: An Experimental Analysis of the Transparent Self. Wiley-Interscience, New York (1971)
33. Kahn, W., Kram, K.: Authority at Work: Internal Models and Their Organizational Consequences. The Academy of Management Review **19**(1), 17–50 (1994)

34. Keelan, P.J., Dion, K.K., Dion, K.L.: Attachment Style and Relationship Satisfaction: Test of a Self-disclosure Explanation. Canadian Journal of Behavioural Science **30**(1), 24–35 (1998)
35. Keller, T.: Parental Images as a Guide to Leadership Sense Making: An Attachment Perspective on Implicit Leadership Theories. The Leadership Quarterly **14**(2), 141–160 (2003)
36. Kim, Y., Sohn, D., Choi, S.M.: Cultural Difference in Motivations for Using Social Network Sites: A Comparative Study of American and Korean College Students. Computers in Human Behavior **27**(1), 365–372 (2011)
37. Kobak, R.R., Sceery, A.: Attachment in Late Adolescence: Working Models, Affect Regulation, and Representations of Self-and Others. Child Development **59**(1), 135–146 (1988)
38. Krasnova, H., Koleva, E., Guenther, O.: It won't Happen to Me! Self-disclosure in Online Social Networks. Proceedings of the 15th Americas Conference on Information Systems, paper 343. San Francisco (2009)
39. Krämer, N.C., Winter, S.: Impression Management 2.0: The Relationship of Self-esteem, Extraversion, Self Efficacy, and Self-presentation within Social Networking Sites. Journal of Media Psychology: Theories, Methods, and Applications **20**(3), 106–116 (2008)
40. Landers, R.N., Lounsbury, J.W.: An Investigation of Big Five and Narrow Personality Traits in Relation to Internet Usage. Computers in Human Behavior **22**(2), 283–293 (2006)
41. Lenhart, A.: Adults and Social Network Web Sites. Pew Internet and American Life Project Report (2009). retrieved from http://www.pewinternet.org/Reports/2009/Adults-and-Social-Network-Websites.aspx
42. Liebermam, M., Doyle, A., Markiewicz, D.: Developmental Patterns in Security of Attachment to Mother and Father in Late Childhood and Early Adolescence: Associations with Peer Relations. Child Development **70**, 202–213 (1999)
43. Loytsker, J., Aiello, J.R.: Internet Addiction and its Personality Correlates. Poster Presented at the Annual Meeting of the Eastern Psychological Association, Washington, DC (1997)
44. Manago, A.M., Graham, M.B., Greenfield, P.M., Salimkhan, G.: Self-presentation and Gender on MySpace. Journal of Applied Developmental Psychology **29**(6), 446–458 (2008)
45. Meade, A.W., Craig, S.B.: Identifying Careless Responses in Survey Data. Psychological Methods **17**(3), 437–455 (2012)
46. Mikulincer, M., Nachshon, O.: Attachment Styles and Patterns of Self-disclosure. Journal of Personality and Social Psychology **61**(2), 321–331 (1991)
47. Miller, L.C., Berg, X.H., Archer, R.L.: Openers: Individuals who Elicit Intimate Self-Disclosure. Journal of Personality and Social Psychology **44**(6), 1234–1244 (1983)
48. Moore, K., McElroy, J.C.: The Influence of Personality on Facebook Usage, Wall Postings, and Regret. Computers in Human Behavior **28**, 267–274 (2012)
49. Muscanell, N.L., Guadagno, R.E.: Make New Friends or Keep the Old: Gender and Personality Differences in Social Networking Use. Computers in Human Behavior **28**(1), 107–112 (2012)
50. Park, N., Kee, K.F., Valenzuela, S.: Being Immersed in Social Networking Environment: Facebook Groups, Uses and Gratifications, and Social Outcomes. CyberPsychology & Behavior **12**(6), 729–733 (2009)
51. Pedersen, D.M., Higbee, K.L.: Personality Correlates of Self-disclosure. Journal of Social Psychology **78**(1), 81–89 (1969)
52. Peter, J., Valkenburg, P.: Adolescents' Exposure to Sexually Explicit Material on the Internet. Communication Research **33**(2), 178–204 (2006)

53. Peters, C.S., Malesky, L.A.: Problematics Usage among Highly-engaged Players of Massively Multiplayer Online Role Playing Games. CyberPsychology & Behavior **11**(4), 481–484 (2008)

54. Preece, J., Nonnecke, B., Andrews, D.: The Top Five Reasons for Lurking: Improving Community Experiences for Everyone. Computers in Human Behavior **20**(2), 201–223 (2004)

55. Raacke, J., Bonds-Raacke, J.: MySpace and Facebook: Applying the Uses and Gratifications Theory to Exploring Friend-Networking Sites. CyberPsychology & Behavior **11**(2), 169–174 (2008)

56. Rice, L., Markey, P.M.: The Role of Extraversion and Neuroticism in Influencing Anxiety following Computer-mediated Interactions. Personality and Individual Differences **46**(1), 35–39 (2009)

57. Ridings, C., Gefen, D., Arinze, B.: Psychological Barriers: Lurker and Poster Motivation and Behavior in Online Community. Communications of the Association for Information Systems **18**, 329–354 (2006)

58. Ross, C., Orr, E.S., Sisic, M., Arseneault, J.M., Simmering, M.G., Orr, R.R.: Personality and Motivations Associated with Facebook Use. Computers in Human Behavior **25**(2), 578–586 (2009)

59. Schau, H.J., Gilly, M.C.: We are What We Post: Self-presentation in Personal Web Space. Journal of Consumer Research **30**, 385–404 (2003)

60. Schouten, A.P., Valkenburg, P.M., Peter, J.: Precursors and Underlying Processes of Adolescents' Online Self-disclosure: Developing and Testing an "Internet-attribute-perception" Model. Media Psychology **10**(2), 292–314 (2007)

61. Special, W.P., Li-Barber, K.T.: Self-disclosure and Student Satisfaction with Facebook. Computers in Human Behavior **28**(2), 624–630 (2012)

62. Troth, A., Miller, J.: Attachment links to team functioning and leadership behavior. In: Sheehan, M., Ramsay, S., Patrick, J. (eds.) Transcending Boundaries: Integrating People, Processes, and Systems, 1999, pp. 385–392. Griffith University, Brisbane (2000)

63. Tonteri, L., Kosonen, M., Ellonen, H.K., Tarkiainen, A.: Antecedents of an Experienced Sense of Virtual Community. Computers in Human Behavior **27**(6), 2215–2223 (2011)

64. Tufekci, Z.: Can You See Me Now? Audience and Disclosure Management in Online Social Network Sites. Bulletin of Science and Technology Studies **28**(1), 20–36 (2008)

65. Van Horn, K.R., Arnone, A., Nesbitt, K., Desilets, L., Sears, T., Giffin, M., Brudi, R.: Physical Distance and Interpersonal Characteristics in College Students Romantic Relationships. Personal Relationships **4**(1), 25–34 (1997)

66. Venkatesh, V., Morris, M.G., Davis, G.B., Davis, F.D.: User Acceptance of Information Technology: Toward a Unified View. MIS Quarterly **27**(3), 425–478 (2003)

67. Vera, E.M., Betz, N.E.: Relationships of Self-Regard and Affective Self-Disclosure to Relationship Satisfaction in College Students. Journal of College Student Development **33**(5), 422–430 (1992)

68. Walther, J.B., Anderson, J.F., Park, D.W.: Interpersonal Effects on Computer-mediated Interaction: A Meta-analysis of Social and Antisocial communication. Communication Research **21**(4), 3–43 (1994)

69. Wang, J.L., Jackson, L.A., Zhang, D.J.: The Mediator Role of Self-disclosure and Moderator Roles of Gender and Social Anxiety in the Relationship between Chinese Adolescents' Online Communication and Their Real-world Social Relationships. Computers in Human Behavior **27**(6), 2161–2168 (2011)

70. Wang, Y.-M.: A Study of Adult Attachment and the Usage of Internet Friend-making Function. Unpublished Master's Thesis, Chinese Culture University, Taipei, Taiwan (2005)
71. Waters, E., Merrick, S., Treboux, D., Crowell, J., Albersheim, L.: Attachment Security in Infancy and Early Adulthood: A Twenty-year Longitudinal Study. Child Development **71**(3), 684–689 (2000)
72. Wehrli, S.: Personality on Social Network Sites: An Application of the Five Factor Model. ETH Zurich Sociology Working Paper No. 7, Swiss Federal Institute of Technology Zurich (2008)
73. Xu, H., Dinev, T., Smith, J., Hart, P.: Information Privacy Concerns: Linking Individual Perceptions with Institutional Privacy Assurances. Journal of the Association for Information Systems **12**(12), 798–824 (2011)
74. Young, K.S., Rodgers, R.C.: The Relationship between Depression and Internet Addiction. CybersPsychology and Behavior **1**(1), 25–28 (1998)
75. Youngblade, L.M., Belsky, J.: Parent-child Antecedents of 5-year-olds' Close Friendships: A Longitudinal Analysis. Developmental Psychology **28**(4), 700–713 (1992)
76. Zimmerman, P.: Attachment Representations and Characteristics of Friendship Relations during Adolescence. Journal of Experimental Child Psychology **88**, 81–101 (2004)

An Innovative Use of Multidisciplinary Applications Between Information Technology and Socially Digital Media for Connecting People

Chutisant Kerdvibulvech[✉]

Graduate School of Communication Arts and Management Innovation,
National Institute of Development Administration, 118 SeriThai Rd., Klong-chan,
Bangkapi, Bangkok 10240, Thailand
chutisant.ker@nida.ac.th

Abstract. Although digital media, social media and new media were found for a while, an overview of integrated multidisciplinary research of new, social and digital media with information technology, such as augmented reality, virtual reality and interactive applications has not been yet presented. This paper discusses the current development of the socially digital media work applying augmented reality for advertising, marketing and display research. These are a very important issue for developing some new social networks, i.e. to connect people through various senses of human (i.e., vision, hearing, somatosensation, olfactory and gustatory). In other words, a literature review of the recent new media and augmented reality works is discussed also. Rather than only augmented reality, this paper also provides a brief concept for future new media research when utilizing and using some other innovative technologies into every human sense. By tackling these complex problems, we can find advanced solutions breeding novel trends in socially digital media and information technology in an integrated and multidisciplinary manner.

Keywords: Socially digital media · Social media · New media · Augmented reality · Interactive experience · Multidisciplinary · Innovative use · Information technology

1   Introduction

In this current digital age, media consumption is not anymore restricted only to radios, books, newspapers, magazines, and television screens. It is now enjoying broader, wider and more interestingly use every year, from computers and smartphones to social media and multiscreen environments. Furthermore, with the rise of social media websites such as Facebook, Instagram, YouTube, Twitter and the emergence of digital television, the way people, especially young people [1], learn and obtain information has been obviously revolutionized over time. The conventional way of media has now become automatically digital media, sometimes known as new media or multimedia in some cases [2]. Some of examples of digital media are video, audio, and blogs. Zigmond and Stipp mentioned well in Harvard Business Review [3] about

© Springer-Verlag Berlin Heidelberg 2015
L. Wang et al. (Eds): MISNC 2015, CCIS 540, pp. 60–69, 2015.
DOI: 10.1007/978-3-662-48319-0_5

the emergence of digital media and the combination of television watching and internet use technologically for advertisers, particularly multitaskers. Moreover, media and advertising industries are increasingly focusing about a new technology to apply beneficially in the context of information technology and communication arts such as socially digital media. Since the influence of the internet is increasingly ever-growing and more and more people get technologically connected to it, the blending of digital media with other innovative technologies and information tools in this 21st century has become a dynamic trend from many leading researchers worldwide, both art and science. Digital media expand increasingly and scientifically horizons into new sensory modalities every day, while social media are not just any individual platform, but also used in more integral platform. Social media are often used in socially advertising and marketing for our today's business.

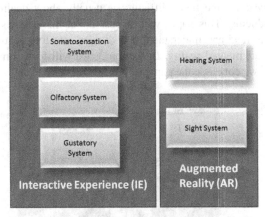

Fig. 1. The term 'interactive experience' is defined for communication in somatosensation, olfactory and gustatory.

In this paper, we provide an overall concept for creative technologies utilizing in new media to connect people socially and digitally. We do this by dividing technological communications into two main categories: augmented reality for visual communication (sight system) and interactive experience (somatosensation, olfactory and gustatory systems). The first category is augmented reality (AR) that was defined long time ago. Augmented reality is a technological filed that integrates the generated virtual world of the computer with the physical world of the human being and user [4]. Second, we define the term 'interactive experience', for other three senses, except only sight and hearing. We shortly called it IE in this paper. Figure 1 depicts our presented definition clearly. Even though it is also important, in this paper we do not focus on the hearing system, as it has been presented long time ago for communication such as 2G mobile phone.

First, augmented reality has recently used in 3D model, digital media, advertising and even some learning tools in school. As described in [5] by Sarracino, it also used to improve the learning process of young students in the museum. Briefly speaking, it is done by overlaying computer graphical information with the camera view of computer devices to help them to learn. Augmented reality is applied to advertising, public relations and integrated product designing systems. Various vision-based advertising and public relations works we found were applied this technology. In [6], Bule and Peer presented interactively an augmented reality advertising system. This system aimed to change marketing strategies by using a camera to insert augmented reality thoughts to the users in front of the real-time camera in the form of entertained graphical aid information (i.e., comic books and speech balloons). Second, we believe that digital media in the future is gone beyond only one individual sense of human. It will shift to the interactive experience, mainly in three senses: somatosensation, olfactory and gustatory systems. However, by connecting people through the aforementioned senses is not completely trivial. Each system for communication has their unique difficulties. For digital somatosensation communication, the main challenge is how to reproduce the perception realistically, naturalistically and correctly. For olfactory and gustatory systems, the main difficulty is usually about dealing with chemical elements for reproducing. This paper will discuss about these issues in the following sections. For easy to read and understand, it is divided into four sections. In section 2, we will talk about augmented reality. In section 3, we will focus on interactive experience. In section 4, we conclude the paper.

Fig. 2. An interactive augmented reality marketing system was built by Bule and Peer [6] to augment comic books speech balloons as the thought of customer service.

2   Augmented Reality

Although all definitions of augmented reality, augmented virtuality (AV), mixed reality (MR) and virtual reality (VR) are quite similar, augmented virtuality and augmented reality are a subset of mixed reality. At the same time, virtual reality is a one hundred percent graphical immersive environment. Each application of augmented reality, augmented virtuality, mixed reality and virtual reality is usually done in real-time, but it is not always required. The concept of augmented reality, augmented virtuality and mixed reality is used not only in socio-scientific issues, but also in advertising, public relations and integrated product designing systems. This is because we find many technological marketers use this technology to promote their products online [7]. It is deemed to change conservative marketing strategies. For example, Bule and Peer [6] built an interactive augmented reality marketing system to overlay thoughts to the customers in the form of comic books speech balloons for marketing purpose. This interactive augmented reality marketing system augments comic books speech balloons as the thought of customer service, as depicted in Figure 2. It uses a vision-based algorithm to detect face position, so that the comic caption is projected accurately above the customer's head. The system is noted that the computation process is in real-time which is important for an interactive augmented reality application. They tested the system in a real environment at two exhibition fairs: Hairdressing exhibition fair and Coinvest investment conference in Slovenia, southern Central Europe. It is noted in their paper that people stopped almost always, particularly a group of people both young and old, when they pass by the booth of experiment. They were entertained when they saw themselves on the large LCD or projector screen and even more so after the comic books speech balloon appeared over their head. The system tracks the head's position automatically while the text can also be changed, so that it is good for marketing strategies.

Fig. 3. The CYBERII financial technology for financial trading in digital media using augmented reality was developed by Maad et al. from Trinity College Dublin [8].

Increasingly, there are numerous research works towards augmented reality in marketing and advertising strategies. For instance, Maad et al. from Trinity College Dublin [8] explained about emerging digital media for electronic capital market trading using augmented reality. They mentioned clearly "*It motivates further ergonomic studies involving the evaluation of utility and usability of augmented reality technologies, including the CYBERII technology, in the field of electronic commerce.*" found in [9]. Figure 3 displays how this CYBERII financial application works. The work features the augmentation of people in a graphical world of capital market indicators to explore. In other words, the system inserts and augments the perception of each individual of the financial services and capital market activity with immersive information for exploration and analysis. This integration between a physical world and a graphical world of capital market indicators reflects some real enviroment through a correspondence closely with immersive experience. It extends the possibility of augmented reality-based tool to show immersive experience more immersive, social and natural. Furthermore, a prototype system [10] of augmented reality-based advertising strategies is presented. It is done by overlaying some paper leaflets of a retailer with one's own immersive content. Figure 4 displays the prototype system in a real environment. On the left side, it is a bundle offer, while the right side is price and quality comparison. By using this prototype, it is easier for a retailer to check comparably both the quality and the price of the products conveniently. Simultaneously, the smartphone can be used to point to leaflet of the competitor, so that he/she is able to see a personal augmented overlay on the smartphone.

Fig. 4. GuerrillAR, developed by Löchtefeld et al. [10], augments competitor's leaflets with graphical aid information, including a bundle offer (left) and price/quality comparison (right).

In this today's world where technologies shape our lives, internet gives more and more people a voice. For this reason, many people such as customers like to have a voice back to the brands they like commercially. It includes that they prefer to be a part of brands. Interestingly, the concept for integrating social media and digital media is recently discussed in some new augmented reality applications. For example, one of very first social augmented reality platforms in the world was commercially presented in [11], called Tagger (Tag the world). This new augmented reality mapping and social platform have made augmented reality more sociable by creating the ability to manage and share mutual innovative information. It is an Android and iOS application for smartphone that allows people to upload and then share their own content such as personal objects, images, and locations with other people (e.g., friends and followers). This means that people can build their unique tags for every sort of media, such as magazines, adverts and posters. After that, they can add buttons and links to many needed images. Next, they can manage and create their own tags with innovative content freely. The content can be such as 3D games and graphics. Also, we can see every tag that other people have left in the digital format of a leaderboard.

The most popular content then rises to the top, so that it is able be voted up by other people with this real-time smartphone application. By doing this, every physical object is gradually able to take on a totally new life when scanned with the application. However, it is important to note that people are able to only access by going to a physical place and using the application. Figure 5 shows an example of this social augmented reality platform for the Michael Jackson experience to celebrate the launch of Xscape. Recently, augmented reality is used for technologically extended clothes and scarf, called Scarfy [12]. This prototype system is used for a scarf. It detects the way it is ties and then augments some interaction between a human and his/her garments. After that, augmented information by shape-change and vibration will be shown and received.

Fig. 5. The Michael Jackson experience on the commercially first social augmented reality platforms to celebrate the launch of Xscape was discussed in [11].

3   Interactive Experience

Computer vision [13] is a technological branch for study and research of acquiring, analyzing, understanding and interpreting images from the physical world. The digital images are usually acquired using both video cameras and infrared cameras. For years, it has been touted as a great technology in related disciplines such as those used in the robotic and augmented reality fields. In other words, computer vision is a tool for studies in other relevant fields. For instance, this technology has been used for motion analysis, head recognition and hand tracking [14]. In addition, it has been used by neuroscientists for years. An interesting review study from Harvard University [15] discusses constructively the integration between biologically-inspired computer vision and experimental neuroscience methodologies. Furthermore, augmented reality is often intersecting and relying on computer vision for application development. Hence, digital media that use augmented reality are, of course, often used primary

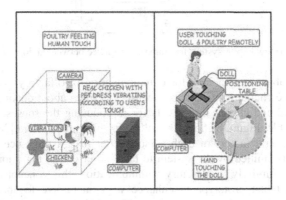

Fig. 6. A system for pet, Pet Internet [17], allows human to touch his/her pet remotely.

information based on (often only and solely) vision. However, now that appears to be changing beyond vision. In fact, communication is currently drifting away from each individual sense of human for connecting people to people. In other words, digital media are able to extend technologically to other particular senses. Those senses are somatosensation, olfactory and gustatory. They are also very important means of communication [16]. We call this communication for these three senses as interactive experience.

To develop robust systems for somatosensation communication remotely, it is not a trivial task. This is because it often deals with so many sensors, so that the main difficulty in designing somatosensation-based tangible user interfaces is how to reproduce the perception realistically and perfectly. Some new tools such as haptic technology and sensor technology are required. Also, this unique communication is not just only human-human interaction, but also includes in animal-human interaction. For instance, Figure 6 shows an overview of the system of Pet Internet for a pet chicken [17]. Also in [18] by using haptic technology, the interactive system makes human and his/her pet to communicate and feel closer to one another via haptic responses by exchange of haptic feedback. Figure 7 depicts an overview haptic system for transmitting force between a pet dog and human on remote human-pet interaction. For human-human interaction, the idea was started in [19] by Tangible Media Group from MIT Media Lab to explore the sense of somatosensation for remote communication. More concretely for human-human interaction, O'Modhrain [20] presented a concept for the senses of somatosensation to hold the main key for designing immersive hand-held mobile systems by using haptic feedback between the interleaved and interdependent spaces.

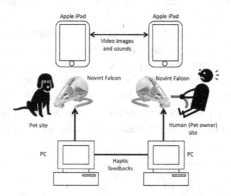

Fig. 7. An overview haptic system for transmitting force between a dog and human [18]

To achieve robust applications for communication in olfactory and gustatory remotely, it is more challenging. This is because it is often about dealing with thermal, electrical, inertial and chemical elements for reproducing the senses. However, it is now not impossible. Figure 8 shows this olfactory wearable system [21] by automatically tracking several factors of each person. After that, a fragrance in various and different forms is emitted by the system to have an effect on the impression of each user differently. Similarly to olfactory communication, this concept is later used in implementing a tablet prototype to enhance with interactive feedback such as air streams and environmental moist, called ambiPad [22]. It enriches mobile new media with ambient feedback used in some innovative movie theatres for 4 dimensional (4D) films. This prototype enriches mobile multimedia with ambient feedback around the display of tablet to allow for any thermal trigger. A good discussion about symbolic olfactory display can be found in [23]. In addition, a preliminary prototype for digital olfactory and gustatory actuation was developed in [24]. Some various tastes are imitated by this system. A spoon and a bottle are also used in the system. By using thermal, electrical and magnetic stimulation technologies [25], it is able to simulate olfactory and gustatory sensations in various forms. Moreover, for transmitting a similar sense of taste, [26] built a system that has a haptic device that can provide a physical interface. This first haptic device is then used with the second device remotely. The amount of the kiss's shape and force by one person of the first haptic device is sent to to another person using the second haptic device. Thus, this realistic kissing simulator can produce the result as nearly as practicable. A similar idea for transmitting a kiss remotely between their loved ones, presented previously by Ames et al. from Stanford University, can be found in [27]. They called this concept as Skype Kiss.

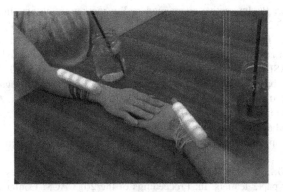

Fig. 8. An olfactory-based interactive wearable application to emit a delicate fragrance, called Light Perfume, was developed by Choi et al. [21] from Keio University.

4   Discussion

Social media, digital media, video-sharing sites, and gadgets such as iOS devices are now fixtures of our social culture, especially for youth [28]. In this paper, we present a preliminary concept of the recent development of the socially digital media work. We describe how to apply augmented reality and interactive experience to connect people innovatively through different senses of human. In augmented reality-based systems, it deals about communication for vision. In interactive experience-based applications, the communications for senses of somatosensation, olfactory and gustatory are explained. A literature review of the recent socially digital media researches is then discussed. By solving the aforementioned complex research problems, it can lead us to breed new trends in socially digital media and information technology in a multidisciplinary context. Part of our future work is aimed at developing a robust system for augmented reality and interactive experience.

Acknowledgments. This research presented herein was partially supported by a research grant from the Research Center, NIDA (National Institute of Development Administration).

References

1. The Asian Media Landscape is Turning Digital. The Nielsen Company (2012)
2. Gee, J.P.: New Digital Media and Learning as an Emerging Area and "Worked Examples" as One Way Forward. MIT Press (November 2009)
3. Zigmond, D., Stipp, H.: Multitaskers may be advertisers best audience. Harvard Business Review, January-February 1-3, 2011
4. Kerdvibulvech, C.: Real-time augmented reality application using color analysis. In: IEEE Southwest Symposium on Image Analysis and Interpretation (SSIAI 2010), pp. 29–32. IEEE Computer Society, Austin (2010) ISBN 978-1-4244-7801-9
5. Sarracino, F.: Can augmented reality improve students` learning? A proposal for an augmented museum experience. Profesorado, Revista de Currículum y Formación del profesorado (2014)
6. Bule, J., Peer, P.: Interactive augmented reality marketing system. In: World Usability Day, Paper ID 2505 (2013)
7. Connolly, P., Chambers, C., Eagelson, E., Matthews, D., Rogers, T.: Augmented reality effectiveness in advertising. In: 65th Midyear Conference on Engineering Design Graphics Division of ASEE, October 3–6, Houghton, Michigan, pp. 109–115 (2010)
8. Maad, S., Garbaya, S., McCarthy, J.B., Beynon, M., Bouakaz, S., Nagarajan, R.: Virtual and Augmented Reality in Finance: State Visibility of Events and Risk, Augmented Reality, Soha Maad (Ed.). InTech (2010)
9. Maad, S., Garbaya, S., Bouakaz, S.: From virtual to augmented reality in financial trading: a CYBERII application. Journal of Enterprise Information Management 21(1), 71–80 (2008)
10. Löchtefeld, M., Böhmer, M., Daiber, F., Gehring, S.: Augmented reality-based advertising strategies for paper leaflets. In: Proceedings of the 2013 ACM Conference on Pervasive and Ubiquitous Computing Adjunct Publication, UbiComp 2013 Adjunct, pp. 1015–1022. ACM (2013)
11. Spire Research and Consulting Pte Ltd, Augmented Reality Bridging the gap between the real and virtual consumer experience (2015)
12. von Radziewsky, L., Krüger, A., Löchtefeld, M.: Scarfy: augmenting human fashion behaviour with self-actuated clothes. In: Proceedings of the Ninth International Conference on Tangible and Embedded Interaction (TEI). ACM, Palo Alto (2015)
13. Klette, R.: Concise Computer Vision. Springer (2014) ISBN 978-1-4471-6320-6
14. Kerdvibulvech, C.: A Methodology for Hand and Fingers Motion Analysis Using Adaptive Probabilistic Models. EURASIP Journal on Embedded Systems (JES) 18, 9 (2014)
15. Cox, D.D., Dean, T.: Neural Networks and Neuroscience-Inspired Computer Vision. Current Biology 24(18), R921–R929 (2014)
16. Hertenstein, M.J., Keltner, D., App, B., Bulleit, B.A., Jaskolka, A.R.: Touch Communicates Distinct Emotions. Emotion, the American Psychological Association 6(3), 528–533 (2006)
17. Teh, J.K.S., Cheok, A.D., Choi, Y., Fernando, C.L., Peiris, R.L., Fernando, O.N.N.: Huggy pajama: a parent and child hugging communication system. In: Proceedings of the 8th International Conference on Interaction Design and Children, pp. 290–291. ACM (2009)

18. Murata, K., Usui, K., Shibuya, Yu.: Effect of haptic perception on remote human-pet interaction. In: Yamamoto, S. (ed.) HCI 2014, Part I. LNCS, vol. 8521, pp. 226–232. Springer, Heidelberg (2014)

19. Chang, A., Kanji, Z., Ishii, H.: Designing touch-based communication devices. In: Proceedings of Workshop No. 14: Universal Design: Towards Universal Access in the Information Society, Organized in the Context of CHI, Seattle, WA (2001)

20. O'Modhrain, S.: Touch and Go — Designing Haptic Feedback for a Hand-Held Mobile Device. BT Technology Journal 22(4), 139–145 (2004)

21. Choi, Y., Parsani, R., Roman, X., Pandey, A.V., Cheok, A.D.: Light perfume: designing a wearable lighting and olfactory accessory for empathic interactions. In: Nijholt, A., Romão, T., Reidsma, D. (eds.) ACE 2012. LNCS, vol. 7624, pp. 182–197. Springer, Heidelberg (2012)

22. Löchtefeld, M., Lautemann, N., Gehring, S., Krüger, A.: ambiPad: enriching mobile digital media with ambient feedback. In: Mobile HCI, pp. 295–298 (2014)

23. Kaye, J.: Symbolic Olfactory Display, M.S. Thesis in Media Arts & Sciences, Massachusetts Institute of Technology (MIT) Media Lab, Cambridge, MA (May 2001)

24. Ranasinghe, N., Karunanayaka, K., Cheok, A.D., Fernando, O.N.N., Nii, H., Ponnampalam, G.: Digital taste & smell for remote multisensory interactions. In: Proceedings of the 6th International Conference on Body Area Networks, pp. 128–129 (2011)

25. Ranasinghe, N., Lee, K.-Y., Suthokumar, G., Do, E.Y.-L.: Taste+: digitally enhancing taste sensations of food and beverages. In Proceedings of the ACM International Conference on Multimedia (ACM Multimedia), pp. 737–738. ACM (2014)

26. Saadatian, E., Samani, H., Parsani, R., Pandey, A.V., Li, J., Tejada, L., Cheok, A.D., Nakatsu, R.: Mediating intimacy in long-distance relationships using kiss messaging. International Journal of Human-Computer Studies 72(10), 736–746 (2014)

27. Ames, M.G., Go, J., Kaye, J.J., Spasojevic, M.: Making love in the network closet: the benefits and work of family videochat. In: Proceedings of the ACM Conference on Computer Supported Cooperative Work, pp. 145–154. ACM, New York (2010)

28. Ito, M., Horst, H.A., Bittanti, M., Boyd, D., Herr-Stephenson, B., Lange, P.G., Pascoe, C.J., Robinson, L.: Living and Learning with New Media: Summary of Findings from the Digital Youth Project. The D. John and C.T. MacArthur Foundation Reports on Digital Media and Learning (November 2008)

Acceptance of Online Social Networks as an HR Staffing Tool: Result from a Multi-country Sample

Chu-Chen Rosa Yeh[1], Karen Castellanos Gossmann[1,2], and Yu-Hui Tao[2(✉)]

[1] National Taiwan Normal University, Taipei, Taiwan
rosayeh@ntnu.edu.tw, karengossmann12@gmail.com
[2] National University of Kaohsiung, 700, Kaohsiung University Rd., Nanzih District, Kaohsiung 811, Taiwan, R.O.C.
ytao@nuk.edu.tw

Abstract. The popular online social networks are increasingly used by business organizations for various business activities. This research focused on the adoption of online social networks by human resources (HR) professionals as a tool in organizational staffing activities. The Technology Acceptance Model (TAM) was the theoretical basis of this study which sought to investigate the issue with a multi-country sample. Online questionnaire invitations were sent to HR practitioners in Taiwan, India, Spain and Guatemala, which generated 101 valid responses. Partial least square-based structural equation modeling was used to fit the proposed model and test the research hypotheses. Results indicated that, as hypothesized, the perceived usefulness, perceived ease of use, and subjective norms in TAM model effectively explained HR practitioners' behavioral intention to use online social networks for staffing activities in our sample. Limitations and future research were also discussed.

Keywords: Online social networks · TAM · HR staffing · HR professionals

1 Introduction

Online social network usage has a wide-spread increase in the last decade. Business organizations have identified numerous ways to add value to various business activities by using online social networks. The Human Resources (HR) community has also beginning to recognize online social networks as a potentially valuable tool, especially for tasks that can be conducted online, such as providing customer services, gathering public opinions, and posting job vacancies. HR professionals have a particular interest in the use of online social networks for staffing activities such as job posting, job referral, and the more controversial applicant screening through an analysis of the job applicant's profile and remarks in the social network account.

System acceptance is increasingly viewed as an important element for the measurement of success of information systems [1-2]. Thus, a study of the acceptance of online social networks in staffing activities will inform the HR community of its potential as a staffing tool in the professional context. The most researched model among the vast body of research on the acceptance and use of information technology has been the Technology Acceptance Model (TAM) developed by Davis [3].

© Springer-Verlag Berlin Heidelberg 2015
L. Wang et al. (Eds): MISNC 2015, CCIS 540, pp. 70–79, 2015.
DOI: 10.1007/978-3-662-48319-0_6

This model states that a highly perceived usefulness (PU) and perceived ease of use (PEOU) will yield through the intention to use an actual usage of a new technology. The work of Davis [3] has been replicated and the results have proven that the model holds for different technologies, persons, settings and time [4].

As technologies have evolved, new studies continue to be conducted in order to see if the model still holds for these newer technologies. In their TAM meta-analysis, King and He [5] confirmed that TAM is a valid and robust model, however, professionals and general users produce quite different results between the type of use. In addition, internet study results should not be generalized to other contexts, such as job task applications, general use, and office application. Since online social network acceptance by HR professionals in staffing activities is one of such contexts raised by King and He [5], it is appropriate to validate the TAM model again in this context. Furthermore, in another more insightful meta-analysis, Schepers and Wetzels [6] further confirmed that subjective norm (SN) has a significant influence in behavioral intention to use, which has an important implication for intra-company setting. Therefore, it is important to include subjective norm in studying HR professionals' behavior intention to use a new technology.

Globalization is today's reality, which forces businesses to compete in a global environment. This creates pressing needs to further investigate how different cultures react to different technologies. In the abovementioned TAM meta-analysis researches, culture is a contextual factor moderating the TAM core model [5], and the technology adoption process differs by culture, Western and non-Western in this case [6]. In particular, in a three-country TAM study, Straub, Keil, and Brenner [4] showed that the same TAM instrument held for the U.S. and Switzerland samples, but not for the Japanese.

The main purpose of this research is to examine whether TAM can be applied to the HR context in the use of the social network technology, focusing specifically on the staffing activities of the organization. Testing cultural difference is not the intention of this research, but by targeting a sample consisting of HR professionals from multiple countries will help validate the TAM model across cultures.

The remaining paper is organized as follows: a theoretical background is provided with the hypotheses generated, followed by the research method with detailed designs of sampling, data collection procedure and measurement. Then, the results based on the corresponding analyses are presented. Finally, the research conclusions are discussed with research limitations and future work.

2 Background Literature and Hypotheses

2.1 The Technology Acceptance Model

Davis [7] developed The TAM as an attempt to improve understanding of user acceptance process of a new technology and to provide theoretical basis for a practical methodology for testing user acceptance. This model incorporates factors such as PU and PEOU to explain both the attitudes toward a new technology and the actual usage of the technology. In Davis [7], PU is defined as "the degree to which a person believes that using a particular system would enhance his or her job performance" while PEOU is defined as "the degree to which a person believes that using a particular system would be free of effort".

The importance of this model is based on the fact that performance gains are often lost to the unwillingness of users to accept available systems that could help enhance their performance on the job [8-9]. The model is based on two main beliefs. First, the belief that a system will help people to improve their job performance will increase their intention to use the system and actual use of the system. Second, the belief that a system is easy to use will increase their intention to use the system and actual use of the system.

TAM was extended to include both social influence processes (subjective norm, voluntariness, and image) and cognitive instrumental processes (job relevance, output quality, results demonstrability and perceived ease of use). This new model is now referred to as TAM2 [10]. The present study uses both the original TAM and one of the new constructs from the extended model TAM2, subjective norm. This construct has been chosen since online social networks are believed to possess a social aspect that could create an effect in the acceptance of this new technology. Subjective norm is defined as "a person's perception that most people who are important to him think he should or should not perform the behavior in question" [10]. In the case of online social networks, the consideration that other people think that the use of these networks will enhance the results of the staffing activities within the organization, may increase the usage of the technology.

The behavioral Intention dimension is based on the Theory of Reasoned Action, it states that intentions lead to action [10]. It is assumed that people who intend to use a system will accept the technology and later perform a behavior of usage. TAM has been widely applied to a diverse set of technologies and users. For a summary on these studies refer to King and He [5] who conducted a meta-analysis on TAM. The meta-analysis revealed that many studies have been conducted to understand the moderating effects of factors such as age, gender, experience, voluntariness and culture in the technology acceptance sphere.

2.2   TAM and the Online Social Networks

Even though the relationships between perceived usefulness, perceived ease of use and the behavioral intention to use have been tested before [5-6], the PU, PEOU of the online social networks in the HR context has never been tested. As the HR professionals perceive the system to be more useful, their behavioral intention to use social networks for staffing activities of the organization will increase because the system will generate a greater value to the HR practitioners. Thus,

Hypothesis 1: Perceived usefulness of online social networks for staffing activities will have a significant influence on behavioral intention to use them.

As the HR professionals perceive social networks to be easy to use for staffing activities, their behavioral intention to use the social networks will increase. Thus,

Hypothesis 2: Perceived ease of use of online social networks for staffing activities will have a significant influence on behavioral intention to use them.

As concluded in the TAM meta-analysis by Schepers and Wetzels [6], subjective norm is a critical factor to intention to use. Therefore, as the HR professionals perceive that the people in the organization and other HR professionals see the use of social networks for the staffing activities of the organization as important, their behavioral intention to use social networks will increase. Thus,

Hypothesis 3: Subjective norms will have a significant influence on behavioral intention to use online social networks for staffing activities in the organization.

3   Research Design

This is an empirical study on Davis' TAM using social network tools in HR staffing context. HR practitioners were asked to complete a questionnaire containing PU, PEOU, SN, and behavior intention to adopt online social networks. Quantitative statistical analysis was applied to test hypotheses.

The population researched in this study is HR practitioners in multiple countries with possible diverse cultures as validated in Straub, Keil, & Brenner [4]. However, the diverse-culture sample will not be tested for the moderating effect, but for re-examining the core TAM with SN. The four countries selected were Guatemala in Central America, Spain in Western Europe, and Taiwan and India in Asia. The contact information of HR professionals in different countries was gathered by visiting the websites of different companies and collecting the information of the HR personnel. Later these personnel were contacted and asked to fill out the survey questionnaire. Social networks such as Facebook and LinkedIn were also used to contact the HR professionals through the professional communities.

The questionnaire used in this research was constructed with validated scales to measure the constructs of the study. This helps the study to ensure higher reliability and validity. For TAM model, PU and PEOU were each measured using 6 items developed by Davis [7]. SN is measured using 3 items adapted from Srite and Karahanna [12]. Behavioral intention was measured using one item adapted from Srite and Karahanna [12]. The questionnaires were translated into three languages, Chinese to fit the Taiwanese sample, English for the Indian sample and Spanish for the Spanish and Guatemalan sample. The pilot test was conducted on 45 HR practitioners in different organizations both in Taiwan and Guatemala. The pilot result confirmed the construct validity of the research instrument before proceeding to a larger-scale data collection effort. A seven-point Likert-scale was used for measurement of study variables where 1 represents strongly disagree and 7 represents strongly agree.

Even though previously validated scales were used in this study, tests were conducted to assure that the instrument effectively measured what it was intended to measure [13]. Internal consistency, indicator reliability, convergent and discriminant validity were all assessed. First, exploratory factor analysis was applied to ensure construct dimensionality. Confirmatory factor analysis was then conducted using partial least square (PLS) structural equation modeling, specifically smartPLS [14]. Both the items' loadings and t-values were evaluated.

According to Fornell and Larcker [15], the AVE should be greater than 0.5 and Composite Reliability (CR) above 0.7. All the CR and AVE values pass the hurdle of 0.7 and 0.5 respectively. As for the factor loadings, according to Hair et al. [16], loadings greater than 0.5 are acceptable. In this study all loadings are above 0.6 which proves convergent validity.

Table 1. Descriptive statistics of the sample

Variable	Category	Count	Percentage (%)
Nationality (N=101)	Indian	20	19.8
	Taiwanese	24	23.8
	Guatemalan	40	39.6
	Spanish	17	16.8
Age (N=94)	20-29	36	38.3
	30-39	38	40.4
	40-49	15	16.0
	50-59	5	5.3
Gender (N=101)	Female	56	55.4
	Male	45	44.6
Education (N=94)	College degree or certificate	8	8.6
	Bachelor degree	43	45.7
	Master degree	43	45.7
Personal Use of Social Networks (N=94)	I don't own a social profile	4	4.3
	I rarely check my profile (less than once a month)	5	5.3
	I seldom check it (once a month)	9	9.6
	I check it frequently (once a week)	24	25.5
	I always check it (every day)	52	55.3
Organizational Size (N=94)	Up to 10 employees	9	9.6
	Up to 50 employees	12	12.8
	Up to 250 employees	13	13.8
	251-500 employees	10	10.6
	501-1000 employees	13	13.8
	More than a 1000 employees	37	39.4
Organizational Social Networks Use (N=94)	The company has strict practices that prohibit the use of social networks	30	31.9
	The company has loose policies that prohibit the use of social networks	23	24.5
	The company encourages the use of social networks	10	10.6
	The company owns a social network profile	22	23.4
	The company demands the use of social networks for staffing activities	9	9.6
Industry (N=99)	Financial Services	16	16.2
	IT/Technology Industry	7	7.1
	Products Manufacturing	15	15.2
	Service Industry	22	22.2
	Telecommunications	8	8.1
	Consulting	6	6.1
	Other	25	25.3

4   Data Analysis Results and Discussion

A total of 101 valid questionnaires were collected. Table 1 shows the demographics of the sample.

The nationalities of the sample are 20, 24, 40, and 17 for Indian, Taiwanese, Guatemalan, and Spanish, respectively. The profile of this multi-country sample is as follows: the majority age between 20 and 39, are female, and have a college degree and above; they work for organizations that are mostly medium and big enterprises, with restricted social network use, and belonging to services and manufacturing industries.

Table 2 demonstrates good reliability and convergent validity of the three constructs. Results of factor loadings, composite reliability and average variance extracted are also presented. Specifically, means of PU, PEOU, and SN are higher than 4, indicating a somewhat positive perception. Factor loadings, composite reliability values, and average variance extracted are all above the suggested value.

Table 2. Descriptive statistics, factor loading, CR, AVE and Items

Variables	Mean/S.D	Items	Factor loading>.6	Composite reliability (CR) >.7	Average variance extracted (AVE) >.5
Perceived Use- fulness	4.51/1.49	PU1	0.92	0.95	0.78
		PU2	0.92		
		PU3	0.91		
		PU4	0.81		
		PU5	0.82		
		PU6	0.89		
Perceived Ease of Use	4.89/1.36	PEOU1	0.87	0.92	0.64
		PEOU2	0.80		
		PEOU3	0.77		
		PEOU4	0.78		
		PEOU5	0.84		
		PEOU6	0.76		
Subjective Norm	4.40/1.48	SN1	0.89	0.90	0.73
		SN2	0.84		
		SN3	0.82		

Table 3 presents results for the study's discriminant validity. Discriminant validity was evaluated by calculating and comparing the correlation coefficients and the square root of the AVE. None of the variable's correlations with other variables are higher than the square root of AVE for that variable, which indicates good discriminant validity.

Table 3. Overview of discriminant validity testing

	Average va-riance extracted	Perceived Use-fulness	Perceived Ease of Use	Subjective Norm
Perceived Use-fulness	0.78	(0.88)[a]		
Perceived Ease of Use	0.64	0.62	(0.80)[a]	
Subjective Norm	0.73	0.69	0.51	(0.85)[a]

[a] Numbers in the parentheses are the square roots of the average variance extracted

Pearson product-moment correlation coefficients were computed to assess the relationship between pairs of study variables and are presented in Table 4. Correlation was found to be significant among different variables, such as between PU and PEOU, PU and SN, PU and behavioral intention, PEOU and SN, PEOU and behavioral intention, and SN and behavioral intention.

Table 4. Means, standard deviations and correlation coefficients

Var	Mean	St. D.	1	2	3	4	5	6	7	8	9
1	0.55	0.5									
2	2.88	0.86	.13								
3	3.37	0.64	.22*	-.04							
4	4.22	1.09	-.15	-.37**	.00						
5	4.24	1.79	.18	-.29**	.17	.09					
6	2.54	1.40	.05	-.05	.05	.15	-.09				
7	4.52	1.49	-.05	-.17	.03	-.10	-.08	.27**	(0.94)[a]		
8	4.9	1.36	-.04	-.26*	-.04	.11	.05	.12	.62**	(0.89)[a]	
9	4.4	1.49	-.08	-.10	.05	-.12	.02	.15	.69**	.51**	(0.81)[a]
10	4.7	1.71	.06	-.09	.04	-.06	.00	.24*	.68**	.57**	.62**

[a] Numbers in parenthesis indicate the construct's Cronbach's alpha.

Variable Coding:

1. Gender; 2. Age; 3. Education; 4. Personal network use; 5. Organizational Size;
6. Organizational network use; 7. Perceived usefulness (PU); 8. Perceived ease of use (PEOU);
9. Subjective norm (SN); 10. Behavioral intention
Gender= Female (1) Male (0)
Age= Under 20 (1), 20-29 (2), 30-39 (3), 40-49 (4), 50-59 (5), 60 or above (6)
Education = High school (1), college or certificate (2), bachelor (3), master (4), doctorate (5)

Results of hypothesis testing are presented for this study. According to Moore, McCabe, Duckworth, and Alwan [17], a significant relationship is found at 90% confidence level when t>1.65, at 95% when t>1.984, and at 99% when t>2.626. Figure 1 summarizes the hypothesis testing results, which will be individually stated below.

| Perceived Usefulness | H1 (0.34)** |

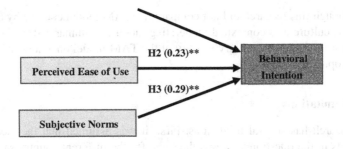

* p < 0.1 with t>1.645, ** p< 0.05 with t>1.984, ***p <0.01 with t>2.626

Fig. 1. Hypothesis testing summary

Hypothesis one stated: Perceived usefulness of social networks for staffing activities will have a significant influence on behavioral intention to use this system. As expected, the perceived usefulness of HR professionals toward a system shows a significant and positive influence (β= 0.34, t=2.55) on the behavioral intention to use online social networks for staffing. This hypothesis is supported.

Hypothesis two stated: Perceived ease of use of social networks for staffing activities will have a significant influence on behavioral intention to use this system. The result shows perceived ease of use does have a significant effect on behavioral intention to use a system (β= 0.23, t=2.31) such that the easier the individual perceives the system, the higher his/her behavioral intention to use it. This hypothesis is also supported.

Hypothesis three stated: Subjective norms will have a significant influence on behavioral intention to use social networks for staffing activities in the organization. The result shows subjective norms having a positive and significant impact on behavioral intention (β= 0.29, t=2.27). This hypothesis is also supported.

5   Conclusions and Limitations

5.1   Conclusions

Through this research we learn that the perceived usefulness, perceived ease of use and subjective norms all have an effect on the behavioral intention of HR practitioners to use the online social networks for organizational staffing activities in our multi-country sample.

Even though TAM has been proven to be a robust model, as newer technologies are emerging it is still important to find out the factors that affect acceptance of them. In the case of the online social networks, the proposed factors from the TAM still show a major impact on the acceptance of this technology in the HR field. This research confirms that even with a new technology these relationships remain valid, which accomplished the research purpose as listed in the introduction.

Another contribution of this research setting is that a multi-country sample was examined, which is similar to the few cross-country study in Straub Keil, and Brenner

[4]. Although this research did not compare the TAM model country by country as in [4] or use culture as a contextual moderating factor as summarized in [5], it is a good setting for testing cross-country validity of the TAM model on a new application with such a population mix.

5.2   Limitations

This research has several limitations. First, it was assumed that the meaning of the constructs in the questionnaire was the same for the different country samples. However, this may not be true; the different samples could perceive the concepts presented in the instrument as different. This is a common concern when conducting cross-cultural research. This does not make the research invalid, but it is important to mention that measurement equivalence may be of concerns.

Another limitation of this study is that the data for all study variables were self-reported by HR professionals. Even though self-report is the only method available for collecting the needed information, the results of the study could be affected by common method variance or the respondents' lack of commitment to respond truthfully to the questionnaire. Accessibility to the sample was another issue; it was difficult to contact HR professionals in other countries because of the lack of connections in these countries. Larger sample would allow a higher level analysis of the data, such as country comparisons.

The data was collected through online questionnaires, so there was no way for the researchers to control who could access the survey, nor who could complete the survey. Finally, the use of single-item measurement for one construct could be seen as a limitation. Even though there is a debate among different research fields, and some researchers support the use of this practice for concrete constructs, it can still be considered a limitation.

Cultural differences may affect the way the respondents view the constructs, thus, direct translation of the instruments might not be the most appropriate way to gather the data. A previous step that was taken in this study before delivering the questionnaire to respondents was an expert review in every language. This might be an important step that other researchers also need to take in order to make sure that the instrument to be used delivers the intended meaning.

5.3   Future Research Suggestions

Cultural constructs can be used to moderate the relationships of the proposed model in TAM [5-6], such as Hofstede's dimensions [18-21]. This could provide an insightful perspective of culture on the acceptance of a new technology. Organizational culture could also affect the acceptance of technology in the workplace, so it can be the focus of future research. Country comparison [4] among the samples is another direction for future research. Larger sample will be needed so that the comparison among different countries can be meaningful. Other methodologies can also be used to conduct these studies, such as a qualitative research approach which allows the respondents to provide more in depth information on the factors that affect their acceptance, and more importantly the reasons why the factors are important. This information cannot be extracted from quantitative studies.

References

1. DeLone, W., McLean, E.: Information systems success: The quest for the dependent variable. Information Systems Research **3**(1), 60–95 (1992)
2. DeLone, W., McLean, E.: The DeLone and McLean model of information systems success: A ten-year update. Journal of Management Information Systems **19**(4), 9–30 (2003)
3. Davis, F.: Perceived usefulness, perceived ease of use, and user acceptance of information technology. MIS Quarterly **13**(3), 319–339 (1989)
4. Straub, D., Keil, M., Brenner, W.: Testing the technology acceptance model across cultures: A three country study. Information & Management **33**, 1–11 (1997)
5. King, W., He, J.: A meta-analysis of the technology acceptance model. Information and Management **43**, 740–755 (2006)
6. Schepers, J., Wetzels, M.: A Meta-analysis of the technology acceptance model: investigating subjective norm and moderation effects. Information & Management **44**(1), 90–103 (2007)
7. Davis, F.D.: A technology acceptance model for empirically testing new end-user information systems: Theory and results. Doctoral dissertation, Massachusetts Institute of Technology, Sloan School of Management (1985)
8. Bowen, W.: The puny payoff from office computers. Fortune, 20–24 (1986)
9. Young, T.R.: The lonely micro. Datamation **30**(4), 100–114 (1984)
10. Fishebein, M., Ajzen, I.: Belief, attitude, intention and behavior: An introduction to theory and research. Addison-Wesley, Reading, Massachusetts (1975)
11. Venkatesh, V., Davis, F.D.: A theoretical extension of the technology acceptance model: Four longitudinal field studies. Management Science **46**(2), 186–204 (2000)
12. Srite, M., Karahanna, E.: The Role of espoused national cultural values in technology acceptance. MIS Quarterly **30**(3), 679–704 (2006)
13. Crocker, L.S., Algina, J.: Introduction to classical and modern Test theory. Harcourt Brace Jovanovich, Fort Worth, TX (1986)
14. Ringle, C., Sarstedt, M., Straub, D.: Editors' Comments: A critical look at the use of PLS-SEM in MIS Quarterly. MIS Quarterly **36**(1), iii–xiv (2012)
15. Fornell, C., Larcker, D.F.: Evaluating structural equation models with unobservable variables and measurement error. Journal of Marketing Research, 39–50 (1981)
16. Hair, Jr., J.F., Anderson, R.E., Tatham, R.L., Black, W.C.: Multivariate data analysis, 5th edn. Prentice-Hall, Upper Saddle River (1998)
17. Moore, D.S., McCabe, G.P., Duckworth, W.M., Alwan, L.C.: The practice for business statistics: Using data for decisions, 2nd edn. W.H. Freeman and Company, New York (2009)
18. Hofstede, G.: Culture's consequences: International differences in work-related values, vol. 5. Sage Publications, Incorporated (1980)
19. Hofstede, G.: Culture's consequences: Comparing values, behaviors, institutions and organizations across nations, 2nd edn. Sage Publications, Incorporated (2001)
20. Hofstede, G., Bond, M.H.: The Confucius connection: From cultural roots to economic growth. Organizational Dynamics **16**(4), 4–21 (1988)
21. Wu, M.: Hofstede's cultural dimensions 30 years later: a study of Taiwan and the United States. Intercultural Communication Studies **15**(1), 33 (2006)

Organization Learning from Failure Through Knowledge Network: Case Study on an Engineering Company in Japan

Sanetake Nagayoshi[1(✉)] and Jun Nakamura[2]

[1] Research Institute of the Faculty of Commerce, Waseda University,
Takada Bokusha Bldg. 4F, 1-101, Totstuka-machi, Shinjuku-ku, Tokyo 169-0071, Japan
snagayoshi@aoni.waseda.jp
[2] Graduate Program in Business Architecture, Research Center for Social and
Industrial Management Systems, Kanazawa Institute of Technology, 7-1,
Ohgigaoka, Nonoichi, Ishikawa 921-8501, Japan
jyulis@neptune.kanazawa-it.ac.jp

Abstract. Making an attempt for new business and innovation to seek better performance is an essential matter for company to keep sustainable growth. Learning from failure is one of the ways to seek better performance, but it is extraordinarily rare for organizations to do it well. This paper reported a successful case study in Sangikyo Corporation in Japan. Since they succeeded in establishing a way to learn from failure, they had accumulated 41 failure case studies and lessons learned from the failures since 2005. In addition, they eventually earned thirty-five fold profits more over the past six years. In this paper, authors explored success factors for organizational learning from failure based on the case in Sangikyo Corporation, from points of willingness to serve, common purpose and communication. This paper however, just showed preliminary hypotheses for further research, and it is necessary to carefully examine them for being concrete hypotheses beyond this paper.

Keywords: Organization learning from failure · Success factor · Willingness to serve · Common purpose · Communication · Knowledge sharing platform

1 Introduction

It is quite important to try new business in the mature economic society of Japan, although it is not so easy to succeed in new business. Some attempts resulted in success, while others did not. According to Mitsubishi UFJ Research and Consulting [1], half of the companies which were challenged new business made some a mistake and/or failure. The reason for them was lack of organizational capability; for instance, problems in employee education and recruiting (25.1%), and low sill (22.9%) [1].

In order to improve organizational capability in this situation, company should invest in human resource development and learn from fault in daily task and failure of the initiative [2]. It is important to learn from failure [3] in order to prevent from repeating a similar fault. Since there were, however, few available case studies of failure, companies were struggling with learning from failure in practice.

© Springer-Verlag Berlin Heidelberg 2015
L. Wang et al. (Eds): MISNC 2015, CCIS 540, pp. 80–93, 2015.
DOI: 10.1007/978-3-662-48319-0_7

There were many companies which try new business, but unfortunately they did not always succeed in it. Since it was significant to continue making such attempt to achieve better business performance, it was important to study a methodology of learning from failure.

Sangikyo Corporation in Japan succeeded in establishing their way to learn from failure. They had accumulated 41 failure case studies and lessons since 2005. And they eventually earned thirty-five fold profits more over the past six years [4]. In this paper, authors employed Sangikyo Corporation to explore the success factors for learning from failure in it.

There were complicated causal relationship among the organizational activity in the learning from failure, success in knowledge sharing, organizational capability improvement, success in new business, and the financial performance. It was impossible to cope with all the relationships together at once. In this paper, authors focused on exploring the reasons why personnel who made a fault/mistake/incident reported and disclosed it in the company. The purpose of this paper is to report a case in Sangikyo Corporation and generate hypothetical success factors for learning of failure.

The remainder of this paper is organized as follows: First, the related studies are reviewed in the following chapter and then the research questions are described in Chapter3. Next the research method is described in Chapter4. The viewpoints for analyzing the case are shown in Chapter5. And then case in Sangikyo Corporation is shown in Chapter 6. The case is analyzed based on the viewpoints in Chapter7, and discussed in Chapter8. Finally, authors conclude this study and the limitations of this study and implications for future research are described in Chapter9.

2   Related Study

According to Hatamura [5], failure is defined as "a human act of not reaching the defined goal," "an unfavorable and unexpected result of human act." And he also indicates that there are an invaluable failure and a non-valuable failure. An invaluable failure is defined as "an unavoidable failure even with extreme caution," which is excursion into unknown. A non-valuable failure is defined as "a failure other than invaluable failure."

Most organizations do a poor job of learning from failure [2, 6]. Carmeli and Schaubroeck [7] indicate that learning from failure is an important facilitator of preparedness for both present and prospective crises. Edmondson [3] suggests

> Wisdom of learning from failure is incontrovertible. Yet organizations that do it well are extraordinarily rare. This gap is not due to a lack of commitment to learning. Managers in the vast majority of enterprises genuinely wanted to help their organizations learn from failures to improve future performance. In some cases they and their teams had devoted many hours to after-action reviews, postmortems, and the like. But time after time I saw that these painstaking efforts led to no real change. Failure and fault are virtually inseparable in most households, organizations, and cultures. Every child learns at some point that admitting failure means taking the blame. This is why so few organizations have shifted to a culture of

psychological safety in which the rewards of learning from failures can be fully realized.

But companies which invested significant money and effort into becoming learning organization with ability to learn from failure struggle with it [8].

When people feel psychologically safe, learning from failures is enabled [9]. Positive feedback loops are critical. Relationship quality is both an outcome and a mediating variable, and procedural issues are critical in fostering a climate for positive reinforcement and building of mutual trust and confidence in relationship [10].

Since it often needs hardship in general, it is important to set suitable environment and system for learning from failure. Executives and managers who are eager to succeed in learning from failure should set suitable environment and system in company. Hence the question in this paper was what and how they do for it in practice, which authors explored with a case study.

3   Research Question

Authors in this paper employed a case in Sangikyo Corporation because they succeeded in learning from failure and they eventually earned more profits. This seemed to be so unique that significant findings were expected.

Authors intended, in general meanings, to study how company can succeed in business and financial performance by learning from failure. Since the relationship however, between the success of learning from failure and the financial performance was complicated, it was impossible to clarify causal relationship at once. Hence proper scope to study should be set in this paper.

Process from occurring failure to resulting in financial performance was; 1) failure occurrence in an activity and/or event, and then recognition of the failure, 2) comprehension of fact, 3) cause and result analysis, 4) development of preventive measures, 5) deployment of the measures, 6) organizational learning, 7) execution of the measures and 8) success in business and financial performance. Since this process included complicated causal relationships, it was ambitious to clarify the all at once. In this paper, authors focused on the first two steps; 1) failure occurrence in an activity and/or event, and then recognition of the failure and 2) comprehension of fact, are focused. And the rests should be studied in other papers.

Research questions in this paper are;

RQ1: Why does the personnel who makes a mistake leading to failure admit the failure and notify other personnel?

RQ2: Why does the personnel who makes a mistake leading to failure disclose the fact in the mistakes without keeping something against them?

Noted these questions are parts of the general research question; How can company succeed in business and financial performance by learning from failure? [11, 12]

4   Research Method

Authors conducted qualitative research with the case in Sangikyo Corporation. The data was collected through ten interviews from July 2014 to January 2015 conducted by authors. The interviews are listed in Table 1. The interviews were recorded with IC voice recorder and the electric voice data files of the interviews were stored by authors. And authors transcribed them, and a manager from the company reviewed it to verify.

Table 1. List of the Interviews

No	Date	Place	Interviewee	Theme
1	July 8, 2014	Waseda University (Tokyo)	Vice President Mr.Sengoku Director Mr.Tokunaga	Overview of the activity for learning from failure
2	August 8,2014	Main Office		Knowledge sharing platform
3	September 5,2014	Sangikyo Corporation (Yokohama City)	Director Mr.Tokunaga Mr.A	Failure in Project A
4			Mr.B,Mr.C	Fault in Project B
5	September 30,2014		Mr.D,Mr.E	Failure in Project A
6			Mr.F,Mr.G	Fault in Project B
7			Mr.H,Mr.I,Mr.J	Failure in Project C
8	November 11,2014		Mr.K,Mr.L	Minor faults in the projects
9			Mr.M,Mr.N	Administrative division
10	January 14,2015	Waseda University	Vice President Mr.Sengoku Director Mr.Tokunaga	Overview of the activity for learning from failure

Authors employed Eisenhardt [13] to approach the research questions. In the next chapter, viewpoints for case analysis were set by employing Barnard [14], Mendelson and Pillai [15], and Mendelson and Ziegler [16]. Next the case data was described, and then analyzed based on the viewpoints. Finally the results were discussed.

5   Viewpoints for Analysis

Barnard [14] suggests that formal organization sets forth three necessary elements for organizations: communication, willingness to serve, and common purpose, and also suggests that an organization that cannot accomplish its purpose cannot survive.

Task force team for learning from failure in Sangikyo Corporation had worked for 10 years regardless difficulties of learning from failure. Since the task force team had rationale to survive as a formal organization, the case was analyzed from the points of communication, willingness to serve and common purpose in this paper.

Internal knowledge and information dissemination has three types of dissemination including vertical dissemination, horizontal dissemination and over-time dissemination [15, 16] in terms of the communication. The vertical dissemination means interaction between superiors and subordinates. The horizontal dissemination means interaction among employees and internal divisions. The over-time dissemination means learning from past case studies including both of successful cases and failure cases.

In terms of the willingness to serve, their motivation to contribute to organization depends on incentives. According to Barnard [14], in all sorts of organizations, affording of adequate incentives becomes the most definitely emphasized task in their existence. Specific inducements range from "material inducements" to "ideal benefactions," while general incentives include personal comfort in social relations.

In terms of the common purpose, employee can and will accept a communication as authoritative only when four conditions simultaneously obtain: he/she can and does understand the communication; he/she believes that it is not inconsistent with the purpose of the organization; he/she believes it to be compatible with his/her personal interest as a whole; and he/she is able mentally and physically to comply with it [14]. The case was analyzed from the point of generating purpose of organization and relationship between communication and incentives.

6   Case Study

Sangikyo Corporation [17] was founded in 1965 primarily as a company which dealt with engineering services for installation and maintenance of microwave communication systems. The company adopted more Western standards and developed a unique corporate mentality and work ethic which was built around being a knowledge-based company.

President thought new challenge sometimes led them failure, but also thought it important to avoid repeating the same failure. Sangikyo Corporation started to organize task force team to learn from failures in 2005, aims of which were improving their organizational capability to establish their position as a prime contractor.

When failure/fault occurred, they organize task force team in order to reflect and learn something from it. They had accumulated 41 failure/fault cases and lessons learned from failures/faults since 2005, when the president made a decision to start it. Mr. Sengoku, vice president, and Mr.Tokunaga, a director at Sangikyo Corporation emphasized that they had never had the same failure/fault they examined in the task forces since they started it.

6.1   Failure/Fault/Incident Discussed in the Task Force

There were two types of issues to be discussed in task force. The first one was incident, such as human error, which causes failure, for example, an employee negligently turned power off, an employee negligently started an auto-truck and broke a cable, and an employee left a personal computer in a train. In such cases, the employee/s who made incidents, as a rule, had to report it to the company as soon as possible through their knowledge sharing platform, and then company executives are notified it through automatically forwarded mobile phone-mail. Serious incident which could cause significant failure was pointed by the board, and were analyzed and discussed to avoid repeating it in a task force team.

The other one was failure like financial loss of project, delay of delivery and/or fault like error reporting in a business. They were also pointed by the board, and were analyzed and discussed to avoid repeating it in a task force team.

6.2   Process

An employee who faulted and/or made an incident causing failure is appointed as leader in a task force team. The task force team is, in general, organized with all employees related to failure, the fault and/or the incidents. In addition to them, the professional personnel in charge of quality assurance sometimes join the task force team.

There were some following opinions and some dissenting opinions for appointing the employee who faulted and/or made the incident causing a failure as leader in a task force team. According to Mr. Sengoku, the vice president and Mr.Tokunaga, a director, he/she was disappointed the most seriously and never made a mistake but someone else can make the same mistake. He/she could be disappointed, if someone else in the company made the same mistake. This was the reason why an employee who faulted and/or made a mistake causing failure played a role of leader in task force team.

Once a task force team was organized, they started to analyze failure/fault/incident and discussed countermeasures to avoid repeating it. After completing verification, cause analysis and countermeasures, the leader of the task force team made a report and deployed it through their knowledge sharing platform to all the employees.

In the meantime of analysis and discussion in the task force team, company executives and supervisors went to the customer to calm down the situation if the failure/fault/incident was significantly related with the customer's business performance.

6.3   Knowledge Sharing

A report of a task force team was released to all the employees through their knowledge sharing platform, as shown in Figure 1. The report included verification, causal analysis and countermeasures of failure/fault/incident, and in which all data including name of leader in task force and customer, and number of lost money was revealed.

At the same time, they also disclosed the report to the customer who was suffered from the failure, and they got trust and further business from the customer again. Since in general, a lot of customers did not have enough knowledge to avoid repeating the same failure/fault/incident. It was appreciated that Sangikyo Corporation volunteers to disclose the reports to customer.

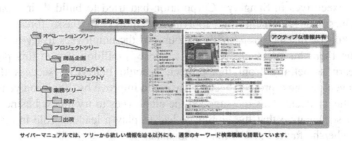

Fig. 1. Knowledge sharing platform in Sangikyo Corporation   (In Japanese Only [18])

6.4   Knowledge Utilization

All the employees were able to browse the reports through their knowledge sharing platform from anywhere. Since not only employees related to task force but also employees from other divisions browse the reports, they reviewed the reports to avoid repeating the same failures/faults/incidents when they did similar project. Project leader in a similar project sometimes asked directly the employee who made a mistake through e-mail and phone call. In case that unexperienced employees were assigned to a similar project, experienced employee and/or their supervisor often asked them to check the reports before doing a task. In the meantime, they avoided repeating the same failure/fault/incident.

In the cases that caused negative impact on the company due to relatively large amount of money lost, the reports were not only disclosed through their knowledge sharing platform but also explained in a corporate event where the president communicated with the employees.

6.5   Appraisal System

Sangikyo Corporation implemented the appraisal system called skill-based competency evaluation, in which employee was graded by skill, annual targets were set with the grade and they were evaluated by the set criteria with transparency. It had four competency criteria like personality, operational skill, human skill and conceptual skill. The employees were evaluated not only their achievement but also the competency criteria. Hence if unfortunately employee faulted, he/she received a low evaluation in terms of achievement. But he/she received a good evaluation if he/she managed a task force team, analyzed and reports well, because he/she was considered as a skilled employee. In addition, employees' bonus relied more on company performance than on individual performance.

Actually, for instance, some employee promoted after he/she played a role of leader in a task force team. And a leader who took responsibility in a task force was rewarded when same failure did not occur for a certain period.

6.6   President Message

Company executives in Sangikyo Corporation had tried to build their capability for new business development by applying learnings in task forces. President Mr.Sengoku, for instance, repeatedly encouraged the employees to try new business by sending them his message, "Even if we lost 50millilon yen in this project, we should learn something from this project. I could believe it success if we lost 30million yen in a next project, because we could improve by 20million yen by learning from this project." This message led the employees to try a new business and not to think it so negative that the employee had to play a leaders' role in a task force team when he/she faulted and/or made a mistake.

Company executives made a proposal to their customer in order to engage new business, in which a learning from failure was applied. The president, for example, not only apologized to the customer but also proposed to the customer to improve their facilities, when an employee made a mistake to turn off power. It was because the task force team concluded that it could set off a similar accident unless the customer improved the facilities. Thus, Sangikyo Corporation received more trust from the customer, and they continuously made an engagement with them.

7   Case Analysis

The process for learning from failure in Sangikyo Corporation was; 1) Occurrence of failure/fault/incident, 2) Activity of task force team, 3) Reporting and 4) Implementation.

As described in Chapter 5, the case was analyzed from the points of (1) communication (2) willingness to serve; and (3) common purpose. In addition, the common purpose was decomposed into three types of dissemination; horizontal dissemination, vertical dissemination and over-time dissemination.

7.1   Communication

Authors analyzed the case from the points of horizontal dissemination, vertical dissemination and over-time dissemination.

When employee made a fault/failure, they immediately notified it through their knowledge sharing platform. And employee honestly disclosed detail situation of the fault/failure in task force team. Reports in task force teams were open to all employees through the platform. Moreover employee received and answered inquiry from other employees after report disclosure.

In the meantime, horizontal dissemination seemed to work well, and they eventually succeeded in avoiding repeat of the same failure.

Employee who made a fault/incident, they also immediately notified it to company executives through their knowledge sharing platform. In the meantime, the fault and accident were not hidden.

In case that unexperienced employees were assigned to a similar project, experienced employee and/or their supervisor asked them to check a report before doing a task. Employees in Sangikyo Corporation recognized detail of a failure, which was so serious that the company lost/could lost a large amount of money, not only through their knowledge sharing platform, but also at a corporate event where the president communicated with the employees. In the meantime, they avoided repeating the same failure/fault/incident.

In addition, the president repeatedly encouraged employees to try new business by learning from failure, and company executives sometimes made a proposal to their customer by using report from task force. These behaviors of company executives motivated them to learn from failure.

Thus, the vertical dissemination, which means interaction between superiors and subordinates, worked well in Sangikyo Corporation.

The over-time dissemination means learning from past case studies including both of successful cases and failure cases. A project leader in a similar project not only reviewed reports of task force teams but also sometimes asked an employee who made a mistake through e-mail, phone call. And they consequently learned from past failures, and they avoided repeating the same failure/fault.

7.2   Willingness to Serve

Since a person who made a fault does not always express all to avoid unfavorable situation, it is sometimes difficult to analyze cause of failure rightly and make a proper countermeasure. Hence a third party who stands on a neutral position, not a person who makes a fault, analyzes a failure.

In Sangikyo Corporation however, an employee who made a fault was appointed as a leader in charge of analysis, counter measurement and reporting. This was because he/she was disappointed the most seriously and never made a mistake, but someone else in the company could make the same mistake, and he/she would be disappointed more, if someone else in the company made the same mistake which could lead the company fail.

They implemented the appraisal system called skill-based competency evaluation. In this system, the employee was highly evaluated when he/she played a leader role in task force team. When he/she was able to manage a task force team well, he/she was highly evaluated. When he/she, to the contrary, was not able to manage, he/she was not evaluated. This was an incentive for an employee to positively play a leader role in task force. Since in Sangikyo Corporation, employee who faults played a role of leader in a task force team, they effectively analyzed cause of failure/fault and make a countermeasure.

In addition, he/she promoted when he/she managed task force team well. He/she rewarded when countermeasures worked well and same failure had not recurred for a certain period.

Since, moreover, their bonus relied more on company performance than on individual performance, they could not receive entire bonus even if their individual performance was satisfied or achieved more than expectation. Since they received full bonus when both of individual performance and company performance were satisfied, they had more incentives to improve company performance by directly contributing to a task force team; as a leader to analyze cause of failure/fault in detail and report result properly.

In the meantime, they immediately notified their fault through their knowledge sharing platform when it occurred. They told fact honestly, and they positively analyzed cause and countermeasure. And they deployed report to all the employees.

7.3   Common Purpose

In terms of employee work style, they did their job by team not individually. Since they shared a task with team members and achieved a goal by team, when a certain team member made a fault, someone else in the team could fault the same. In the interviews for the managers in Sangikyo Corporation, a manager Mr.A and Mr.B said, "Many of the faults being committed today are ones that any one of us could commit."

Since, in Sangikyo Corporation, many employees did similar tasks, other employees could the made the same mistake. This was why they notified fault immediately when they made a fault, and they told fact honestly, positively analyzed cause and countermeasure, opened report to all employees.

Authors concluded the above analysis in Table 2.

Table 2. Analysis of Task Force in Sangikyo Corporation

		Failure/Fault /Incident	Task Force Activity	Reporting	Execution
Communication	Horizon-tal	Knowledge sharing	Honest failure discloser	Open to all the employees	Q&A with other employees
	Vertical	Notification to executives	Encouragement by executives	Advice by supervisors	Proposal to customer
	Over-Time	–	–	–	Review past reports
Willingness to Serve		Skill Based Competency, Company Performance weighted Bonus, Reward for task force activity			
Common Purpose		Sharing tasks with colleagues	No willing for colleagues' fault	Same faults to everyone	Sharing tasks with colleagues

8   Discussion

Authors discuss the research questions, RQ1 and RQ2 respectively, in this chapter.

8.1   RQ1: Why Does the Personnel Who Makes a Mistake Leading to Failure Admit the Failure and Notify Other Personnel?

In Sangikyo Corporation, employees who made a fault/mistake/incident notified not only company executives but also all employees through their knowledge sharing platform. This was one of the rules as a code of conduct in Sangikyo Corporation. If he/she missed a notification of fault/mistake/incident, he/she could be punished. Since employee in general, did not want to receive negative evaluation, he/she notified such a fault/mistake/incident through their knowledge sharing platform as soon as possible. Completing his/her responsibility to notify his/her fault/mistake/incident to company executives, he/she might think that the responsibility to settle it was transferred to the company executives, and he/she felt a ease since corporate executives and/or superiors helped them to settle it.

Since in addition, many of employees did similar tasks, other employees could make a same mistake. When he/she notified fault/mistake/incident immediately, the fault/mistake/incident could be avoided to occur.

These were why they notified the fault immediately after the fault occurred, as Figure 2 shows:

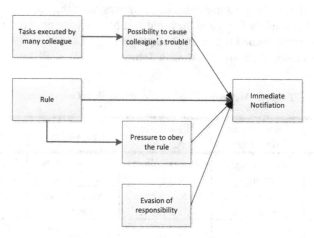

Fig. 2. Framework on Immediate Notification

8.2   RQ2: Why Does the Personnel Who Makes a Mistake Leading to Failure Disclose the Fact in the Mistakes Without Keeping Something against Them?

In Sangikyo Corporation, employees who made a fault/mistake/incident played a leader role in task force team. From a third party's point of view, he/she himself/herself was disappointed the most seriously and never made the same mistake but someone else in the company could make the same mistake. He/she could be disappointed more, if someone else in the company made the same mistake which could lead the company fail. On the other hand, from an employee's point of view, he/she had a strong pressure from company executives or other employees, because they could lose their profit due to the failure/fault/mistake. In this sense, being a leader in task force team meant a kind of atonement for themselves.

They also had positive incentives to be a leader of task force team, although they unwillingly became a leader. They promoted under the skill based competency system and/or could get a prize when they managed task force well. In addition, the president repeatedly encouraged the employees to learn from failure by sending his message to them. This message motivated and forces the employees to be a leader of task force team, even when they made a mistake. Framework on the leader undertakes is shown in Figure 3.

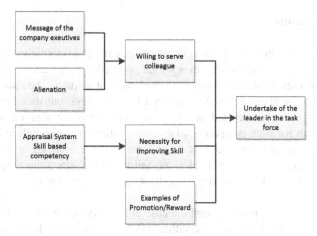

Fig. 3. Framework on Leader Undertakes

An employee who made a fault managed task force to analyze cause of the fault, make measures and report to all the employees as a leader. This was also a kind of atonement. He/she felt so disappointed that he/she might struggle with a sense of alienation due to failure/fault/mistake. He/she did task force job in order to cope with the sense of alienation. Thus, he/she felt that he/she contributed the company as a member of the company.

Since they had to disclose all fact in order to make task force effective, they honestly disclosed the fact without hiding even if it was unfavorable matter for them. The rewarding system and appraisal system called skill based competency also made them be a leader of task force team. When he/she managed it well, they were highly evaluated, promote and be rewarded.

These were why a personnel who made a mistake leading to failure disclosed fact in the mistake without keeping anything against them, as Figure 4 shows.

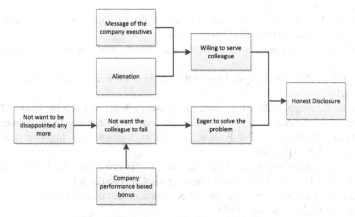

Fig. 4. Framework of the Honest Disclosure

9   Conclusion

Many companies try new business in order to seek better performance and kept sustainable growth. Since even companies which succeeded in a current business could often lose their competitive advantages in environment change. They sustained competitive advantages when they succeeded in a new business, but they lost when they failed. Failure generally occurred in a new business because it was a new challenge, but it was crucial for such companies to avoid repeating the same failure/fault. Hence Sangikyo Corporation organized task force team in order to avoid repeating failure. They had succeeded in avoidance of same failures/faults since they organized task force team. And then they could eventually enjoy more profit from the task force team.

In this paper, authors just explored the case in Sangikyo Corporation for hypothetical success factors for learning from failure. There are therefore limitations. Since it depends just on the single case in Sangikyo Corporation, the findings may not be applicable to other industries and other companies. It is needed to study in further research to qualify. In addition, since in this paper, authors conducted qualitative study, it is necessary to collect quantitative study to further support the findings. To this end, more studies are needed to generalize the findings beyond this study.

Acknowledgement. The authors thank the vice president Mr. Yasukazu Sengoku, a director Mr. Masashi Tokunaga, and the employees in Sangikyo Corporation, who kindly dedicated to our research works. In addition, this research was supported by a grant from JSPS (JSPS KAKENHI Grant Number 15K00319).

References

1. Mitsubishi UFJ Research and Consulting: Research on SME new business in Japan (in Japanese) (2012)
2. Kanno, H.: Management with Learning from Failure (Shippai No Keieigaku) (in Japanese). Nihon Keizai Shinbunsha, Tokyo (2014)
3. Edmondson, A.C.: Strategies for learning from failure. Harvard Business Review (April 2011)
4. Sengoku, M.: Shain No Ichigyo Houkoku ga Kaisha wo Kaeru (in Japanese). Kanki Publishing (2007)
5. Hatamura, Y.: Shippaigaku No Susume (in Japanese). Kodansha Bunko, Tokyo (2005)
6. Cannon, M.D., Edmondson, A.C.: Failing to Learn and Learning to Fail (Intelligently): How Great Organizations Put Failure to Work to Innovate and Improve. Long Range Planning **38**, 299–319 (2005)
7. Carmeli, A., Schaubroeck, J.: Organizational Crisis-Preparedness: The Importance of Learning from Failures. Long Range Planning **41**, 177–196 (2008)
8. Edmondson, A.C.: The local and variegated nature of learning in organization. Organization Science **13**(2), 128–146 (2002)
9. Carmeli, A.: Social Capital, Psychological Safety and Learning Behaviors from Failure in Organizations. Long Range Planning **40**, 30–44 (2007)

10. Ariño, A., de la Torre, J.: Learning from Failure: Towards an Evolutionary Model of Collaborative Ventures. Organization Science **9**(3), 306–325 (1998)
11. Nagayoshi, S.: How can company improve financial performance by learning from failure? In: Uden, L., Oshee, D.F., Ting, I.-H., Liberona, D. (eds.) KMO 2014. LNBIP, vol. 185, pp. 333–336. Springer, Heidelberg (2014)
12. Nagayoshi, S., Nakamura, J.: Success Factors for Learning of Failure - Case Study on Building Capability by "Reflection and Learning" in Sangikyo Corporation. Waseda Bulletin of International Management, No. 45 (2015)
13. Eisenhardt, K.M.: Building Theories from Case Study Research. Academy of Management Review **14**(4), 532–550 (1989)
14. Barnard, C.I.: The Functions of the Executive. Harvard University Press (1968)
15. Mendelson, H., Pillai, R.: Information Age Organizations, Dynamics and Performance. Journal of Economics Behavior & organization **38**, 253–281 (1999)
16. Mendelson, H., Ziegler, J.: Survival of the Smartest. John Wiley & Sons, Inc. (1999)
17. Sangikyo Corporation: History of Sangikyo Corporation. http://www.sangikyo.com/article_service/article/about.html (retrieved on March 24, 2015)
18. http://www.sangikyo.co.jp/article_service/article/index.html (retrieved on March 6, 2015)

The Integration of Nature Disaster and Tourist Database: The Effect of Extreme Weather Event on the Seasonal Tourist Arrival in Taiwan

Ching Li and Ping-Feng Hsia[✉]

Graduate Institute of Sport, Leisure and Hospitality Management,
National Taiwan Normal University, Taipei 106, Taiwan
60031011a@ntnu.edu.tw

Abstract. The study integrated nature disaster and tourist databases to understand how the impact of extreme weather event on tourist arrival. Through multilayer perceptron analysis, the relationship between extreme weather events and the tourist arrival was estimated. Data were collected from National Fire Agency and Tourism Bureau for the period 1990–2010. The results of the investigation show that the key extreme weather event in spring is heavy rain; in summer and autumn, the threat of typhoons is the major factor, and in winter, it is the cold air incursion. In conclusion, the seasonal variation in the tourist arrival is affected by different types of extreme weather events. Furthermore, the scale of the extreme weather event was the most important factor that affects the change in the tourist arrival.

Keywords: Extreme weather events · Multilayer perceptron analysis · Tourist arrival · Seasonal impact

1 Introduction

Long-term government monitoring data can provide useful information. The transparency, accuracy and timeliness in information transmission direct influence the process of restoration and the revival of tourist destination industry. On the other hand, tourists consider the weather conditions of their intended destination when they plan a trip because a weather event might threaten their travel experience and themselves or their belongings [1] [2]. However, it is difficult to assess how a weather forecast or change in weather conditions affects tourists' preferences over destinations. If the tourist arrivals cannot be predicted efficiently, tourism operators cannot implement development plans. Therefore, the study integrated nature disaster database provided by the National Fire Agency, Ministry of the Interior, and tourist database provided by the Tourism Bureau, Ministry of Transportation and Communications, Republic of China.

Owing to regional interaction between the geographical environment and climatic characteristics, there are different types of extreme weather events in each area; all of which may be destructive in their own right [3] [4]. Taiwan is an area that is exposed

© Springer-Verlag Berlin Heidelberg 2015
L. Wang et al. (Eds): MISNC 2015, CCIS 540, pp. 94–105, 2015.
DOI: 10.1007/978-3-662-48319-0_8

to one of the highest levels of global climate crisis [5] [6]. The proportion of population live in the urban is 80.4% [7]. The population in 73.1% of the area faces, on average, more than three natural disasters each year, and 90% of these disasters are related to climate, weather, and water [8].

The study focused on the extreme weather event, because the weather condition became more and more extreme. When facing agricultural issue, people began to notice that climate changed in an extreme way [9]. Especially, precipitation extreme is on a global scale [10]. Extreme weather events are also important factors that influence the tourist arrival in its selection of travel destinations [11].

The two principal factors in this study are the frequency and scale of extreme weather events. The frequency of extreme weather events has been discussed previously and considered as a factor in weather confidence of travel destinations [1]. The overall cost of the damage caused by extreme weather events is reported in Taiwan's Disaster Responses Reports [8], and it reflects the scale of the extreme weather events. The aim of this study is to determine the extreme weather event frequency and scale to predict the tourist arrival, and to assess when the general estimation should be corrected.

2   Nature Disaster and Tourist Databases

The study used two databases: (a) nature disaster database: Disaster Responses Reports published by Taiwan National Fire Agency; (b) tourist database: Annual Tourist Report published by Taiwan Tourism Bureau. The brief introduction as followed. Besides, in private sector, there were relative tool provided to public, such as seasonal weather forecast.

2.1   Nature Disaster Database: Disaster Responses Report

Disaster Responses Report was published by Taiwan National Fire Agency. The purpose of the Disaster Responses Report was to report the yearly affair of National Fire Agency in Taiwan, including fire control human resources, facilities, equipment, and policy of disaster prevention, natural disaster statistic, conflagration report, and first aid of disaster report. The natural disaster statistic included casualties and loss of typhoon, flooding, earthquake and other natural disaster [8]. Tourist destination manager could use the statistic of natural disaster to help them making the prevention strategy.

2.2   Tourist Database: Annual Tourist Report

Annual Tourist Report was the tourism statistics conducted by Taiwan Tourism Bureau. It contained three different statistic data. First was visitor arrivals, second was outbound departures of nations of the Republic of China, and the last one was visitors to principal tourist spots in Taiwan. The study focus on visitors to principal tourist spots in Taiwan. The data of visitors to principal tourist spots were collected from Taiwan principal tourist spots by month, which could be used as reference for tourist destination management [12].

2.3   Seasonal Weather Forecast

Government sometimes will ask private sector for assistant and operation in order to gather more information and improve the state-of-the-art technique more efficiently. Atmospheric and Environmental Research (AER) helps US government agencies better anticipate and manage climate and weather related risks. It provides several climate and weather forecast information, including arctic oscillation and hurricane. There is also seasonal weather forecast provided by AER, focus on snow cover, polar vortex, and blizzard [13]. However, the climate and weather condition is different because of different geographical features, so it is necessary to confirm the key extreme weather event in Taiwan.

3   The Extreme Weather Event in Taiwan

In Taiwan, the frequency and scale of typhoon and hot wave occurred more often and the days of extreme high temperature and drought increased. Besides, those damage events such as the flood and mudslide happened frequently as the increase in heavy rainfall incidents [14] [15] [16] [17]. Recent years, (1985-2009) the damage weather types in Taiwan caused the loss of 16 billion and 3 hundred million NT dollars per year in average. Typhoon (87%) and heavy rainfall (10%) were the main factors [18]. The study referred to "Taiwan's climate extremes and the catastrophic weather alert's standard", issued by Central Weather Bureau(CWB), to define the various types of extreme weather events and its scale in Taiwan [19]. The most common extreme weather events in Taiwan include typhoons, strong winds, hot waves, cold air incursion, drought, heavy rain and flooding. The definition and trend of those extreme weather events in Taiwan described as follow.

3.1   Extreme Wind: Typhoon and Strong Winds

The definition of strong wind is that the wind speed is higher than level 6 (10.8 - 13.8 m s–1) [20]. When average wind speed reaches level 6 and wind gust speed reaches level 8, Central Weather Bureau in Taiwan will announce strong wind warning. The strong wind most occurs during the period of typhoon attacking Taiwan [21]. Typhoon was the most devastating meteorological disaster in Taiwan, it usually combined with strong wind and heavy rain to easily cause casualties, industry's loss and the damaged of building and public facilities. Invaded typhoon with the wind speed scale up to seven or more, the Central Weather Bureau would issue a typhoon warning. It was usually occurred from July to September [22]. After 2000, according to statistics, the average annual number of typhoons invading Taiwan has increased from 3.3 to 5.7, in which the proportion of light typhoon decreased, while the number of moderate and severe typhoon significantly increased [23]. The frequency of heavy rainfall typhoons and the intensity of rainfall both had the increasing trends [24].

3.2   Extreme Temperature: Hot Waves and Cold Air Incursion

The definition of extreme high temperature in Taiwan is over 35°C. When the extreme high temperature prolongs several days, it is called hot wave. In summer of Taiwan, the days of highest temperature and the frequency and intensity of hot wave has increased, and the days of extreme low temperature has decreased [25]. Taipei city is the most remarkable among all. It increased 1.36 days every year [26]. The survey found that the frequency of hot wave incidents occurred more often after 2000. The average high temperature days were up to 32 days in 1999-2008 [27]. Previous study predicted that the highest temperature over 32°C in summer days would increase [28]. On the other hand, the extreme low temperature is below 10°C, and it causes cold air incursion. The days of extreme temperature occurred significantly reduce over the past century in Taiwan. A year's low temperature days in Taipei has been less than 10 days [27].

3.3   Drought

The rainfall intensity increased, but the hours of rainfall decreased in Taiwan for century [17]. The drought occurred sometimes due to the lack of rainfall, uneven distribution of rainfall time, the amount of water can only preserve 20.5% constrained by topography and the need of water increase. In Taiwan's standards, more than 20 continuous days without rain is drought events [24]. Every year in Taiwan, November to April next year is the dry season, rainfall is about 20% and the southern region has less rainfall among all [29]. Previous study predicted that the number of days without rain will increase in future, but the number of days of heavy rain will not have a significant change. Although the rainfall of extreme rainfall events might climb, the drought period in central and southern region is expected to be longer [28]. The north area might also face the problem that frequency and intensity of drought rise in winter and spring because of insufficient rainfall [25].

3.4   Extreme Rainfall: Heavy Rain and Flooding

The days of torrential rain occurred with an increasing trend, while the days of small rain significantly reduced [25] [26]. The frequency and intensity of flooding, landslides and other disasters may increase. When a single area accumulates the rainfall between 200-350 mm in 24 hours is torrential rain event. The rainfall accumulates more than 350 mm is super torrential rain event [22]. Short duration rainfall (less than a day) will influence slope land disaster. Long duration rainfall (1-3 days) may cause flood disaster [30]. In 2006, 609 floods in four days with 14 observatories accumulated rainfall up to 1,000 mm. It caused flooding, hillside collapse, rail and road traffic disruption and other disaster in central and southern regions. Agricultural losses were more than 1.1 billion Taiwan dollars [31].

4   Materials and Methods

This study used multilayer perceptron analysis to estimate the frequency and scale of different type extreme weather events and their effects on the tourist arrival. The data of extreme weather events for 1990–2010 are collated in the Disaster Responses Reports published by the National Fire Agency, Ministry of the Interior, Taiwan. In these reports, extreme weather events are classified into seven items: typhoons, strong winds, heat waves, cold air incursion, drought, heavy rain and flooding. The reports also include the annual frequency and damage costs of extreme weather events. This study used the damage costs of extreme weather events to reflect the scale of the extreme weather events in Taiwan. Data on the tourist arrival from 1990 to 2010 were collected from Annual Tourist Reports published by the Tourism Bureau, Ministry of Transportation and Communications, Republic of China.

To avoid the impact of inflation of numbers and extreme values, the difference between the actual tourist arrival and the expected tourist arrival was used. The process of calculation was as follows: (1) the seasonal tourist arrival for each year was aggregated, (2) the average growth rate of the tourist arrival was computed, (3) using the growth rate, the expected seasonal tourist arrival was computed for each year, and (4) the difference between the actual and the expected tourist arrival was computed.

$$S_n = \sum M_m . \tag{1}$$

$$R_n = (S_{n+1} - S_n)/S_n . \tag{2a}$$

$$R' = (\sum\nolimits^{1990^{2010}} R_n)/N . \tag{2b}$$

$$E_n = [S_{n-1}(1+R')]/3 . \tag{3}$$

$$P_i = (M_m - E_n)/1000 . \tag{4}$$

m: month, spring is from March to May; summer is from June to August; autumn is from September to November; winter is from December to February
i: season
S_n: total of seasonal tourist arrival each year
M_m: tourist arrival each month
R_n: growth rate of tourist arrival each year
R': average growth rate of tourist arrival in 1990–2010
E_n: expected monthly tourist arrival
P_{spring}: difference between actual and expected tourist arrival (per thousand)

For example, the step of calculating tourist arrival in 2010 spring was followed. Firstly, the tourist arrivals, from March to May in 2010, were added up to represent the tourist arrival in 2010 spring. Secondly, the growth rate of tourist arrival was counted each spring and summed up to rid of 20, then the average growth rate of tourist arrival in spring from 1990 to 2010 was figured out. Thirdly, the expected 2010 spring monthly tourist arrival could be calculated according to the growth rate. Finally, using the actual monthly tourist arrival to subtract the expected monthly tourist arrival, then counted tourist arrival in thousand. The difference between actual and expected tourist arrival in thousand was used as dependent variable.

Through multilayer perceptron analysis, the examination of the relationship between the tourist arrival and extreme weather events comprised the following two parts:

4.1   Identify Key Seasonal Extreme Weather Events

The frequency and scale of seven types of extreme weather events were analyzed and the key extreme weather events for each season determined.

4.2   Define the Order of Importance of the Extreme Weather Events for Each Season

In multilayer perceptron analysis, the importance chart was drawn and the importance was calculated, which showed that the results were dominated by type, frequency, or scale of extreme weather events. Moreover, the type, frequency, and scale of extreme weather events were assessed for their order of importance in each season.

5   Results

The description of sample was as follow. In spring, 45 cases were assigned to the training sample and 12 to the testing sample. The 3 cases excluded from the analysis. In summer, 43 cases were assigned to the training sample and 6 to the testing sample. The 11 cases excluded from the analysis. In autumn, 38 cases were assigned to the training sample and 7 to the testing sample. The 12 cases excluded from the analysis. In winter, 36 cases were assigned to the training sample and 14 to the testing sample. The 5 cases excluded from the analysis (Table1.).

Table 1. Case processing summary

		Spring		Summer		Autumn		Winter	
		N	Percent	N	Percent	N	Percent	N	Percent
Sample	Training	45	78.9%	43	87.8%	38	84.4%	36	72.0%
	Testing	12	21.1%	6	12.2%	7	15.6%	14	28.0%
Valid		57	100.0%	49	100.0%	45	100.0%	50	100.0%
Excluded		3		11		12		5	
Total		60		60		57		55	

The model summary displayed information about the results of training sample. Sum-of-squares error is displayed because the output layer has scale-dependent variables. This is the error function that the network tries to minimize during training. The average overall error is the ratio of the sum-of-squares error for all dependent variables to the sum-of-squares error for the "null" model, in which the mean values of the dependent variables are used as the predicted values for each case. The average overall relative error and relative errors are fairly constant across the training and testing samples, which gives you some confidence that the model is not overtraining and that the error in future cases scored by the network will be close to the error reported in this table (Table 2.)

Table 2. Model summary

		Spring	Summer	Autumn	Winter
Training	Sum of Squares Error	15.73	25.69	17.52	10.30
	Average Overall Relative Error	0.71	1.22	0.95	0.59
	Stopping Rule Used	1 consecutive(s) step with no decrease error[a]			
	Training Time	00:00:00.048	00:00:00.564	00:00:00.052	00:00:00.034
Testing	Sum of Squares Error	7.91	2.66	9.94	5.26
	Average Overall Relative Error	1.00	1.56	1.00	1.04

Dependent: tourist arrival
a. Error computations are based on the testing sample.

5.1   Key Seasonal Extreme Weather Events

The multilayer perceptron analysis revealed that the tourist arrival in different seasons was affected by different types of extreme weather event. The key extreme weather events in spring were typhoons, cold air incursion, drought, heavy rain, and flooding, both in frequency and in scale. In summer, they were typhoons, strong wind, drought, heavy rain, and flooding, both in frequency and in scale. In autumn, they were typhoons, heat waves, and heavy rain, both in frequency and in scale, and cold air incursion in frequency. In winter, they were cold air incursion and heavy rain, both in frequency and in scale, and drought in frequency (Table 3.).

Table 3. Key seasonal extreme weather events

	Spring	Summer	Autumn	Winter
Typhoon	F/S	F/S	F/S	
Strong wind		F/S		
Heat wave			F/S	
Cold air incursion	F/S		F	F/S
Drought	F/S	F/S		F
Heavy rain	F/S	F/S	F/S	F/S
flooding	F/S	F/S		

F: Frequency of extreme weather event
S: Scale of extreme weather event

5.2   The Importance of Extreme Weather Event in Each Season

The most important extreme weather event was not exactly the same in each season. Observing the normalized importance of more than 50%, revealed that the top two important extreme weather events were the scale of heavy rain and the frequency of flooding in spring. In summer, they were the scale of typhoons, heavy rain, flooding, and drought. In autumn, it was the scale of typhoons, and in winter, it was the scale of the cold air incursion (Table4.).

Table 4. Importance of extreme weather event in each season

		importance	normalized importance
Spring	scale of heavy rain	0.41	100.00%
	frequency of flooding	0.22	55.30%
Summer	Scale of typhoon	0.21	100.00%
	Scale of heavy rain	0.19	93.90%
	Scale of flooding	0.17	84.70%
	Scale of drought	0.13	61.90%
Autumn	Scale of typhoon	0.41	100.00%
Winter	Scale of cold air incursion	0.68	100.00%

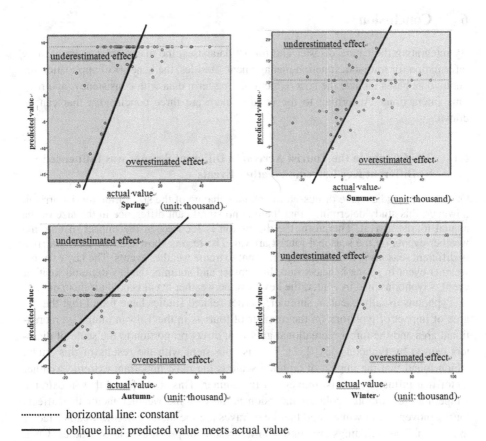

.... horizontal line: constant

────── oblique line: predicted value meets actual value

Fig. 1. Predicted-by-observed chart for tourist arrival

5.3   Adjustment of Effect of Extreme Weather Event on Tourist Arrival Prediction

When there was no extreme weather event occurring, predicted value as a constant, located on horizontal line, was computed by multilayer perceptron model. The oblique line indicated that predicted value meets actual value. In the left area of oblique line showed the effect of extreme weather event was underestimated; on the other hand, the right part indicated the effect of extreme weather event was overestimated. To discuss the effect of extreme weather event, the study observed distribution of observation excluding constant. In autumn and winter, most of observation located in right area of oblique line, which means the effect of extreme weather event was overestimated. In spring and summer, the observation spread in left and right area equally, this indicated that the effect of extreme weather event was irregular (Fig. 1.).

6   Conclusion

By integrating the nature disaster and tourist databases, the managers can apply more information for tourist destination management. Besides, the benefits of application the public data bases include the cost reduction, long term data with consistency, and real-time interactions. According to the results, there are three conclusions that can be drawn:

6.1   The Change in the Tourist Arrival in Different Seasons was Influenced by Different Key Extreme Weather Events

Other recent studies have discussed the annual change of the size of the tourist arrival; however, this study determined that there is no significant difference in the size of the annual tourist arrival. Therefore, the discussion is focused on the impact of extreme weather events on the seasonal tourist arrival. The results show that the tourist arrival in different seasons is influenced by different extreme weather events. The key extreme weather event in spring is heavy rain. In summer and autumn, the key extreme weather event is typhoons, and in winter, the key extreme weather event is cold air incursion.

Typhoons usually occur in summer. Earlier related studies have found that the degree of impact of typhoons on the number of tourists in the Taiwan mountain recreational area and the forest recreational area is in direct proportion to the scale of disasters caused by typhoons [24] [32]. This is consistent with the results of this study, which found that the impact of typhoon scale is the most important extreme weather factor that influences tourist movement in summer. This study found that in extreme temperature conditions, cold air incursion is the extreme weather factor that affects tourist movement in winter, and that heat waves are the major extreme weather factor in autumn. These findings are similar to the results of previous related studies. Chen found that colder or hotter weather affects tourists' choice of travel destination [33]. Chi et al. conducted research on Taiwan's forest recreational area in winter and found that tourist movement would decrease because of the clammy weather [34]. This study found that heavy rainfall also has an influence on tourist movement in all four seasons, especially in spring and summer. However, past studies have focused mostly on heavy rainfall in summer, and identified that heavy rainfall in summer caused a reduction in tourist movement [34] [35] [36], but little mention was made regarding the impact of heavy rainfall in the other seasons.

6.2   The Scale of Extreme Weather Events Was the Important Effect on the Change of Tourist Arrival

The impact of the frequency and scale of extreme weather events was discussed. The results show that the scale of extreme weather events has greater significance to the tourist arrival than the frequency. In spring, the most important extreme weather events is scale of heavy rain; in summer and autumn, it was scale of typhoon; in winter, it was scale of cold air incursion.

6.3   The Effect of Extreme Weather Events Needed to Be Adjusted in Autumn and Winter

The influence of extreme weather event was unpredictable in spring and summer, because the tourist arrival was influenced by more than key extreme weather event. In autumn, the most important extreme weather event was typhoon. And it was cold air incursion in winter. However, based on result, it is obvious that the effect of extreme weather event was overestimated in autumn and winter, which means that the prediction of tourist arrival needs to be adjusted higher when extreme weather event occurring in autumn and winter.

7   Recommendation

In practical application, tourism destination manager should pay attention the weather forecast about heavy rain and typhoon in spring and summer separately. Besides, the impact of extreme weather events in autumn and winter was minor, which meant the manager should make preparation for risk management and make sure the tourist safety. In future study, the result and model of multilayer perceptron analysis should be verified further. For example, using the data from 2011 to 2014 test the model and verify the result of the study. Moreover, the future study could focus on the key extreme weather event each season to do more research.

Acknowledgment. The study is part of research finding in the study project (NSC101-2410-H-003-136) which is supported by the Ministry of Science and Technology, Taiwan, R.O.C.

References

1. de Freitas, C.R.: Tourism climatology: evaluating environmental information for decision making and business planning in the recreation and tourism sector. Int. J. Biometeorol. **48**, 45–54 (2003)
2. Hartz, D.A., Brazel, A.J., Heisler, G.M.: A case study in resort climatology of Phoenix, Arizona. USA. Int. J. Biometeorol. **51**, 73–83 (2006)

3. Francis, D., Hengeveld, E.: http://www.ec.gc.ca/default.asp?lang=En&n=FD9B0E51-1
4. Vellinga, P., Verseveld, W.J.V.: Climate Change and Extreme Weather Events. World Wide Fund For Nature (WWF), Gland, Switzerland (2000)
5. World Bank. http://www-wds.worldbank.org/external/default/WDSContentServer/WDSP/IB/2005/11/23/000160016_20051123111032/Rendered/PDF/344230PAPER0Na101officia l0use0only1.pdf
6. Harmeling, S.: Climate change risk Index 2011. Report ISBN 978-3-939846-74-1 (2011)
7. Urban and Housing Development Department. http://www.ndc.gov.tw/encontent/m1.aspx?sNo=0007231
8. National Fire Agency, Ministry of the Interior. http://www.nfa.gov.tw/main/index.aspx
9. Van Oort, P.A.J., Timmermans, B.G.H., Meinke, H., Van Ittersum, M.K.: Key weather extremes affecting potato production in The Netherlands. Eur. J. Agron. **37**(1), 11–22 (2012)
10. Zwiers, F.W., Alexander, L.V., Hegerl, G.C., Knutson, T.R., Kossin, J.P., Naveau, P., Nicholls, N., Schär, C., Seneviratne, S.I., Zhang, X.: Climate extremes: challenges in estimating and understanding recent changes in the frequency and intensity of extreme climate and weather events. In: Climate Science for Serving Society, pp. 339–389. Springer, Netherlands (2013)
11. Murphy, P.E., Bayley, R.: Tourism and disaster planning. Geogr. Rev. **79**, 36–46 (1989)
12. Tourism Bureau. http://admin.taiwan.net.tw/auser/B/annual_statistical_2012_htm/Chinese/htm/p01.htm
13. Atmospheric and Environmental Research (AER). https://www.aer.com/about-us
14. Chen, Y.L.: The climate changes in Taiwan during recent 100 years. Sci. Dev. **424**, 6–11 (2008)
15. Tung, C.P., Lin, C.Y.: The challenge and response for climate change. Sci. Devel. **424**, 28–33 (2008)
16. Environmental Protection Administration: http://unfccc.epa.gov.tw/unfccc/english/_uploads/downloads/01_Extreme_Events_and_Disasters_from_Typhoon_Morakot-the_Biggest_Threat_ever_to_Taiwan.pdf
17. Central Weather Bureau (CWB): 1897-2008 Statistics on Climate Chang in Taiwan. Report ISBN 978-986-02-1702-5, p. 77 (2009)
18. Chen, C.K.: Looking back and looking into future on meteorological hazard mitigation in Taiwan. In: 2011 Conference on Weather Analysis, Forecasting and Seismic Observation. Taipei, Taiwan, Central Weather Bureau (2011)
19. CWB. http://www.cwb.gov.tw/V7/prevent/plan/prevent-faq
20. World Meteorological Organization. https://www.wmo.int/pages/prog/www/IMOP/publications/CIMO-Guide/Prelim-2014Ed/Prelim2014Ed_P-I_Ch-5.pdf
21. CWB. http://www.cwb.gov.tw/V7/prevent/warning.htm
22. CWB. http://www.cwb.gov.tw/V7/knowledge/encyclopedia/me028.htm
23. Lin, C.C.. http://www.libertytimes.com.tw/2011/new/sep/21/today-life10.htm
24. Chiou, C.R., Dai, J.S., Tsai, W.L., Chan, W.H.: Influences of Natural Disasters on Tourist Arrivals at National Forest Recreation Areas in Taiwan. Q. J. Chin. For. **44**(2), 249–264 (2011)
25. Hsu, H.H., Wu, Y.C., Chou, C., Chen, C.D., Chen, Y.M., Lu, M.M.: Taiwan climate change science report 2011. National Science and Technology Center for Disaster Reduction, Taipei, Taiwan, p. 67 (2011)
26. Lu, M.M., Cho, Y.M., Hsu, T.C., Li, C.T., Lin, Y.C., Li, S.Y.: Characteristic of Taiwan hundred-year climate change. In: 2011 Conference on Weather Analysis, Forecasting and Seismic Observation, Taipei, Taiwan, Central Weather Bureau (2011)

27. Wu, Y.C., Chen, Y.M., Chu, R.L.: Trend of Taiwan climate change. Natl. Appl. Res. Lab. Q. **25**, 40–46 (2010)
28. Liu, C.M., Wu, M.C., Lin, S.H., Chen, Y.C., Yang, Y.T., Lin, W.H., Tseng, Y.H., Chen, C.D.: The prediction of Taiwan future climate change. Global Change Research Center, National Taiwan University, Taiwan (2008)
29. Lu, C.J.: Long-term trend analysis of drought in Southern Taiwan. Doctoral dissertation, National Cheng Kung University (2006)
30. Lu, M.M., Chen, C.J., Lin, Y.C.: Long-term Variations of the Occurrence Frequency of Extreme Rainfall Events during the Period of 1951-2005. Atmo. Sci. **35**, 87–102 (2007)
31. Lien, W.Y., Fu, C.C., Hsieh, L.S., Yu, B.S.: Taiwan natural disaster - typhoon. Natl. Appl. Res. Lab. Q. **13**, 12–14 (2007)
32. Lee, Y.F., Chi, Y.Y.: Impact assessment of natural disaster on annual visitor number of recreational district. In: The 14th Leisure, Recreation and Tourism Research Symposium and International Forum, Hualien, Taiwan, National Dong Hwa University (2012)
33. Chen, C.C.: A study on the application and visiting to Fushan Botanical Garden and their relationships to climate factors. Doctoral dissertation, National Taiwan Normal University, p. 53 (2013)
34. Chi, Y.N., Wu, J.Y., Lin, S.T.: Study of trend of tourist flow in National Park. Taiwan For. J. **28**, 23–30 (2002)
35. Yang, W.G., Ho, P.T., Lee, S.S.: The Analysis of Taiwan Theme Park Development. J. Chaoyang Univ. Technol. **13**, 247–269 (2008)
36. Chong, F.: http://news.sina.com.tw/article/20130806/10330324.html

Evolution of Social Networks and Body Mass Index for Adolescence

Hsieh-Hua Yang[1] and Chyi-In Wu[2(✉)]

[1] Department of Health Care Administration, Oriental Institute of Technology,
58, Sec 2, Sihchuan Rd., Pan-Chiao Dist., New Taipei City 22061, Taiwan
FL008@mail.oit.edu.tw
[2] Institute of Sociology, Academia Sinica, Nankang, Taipei 11529, Taiwan
ssslciw@gate.sinica.edu.tw

Abstract. We hypothesized that social norm for BMI has effect on the evolution of social networks. The goal of the present study is to analyze the evolution of friendship network and body mass index (BMI) for adolescent in 3 different classes—boys' and girls' class, boys' class, and girls' class. A network survey was carried out in classrooms of high schools. The participants came from 3 classes. Sociometric data were collected by having each student nominate up to 16 intimate classmates. Panel data was collected 7 times across 2 semesters from Sep. 2008 to Jun. 2009. The program SIENA was applied to estimate the models for the evolution of social networks and BMI. The result showed that BMI has effect on girls' class, but not boys'. In conclusion, the different composition of gender in a class will construct social norm about BMI and exerts different effect on network evolution. It is helpful for intervention plan focused on norms.

Keywords: Social networks · BMI · Adolescent · Evolution

1   Introduction

Social network formation is an evolution process. Many theories are applied to explain the mechanism of social network evolution. Social comparison theory [1] holds that people have a drive to make comparisons that result in favorable outcomes and individuals will compare themselves to those with whom they are most similar rather than with individuals who are perceived to be more dissimilar. Social learning theory [2] proposes that individuals' behaviors may arise and be reinforced through observing and imitating others, including peers. And social control theory [3] posited that social context and network norms determine individual behavior, and suggested that peers have a strong influence on individual behavioral patterns. Fredrickson and Roberts [4] propose objectification theory to illuminate that girls and women are typically acculturated to internalize an observer's perspective as a primary view of their physical selves.

Prior adolescent BMI related studies indicated the importance of social norms in affecting health behaviors [5]. Brault et al. [6] indicated that body image concerns and

© Springer-Verlag Berlin Heidelberg 2015
L. Wang et al. (Eds): MISNC 2015, CCIS 540, pp. 106–115, 2015.
DOI: 10.1007/978-3-662-48319-0_9

weight stigmatization stand out as important factors in the prediction of developmental weight trajectories. A review by Robinson et al. [7] indicated that information about eating norms influences choice and quantity of food eaten. Even the biases of self-reporting weight were due to the social norms regarding "ideal" weight [8]. Social norms have been included in models of health behavior [9, 10]. Perceived norm exist at the individual, psychological level and represent a person's interpretation of the prevailing collective or cultural norm [11]. Social norms are mechanism through which peer influence may operate without any direct contact. A high prevalence of a behavior among young people in a setting may send a subtle message that such behavior is accepted and expected, which may encourage adoption of that behavior throughout that social setting.

Social norm is measured by friends' behavior. Eisenberg et al. [12] calculated social norm by the prevalence of respondents' dieting behaviors. Font, Fabbri, and Gil [13] measured social norms by the behavior of a reference group. Eisenberg et al. [14] measured social norm effects by asking respondents whether they had friends who dieted to lose of keep from gaining weight. Thompson et al. [15] used multifaceted scales to measure peer influence, such as the degree of teasing that respondents received from friends about weight and appearance. Shomaker and Furman [16] measured the social reinforcement of thinness from close social group members. Christakis and Fowler [17] examined obesity spreading through social networks.

Comparing one's BMI to those of individuals perceived as more attractive is more common among females than males [18]. It has been associated with increases in body dissatisfaction for girls. Forney and Ward [19] examined the moderating effect of perceptions of social norms and found the effect had gender difference. The findings suggested that women who perceived a social environment that values thinness and approves of disordered eating are more likely be at risk for eating disorders.

In this research, BMI is the focus to be tested during friendship network evolution. The argument is that social norm about BMI in a class will exert effect on network evolution, and the effect has gender difference.

2   Method

2.1   Participants

Studying the friendship network evolution, it must be assumed that the group is closed, the boundary of the friendship network is clear. The data was collected from high school in Taiwan. The participants were from 3 classes. There were 15 boys and 31 girls in class I, 49 boys in class II, and 47 girls in class III.

2.2   Measures

Panel data was collected during 2 semesters from Sep. 2008 to Jul. 2009. Sociometric data were collected 7 times by having each student nominate up to 16 intimate classmates. Their height and weight of the participants were collected in the beginning of the 1st semester. And BMI was calculated. The BMI for a person is defined as their body mass divided by the square of their height—with the value universally being given in units of kg/m^2. Then BMI was recoded as 1 (BMI<18.5), 2 (18.5≤BMI<24), and 3 (BMI≥24).

2.3   Data Analysis

The program SIENA (Simulation Investigation for Empirical Network Analysis) was applied to estimate the models for the evolution of social networks according to the dynamic actor-oriented model of Snijders [20, 21, 22]. For the estimation, sociometric data was transformed into adjacency matrices. Boy was coded as 1 and girl as 2.

In the study, the evolution model components include density, reciprocity, transitivity, gender similarity, BMI ego, BMI alter and BMI similarity. Density effect is defined as the number of outgoing ties. Reciprocity effect is defined the number of reciprocated ties. A transitive triplet for actor is an ordered pairs of actors (j, h) for which i→j→h and also i→h. The program can estimate all possible effects simultaneously, and is suitable to test our hypotheses. The SIENA program is included in the StOCNET system, and can be downloaded from http://stat.gamma.rug.n1/stocnet/ [23].

3   Result

The analyses were run on the 46 (15 boys and 31 girls), 49 (boys), and 47 (girls) students of 3 classes separately. As table 1, the probability of lean was almost the same for class II and III. The probability of normal BMI was the highest for class III. And obviously, the probability of overweight was higher for boys than girls.

Table 1. Description of BMI

Wave	Lean (BMI<18.5) n (%)	Normal (18.5≤BMI<24) n (%)	Overweight (BMI≥24) n (%)
Class I (n=46)			
Boys	4 (8.70)	5 (10.87)	6 (13.04)
Girls	10 (21.74)	17 (36.96)	4 (8.70)
Class II (n=49)			
All boys	11 (22.45)	31 (63.27)	7 (14.28)
Class III (n=47)			
All girls	10 (21.28)	34 (72.34)	3 (6.38)

As table 2, the average degree at beginning is between 4.889 and 6.800 for class I, 6.292 and 7.854 for class II, and 5.804 and 7.022 for class III. As Fig. 1, the density of class II was the highest among these classes almost during all stages except w2 and w7. The density of class I was the lowest among 3 classes except w3 and w7.

Table 2. Density and degree

Wave	w1	w2	w3	w4	w5	w6	w7
Class I							
Density	0.120	0.123	0.143	0.148	0.106	0.117	0.142
Average degree	5.511	5.644	6.578	6.800	4.889	5.400	6.533
Class II							
Density	0.160	0.128	0.162	0.151	0.158	0.147	0.132
Average degree	7.854	6.292	7.958	7.396	7.729	7.208	6.479
Class III							
Density	0.145	0.136	0.139	0.149	0.132	0.139	0.123
Average degree	6.804	6.370	6.543	7.022	6.196	6.543	5.804

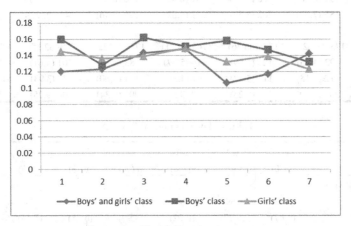

Fig. 1. Density

The possible relational ties between students were 2070, 2352, and 2162 for class I, II, and III respectively. As table 3, around 80% of them had no relations (0 to 0). About 2-7% of them were newly built relations presented as 0 to 1. About 3-8% was dissolved relations presented as 1 to 0. And less than 10% were continuing relations presented as 1 to 1. Newly built relational ties in class I were lower than the other classes except at the end of observation. And the dissolved ties in class I were also the lowest during all stages except w4-5.

As table 4, possible dyads were 1035, 1176, and 1081 for 3 classes respectively. Of these dyads, about 5-9% was mutual dyads and 8-16% was asymmetric dyads. Obviously, the probability of asymmetric dyads was the highest for class II (13.01%-16.67%). And the probability of asymmetric dyads was the least for class I.

The result of the evolution model is presented as Table 5. For network evolution, the rate function describes the average number of changes in network ties between measurement points. The patterns of rate change of these 3 classes were almost similar in the first semester, as Fig. 2. The rate was increased during w2-3, then decreased during w3-4. And in the 2nd semester, their patterns were different. For class II, the rate was increased then decreased. For class I, the rate was increased to the end of observation. And for class III, the rate was kept steady in the 2nd

Table 3. Change frequencies of relational ties

Wave	w1-2	w2-3	w3-4	w4-5	w5-6	w6-7
	n (%)	n (%)	n (%)	n (%)	n (%)	n (%)
Class I						
0 to 0	1745(84.30)	1710(82.61)	1691(81.69)	1719(83.04)	1770(85.51)	1715(82.85)
0 to 1	77 (3.72)	106 (5.12)	83 (4.01)	45 (2.17)	80 (3.86)	112 (5.41)
1 to 0	71 (3.43)	64 (3.09)	73 (3.53)	131 (6.33)	57 (2.75)	61 (2.95)
1 to 1	177 (8.55)	190 (9.18)	223 (10.77)	175 (8.45)	163 (7.87)	182 (8.79)
Class II						
0 to 0	1858(79.00)	1870(79.51)	1846(78.49)	1825(77.59)	1877(79.80)	1905(80.99)
0 to 1	117 (4.97)	180 (7.65)	124 (5.27)	172 (7.31)	104 (4.42)	101 (4.29)
1 to 0	192 (8.16)	100 (4.25)	151 (6.42)	156 (6.63)	129 (5.48)	136 (5.78)
1 to 1	185 (7.87)	202 (8.59)	231 (9.82)	199 (8.46)	242 (10.29)	210 (8.93)
Class III						
0 to 0	1757(81.27)	1746(80.76)	1745(80.71)	1756(81.22)	1776(82.15)	1776(82.15)
0 to 1	92 (4.26)	123 (5.69)	116 (5.37)	83 (3.84)	101 (4.67)	85 (3.93)
1 to 0	112 (5.18)	115 (5.32)	94 (4.35)	121 (5.60)	85 (3.93)	119 (5.50)
1 to 1	201 (9.30)	178 (8.23)	207 (9.57)	202 (9.34)	200 (9.25)	182 (8.42)

Table 4. Dyad counts of 7 waves

	w1	w2	w3	w4	w5	w6	w7
	n (%)	n (%)	n (%)	n (%)	n (%)	n (%)	n (%)
Class I (1035)							
Mutual	69	69	83	93	65	74	93
	(6.67)	(6.77)	(8.02)	(8.99)	(6.28)	(7.15)	(8.99)
Asymmetric	110	116	130	120	90	95	108
	(10.63)	(11.21)	(12.56)	(11.59)	(8.70)	(9.18)	(10.43)
Class II (1176)							
Mutual	112	61	93	83	88	79	72
	(9.52)	(5.19)	(7.91)	(7.06)	(7.48)	(6.72)	(6.12)
Asymmetric	153	180	196	189	195	188	167
	(13.01)	(15.31)	(16.67)	(16.07)	(16.58)	(15.99)	(14.20)
Class III(1081)							
Mutual	89	81	72	98	72	78	68
	(8.23)	(7.49)	(6.66)	(9.07)	(6.66)	(7.22)	(6.29)
Asymmetric	135	131	157	127	141	145	131
	(12.49)	(12.12)	(14.52)	(11.75)	(13.04)	(13.41)	(12.12)

semester. Among these 3 classes, class II had the highest rate parameter. The endogenous network effects are all statistically significant. The density, reciprocity and transitive triplets have effect during all stages. The reciprocity effect indicates a preference for reciprocating relationships. The transitivity effect indicates a preference for being friends with friends' friends. The rate function was dependent on the outdegree.

The effect of BMI was significant for network evolution in girls' class and in heterosexual class. In class I, the positive parameter of BMI alter implies that actors with higher value of BMI are more popular than others. In class III, the negative parameter of BMI alter implies that actors with higher value of BMI are less popular than others, and the positive parameter of BMI ego implies that actors with higher value of BMI are more active in sending friendship nominations. In the boys' class, BMI had no effect.

Table 5. Estimates of the evolution model

Variable	Boys and Girls' Estimate(se)	Boys' Estimate(se)	Girls Estimate(se)
Rate parameter			
w1-2	2.871 (0.275)	4.395 (0.412)	3.082 (0.284)
w2-3	3.345 (0.332)	5.199 (0.441)	4.435 (0.411)
w3-4	2.537 (0.235)	3.678 (0.300)	3.381 (0.309)
w4-5	3.033 (0.299)	5.728 (0.514)	3.117 (0.281)
w5-6	3.058 (0.338)	2.887 (0.232)	3.025 (0.280)
w6-7	4.021 (0.422)	3.207 (0.256)	3.300 (0.288)
out-degree effect on rate	0.090 (0.015)*	0.099 (0.011)*	0.103 (0.018)*
Density	-1.852 (0.059)*	-1.539 (0.040)*	-1.676 (0.047)*
Reciprocity	1.239 (0.080)*	0.829 (0.057)*	1.186 (0.065)*
Transitive triplets	0.125 (0.009)*	0.088 (0.005)*	0.112 (0.009)*
Gender similarity	0.496 (0.094)*	-	-
BMI alter	0.146 (0.045)*	-0.058 (0.040)	-0.123 (0.57)*
BMI ego	-0.080 (0.049)	0.033 (0.041)	0.217 (0.063)*
BMI similarity	-0.085 (0.098)	-0.079 (0.078)	0.012 (0.105)

* $p < .05$

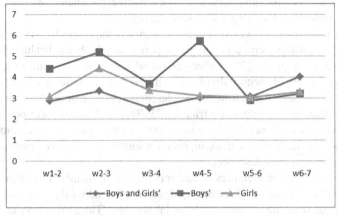

Fig. 2. Rate parameter

4   Discussion

Longitudinal network data can yield important insights into social processes. Methodologically, the assessment of friendship evolution needs to be based on longitudinal designs that include multiple measurement waves. Poulin and Chan [24] indicated that such detailed analysis of stability will allow a better understanding of the dynamic processes by which friendships change over time and affect children's and adolescents' psychosocial development. Two semesters' observations may offer rich materials for understanding adolescents' social networks development.

4.1   Friendship Network Evolution

The rate parameter is up and down, then increasing or decreasing at the last observation. It represents the friendship network is dynamic but not steady. Further, the observation showed that boys' class had the highest rate parameter, especially during w4-5. Obviously, boys make more new friends in the beginning of the 2nd semester. However, the other 2 classes did not have the same result. After w4-5, the rate parameter is rather steady for girls' class and slightly increasing for heterosexual class. Both the frequency of relational ties and mutual dyads showed that girls' friendship is kept steadier than boys. The reason may due to the intimate groups. Girls are more likely than boys to spend their time in small intimate groups, usually made up of two or three people [25, 26, 27]. Besides, Girls' language is relatively more likely to be collaborative and affiliative [28, 29], while boys' language is relatively more assertive, controlling, and competitive and incorporates more demands [30]. Thus, girls have better interpersonal skills than boys and maintain intimate friendship.

The network effects, including density, transitive triplets, and reciprocity are important during all stages. The results are the same as the results of other researches [20-22, 31]. The positive parameter of reciprocity represented the participants had preference for reciprocating relations. The positive parameter of transitive triplets meant the preference for being with friends' friend.

In class I, heterosexual class, gender similarity is important for network evolution. Heterosex friendships represent unique opportunities both for healthy development and adverse development [32]. Other-sex friendships may aid youth in learning how to interact productively with individuals with interests, experiences, and backgrounds that differ from their own. However, adolescents report feeling closer to their same-sex friends than to their other-sex friends [33, 34]. And friends who provide psychological closeness are beneficial for adolescents' socio-emotional development [5, 35]. Thus, same-sex friendships are an important resource for the development of psychological health during adolescence.

Social networks link members of one relationship to members of others in a structure of ties. The formation of friendship dyads is affected by the structure of triads as well as by the structure of the overall social network. The analysis on the social network level aids in bridging the gap between the micro and the macro level [36].

4.2   BMI and Social Network Evolution

The similarity of BMI had no effect, but BMI ego and BMI alter had effect in class I and III, but not II. The phenomena meant that the students perceived norm about BMI are different for 3 classes, and girls are more sensitive to norms of BMI.

For adolescence, weight and shape become increasingly salient and important in girls' everyday lives. And a girls' understanding of her body is based not only on her own views of it, but also on how she believes others view it [32]. It is possible that sociocultural differences in the relatively greater emphasis placed on appearance for women than men [37, 38, 39] explain why boys are more attuned to their BMI than girls.

In girls' class, actors with higher value of BMI are less popular, while in heterosex class, actors with higher value of BMI are more popular. A friendship is more likely to come into existence if each individual perceives the other as attractive, responsive, and in particular, similar in a variety of ways [40, 41]. An actor of higher BMI is attractive or not depends on the context. Due to the prevalence of the obesity stigma in the general population [42], the experiences of being avoided or excluded may be a source of interpersonal strain. In girls' class, girls of higher BMI may experience interpersonal strain, but they still try to sending friendship nominations.

Zhang et al. [43] indicated that the effect of peer influence varied based on the underlying distribution of BMI. The probability of overweight in a class may exert dose effect on social network change. It needs further study.

4.3   Conclusion

The purpose of this article is testing the effects of BMI on friendship network evolution. A longitudinal network behavior models indicated that gender similarity is influential in the formation of social ties, and the associations between BMI and friendship network have been tested in girls' class. Social network analysis is uniquely suited for measuring and understanding the behavior of peers because it provides a formal means for "mapping" friendships and measuring properties of those friendships [44]. The results support that female adolescents are more susceptible to social norm about BMI and are helpful for intervention plan focused on norms.

Acknowledgments. The work is supported by the National Science Council (NSC101-2410-H-161-002).

References

1. Festinger, L.: A Theory of Social Comparison Processes. Hum. Relat. **7**, 117–140 (1954)
2. Bandura, A.: Social Learning Theory. Prentice Hall, Englewood Cliffs (1977)
3. Hirschi, T.: Causes of Delinquency. University of California, Berkeley (1969)
4. Fredrickson, B.L., Roberts, T.-A.: Objectification Theory toward Understanding Women's Lived Experiences and Mental Health Risks. Psychol. Women Quart. **21**, 173–206 (1997)

5. Wang, Y., Xue, H., Chen, H.-J., Igusa, T.: Examining Social Norm Impacts on Obesity and Eating Behaviors among US School Children Based on Agent-based Model. BMC Public Health **14**, 923 (2014). doi:10.1186/1471-2458-14-923

6. Brault, M.-C., Aime, A., Begin, C., Valois, P., Craig, W.: Heterogeneity of Sex-stratified BMI Trajectories in Children from 8 to 14 Years Old. Physiol. Behav. **142**, 111–120 (2015)

7. Robinson, E., Thomas, J., Aveyard, P., Higgs, S.: What Everyone Else Is Eating: A Systematic Review and Meta-analysis of the Effect of Informational Eating Norms on Eating Behavior. J. Acad. Nutrition and Dietetics **114**, 414–429 (2014)

8. Gil, J., Mora, T.: The Determinants of Misreporting Weight and Height: The Role of Social Norms. Econ. Hum. Biol. **9**, 78–91 (2011)

9. Fishbein, M., Ajzen, I.: Belief, Attitude, Intention and Behavior: An Introduction to Theory and Research. Addison-Wesley, Reading (1975)

10. Ajzen, I.: The Theory of Planned Behaviors. Organ. Behav. Hum. Dec. **50**, 179–211 (1991)

11. Kallgren, C.A., Reno, R.R., Cialdini, R.B.: A Focus Theory of Normative Conduct: When Norms Do and Do Not Affect Behavior. Pers. Soc. Psychol. B. **26**, 1002–1012 (2000)

12. Eisenberg, M.E., Neumark-Sztainer, D., Story, M., Perry, C.: The Role of Social Norms and Friends' Influences on Unhealthy Weight-control Behaviors among Adolescent Girls. Soc. Sci. Med. **60**, 1165–1173 (2005)

13. Font, J.C., Fabbri, D., Gil, J.: Decomposing Cross-country Differences in Levels of Obesity and Overweight: Does the Social Environment Matters? Soc. Sci. Med. **70**, 1185–1193 (2010)

14. Eisenberg, M.E., Neumark-Sztainer, D.: Friends' Dieting and Disordered Eating Behaviors among Adolescents Five Years Later: Findings from Project EAT. J. Adolescent Health **47**, 67–73 (2010)

15. Thompson, J.K., Shroff, H., Herbozo, S., Cafri, G., Rodriguez, J., Rodriguez, M.: Relations among Multiple Peer Influences, Body Dissatisfaction, Eating Disturbance, and Self-esteem: a Comparison of Average Weight, at Risk of Overweight, and Overweight Adolescent Girls. J. Pediatr. Psychol. **32**, 24–29 (2007)

16. Shomaker, L.B., Furman, W.: Interpersonal Influences on Late Adolescent Girls' and Boys' Disordered Eating. Eat Behavior **10**, 97–106 (2009)

17. Christakis, N.A., Fowler, J.H.: The Spread of Obesity in a Large Social Network over 32 Years. New Engl. J. Med. **357**, 370–379 (2007)

18. Pinkasavage, E., Arigo, D., Schumacher, L.M.: Social Comparison, Negative Body Image, and Disordered Eating Behavior: The Moderating Role of Coping Style. Eating Behaviors **16**, 72–77 (2015)

19. Forney, K.J., Ward, R.M.: Examining the Moderating Role of Social Norms between Body Dissatisfaction and Disordered Eating in College Students. Eating Behaviors **14**, 73–78 (2013)

20. Snijders, T.A.B.: Stochastic Actor-oriented Models for Network Change. J. Math. Sociol. **21**, 149–172 (1996)

21. Snijders, T.A.B.: The Statistical Evaluation of Social Network Dynamics. In: Sobel, M.E., Becker, M.P. (eds.) Sociological Methodology. Basil Blackwell, Boston (2001)

22. Snijders, T.A.B., Steglich, C., Schweinberger, M.: Modeling the Coevolution of Networks and Behavior. In: van Montfort, K., Oud, J., Satorra, A. (eds.) Longitudinal Models in the Behavioral and Related Sciences. Lawrence Erlbaum Associates, New Jersey (2007)

23. Snijders, T.A.B., Steglich, C., Schweinberger, M., Huisman, M.: Manual for SIENA version 3, University of Groningen: ICS. University of Oxford, Department of Statistics, Oxford (2007). http://stat.gamma.rug.nl/stocnet

24. Poulin, F., Chan, A.: Friendship Stability and Change in Childhood and Adolescence. Dev. Psychol. **30**, 257–272 (2010)
25. Fabes, R., Martin, C., Hanish, L., Anders, M., Madden-Derdich, D.: Early School Competence: The Roles of Sex-segregated Play and Effortful Control. Dev. Psychol. **39**, 848–858 (2003)
26. Maccoby, E.E.: Gender and Group Process: A Developmental Perspective. Current Directions in Psychological Science **11**, 54–58 (2002)
27. Moller, L., Hymel, S., Rubin, K.: Sex Typing in Play and Popularity in Middle Childhood. Sex Roles **26**, 331–353 (1992)
28. Leaper, C.: Influence and Involvement in Children's Discourse: Age, Gender, and Partner Effects. Child Dev. **62**, 797–811 (1991)
29. Strough, J., Berg, C.A.: Goals as a Mediator of Gender Differences in High-affiliation Dyadic Conversations. Dev. Psychol. **36**, 117–125 (2000)
30. Leaper, C., Smith, T.: A Meta-analytic Review of Gender Variations in Children's Language Use: Talkativeness, Affiliative Speech, and Assertive Speech. Dev. Psychol. **40**, 993–1027 (2004)
31. van Duijn, M.A.J., Zeggelink, E.P.H.: Huisman, M, Stokman, F.N. and Wasseur, F.W., Evolution of Sociology Freshmen into a Friendship Network. J. Math. Sociol. **27**, 153–191 (2003)
32. Kiuru, N., Burk, W.J., Laursen, B., Nurmi, J.-E., Salmela-Aro, K.: Is Depression Contagious? A Test of Alternative Peer Socialization Mechanisms of Depresssive Symptoms in Adolescent Peer Networks. J. Adolescent Health **50**, 250–255 (2012)
33. Centola, D., van de Rijt, A.: Choosing Your Network: Social Preferences in an Online Health Community. Soc. Sci. Med. **125**, 19–31 (2015)
34. Cruwys, T., Bevelander, K.E., Hermans, R.C.J.: Social Modeling of Eating: A review of When and Why Social Influence Affects Food Intake and Choice. Appetite **86**, 3–18 (2015)
35. Davision, T.E., McCabe, M.P.: Relationships Between Men's and Women's Body Image and Their Psychological, Social, and Sexual Functioning. Sex Roles **52**, 463–475 (2005)
36. Granovetter, M.S.: The Strength of Weak Ties. Am. J. Sociol. **78**, 1360–1380 (1973)
37. Strahan, E.J., Wilson, A.E., Cressman, K.E., Buote, V.M.: Comparing to Perfection: How Cultural Norms for Appearance Affect Social Comparisons and Self-image. Body Image **3**, 211–227 (2006)
38. Wilson, J.M.B., Tripp, D.A., Boland, F.J.: The Relative Contributions of Subjective and Objective Measures of Body Shape and Size to Body Image and Disordered Eating in Women. Body Image **2**, 233–247 (2005)
39. Wilson, J.M.B., Tripp, D.A., Boland, F.J.: The Relative Contributions of Waist-to-hip Ratio and Body Mass Index to Judgements of Attractiveness. Sexualities, Evolution & Gender **7**, 245–267 (2005)
40. Cramer, D.: Close Relationships: The Study of Love and Friendship. Arnold, London (1998)
41. Duck, S.W.: Friends for Life: The Psychology of Personal Relationships. Harvester, New York (1991)
42. Hilbert, A., Rief, W., Braehler, E.: Stigmatizing Attitudes toward Obesity in a Representative Population-based Sample. Obesity **16**, 1529–1534 (2008)
43. Zhang, J., Tong, L., Lamberson, P.J., Durazo-Arvizu, R.A., Luke, A., Shoham, D.A.: Leveraging Social Influence to Address Overweight and Obesity Using Agent-based Models: The Role of Adolescent Social Networks. Soc. Sci. Med. **125**, 203–213 (2015)
44. Ennett, S.T., Bauman, K.E.: Adolescent Social Networks: School, Demographic, and Longitudinal Considerations. J Adolescent Res. **11**, 194–215 (1996)

Effect of Task-Individual-Social Software Fit in Knowledge Creation Performance: Mediation Impact of Social Structural Exchange

ChienHsing Wu[✉], Shu-Chen Kao, and Cheng-Hua Chen

National University of Kaohsiung, 700 Kaohsiung University Rd.,
Nanzih District, Kaohsiung 811, Taiwan, R.O.C.
chwu@nuk.edu.tw, kaosc@mail.ksu.edu.tw, ron79913@yahoo.com.tw

Abstract. The current study discloses empirically the role of task-individual-social software fit in knowledge creation in the context of manufacturing industry, service industry, and research institute. A salient consideration is the mediation effect of social structural exchange on the role toward knowledge creation performance. Results of data analysis from 279 valid samples reveal findings as follows. First, effect of task-individual-social software fit is confirmed. Second, structural exchange does not mediate the role of task-individual-social software fit toward creation performance. Third, task-individual-social software fit is associated significantly with variables represented by social software, creation task, and individual cognition, respectively. Finally, both goal-free and goal-frame creation mode significantly influence fit of features of creation task, individual cognition, and social software, and so do both analytical and intuitive cognition style. Discussion and implications are also addressed.

Keywords: Knowledge creation · Social software · Individual cognition · Creation task · Social structural exchange

1 Background and Motivation

For the past few decades, advanced social software (e.g., Facebook and Google+) has been developing its distinct capabilities to bring innovative ways individual and organizations establish new mindsets in almost every field [1-7]. This also propels the effects of creation, raising our expectation levels towards it. For example, in the study of the social applications von Krogh [4] reported that Facebook likely affects employers' innovative work characteristics and in consequence influences company's ability to innovate. A literature survey indicated that knowledge creation is a multidimensional issue mainly linked to leadership, process, environment, and strategy. Although studies depicted in literature have reported the important arguments, aspects, comments, and factors that are associated significantly with knowledge creation, they also pay limited attention to the role of fit that concurrently considers such players' features in knowledge creation as task features (e.g., goal is defined or not) [8], individual features (e.g., analytical style), and social software features (e.g., used for communication and collaboration) [9]. The current study presents a research model

© Springer-Verlag Berlin Heidelberg 2015
L. Wang et al. (Eds): MISNC 2015, CCIS 540, pp. 116–130, 2015.
DOI: 10.1007/978-3-662-48319-0_10

and empirically examines the role of task-individual-social software fit model (TISF) in knowledge creation. A salient consideration is the mediation effect of social structural exchange on the role toward knowledge creation performance.

2 Related Concepts

2.1 Knowledge Creation

Knowledge creation is a value-added activity and process [1, 8, 10]. Therefore, whether the creation is successful or not needs to go through a process of visualization and accomplishment [11]. Based on the argument that human is the main creation resource, the IAM evaluates human's ability and competiveness to create profit. BS focuses on the learning factors and expectations for growth. To address this issue, index to measure non-financial variables for creation (or innovation) including quality, customer relationship, management efficiency, association profits, techniques, brand value, employer relationship and environmental problems are proposed. On the one hand, this indix focuses on the intangibles' impact towards the market. On the other hand, some studies placed emphasis on the creation outcomes, such as (1) product innovation achievements (e.g., (proportion of successful products) [1], (2) manufacturing/service process innovation achievements (e.g., quality and reliability)[12], (3) management innovation achievements (e.g., planning and preocedural) [12, 13], (4) organizational innovation achievements, and (5) strategic innovation achievements.

Although there are a lot of variables and since the creatin task is diverse and dynamic, the current study adopts arguments by Shin [1, 8, 14] as the variables to represent knowledge creation evaluators, including (1) product creation achievements, (2) manufacturing/service process creation achievements, and (3) management creation achievements, but excludes both strategy and organization creation achievements mainly due to the difficulties of measurement (e.g., top level involved and long period of time to observe).

2.2 Fitness

There have been various models regarding the analysis of fitness. Among models are individual-environment [15], individual-team [16], individual-jobs [17], and task-individual-technology [18]. Upon the continuous renewal of the ICT, the fitness between individual and environment (virtual space) has been highly discussed [12, 13, 19]. The fit theory indicates that task-technology-individual fit model can be applied to knowledge management systems. It suggests that task, technology and individuals are the three fitness factors that influence work efficiency. If everything fits well, the achievements of the work can be improved. However, the usage of technology and the interaction has not been addressed. Junglas et al. [12] applied the task-technology fitness model to investigate the information system of mobile phones; yet, personal cognitive behavior was not included as well. On the other hand, based on the perspective of use intention of social communication technology. (SCT), Koo et al. [20]

confirmed a TTF fitness model showing that SCT use intention will be affected by the attributes of the task. However, the research did not address the role of social network attributes in their model.

In general, the analysis of fitness is a systematic outcome based on the concept of interplay. To be effective, all the contributors must cooperate with one another. The current study argues that knowledge creation is a systematic concept; the creation outcome can be enhanced by concurrently considering the creation task characteristics (e.g., goal is defined or not defined), social media attributes (e.g., support of interaction and maintenance of positive development), and personal characteristics (e.g., analytical or intuitive style). The task-individual-social software fit model is proposed and the first hypothesis is then defined as follow.

H1: Task-individual-social software fit model significantly influences knowledge creation performance

2.3   Social Structural Exchange

Cropanzano and Mitchell [21] argued that social structural exchange theory can be used to represent sociology and psychology when discussing social behaviors. Social structural exchange theory is also called structural exchange theory, and it is initially originated social exchange theory. It stresses the objective to exchange under an environment and that the transaction has values to share knowledge and information. However, the prerequisite is that it has to have a scale. In other words, social exchange will share values through the transaction to create belief and know the behaviors of the social groups. Furthermore, the more exchange, the better development of the structure which leads to higher value. Above all, it will bring positive effects towards knowledge creation. Literature has shown that the structural exchange theory has been adopted in various fields to describe its impact on the exchange merits [22]. For example, Liang et al. [22] employed meta-analysis technique to examine and confirm that sharing will be likely affected by variables in social exchange (e.g., cognitive benefits and activities).

To further address this concept, Lyons & Scott [23] also reported that individual sharing behavior is associated with social structural theory. However, literature paid limited attention to the antecedents of social exchange behavior, in addition to trust, relationship, reputation, commitment and benefit. The current study argues that whether the factors represented by the environment in which the social exchange is taking place have effects on the outcomes derived by the exchange behavior towards creation merits. For instance, an multi-facet environment that is continuously interacted and adjusted in its fit will be increasing trust, relationship, and commitment towards better creation merits. This implies that social exchange may be alternating the effect of aforementioned fit on the creation merits. Therefore, the second hypothesis is defined as follow.

H2: Structural social exchange significantly mediates the effect task-individual-social software fit on knowledge creation performance

2.4   Social Software

With the renewal and updates in internet technology, social software (e.g., FaceBook, Google +) has demonstrated its success impact on our daily lives and works [5]. Yu et al., [24] compared social network media (blog and online forum) with traditional media (television press) to investigate their effects towards company achievements, and reported that social media has a significant effect on corporation success, especially when searching or igniting creativity, leading to more space for development. When discussing the social software and development, Carpenter et al. [25] suggests that social websites contributed a lot to the social network. This concept implies that social media can affect government policy, and that information management, safety, crime and privacy are all issues that need to be addressed. Obviously, to develop and maintain the value social software creates, technological development and applications, and big data analysis as well, are not the only issues, social relevance-oriented software development is also an important issue for the new networking society development. In fact, virtual society formed by social software has shown positive and negative effects on social values [26]. For example, satisfying our demands, increase interactive efficiency, increase novelty, accelerate information sharing, and flourish our daily lives are all positive merits while invasion of privacy, business fights, trustworthy issues, weak loyalty, energy consumption, Internet crime, and the internet addiction are all negative.

Social software can be viewed as a tool to network social entities. There are two different approaches regarding sociology theories: macro social theory and micro social theory. Particularly, the macro theory points out that the development of social software should follow the functions in our society (e.., knowledge sharing and opinion expression), and conflicts are considered normal (e.g., asking for unfair treatment to be revised) [27]. In fact, core value of social software is the functionality of socialization [27]. It should bring positive values towards social network, which can be divided into two parts: developing interactive relationship and retaining positive operation. To have the former possible, functions of management and analysis in social software is utilized for supporting activities management [9, 28, 29] while those of interaction and collaboration utilized for supporting interaction and collaboration [28, 30]. For the later, regulation and trust is used to maintain social order [26, 31] while privacy and safety can be used to assure perseverance and effectiveness [23, 31].

Social software is creating a new system that all the actions within it are all related. Any movement in a section will change the dynamics of the whole system. For instance, medical website will affect the privacy of people during the process of the transaction, sometimes weakening the value of online systems. If the benefit is greater than the cost, positive effects will occur. Therefore, the development of social software and the social network it created will change constantly, expectedly, leading to a suitable and positive social platform for entities. As a result, the current research argues that social software is one of the important contributors that affect the aforementioned fit development toward creation merits. The third hypothesis is then defined as follow.

H3: Social software significantly influences task-individual-social software fit

2.5   Creation Task

Schulze & Hoegl [1] reported that the SECI model developed by Nonaka & Takeuchi [32] are composed of four different steps including socialization, externalization, combination and internalization. Kao et al. [8] based on their data analysis divided creation task characteristics into three modes which are goal-free, goal-driven, and goal-framed. The current study adopted the three modes as the creation task features. The main argument for this is that the way that a goal is defined, not defined, or framed will be influencing the SECI operation towards the fit. For instance, Lyles [33] stressed that innovators are reliant on the past, making them hard to accept the new concepts, and in consequence whether a goal is defined will be likely to affect the way people interact.

In fact, knowledge creation is similar to chasing a target. The target will change due to the new concepts, interaction, environments, and background as well [34]. Martins & Terblanche [35] argued that freedom can be a stimulant in new concept development and that anyone can achieve the goal under freedom. Goal-free will not limit the goals from the past, making it more applicable to the knowledge creation [33]. Kao et al. [8] analyzed the relationship between creation modes and achievements, and reported that it is easier to have creation ideas given the non-target mode. However, they did not address the impact when innovators' personality and the environment under social media are considered.

As aforementioned, the knowledge creation under the environment presented by social software is an outcome of the dynamic process. The current study argues that whether the target is defined or not would influence the exchange process or fitness toward creation merits, and therefore the forth hypothesis is defined as follow.

H4: Creation task significantly influences task-individual-social software fit

2.6   Individual Cognition

Generally, the thinking behavior is associated with new knowledge development from the past experience and existing knowledge via a production process. Basically, this production process is guided by the cognitive system that includes thinking behavior, memory, learning, information processing, and problem solving skills. To address this concept, Garavelli et al. [36] reported that that knowledge development is a cognitive process through social interaction to create ideas. Interaction and the flow of knowledge depend on individual cognitive attributes which are affected by the interpretation process. They can re-organize with the original elements to create new ones as the change will become the source for knowledge creation [40].

Chalmers et al. [37] also stated that the choice of information and construction is an important part of the cognitive process. Choice, construction and the method of invention are all affected by internal and external facts including goals, beliefs, context and behavioral mode. However, although interpretation is essential, the content is only the subordinate of the creation process. Also, choice, construction and method of invention are affected by the cognitive process which makes it important to have a reliable cognitive system [37]. In addition, the complex process includes the mode of how it operates (e.g., goal is set) and the environment under social software. Most importantly, the different combinations of these factors will affect the achievement level in the end. For instance, for intuitive people, they may be successful under well-developed social software. Of course, this is just a hypothesis and this study will have deeper discussions regarding this topic. The cognitive attributes (e.g., cognitive style) is the behavioral instinct shown when people are faced in dilemmas or in special conditions.

In the classification of cognitive styles, Allinson & Hayes [38] developed cognitive style index, CSI as to differentiate intuitive and analytic. Kolfschoten et al. [39] stated that intuitive have a significant difference in learning. They also found out that pre-setting the learning environment can be effective towards building up knowledge development space. Literature paid limited attention to the relationship between cognitive styles and knowledge creation. The current study argues that cognitive style will be contributing to the aforementioned fit toward creation merits, and therefore the fifth hypothesis is defined as follow.

H5: Individual cognition significantly influences task-individual-social software fit

3   Method

3.1   Research Model

The research model is illustrated in Figure 1 based on the review of literature in general and research argument in particular. Three independent variables including social software feature, creation task feature, and individual feature and dependent, and their dependent one, task-individual-social software fit are presented. The creation merit is the dependent variable of task-individual-social software fit while structure exchange is the mediator between them. Details are described below.

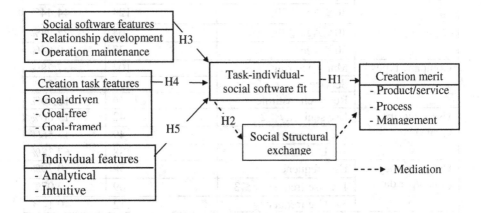

Fig. 1. Conceptual research model

3.2 Sampling and Measure

The questionnaire was designed based on the literature and research arguments as the data collection instrument (Table 2 in extracted form). Each question item was based on a 7-point Likert scale. Bi-polar descriptors were used for each question. Domain experts and specialists in social science, social software, human behavior, and fit theory were consulted to improve the understandability and readability of the questionnaire. A pilot test with a number of individuals was conducted to again ensure understandability and readability. There were 32 items developed to describe variables. The basic information of subjects includes gender, age, company type (e.g., service industry), OSN use seniority (years), use frequency (hours per day), and job seniority (years). The sample was targeted in the manufacturing industry, service industry, and research institute who is experienced in using social network software. The current study did not limit the top N companies due to the concern of wide range of creation in product, process, and management. Therefore, online questionnaire was developed and released in the online community (e.g., Facebook and Linkedin). A special statement to increase valid returned rate was included that manufacturing industry, service industry, and research institute were welcome. A voucher was sent to the samples who were checked valid. The expected sample collected was around 400. Of 579 returned samples, there were 279 valid, indicating 48.19% valid rate. Any returned questionnaire was considered invalid either incomplete, not within the domains, or single answer. The sample statistics is presented in Table 1. There was no particular majority in the collected sample, except the 75.27% for those whose job seniority is more than 4 years.

Table 1. Descriptive statistics

Number of valid sample: 279			
Basic data		Number of sample	Percentage
Gender	Male	142	50.90%
	Female	137	49.10%
Age	Age $\leq$ 29	64	22.94%
	30 $\leq$ Age $\leq$ 39	124	44.44%
	40 $\leq$ Age $\leq$ 49	64	22.94%
	50 $\leq$ Age	27	9.68%
Company types	Manufacturing	104	37.28%
	Service	129	46.23%
	Research institute	46	16.49%
Use seniority (years)	Use seniority $\leq$ 2	42	15.06%
	2 < Use seniority $\leq$ 4	49	17.56%
	4 < Use seniority	188	67.38%
Use frequency (hours/per day)	Use frequency $\leq$ 1	58	20.79%
	1 < Use frequency $\leq$ 3	109	39.07%
	3 < Use frequency	112	40.14%
Job seniority	Job seniority $\leq$ 2	38	13.62%
	2 < Job seniority $\leq$ 4	31	11.11%
	4 < Job seniority	210	75.27%

4   Data Analysis Results and Discussion

4.1   Reliability and Factor Analysis

The results of reliability and factor analysis was presented in Table 2. The Cronbach's Alpha was used to confirm the internal consistency for variables. The items grouped in a factor or sub-factors were adjusted based on the criteria that item to total was less than or equal to 0.35 The Cronbach's Alpha value increases if any item that suited the criteria is deleted. The exploratory factor analysis was employed to derive actual factors. Table 2 shows that there is only on factor in the variable of social software. The items used to represent creation task features, goal-free, goal-driven, and goal-framed, were not as consistent as expected. They were finally grouped into tow sub-factors, goal-free and goal-framed. The variable of cognitive style was grouped into two sub-factors, analytical and intuitive, which is according to our expectation, although CS5 and CS7 were deleted. In particular, the variable of creation performance was grouped into a single factor, which is not according to our argument. The validity test results obtained satisfied the following requirements: (1) item loading greater than 0.35, (2) composite reliability (i.e., Cronbach's α) greater than 0.6, (3) average variance extracted (AVE) greater than 0.5, and (4) square root of AVE for a factor is greater than the correlation coefficient of other factors. The correlation coefficient of factors with AVE validity is shown in Table 3.

Table 2. Reliability and factor analysis results

Vari-able	Sub-factors	Items	Question items	ITT	C's α
SS	-	SS1	Social software benefits social norm toward creation	0.792	
		SS2	Social software builds relationship toward creation	0.857	
		SS3	Social software boosts interaction toward creation	0.840	
		SS4	Social software operates well toward creation	0.830	0.906
		SS6	Social software improves common value toward creation	0.861	
		SS7	Social software benefits social development toward creation		
CT	GFe	CT4	Given a goal defined, it is changeable during creation process	0.494	
		CT5	A basic creation outcome is acceptable	0.384	
	GFa	CT7	Goal is defined in a context	0.799	
		CT8	Given a defined goal context, it is freely changeable within the context	0.774	
		CT9	Given a defined goal context, I can reach it according to my own way	0.765	0.907

Table 2. (*Continued.*)

Vari-able	Sub-factors	Items	Question items	ITT	C's α
IC	AS	IC1	I study issues in detail to solve the problem	0.817	
		IC2	I need theoretical background to solve the problems	0.742	
		IC3	I am categorized by persons into a logical-thinking person	0.774	
		IC4	I like the job needed logical thinking and progressive analysis	0.772	
	IS	IC6	I look briefly at reports, rather than reading with care.	0.553	0.885
		IC8	I am good at idea generation, rather than interpret accurately.	0.578	
TISF	-	TISF1	Social network contains lots of creative sources	0.869 0.929	
		TISF3	Social network helps creative tasks	0.921	
		TISF4	Social network suits creative tasks	0.925	0.932
		TISF5	Social network improves creative ideas		
SE	-	SE1	Sharing frequently with creative opinions in social network	0.917	
		SE2	Collaborating in team with creative activities in social network	0.932	
		SE4	Exchanging frequently with ideas in social network	0.921	0.913
CP	-	CP1	Market share of products or services is increasing	0.872 0.899	
		CP2	Profit of products or services is increasing	0.885	
		CP3	Sales of new products or services are increasing	0.855 0.872	
		CP4	Creative processes are improved frequently	0.851	
		CP6	Creative processes reduces cost	0.885	
		CP7	Creative processes improve production efficiency	0.908	
		CP8	Creative management improves communication and negotiation among employees	0.893	
		CP9	Creative management improves response adjustment to changing environment		0.915
		CP10	Management efficacy is increasing		

ITT: Item to total; SS: Social software; CT: Creation task; GFe: Goal free; GFa: Goal frame; IC: Individual cognition; AS: Analytical style; IS: Intuitive style; TISF: Task-individual-social software fit; SE: Structural exchange; CP: Creation performance

Table 3. Correlation coefficient of factors with AVE validity

Factors	AVE	SS	G-Fe	G-Fa	AS	IS	TISF	SE	CP
SS	0.805	**0.897**							
G-Fe	0.673	0.592	**0.820**						
G-Fa	0.781	0.778	0.614	**0.884**					
AS	0.814	0.741	0.589	0.782	**0.902**				
IS	0.717	0.554	0.496	0.564	0.547	**0.847**			
TISF	0.629	0.783	0.521	0.748	0.756	0.549	**0.793**		
SE	0.852	0.680	0.486	0.584	0.622	0.475	0.720	**0.923**	
CP	0.775	0.711	0.566	0.753	0.747	0.539	0.752	0.730	**0.880**

*p<0.05, **p<0.01, ***p<0.001; The number in the bold diagonal course is the square root of AVE.

4.2   Test Results

The results of hypothesis tests are presented in Table 4. It shows that the proposed variable, task-individual-social software fit, significantly influences the creation performance from the perspective of manufacturing industry, service industry, and research institute (in Model (1)). Thus, H1 (t=18.812, p=0.000) is supported. This finding indicates that the creation outcome is associated with that of concurrently considering the creation task feature, social software feature, and individual features. However, when the mediator (structural exchange) comes into the model (model (2)), the t value reduces from 18.812 to 8.963 and R2 increases from 0.561 to 0.634, yet its effect still remains significant (p=0.000). This implies that there is no mediation effect presented by the structural exchange. Thus, the H2 is not supported. This finding is not according to the research expectation that the interactive environment in which the social exchange is taking place has an effect on the creation merits. It should be noticed that the effect of task-individual-social software fit on structural exchange is significant (t=17.198, p=0.000, R2=0.516).

Table 4 also shows that the three independent variables (i.e., social software, creation task, and individual cognition) significantly influence task-individual-social software fit. Thus H3 (t=6.946, p=0.000), H4 (t=2.741, p=0.007), and H5 (t=4.525, p=0.000) are supported. These findings are exactly according to our research argument that task-individual-social software fit is affected by three variables represented by social software, creation task, and individual cognition, respectively. Moreover, the creation task feature was grouped into two sub-factors, goal-free and goal-frame, which is not consistent with the previous studies that reported goal-driven, goal-free, and goal-frame, respectively (Kao et al., 2011). The research finding shows that both goal-free and goal-frame are significantly related to the task-individual-social software fit (t=2.024, p=0.044 and t=13.719, p=0.000, respectively). Furthermore, the individual cognition was extracted to two sub-factors, analytical and intuitive, and both sub-factors are significantly related to the task-individual-social software fit (t=12.942, p=0.000 and t=4.314, p=0.000, respectively). This finding clarifies the research question that individual cognition is associated with the task-individual-social software fit.n the first level as seen in Table 2, Functional Value (FV), Emotional Value (EmV), and Epistemic Value (EpV) among the five consumption value in both printed and online music product are shown positive and significant at the 0.01 level, except Epistemic Value in online type. Therefore, the factors leading to customers' satisfaction toward printed music consumption support hypotheses 2, 4, and 5, while Social Value (SV) and Conditional Value (CV) are not statistically significant in relation with consumer's satisfaction, thus reject hypotheses 3 and 6. It also shows that satisfaction significantly influenced repurchasing intention for both product types. In the second level, the three sub-factor of Functional Value and two of Emotional Value are taken into exam, respectively. As seen in Table 3, all sub-factors of Functional Value and those of Emotional Value are all having a significant positive relation with the satisfaction.

Table 4. Results of hypothesis tests

Dependent variables	Creation performance				TISF	
Model	Model (1)		Model (2)			
Independent	t	p-v	t	p-v		
TISF (H1)	18.812***	0.000	8.963***	0.000		
Mediator: Social Structural exchange (H2)	-		7.437***	0.000		
R^2		0.561		0.634		
ΔR^2				0.073	t	p-v
Social software (H3)					6.946***	0.000
Creation task (H4)					2.741***	0.007
Individual Co. (H5)					4.525***	0.000
					R^2=0.668	
Creation task Goal-free (H4-a)					2.024**	0.044
Goal-frame (H4-b)					13.719***	0.000
					R^2=0.555	
Individual Co. Ana. style (H5-a)					12.942***	0.000
Int. (H5-b)					4.314***	0.000
					R^2=0.560	

*p<0.1; **p<0.05; ***p<0.01; TISF: Task-Individual-Social Software Fit

4.3   Discussion and Implications

The current research confirms the significance of task-individual-social software fit, significant of social software, significance of creation task presented by goal-free and goal-frame mode, and significance of individual cognition presented by analytical and intuitive style; yet does not confirm the significance of structural exchange as the mediator.

First, task-individual-social software fit has a significant impact on creation performance (in Table 4). This finding supports our argument and the findings by Smith & Mentzer [10, 13, 20] that variable features with a good fit form will likely improve the creative work. Basically, the outcome of fitting analysis is dynamic and systematic. The interplay is taking place while a number of players interact, and thereafter influences the outcomes. To be efficient and effective, all players (e.g., social software) should cooperate appropriately with one another in a quite fitting form. As aforementioned, the knowledge creation is a multidimensional issue that involves such as variables as technology (social software), personal thinking behavior (individual cognition), and creation task features (goal-free). The research finding confirms that the fit among three variables is significantly important to the merit of knowledge creation. It is suggested that knowledge creation researchers, consultants, and agents should concurrently consider to enhance creation performance the creation task characteristics (e.g., goal is defined or not defined), social media attributes (e.g., support of interaction and maintenance positive development), and personal characteristics (e.g., analytical or intuitive thinking style).

Second, the in-depth analysis presented in Table 4 reveals that both foal-free and goal-frame creation mode significantly influences the fit. This implies that both goal-free and goal-frame creation strategy likely influences the degree of fit toward creation performance. In the current study, the goal-free is non-definable target while goal-frame is definable target, but within a range, and both strategies ignore the process of reaching any target. Although previous studies underline that there are different effects of creation modes towards creation outcomes, the knowledge creation created by social media users is a dynamic process. The finding in the current research reveals that freely developing creation idea with or without a range will be likely to enhance the engagement from the micro view of fit analysis to enhance creation outcomes. Moreover, it is found that both analytical thinking style and intuitive thinking one are important to the degree of fit toward creation outcomes. We suggest based on these findings that the fit strength comes from creation modes determined and thinking behavior styles the creators take on.

Finally, we argue that structural exchange may play an important role to alternate the place of fit of social software, creation task, individual cognition. However, the finding does not support this and one of the concepts by Lyons & Scott (2013) that the power of social structural theory does not change the effect of variable fit on creation performance. This implies when antecedents of knowledge creation incorporates appropriately into a quite fit form an environment in which exchange of opinions, comments, and ideas will not be likely necessarily important. One of the possible reasons may be that the social software is successfully utilized to provide users with such functions as presentation, interaction, collaboration, and cooperation. Under such a platform, what is mostly needed is whether creation task and personal thinking style work well together. Our research finding confirms that the subjects perceive that no matter that goal is defined or not defined should better consider the thinking behavior to adjust to a proper fit form, and doing so will likely foster creation merits. For example, when a goal-free strategy is determined which thinking style should be used to lead to a better outcome. However, the current study did not cover this issue and will put it in one of the future research focuses.

5   Concluding Remarks

The exploratory study focused on the importance of knowledge creation performance with respect to the perspective of manufacturing industry, service industry, and research institutes. The task-individual-social software fit was highlighted that influenced creation performance. Potential determinants (e.g., social software, creation mode, and individual cognition) that contributed to the fit were defined and examined. The structural exchange, the mediator that may take the place of the fit, was also examined; yet was confirmed not supported. The exploratory study serves as groundwork for developing research on knowledge creation for both theory and practice. The research findings suggest that social software, personal thinking behavior, and creation mode should be considered concurrently to develop a proper fit, so creation outcomes may be enhanced. Combination of creation mode presented by goal-free and goal-frame and individual thinking styles presented by analytical and intuitive that fosters creation merits is yet known and needs to be disclosed. Further research may also include case or multi-case studies on a specific knowledge context to extend our findings and develop suggestions.

References

1. Schulze, A., Hoegl, M.: Organizational knowledge creation and the generation of new product ideas: A behavioral approach. Research Policy 37, 1742–1750 (2008)
2. Yang, C.-W., Fang, S.-H., Lin, J.L.: Organizational knowledge creation strategies: A conceptual framework. International Journal of Information Management 30, 231–238 (2010)
3. Camison, C., Fores, B.: Knowledge creation and absorptive capacity: The effect of intra-district shared competences. Scandinavian Journal of Management 27, 66–86 (2011)
4. Esterhuizen, D., Schutte, C.S.L., du Toit, A.S.A.: Knowledge creation processes as critical enablers for innovation. International Journal of Information Management 32(4), 354–364 (2012)
5. von Krogh, G.: How does social software change knowledge management? Toward a strategic research agenda. Journal of Strategic Information Systems 21(2), 154–164 (2012)
6. Shan, S., Zhao, Q., Hua, F.: Impact of quality management practices on the knowledge creation process: The Chinese aviation firm perspective. Computers & Industrial Engineering 64(1), 211–223 (2013)
7. von Krogh, G., Geilinger, N.: Knowledge creation in the eco-system: Research imperatives. European Management Journal 32, 155–163 (2014)

8. Kao, S.C., Wu, C.H., Su, P.J.: Which mode is better for knowledge creation? Management Decision **49**(7), 1037–1060 (2011)

9. McAfee, A.: Mastering the three worlds of information technology, Harvard Business Review, pp. 132–144, November 2006

10. Rohrbeck, R.: Exploring value creation from corporate-foresight activities. Futures **44**(5), 440–452 (2012)

11. Popadiuk, S., Choo, C.W.: Innovation and knowledge creation: How are these concepts related? International Journal of Information Management **26**, 302–312 (2006)

12. Junglas, I., Abraham, C., Watson, R.T.: Task-technology fit for mobile locatable information systems. Decision Support Systems **45**, 1046–1057 (2008)

13. Smith, C.D., Mentzer, J.T.: Forecasting task-technology fit: The influence of individuals, systems and procedures on forecast performance. International Journal of Forecasting **26**, 144–161 (2010)

14. Shin, Y.Y.: A Person-environment fit model for virtual organizations. Journal of Management **30**(5), 725–743 (2004)

15. Posner, B.Z.: Person-organization values congruence: No support for individual differences as a moderating influence. Human Relations **45**, 351–361 (1992)

16. Werbel, J.D., Gilliland, S.W.: Person-environment fit in the selection process. Research in Personnel and Human Resource Management **17**, 209–243 (1999)

17. Edwards, J.R.: Person-job fit: A conceptual integration, literature review, and methodological critique. In: Cooper, C.L., Robertson, I.T. (eds.) International review of industrial/organizational psychology, pp. 283–357. Wiley, New York (1991)

18. Goodhue, D.L., Thompson, R.L.: Task-technology fit and individual performance. MIS Quarterly **19**(2), 213–236 (1995)

19. Parkes, A.: The effect of task – individual – technology fit on user attitude and performance: An experimental investigation. Decision Support Systems **54**(2), 997–1009 (2013)

20. Koo, C., Wati, Y., Jung, J.: Examination of how social aspects moderate the relationship between task characteristics and usage of social communication technologies (SCTs) in organizations. International Journal of Information Management **31**, 445–459 (2011)

21. Cropanzano, R., Mitchell, M.S.: Social exchange theory: An interdisciplinary review. Journal of Management **31**, 874–900 (2005)

22. Liang, T. P., Liu, C. C., Wu, C. H.: Can social exchange theory explain individual knowledge-sharing behavior? a meta-analysis, In: Proceedings of the 29th International Conference on Information Systems (ICIS), Paris (2008)

23. Lyons, B.J., Scott, B.A.: Integrating social exchange and affective explanations for the receipt of help and harm: A social network approach. Organizational Behavior and Human Decision Processes **117**(1), 66–79 (2013)

24. Yu, Y., Duan, W., Cao, Q.: The impact of social and conventional media on firm equity value: A sentiment analysis approach. Decision Support Systems **55**(4), 919–926 (2012)

25. Carpenter, J.M., Green, M.C., LaFlam, J.: People or profiles: Individual differences in online social networking use. Personality and Individual Differences **50**(5), 538–541 (2011)

26. He, W.: A review of social media security risks and mitigation techniques. Journal of Systems and Information Technology **14**(2), 171–180 (2012)

27. Fournier, S., Lee, L.: Getting brand communities right. Harvard Business Review, pp. 105–111, April 2009

28. Chen, Y.J.: Knowledge integration and sharing for collaborative molding product design and process development. Computers in Industry **61**, 659–675 (2010)

29. Vezzetti, E., Moos, S., Kretli, S.: A product lifecycle management methodology for supporting knowledge reuse in the consumer packaged goods domain. Computer-Aided Design **43**, 1902–1911 (2011)
30. Vuori, M.: Exploring uses of social media in a global corporation. Journal of Systems and Information Technology **14**(2), 155–170 (2012)
31. Linke, A., Zerfass, A.: Social media governance: regulatory frameworks for successful online communications. Journal of Communication Management **17**(3), 270–286 (2013)
32. Nonaka, I., Takeuchi, H.: The Knowledge-Creating Company. Oxford University Press, Inc., New York (1995)
33. Lyles, M.A.: Organizational learning, knowledge creation, problem formulation and innovation in messy problems. European Management Journal **32**, 132–136 (2014)
34. Hellström, T., Jacob, M.: Boundary organizations in science: From discourse to construction. Science and Public Policy **30**(4), 235–238 (2003)
35. Martins, E.C., Terblanche, F.: Building organizational culture that stimulates creativity and innovation. European Journal of Innovation Management **6**(1), 67–74 (2003)
36. Garavelli, C., Gorgoglione, M., Scozzi, B.: Manage Knowledge Transfer by Knowledge technologies. Technovation **22**, 269–279 (2002)
37. Chalmers, P.A.: The role of cognitive theory in human – computer interface. Computers in Human Behavior **19**, 593–607 (2003)
38. Allinson, C.W., Hayes, J.: The Cognitive Style Index: A measure of intuition-analysis for organizational research. Journal of Management Studies **33**(1), 119–136 (1996)
39. Kolfschoten, G., Lukosch, S., Verbraeck, A., Valentin, E., de Vreede, G.J.: Cognitive learning efficiency through the use of design patterns in teaching. Computers & Education **54**, 652–660 (2010)
40. Wu, C.H., Kao, S.C., Shih, L.H.: Assessing the suitability of process and information technology in supporting tacit knowledge transfer. Behavior and Information Technology **29**(5), 513–525 (2010)

Evaluating PEVNET: A Framework for Visualization of Criminal Networks

Amer Rasheed[1(✉)], Uffe Kock Wiil[1], and Mahmood Niazi[2,3]

[1] The Maersk Mc-Kinney Moeller Institute,
University of Southern Denmark, Odense, Denmark
{amras,ukwiil}@mmmi.sdu.dk
[2] Department of Information and Computer Science,
King Fahd University of Petroleum and Minerals, Dhahran, Saudi Arabia
[3] Facutly of Computing, Riphah International University, Islamabad, Pakistan
mkniazi@kfupm.edu.sa

Abstract. Information visualization has been a burning topic among the researchers in the recent decade. Getting targeted information, which is everyone's desire, is becoming difficult with the abundance of data. In this research, we have made an evaluation of our proposed framework PEVNET by conducting an experiment. Thirty two participants evaluated the system. The experiment was performed in two phases. In the first phase, a usability evaluation and qualitative feedback was carried out to check whether the PEVNET framework provided adequate results to the users. The qualitative feedback was performed by considering two aspects: the ease of use and the functionality. In the second phase, the comparison of the PEVNET had been performed against another state-of-the-art tool. Locating the central person, detecting the hidden interaction patterns between the sub-clusters, and detecting temporal activity were among the main tasks that were to be achieved by the participants. These tasks were to be performed in the groups of participants. The case study of Chicago Narcotics datasets was used. We found that the participants, of the PEVNET group, performed the tasks faster as compared to the other techniques used in the experiment. Among the participants, there were a few domain experts who appreciated our novel visualization features. Anecdotally, we believe that by evaluating the PEVNET in this research paper, we will be able to get the confidence of the crime analysts. We have found that the network visualization of the PEVNET framework, based on the experimental results, has gotten satisfactory feedback from the majority of the participants.

Keywords: Information visualization · Criminal networks · Investigative analysis · Sub-cluster detection

1 Introduction

In the modern era when there is an extensive discussion about the information visualization [1], there is a scarcity of visualization tools that support decision making. Apart from the variety of network visualization tools [2, 3, and 4], the analysts can hardly find the investigative analysis (IA) [4] application which can directly address their

© Springer-Verlag Berlin Heidelberg 2015
L. Wang et al. (Eds): MISNC 2015, CCIS 540, pp. 131–149, 2015.
DOI: 10.1007/978-3-662-48319-0_11

requirements and that they get their targeted solution readily available. There is need for an authority platform with which it is permissible to make a break through with respect to the prevailing challenges to information visualization as these challenges are not making pace with the technology demands.

There has been a widespread dis-satisfaction with the current evaluations on the information visualization tools. There are numbers of reasons for this, for instance, the usage of small datasets, the data sets extracted is not real time, problems in finding domain experts [5], problems in arranging the infra-structure to conduct qualitative evaluations etc. To quote Plaisant et al. [5], 'there is a dire need of systematic development of benchmarks'. It will help in providing different ways of strengthening by making comparisons [5], which is commonly performed in conducting evaluation of some tools. For this purpose, Richard Heurer in his book mentioned the need for the tools and techniques to support intelligence analysts with information structuring, challenging assumptions, and exploring alternatives [6]. Moreover, there has been continuous and sometimes collective efforts by information visualization experts to promote vigilance as to how information visualization can do havoc. In this regards, the infovis community conducts IEEE Visual Analytics Science and Technology (VAST) symposiums every year [5]. It is an effective measure to promote the information visualization sense among the new researchers.

Different ways of visualization have been proposed in the literature and there is a variety of existing visualization techniques, such as Gephi[1], Tomsayer[2] etc. Moreover, there are numerous examples of different objects, their interactions, and their collaborative networks [7, 8] that are based on different methods of visualization but PEVNET, our proposed framework for the visualization of criminal networks [10, 25], has an edge in proposing some novel visualization features.

In this study, we have made the evaluation of the PEVNET framework with the help of an experiment. We are of the view that by evaluating the system we can assess its performance and for this purpose we have studied a number of evaluation techniques [11, 12, 13, 14, and 15]. We have considered the ingredients to conduct an effective evaluation with the perspective of information technology, i.e., 'full datasets, domain specific tasks, and domain experts as participants' [9].

The organization of the remainder of this paper is as follows. In the next section, challenges from the field of criminal network visualization are described. Section 3 describes related work. In Section 4, the PEVNET framework is described. Section 5 describes the experimental settings for conducting the evaluation of the framework. Tasks, datasets and experiment described in the sections 6 and 7. Analysis and results are shown in Section 8. Section 9 provides a discussion and motivation for the readers is described in Section 10. Finally, section 11 concludes the paper followed by the future works.

---

[1] http://gephi.github.io/
[2] https://www.tomsawyer.com/products/visualization/

2   Challenges

Numerous studies [5, 6, and 9] have found that there is a variety of challenges confronting existing visualization tools and techniques. Some of the challenges are, for example, to cope with the computationally interactive visualizations, to create a balance between usability and completeness, to show the huge datasets on a single screen, to solve the problems with the display complexity of the network visualizations, such as cluttering of data [11] etc. Besides that, the problem of accessing the domain experts and finding the appropriate datasets have given us a tough time. Even if one finds the datasets, the conversion of the datasets to the acceptable file format is also one of the big challenges. Last but not the least is a need of a yard stick [5] which can be taken as an authoritative tool in the field of network visualization.

3   Related Work

We have studied numbers of evaluation techniques. The nature of the work carried out in this study, is both qualitative and quantitative. The work closest to our work was carried out by Chi and Reid [16]. The operator framework by Chi and Reid [16] is a generalization of the operations for performing the different visualization techniques. The authors carried out their work by slightly modifying the Card et al. [17] model. The operator framework has manipulated the datasets by using the concept of visualization pipelines. The demonstration was carried out by using the state diagram. The author applied a number of applications, such as Perspective wall [18] and Dynamic querying, [20] for evaluating the framework. The verification and validation of the network analysis and visualization system carried out by Jennifer et al. in CrimeNet explorer [12], a worth quoting evaluation work. They evaluated the system using the cluster recall and clustering precision [19]. They performed an experiment in which they focused on subgroup detection and tried to validate CrimeNet explorer with their proposed structural analysis functionality. Moreover, the authors developed some hypotheses to evaluate the systems' performance. With their proposed features, the subjects had some problems of visual cluttering and also the effectiveness was not supported by the subjects. But there were some supporting arguments as well; for instance, the layman users could be more convenient with their proposed techniques as he/she does not have to go into numbers of case studies, provided to the experiment participants, and he/she could directly adapt himself/herself with the flexibility of CrimeNet. Another work has been performed by Youn-ah-Kang et al. [29]. They also followed the VAST symposium [5] procedures. In their experiments, in which there were a number of student participants, they supplied the participants with numbers of small pieces of documents containing some relevant or irrelevant information about some big event. The participants were required to find out the major event by gathering evidences from the papers provided. Further, they performed their tasks in four strategies, explained to the participants before the start of the experiment. In their subsequent paper [30], the authors provided the examples along with their proposed strategies.

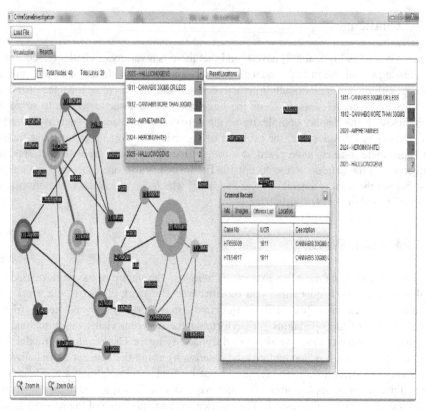

Fig. 1. Clustering using proposed novel clustering algorithm require selection of crime type from the list of values on the PEVNET desktop.

There are a number of tools that are employed for evaluating interactive visualizations, such as stacked histogram, Pie charts, highlighting scatter plots, line graphs[21], Hyperbolic space [22] etc. The self-organizing maps (SOM) was used to evaluate similar objects by demarcating them into clusters so that the users could easily assess the interactions between the different objects.

4   PEVNET-The Proposed Framework

The PEVNET [10, 25] layout is usually comprised of a node and link diagram, depicting the network visualization as shown in Figs. 1 and 2. We have combined network data manipulation techniques with data mining of large datasets by way of visualization. The framework comprises three major parts: network visualization features, visual features based upon temporal patterns detection, and composite features. These three parts exist independently. In the network visualization part, data retrieval is carried out in a simple manner. It is performed by using different techniques; for instance in PEVNET there is a novel clustering algorithm [25] with which the analysts are able to filter out data in a novel way as shown in Fig. 2. Two clusters, comprising of three and four nodes are also shown. The crime type filter displays the cluster of the selected crime type. The nodes with the selected crime type are displayed in such a way that the cluster seems to pop up from the original network snapshot as shown in Fig. 2 [25]. There is a unique and novel feature in PEVNET that makes selection of similar nodes that are linked with same crime type, found in the network. The detection of similar crime is of vital importance. Occasionally, it is needed during crime investigation. The analyst is required to search persons (nodes) that are dealing with some crime types. He will select any node that is involved in for instance detecting the crime type Heroin in the criminal network by watching the legend panel. The system will automatically make those nodes appear prominently which are linked with 'Heroin' crime type as shown in the location window in the Fig. 3. Besides collaborative sub-cluster detection feature, criminal window feature as shown in Fig. 1 etc. are other examples of network visualization features in PEVNET.

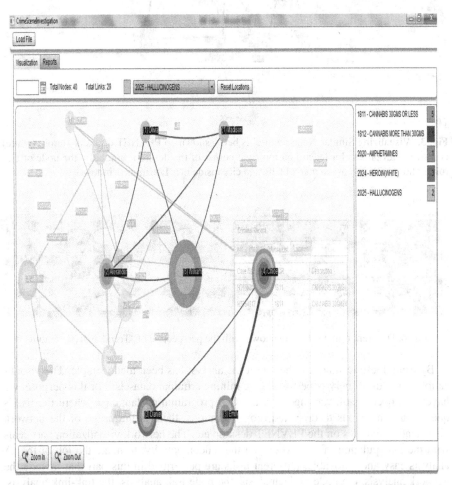

Fig. 2. On selecting the crime type Hallucinogens, the sub-clusters with the selected crime types are displayed. On close observation, the original criminal network is also evident in the background.

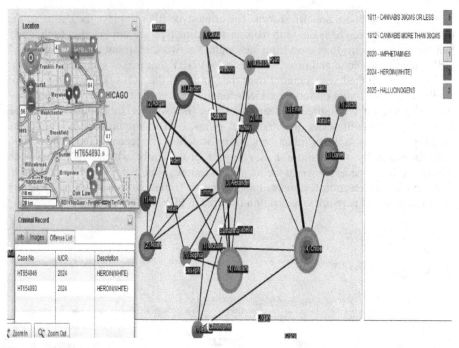

Fig. 3. 'Visualizing Similar Node feature' is being shown on PEVNET desktop. Heroin (white) is shown with pink color legend on top right corner of the desktop. Similarly, the node of Heroin (white) is shown in the map of Chicago city inside the 'Location' window.

Fig. 4. Different crime types are shown with the perspective of 'Trend analysis feature.

By using these techniques, the job of the analyst has been made simple. The visualization of the user query is performed by mining criminal datasets. For this purpose, we have developed a software agent in the C # programming language where the user's queries are analogous to drag and drop operations, there is the usage of the network visualization features on the PEVNET desktop etc. The network visualization part deals with the manipulation of data to extract the hidden activity, to make the job of the IA analysts easy and enjoyable. Different tasks are performed in this part that include the network analysis, the node-node analysis, the node-link analysis, the link-link analysis, and last but not the least, the analysis between the sub-group and group interaction.

The second part of PEVNET framework includes criminal pattern generation based on time. The criminality pattern generation part deals with the generation of the patterns that are evolved over some span of time, for instance the trend analysis feature [26], temporal patterns feature [10] etc. The trend analysis feature [25] as shown in the following Fig. 4 is used to detect variations in criminal activity over some span of time. It helps the analyst in the prediction of certain criminal activities. For instance, the variation of various crime types is shown by Illinois uniform crime type (IUCR) number of the respective crime types in the following Fig 4. Some of the features in PEVNET have already been implemented but those features have been re-examined to discover the new horizons which were over-looked in the past. By making use of the stored procedures and clustering algorithms, many patterns in the network data have been discovered.

Composite network visualization features have been proposed in the third part of the PEVNET. We have extended the composite research [31, 32] by introducing novel visualization techniques and have proposed some composite features; for instance, grouping the selected nodes, merging group into another group, un-grouping group and merging node into group. Composite is a feature that is based upon collapsing and expanding of data. It is difficult to handle visualization of member information, which are nodes; inside the composite. Besides that, other issues that include interactions between different groups of nodes and the members inside one group or another.

We used the case study based on the data from the Chicago narcotics data-sets used by the Chicago Police Department (CPD) to evaluate the framework. It is available in excel file format. The format in Microsoft Excel is more understandable to the majority of users. The format helped us a lot during the system evaluation since the users could validate their findings by comparing their results from the excel sheet.

The two main objectives that were considered while conducting the evaluation in this paper, were as follows:

- 'Evaluation of novel network visualization features'.
- To detect the 'evolution of temporal patterns between sub-clusters in the criminal networks'.

5   Experiment Settings

The experiment was conducted in two phases. There were a total of thirty two participants in the experiment. Both the sessions were conducted in a computer lab at the software engineering department in King Fahd University of Petroleum and Minerals (KFUPM). Each participant had two hours for making his findings. There was a break of fifteen minutes in the session. The participants were given grades for this session. The session counted as a class assignment for the students. Luckily, we managed to get the inputs of the four IA professionals as participants.

In the first phase of experiment, the datasets were extracted from an excel file. The choice of such selection was due to the fact that it is understandable to nearly all the users. In this way, the users could easily validate their queries from the excel file. We got input of one of the four domain experts during the first phase of the experiment. It was served as a yard stick for processing the other participant's scores. User training was given to the participants since they were not experts in the field of network visualization. The user background and the personal information was gathered in the same session. The questionnaire forms were given to each of the participants. There were two types of questionnaire forms, i.e., 'Feature's usability evaluation form' and 'Feature's qualitative survey form'. However, in the 'Feature's qualitative survey' form, there is only an addition of two extra fields against each entry of the first form. These two fields were 'System's functionality' and 'System-ease-of-use'. There were four choices; for instance, satisfied, fair, not satisfied, and not applicable (N/A).The users had to make his/her inputs in any of the choices. The main aim of conducting phase one was to make the participants familiar with the functionalities of the different features in the PEVNET and provide qualitative feedback, such as system functionality and the ease of use. In this way, we also validated the PEVNET functionality.

In the second phase of the experiment, we divided the participants into three teams. The PEVNET user group, the Gephi user group and the desktop user group were the names of the three teams. There were twelve participants in one group and ten participants each in the second and the third groups. Each group included one demonstrator. We gave the user training session to the demonstrators of each group one day before the experiment in another session.

6   Task and Datasets

The datasets were taken out of the total of around 40,000 datasets, known to have been committed in the whole year, 2011. For better understanding, we provided the excel sheet that contained 600 data records of the reported crimes of the IUCR. In Table 1, there is an instance of some of the data records extracted from the IUCR datasets. In the first phase of the experiment; the participants used the data records from the excel sheet for evaluating PEVNET.

In the second phase of the experiment, the participants were to follow the procedure of the VAST symposium [5] in which the participants are usually provided with datasets, scenarios, tasks, definitions, and some procedural information. The excel sheet contained datasets, scenario, tasks, and information about some key players, activities, locations and brief definitions. Regarding the tasks and the key questions for the participants, we were inspired by the important structural characteristics of the criminal networks by McAndrew and Sparrow [27, 28]. So, we set the participant's tasks as, for instance, locating

- 'the central person' [28],
- 'the sub-groups' [28],
- 'the patterns of interaction between the clusters of different crime types' [28],
- 'the criminal activity over some span of time',
- 'the similar crime types across some geographical locations',

- 'the information flux across the network' [28],
- 'the criminal peripheral information', and
- 'the co-offending groups information'.

The questions were the same for all the participants to check the performance in each group; for instance, the time taken, the number of correct answers, overall satisfaction with the methods in the feedback section etc.

We developed a solution key along with the necessary explanation, where needed. The score scale was from '2' to '-1'. In the scale, there were two points for 'satisfied', one for 'fair', 0 for 'not satisfied', and '-1' stands for 'not applicable'. The data from the participants was collected and the score compilation was performed based on the developed key, along with the experts. The scores of each participant were used in the analysis as shown in Table 1. There was a defined criteria to check the performance; for instance, the time taken, number of correct answers, overall satisfaction with the methods, qualitative feedback etc.

The main purpose of conducting this research paper is to evaluate the PEVNET. So we have made the research questions (RQ) which are as follows:

RQ 1. Whether the PEVNET provides adequate support with respect to the system usability, functionality and the ease of use?

RQ 2. Will the PEVNET perform well against the tools and techniques used in the experiment?

7   Experiment

The main tasks were divided into three main sections, i.e., visualization, temporal, and composite in both parts of the experiment. In part one, there were a total of 18 tasks as described in the 'usability evaluation' and the qualitative feedback. The input into these tasks was to be performed in the documents as shown in Table 1.

The usability evaluation was actually the validation of the PEVNET system. The two aspects of the qualitative evaluation, i.e., systems ease-of-use and functionality have been investigated, by the participants, in the qualitative feedback section. After getting the inputs from the participants the documents were collected. Table 1 was formulated after processing the input documents. We also formulated a solution key for the input documents. It was formulated with the help of the domain experts in the field of IA.

The 'experimental score' was actually the percentage coincidence with the exact answers, provided in the solution key. It was calculated by using the traceability matrix technique. According to the author's best knowledge, the performance or efficiency of the system can be assessed by calculating the ease with which the system can be used and secondly, the functionality of the system as to how optimized the system design is. We have calculated 'system ease of use' and 'system functionality' based on the 'qualitative feedback' document. The total numbers of tasks, while performing qualitative feedback, were same as in case of usability evaluation. For assessing these two qualitative parameter, the participants input has been divided in four parts as satisfied, fair, not satisfied, and not applicable.

Table 1. Experiment score by all participants in phase-I.

Participants #	Experiment Score
Participant 1	94
Participant 2	100
Participant 3	20
Participant 4	100
Participant 5	94
Participant 6	98
Participant 7	100
Participant 8	94
Participant 9	94
Participant 10	100
Participant 11	100
Participant 12	94
Participant 13	94
Participant 14	94
Participant 15	94
Participant 16	100
Participant 17	94
Participant 18	100
Participant 19	45
Participant 20	100
Participant 21	94
Participant 22	25
Participant 23	94
Participant 24	94
Participant 25	100
Participant 26	100
Participant 27	94
Participant 28	94
Participant 29	94
Participant 30	94
Participant 31	100
Participant 32	10

Participants #	Performance/Efficiency	
	System ease of use	System's Performance
Participant 1	13.2.2.0	13.3.1.0
Participant 2	11.5.2.0	15.1.2.0
Participant 3	2.0.13.3	13.0.2.3
Participant 4	14.6.0.0	14.4.0.0
Participant 5	12.4.2.0	12.4.2.0
Participant 6	11.5.2.0	11.5.2.0
Participant 7	12.3.2.1	14.3.1.0
Participant 8	13.5.0.0.	13.5.0.0.
Participant 9	11.0.1.6	13.3.2.0
Participant 10	11.5.2.0	11.5.2.0
Participant 11	14.3.1.0	14.3.1.0
Participant 12	12.6.0.0.	12.6.0.0.
Participant 13	13.3.2.0	13.3.2.0
Participant 14	11.5.2.0	11.5.2.0
Participant 15	14.3.1.0	14.3.1.0
Participant 16	13.5.0.0.	13.5.0.0.
Participant 17	13.3.2.0	13.3.2.0
Participant 18	11.5.2.0	11.5.2.0
Participant 19	14.3.1.0	14.3.1.0
Participant 20	13.5.0.0.	13.5.0.0.
Participant 21	13.3.2.0	13.3.2.0
Participant 22	11.5.2.0	11.5.2.0
Participant 23	14.3.1.0	14.3.1.0
Participant 24	12.6.0.0.	12.6.0.0.
Participant 25	13.3.2.0	13.3.2.0
Participant 26	11.5.2.0	11.5.2.0
Participant 27	14.3.1.0	14.3.1.0
Participant 28	13.5.0.0.	13.5.0.0.
Participant 29	13.3.2.0	13.3.2.0
Participant 30	11.5.2.0	11.5.2.0
Participant 31	14.3.1.0	14.3.1.0
Participant 32	13.5.0.0.	13.5.0.0.

In the second part, the details of the group activities are as follows:

7.1 The Desktop User Group

The participants of this group were provided with the excel sheet containing the datasets from the IUCR. These participants were to perform the tasks as described above. The participants who were in this group had a tough time as they had to perform a huge computation before they performed the tasks; for instance,

- to track the identities of the persons involved in the co-offending crimes.
- to count the number of times, each person was involved in the crime and then repeat it for each new crime type.
- to calculate the weight of the node based on the crime count.

The problem got worst when there was counting of crimes especially in the same instance of time; for instance, crimes made in different locations but at the same instance of time.

7.2   The Gephi User Group

The participants of this group used Gephi, a state of the art network visualization software. The participants in this group were provided a desktop with Gephi software installed. The participants had to carry out the tasks described above. We have used Gephi, as a good example representing the existing techniques, in our experiment due to some particular reasons. It is a widely used network visualization tool and is a freely available software tool. Many of our participants had information about it or at least they had heard about it.

7.3   The PEVNET User Group

The participants in this group were provided the PEVNET framework on their computers. The participants had to carry out the tasks described above. The participants in this group were looking relaxed due to the tool support as per requirements of the users. Since the solutions of the majority of the questions were readily available; for instance, locating the central person.

8   Analysis and Results

In phase one of the experiment, the results of the usability evaluation of the PEVNET system are shown in Table 1. In the quantitative part, the values of the 'experimental score' represent the participant's inputs. The 'experimental score' represents the percentage of similar values between the participant's feedback and the solution key. It is actually the percentage of the participant's success in using the PEVNET system. The inputs, by the participants, ranged from 10 to 100 in the column. The participants with the score 10 had the worst grip over the system or his/her assessment was not positive. Either he/she could not understand the instructions during the tutorial or he/she rejected the PEVNET system. The participants with the score 100 had the complete grip over the system or his/her assessment was fully positive and he/she got all the instructions appropriately. In a similar fashion, the other participants scored in between these values. Fig. 6 shows the participant's responses in the usability evaluation. The data is retrieved from Table 1. The blue line shows the participant's confidence, which was outstanding, i.e., around 80 percent, over the PEVNET system. There were three to four participants out of the total of thirty two had some reservations regarding the system layout, system design etc. In the qualitative part, after getting the participant's input, we assigned a score in the scale. Again, the scale was from '2' to '-1' for usability evaluation and qualitative feedback documents. There were four criteria 'satisfied', 'fair', 'not satisfied', and 'not applicable'. The scores were assigned against each criterion. First of all, the percentage scores were calculated as shown in the 'usability evaluation percentage score' column in Table 1 and shown in the shape of the graph in Fig. 6. By comparing the individual participant's input with the solution key, the percentage score was calculated. Regarding the calculation of the efficacy, we can consider an example as shown in the first row, in Table 1, under the 'system ease of use' field, i.e., the value of '13.2.2.0'. In the value, 13 represents the number of 'satisfied' hits, 2 represents number of 'fair' hits, the next 2 represents the number of 'not satisfied' entries, and 0 shows the number of 'not applicable' entries. We plotted the graph against these values, i.e., 'performance/efficiency' of Table 1. The values of the PEVNET system's ease of use and system functionality against the participants have been displayed in the shape of the graphs shown in Figs. 5 (a) and (b), respectively. Again, the blue lines show the

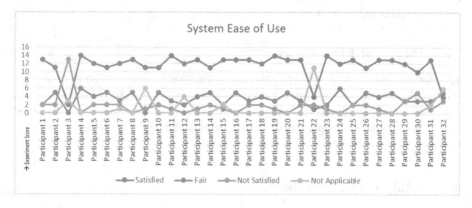

Fig. 5(a). The performance evaluation criteria for PEVNET, i.e., system ease of use of the PEVNET framework.

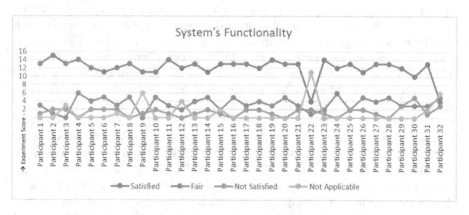

Fig. 5(b). The performance evaluation criteria for PEVNET i.e. functionality of the PEVNET framework

participant's confidence, which was outstanding, over the PEVNET system. The qualitative feedback was around 80 percent in the case of the system's ease of use and 90 percent in the case of the PEVNET functionality as indicated by Figs. 5 (a) and (b), respectively. The red, green, and violet lines, in Figs. 5 (a) and (b), are represent 'fair', 'not satisfied', and 'not applicable', respectively. The reason as to why the outcome was not a hundred percent was due to the dis-satisfaction of two to three participants regarding the system's flexibility, user friendliness, color combinations of the crime type legends, system response when the number of nodes increased etc. In the second part, if we compare the performance and the score of the participants, it is clear that the participants using the PEVNET outdistanced those in other groups. The participants in group 1 had to make more computations due to the involvement of statistical tables.

After getting our results, we addressed our research questions as follows

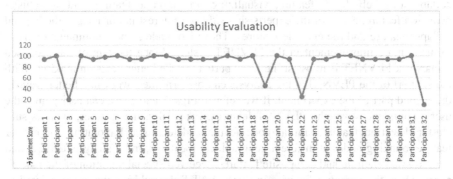

Fig. 6. PEVNET performance evaluation.

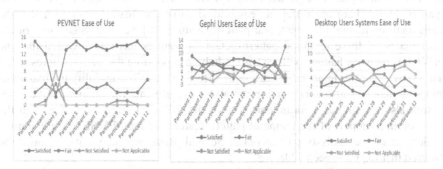

Fig. 7. (a, b, c) The system's ease of use graphs for the PEVNET users, Gephi users, and the desktop users.

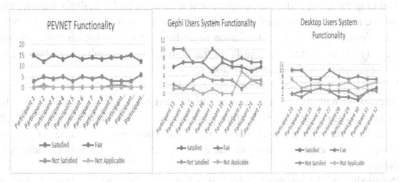

Fig. 8. (a, b, c) The system's functionality graphs for PEVNET users, Gephi users, and the desktop users.

In the above experiment, if we refer to the performance evaluation of the PEVNET in Figs. 5 (a), 5 (b) & 6, it can easily be judged that the majority of the participants, that also include the domain experts, were fully satisfied in all the aspects; for instance, system ease of use, functionality, and usability evaluation.

All the domain experts appreciated the PEVNET system's functionality and the ease of use. Especially, the novel features of the PEVNET that include the detecting collaborating sub-cluster feature, visualizing similar node feature, and sub-cluster detection feature. Some of the experts showed keen interest in our idea of the expand collapse feature and the encircle features. They also made some recommendations as to make more improvement in the PEVNET system. Since the aim of the proposed framework PEVNET is to assist the IA experts to make better decision making. With the support of the PEVNET,   the analysts can perform their work in a better way. In the second part of the experiment, if we refer to Figs. 7 and 8, we can easily infer the user's confidence over the PEVNET features. Especially, the features, such as sub-cluster detection feature, similar node features, detecting collaborating sub-cluster feature, blinking nodes at different geographical locations, trend analysis feature, and the proposed encircle and composite features, received special attention of the participants. Then, the temporal features in the PEVNET that address the issues related to the time taken for instance, in detecting any crime activity that popped up over some span of time. There are different temporal visualization features, for instance, the trend analysis feature [26], temporal patterns feature [10] etc. The satisfaction of the participants on these temporal pattern detection features has also revealed that the PEVNET temporal features will be a great help in supporting the analysts in resolving the issues related to temporal crimes in an effective way. The usability evaluation of temporal data was also conducted in the above experiment. The temporal pattern detection also constitutes a formidable part of the PEVNET framework. So, with the collaborative support of the temporal pattern features of the PEVNET, the cognitive tendencies of the analysts will increase. So with these arguments, it can easily be inferred that RQ1 has been addressed.

From the second part of the experiment, it is evident that the novel features of the PEVNET performed well against the respective features of Gephi in the comparative study. Those novel features, for instance, the detecting collaborating sub-cluster feature, blinking nodes at different geographical locations, trend analysis feature, and the proposed encircle and composite features, provide a sound support in the decision support. We received much appreciation regarding our clustering algorithm technique and the way we presented the details on the demand feature, which are two examples of the re-examining of the existing features.

We have also gotten some criticism, from some of the participants, regarding the implementation with the view to re-examine the existing features, such as locating the central person, thick nodes [10], clustering the sub-group based on the crime types [25], etc. The participants were of the view that there was no marked contribution with the perspective of re-examining the existing network visualization features in the literature.

After analyzing the feedback, which was both positive and negative, we can easily conclude that the PEVNET performed satisfactorily against the other tools and techniques in the experiment. Also, we got some strengths of the PEVNET, i.e., the novel features are a useful contribution to the existing visualization techniques and the features of the PEVNET, which we have re-examined, are enough to satisfy our second research question.

9   Discussion

First of all, the usability evaluation was carried out by the participants with the perspective of a quantitative evaluation. After that, the participant's feedback with respective to the qualitative evaluation was carried out by considering the system ease of use, look and feel, and the system's functionality. As described earlier, both the qualitative and quantitative aspects have been considered whilst making the evaluation of the PEVNET system. We received excellent remarks from the participants in both parts since one of the main aims of conducting this evaluation was to make the participants familiar with the functionalities of the different features in the PEVNET and getting the feedback, such as system ease of use, functionality etc. Figs. 5(a), 5(b), 6, 7 and 8 show the participants responses. The majority of the participants and the domain experts appreciated the usability evaluation and put their confidence in the PEVNET features. The PEVNET features include the node link color feature, locating central person, node size feature, node link details on demand feature, visual filtering, network of clusters, detecting collaborating sub-cluster feature, visualization features based on temporal data, sub¬-cluster detection feature, trend analysis feature, encircle feature, temporal pattern feature, expand collapse feature, and visualizing similar node feature [9, 25, 26].

There were some participants who had some general reservations about the novelty of the features, functioning of the system, locating certain features, validation steps, generality of the PEVNET system, market impact, number of nodes, and time complexity. There were three participants who criticized the system in the quantitative evaluation. They had concerns regarding the diversity of the features. One of them pointed out that there should be a single menu, for instance, like a dashboard. Similarly, in the qualitative evaluation, two to three participants had recommendations regarding the user friendliness of the user interface, dark color combinations of the crime type legends, long system response when the number of nodes increased etc.

At the start, we got some problems as how to make the evaluation of the system, since there was no assessment as to whether the evaluation would be qualitative or quantitative. Since the visual display of the PEVNET demands some qualitative evaluation. The ideal situation for doing this was in a place where there was some IA going on and one could find an adequate number of domain experts who could carry out the verification and validation of the system since they had working knowledge of different visualization tools and techniques. But practically, conducting interviews with the domain experts was a difficult task. The domain experts were required to be traced globally. Thanks to some events, for instance, research conferences, summer schools, symposiums, etc., we had interaction with some domain experts.

We got many problems in getting the data since the access to the criminal datasets was difficult. There were number of formalities; for instance, going through the security check, official verification, license, usage purpose etc. The second issue was having the appropriate datasets. If someone was lucky enough to get some datasets, they were sometimes found to be in an alien format. Often, it was difficult to convert the datasets that conformed to one's system. Luckily, we got datasets from a reliable source. But we had to develop a software agent to convert the datasets to our desired format, for instance, the IUCR datasets were in the excel file format but the system in which the PEVNET was developed supported only the XML file format. We had some negative comments from the participants with respect to re-examining the existing features. They were of the view that there was no distinct improvement in the existing features, which was our claim.

But we have gotten a great deal out of this evaluation. While making the experiments, we came across numbers of thoughts and ideas. Moreover, out of the thirty two participants in the experiments twenty eight were information technology literates and four were domain experts in the field of network visualization and social network analysis. We got some unique ideas from a few of the students, who enjoyed the PEVNET system's unique features. In the PEVNET team, there was a participant that had some reservations regarding the system's ease of use and the functionality. In the Gephi team there was a participant, who was satisfied regarding Gephi system's functionality but he was not satisfied with the ease of use of the Gephi system. He appreciated the PEVNET's system ease of use in locating the features.

So, we will definitely benefit from this activity as it has provided us with a good chance to make improvements in the system based on the reservations, criticisms, recommendations, comments and appreciations of the participants. Naturally the appreciations from the participants, especially the IA experts, also act as a mental tonic for us.

10 Motivation

We have followed nearly all the tasks, by McAndrew and Sparrow [27, 28], in this research evaluation for analyzing the criminal networks. They are the leading lights in the field of IA research. The PEVNET user's group performed good as there was tool support for conducting nearly all the tasks as described earlier. The PEVNET users not only provided the comprehensive inputs but also provided a variety of solutions which, naturally, could be an extra support for the decision makers. This variety of solutions was as follows:

- The PEVNET users not only detected the central person but also provided the five alternatives that could become candidate nodes for becoming the central persons with respect to the nodes' weight.
- The clustering algorithm [26] of the PEVNET is a great support while locating sub-groups, The PEVNET not only provides information regarding sub-groups but also the information of the whole network in the background. In this way, the users are in a better position to make comparisons with other nodes in the network.

Again, it is a good example of support to the decision maker. In PEVNET, we have described another way of collecting the hidden information in the 'sub-groups' [28] with the 'detecting collaborative sub-cluster feature' [26], shown in Figs. 1 and 3. The said feature extracts the sub-group interaction from the datasets and visualizes the internal crime collaboration by using colorful legends.

- Finally, the 'similar node feature' [26] addresses one of the tasks [28], i.e., 'information flux across the network'. With the support of this feature the analysts can get a geographical information flux across the network. The information of the sub-groups having similar crime types can be detected with this feature.

11   Conclusion and Future Works

In this research paper, we have made an evaluation of our proposed framework for the visualization of criminal networks, PEVNET. An experiment has been conducted in which there were thirty two participants and the performance of the PEVNET system has been assessed by way of the usability evaluation and qualitative feedback. Two research questions have been formulated and were successfully tested based on the experimental results. In a nutshell, we have proposed some novel visualization features with an attempt to facilitate IA experts and have made an attempt to add our contribution to the research. We have a firm belief that the PEVNET will create a 'visual literacy' [23], influence user's ideas, and sharpen their critical thinking. Also with this evaluation, we have gotten the confidence of the majority of the participants and especially the domain experts.

We have planned to conduct the comparative study of the PEVNET framework with other network visualization tools as part of our future works. Further, we will also evaluate our recently proposed novel composite feature.

Acknowledgements. First of all, we would like to pay a special thanks to the IT department at King Fahd University of Petroleum and Minerals (KFUPM) in the kingdom of Saudi Arabia. The crucial task in front of the IT department was to equip the relevant computer with the specified software. Secondly, a special thanks to Dr. Niazi-Associate professor in the Software Engineering department, Dr. Aziz Arshad and Mr. Waheed Aslam Lecturers in the ICS department in facilitating our work at KFUPM.

References

1. Information visualization & visual analytics, Pacific Northwest National Laboratory, U.S. Department of energy. http://vis.pnnl.gov/
2. Chernoff, H.: The use of faces to represent points in k-dimensional space graphically. Journal of the American Statistical Association **68**, 361–368 (1973)
3. Furnas, G.W., Buja, A.: Prosection views: dimensional inference through sections and projections. With a discussion by John F. Elder IV, ShingoOue and Daniel B. Carr and a rejoinder by the authors. Journal of Computational and Graphical Statistics **3**(4), 323–385 (1994)

4. Ebel, H., Davidsen, J., Bornholdt, S.: Dynamics of social networks. Complexity **8**(2), 24–27 (2002)
5. Grinstein, G., O'Connell, T., Laskowski, S., Plaisant, C., Scholtz, J., Whiting, M.: VAST 2006 contest: A tale of alderwood. In: Proc. First IEEE Int'l Symp. Visual Analytics Science and Technology (VAST 2006), pp. 215–216 (2006)
6. Heuer, R.J.: Psychology of intelligence analysis
7. Freeman, L.C.: Centrality in social networks conceptual clarification (1978)
8. Brantingham, P.L., Ester, M., Frank, R., Glässer, U., Tayebi, M.A.: Co-offending network mining
9. Carpendale, S.: Evaluating Information Visualizations, Department of Computer Science, University of Calgary, 2500 University Dr. NW, Calgary, AB, Canada T2N 1N4.3
10. Rasheed, A., Wiil, U.K.: PEVNET: A framework for visualization of criminal networks. In: The 2014 IEEE/ACM International Conference on Advances in Social Networks Analysis and Mining (ASONAM 2014), pp. 876–881 6 p. IEEE Computer Society Press (2014)
11. Jordan, P.W.: An Introductin to Usability. Taylor and Francis, Bristol (1998)
12. Xu, J.J., Chen, H.: University of Arizona, Tucson, AZ, CrimeNet explorer: a framework for criminal network knowledge discovery (2005)
13. Park A.J., Tsang, H.H., Brantingham, P.L.: Dynalink: A framework for dynamic criminal network visualization. In: European Intelligence and Security Informatics Conference (2012)
14. Petersen, R.R., Wiil, U.K.: Crimefighter investigator: A novel tool for criminal network investigation. In: European Intelligence and Security Informatics Conference (EISIC), pp. 197–202, September 2011
15. Barlow, T., Neville, P.: A comparison of 2-d visualization of hierarchies. In: Proceedings of Information Visualization, pp. 131–138 (2001)
16. Chi, E.H., Reidl, J.T.: An operator interaction framework for visualization systems. In: Proceedings of the IEEE Symposium on Information Visualization. IEEE (1998)
17. Card, S.K., Mackinlay, J.D., Shneiderman, B.: Readings in information visualization: using vision to think. Morgan Kaufmann (1999)
18. Mackinlay, J.D., Robertson, G.G., Card, S.K.: The perspective wall: detail and context smoothly integrated. In: Proceedings of the SIGCHI Conference on Human Factors in Computing Systems (1991)
19. Roussinov, D.G., Chen, H.: Document clustering for electronic meetings: an experimental comparison of two techniques, Department of MIS, Karl Eller Graduate School of Management, University of Arizona, McClelland Hall 430ww, Tucson, AZ 85721, USA
20. Shneiderman, B.: The Eyes Have It: A Task by Data Type Taxonomy for Information Visualizations
21. Jain, A.K., Dubes, R.C.: Algorithms for clustering data. Prentice Hall (1998)
22. Lampling, J., Rao, R.: Laying out and visualizing large trees using a hyperbolic space. In: Proceedings of the 7th annual ACM Symposium on user Interface Software and Technology. ACM (1994)
23. Li, X., Juhola, M.: Country crime analysis using the self-organizing map, with special regard to demographic factors. SpringerLink (2013)
24. Bleed, R.: Visual literacy in higher education, by Ron Bleed, Maricopa Community Colleges, ELI explorations, August 2005
25. Rasheed, A., Wiil, U.K.: Novel visualization features of temporal data using PEVNET. In: Wang, L.S.-L., June, J.J., Lee, C.-H., Okuhara, K., Yang, H.-C. (eds.) MISNC 2014. CCIS, vol. 473, pp. 228–241. Springer, Heidelberg (2014)

26. Rasheed, A., Wiil, U.K.: Novel Analysis and Visualization Features in PEVNET, the Maersk Mc-Kinney Moeller Institute University of Southern Denmark Campusvej 55, 5230 Odense M, Denmark (2014) (submitted for acceptance)
27. McAndrew, D.: The structural analysis of criminal networks. In: The Social Psychology of Crime: Groups, Teams, and Networks
28. Sparrow, M.K.: The application of network analysis to criminal intelligence: An assessment of the prospects (1991)
29. Kang, Y., Gorg, C., Stasko, J.: Member IEEE, How can Visual Analytics Assist Investigative Analysis? Design Implications from an Evaluation
30. Yi, J.S., Kang, Y., Stasko, J., Jacko, A.J.: Toward a Deeper Understanding of the Role of Interaction in Information Visualization
31. Wiil, U.K.: Issues for the next generation of criminal network investigation tools. In: European Intelligence and Security Informatics Conference (2013)
32. Peterson, R.P.: Criminal Network Investigation: Processes, Tools, and Techniques. Diss. SDU, Faculty of Engineering, The Maersk Mc-Kinney Moller Institute (2012)

Effects of Human Communication on Promoting the Use of a Web-Mediated Service

Research Study on the Japan Local Network System

Norihito Seki[✉]

Hokkai-Gakuen University, 4-1-40, Asahi-Machi, Toyohira-Ku, Sapporo, Japan
nseki@ba.hokkai-s-u.ac.jp

Abstract. In this study, the authors investigated the effects of human communication on promoting use of a web-mediated service by exemplifying a load
matching system for intermediating transportation business. In Japan, the majority of load matching systems is scarcely used and has been inactive other
than few exceptions. Japan Local Network System has been one of the few successful load matching services. The association that operates the system is
actively encouraging its members to mutually communicate and share information. We conducted a questionnaire survey on the member companies and constructed structural equation models based on a covariance structure analysis of
the survey results. The models showed that the metadata shared by the members
via human communication has promoted the use of the service.

Keywords: Logistics · Information sharing · Human communication · Metadata ·
Covariance structure analysis · Intermediary service site · Load matching system

1 Introduction

Web-mediated intermediary services intermediate information on the Internet web
[1]. Examples of web-mediated intermediary services include the NC Network [2] of
the manufacturing industry for intermediating orders of parts and molds and a load
matching system, which intermediates information of loads and available carriers in
transportation business. However, not all such services have been successful, requiring effective means for promoting use of the systems, information distribution and
intermediation among businesses entities.

In this study, we focused on a load matching system. The system intermediates
load and load board information and has been expected to greatly contribute to improving the transportation efficiency of small to medium-sized carriers, which do not
have a nationwide transportation network or physical distribution bases. Unlike big
forwarding agents, which have built a nationwide base network to increase loads, it is
difficult for a small carrier with only one office to increase loads by itself. Small carriers have thus tried to increase the transportation efficiency, load factor and working
rate of their rolling stock by forming a link with several other carriers so that each
acts as a branch of the others.

L. Wang et al. (Eds): MISNC 2015, CCIS 540, pp. 150–160, 2015.
DOI: 10.1007/978-3-662-48319-0_12

They have implemented load matching via telephone, facsimile and computer communication. As the Internet web has expanded nationwide, focus has shifted to web-mediated load matching. The use of the web was expected to reduce the introduction cost and increase the amount of information exchange and thus business chances. In the early 2000s, many web-mediated load matching systems were constructed and increased their members.

However, the results have been so far disappointing. Today, the majority of web-mediated load matching systems in Japan are suspended or closed [3]. The cause was likely to have been that not sufficient carrier and load requests were made. Japan Local Network System has been one of few successful exceptions. The association in charge of operating the system (hereinafter referred to as the "association") is actively holding meetings[1] to encourage interaction among its members. The members communicate with each other not only on the web but also face-to-face. Such real interaction is likely to have some effect. In this study, the authors investigated the roles of face-to-face human communication in promoting the use of a web-mediated service by performing a questionnaire survey on members of the association and conducting a positive analysis of the results.

2   Load Matching System and Actions of Japan Local Network System Association

2.1   Load Matching System

A web-mediated load matching system matches load requests and load board requests via the Internet. Transactions are made by members posting up their requests on the web page and searching for the match.

Load and load board requests are posted together with the data, hours, destination, charges desired, etc. A member requesting for load also registers the information of the truck and wait for a member(s) that want its load to be transported. Similarly a member requesting for load board posts up the information of the load and wait for a member(s) that wants load. This entire set of information is called load matching information and is consumed in a single transaction [4].

2.2   Overview of Japan Local Network System Association

Japan Local Network System has 1,645 business cooperative members (as of the end of fiscal 2008). The main objective is to make transactions via the load matching system. The service makes a total annual turnover of about 58 billion JPY, about 400 thousand transaction contracts a year and about 650 thousand requests posted a

---

[1] In the transport industry, most frequently talked topics at face-to-face communication are those on safe driving and operation. Japan Local Network System is an example of a load matching system for promoting transactions.

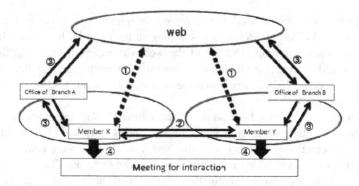

Fig. 1. Flows of information in Japan Local Network System. The dashes lines denote information exchange on the Internet. The thin solid lines denote settlements of accounts, and the bold arrows denote participation to a meeting.

year[5]. Based on its basic principles of "being faithful and respecting the art of commerce" and building a human network, the association attach much importance to meetings in which people interacts with each other.

Japan Local Network involves the following load matching procedures:

① A carrier member search for load on the web page.

② The parties negotiates with each other.

③ The accounts are settled via the association.

④ Members participate to a meeting voluntarily[2].

2.3   Roles of Real Interaction in the Japan Local Network System

The association holds meetings for member to interact with each other in various levels including sectional, regional (7 regions such as Hokkaido, Tohoku and Kanto), and national. For example, a regional meeting consists of a general meeting, sectional meetings and a social. Members exchange information mainly at a sectional meeting. Eight sectional meeting were held during the Kanto regional meeting held on August 23, 2007, to which the authors attended. About 40 cooperatives (possessing 434 trucks in total) were at the sectional meeting for flat body truck owners. They were exchanging information, for example on what trucks should be used for what load and other information for deepening their understanding on load matching. The information exchanged and shared is the carriage infrastructure information [4] and is metadata[3] for load matching information, such as vehicle type, load, and destination.

---

[2] In industrial engineering, voluntary activities have been deemed important. For example, Yoshida [14] mentioned that a compulsory small group activity (called the "QC circle") is a cause of decline and should be substituted by voluntary and creative activities, such as the Creative Dynamic Group Method (GDGM). Saito [15] also mentioned the need of voluntary human activities. In this study, the models shown in Figs. 1, 3 and 4 were constructed also based on their theories.

Table 1. Carriage infrastructure information (metadata for load matching data)

Content	Description
Rolling stock (number and kind)	Metadata for vehicle data in load matching
Favorite cargo	Metadata for load data in load matching
Main destination	Metadata for destination data in load matching
Busy and low seasons	Supply and demand is better adjusted by grasping the changes in load in a year.

The role of a real meeting in the Japan Local Network System can be investigated as a relationship between a virtual community (web forum) and real community (meeting). In an empirical study, Yoshida [7] concluded that the relations continue between a web forum and its off-line meeting (real community). He mentioned that a virtual community complemented and enhanced the roles of a real community [7]. From this viewpoint, the meeting held by the association is connected to its load matching system (bold arrows in Fig. 1).

Meetings are sites for face-to-face communication. The main objective of personal exchange has been reported to be information acquisition [8]. According to media richness theory, face-to-face communication can effectively process ambiguity [9] and is thus appropriate for summarizing opinion and sharing information. Actually many corporation organizations in Japan attach importance to face-to-face conversation [10].

Toyama [11] reported that transactions stopped in a closed e-commerce network that had held meetings but exchanged almost no information whilst transactions increased in another network in which members exchanged information and deepened understanding. This shows that a role of holding a meeting is to promote information sharing and transactions. In a load matching system, an interview investigation by Seki [4] showed that the information obtained via personal exchange was metadata.

As exchange meetings play a big role in Japan Local Network System, it is likely to be a case in which human communication is promoting transactions. To investigate the effects of human communication on promoting transactions, we surveyed the actual utilization state of the System and verified the effects of personal exchange meetings on promoting the use of the load matching system.

3   Positive Analysis

3.1   Hypothetical Model

A hypothetical model was constructed for positive analysis. Use of a load matching system can be expressed with the number of load and board request entries and the number of transaction contracts and thus can be quantitatively analyzed.

---

[3] Metadata is also expressed as "occasion for building relationship", "context shared", "5W1H", and "TPO" information [10],[16]. Metadata is information for understanding information.

A questionnaire survey was conducted on member corporations of Japan Local Network System. Based on the results, the relationships were investigated among the number of personal exchange meetings, metadata and the number of entries. The variables shown in Table 2 were established, and a hypothetical model was created (Fig. 2).

The model was created based on the following idea. Members of Japan Local Network System attend meetings actively to share metadata for better use of load matching information. They introduce information technologies and sub-information systems, such as for producing, distributing and accumulating information. In other words, it is a management information system [12] in which three system morphologies of human, information and information technology are integrated in an organic manner. The model in Fig. 2 depicts such a system.

In a load matching system, the number of entries is believed to determine the number of transaction contracts. In this study, sufficient information distribution was assumed. It was also assumed that the number of entries increased by increasing the number of accesses and that an increase in the number of entries increased the transaction contracts. Variables set were the "number of accesses", "number of entries" and "number of transaction contracts". In Fig. 2, the part enclosed within a dashed-line rectangle shows information exchange on the web.

Metadata sharing among members at personal exchange meetings were also incorporated in the model. Variables "the number of meetings attended" and "metadata sharing" were assumed to affect the "number of accesses", "number of entries" and "number of transaction contracts". As shown in Fig. 2, the objective of this study is to investigate the effects of a personal exchange meeting and metadata sharing on the web-mediated transactions.

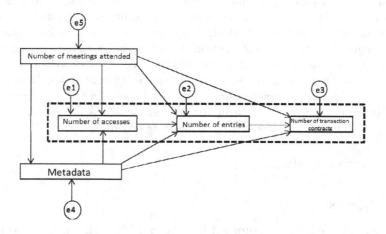

Fig. 2. Hypothetical model in this study. The part in the dashed-line rectangle shows information exchange on the web.

Table 2. Variables

Variables	Explanation
The number of accesses	The number of accesses made to the load matching system in a month
The number of entries	The number of entries made to the load matching system
The number of transaction contracts	The number of contracts made per month via the load matching system
Number of meetings attended	The number of real regional meetings attended per year
Metadata	The number of member corporations for which carriage infrastructure information has been acquired

3.2   Methods and Results

Survey Methods. A questionnaire survey was conducted by sending a form to 663 member corporations of 51 branches (in Hokkaido, Tohoku and Kanto) of Japan Local Network System. The period of the survey was from September 17, 2009, to November 24. Valid responses were made by 145 (response rate: 21.9%). However, 44 were excluded from the analysis due to missing value, and data for 101 were analyzed.

Answering Method. For the five variables, including the number of meetings attended and metadata, the respondents were to enter the actual numbers.

Descriptive Statistics and Correlation Analysis. The descriptive statistics and correlation analysis of the data collected via the questionnaire are shown in Table 3. Because the values of the variables varied greatly, they were transformed by $y^* = \sqrt{x} + \sqrt{x+1}$ [13]. Correlation coefficients between transformed data showed high positive correlations between the "number of entries" and the "number of transaction contracts" (r=.913), between the "number of accesses" and the "number of entries" (r=.636), between the "number of accesses" and the "number of transaction contracts" (r=.625), between "metadata" and the "number of entries" (r=.477) and between "metadata" and the "number of transaction contracts" (r=.495).

Although not high, there were also significant positive correlations between the "number of meetings attended" and "metadata" (r=.269), between the "number of meetings attended" and the "number of entries" (r=.213), and between the "number of meetings attended" and the "number of transaction contracts" (r=.227).

Covariance Structure Analysis (Path Analysis). By using the transformed variable data and the hypothetical model shown in Fig. 2, structural equation models were constructed by covariance structure analysis. Here, latent variables were not introduced, but the variables shown in Table 2 were regarded to be observation variables. Covariance structure analysis was conducted by using Amos19, and the maximum likelihood method was used for estimation. Hereinafter the hypothetical model is referred to as "Model A".

The results for Model A are shown in Fig. 3. Model A showed high conformity values as shown in Table 3, but there were several paths that were not significant (Fig. 3). The paths from the "number of meetings attended" to the "number of accesses", from the "number of meetings attended" to the "number of entries", and from the "number of meetings attended" to the "number of transaction contracts" showed insignificant and small values. The value from the "number of meetings attended" to the "number of accesses" was negative. Model B was constructed by excluding the insignificant paths and also based on the correlation coefficients shown in Table 5 (Fig. 4).

In Model B, all paths were significant. As shown in Table 3, the conformity indices were satisfactory for both Models A and B. Compared to Model A, Model B showed a smaller difference between GFI and AGFI and a lower AIC value, suggesting higher conformity of Model B. Therefore, Model B was used for subsequent analysis.

The conformity indices of Model B were satisfactory ($\chi2$=3.428; GFI=.987; AGFI=.951; CFI=1; REMSEA=0), showing good conformity between the data and model. The conformity value of the paths was .546 from the "number of accesses" to the "number of entries", and was as high as .877 from the "number of entries" to the "number of transaction contracts". The value was .273 from "metadata" to the "number of accesses", .327 from "metadata" to the "number of entries", .077 from "metadata" to the "number of transaction contracts", and .317 from the "number of meetings attended " to "metadata". The indirect effect of the "number of accesses" on the "number of transaction contracts" was .479, and that of "metadata" on the "number of transaction contracts" was .418 (Table 5).

Table 3. Comparison of model conformity

	Model A (Hypothetical model)	Model B (Model for analysis)
χ^2	2.367(p=.124)	3.428(p=.489)
GFI	0.991	0.987
AGFI	0.861	0.951
CFI	0.995	1
RMSEA	0.117	0
AIC	30.367	25.428

Table 4. Standardized indirect effects in Model B

	Number of meetings attended	Metadata	Number of accesses	Number of entries
Metadata	0	0	0	0
Number of accesses	0.087	0	0	0
Number of entries	0.151	0.149	0	0
Number of transaction contracts	0.157	0.418	0.479	0

Table 5. Descriptive statistics and correlation coefficients

	Mean	Standard deviation	Number of meetings attended	Metadata	Number of accesses	Number of entries	Number of transaction contracts
Number of meetings attended	**2.783**	**1.464**	1				
	2.059	2.473	1				
Metadata	**5.331**	**3.825**	**0.269****	**1**			
	10.327	15.362	0.317**	1			
Number of accesses	**9.821**	**6.167**	**0.084**	**0.273****	**1**		
	33.139	45.801	0.071	0.264**	1		
Number of entries	**6.882**	**5.541**	**0.213***	**0.477****	**0.636****	**1**	
	19.059	31.482	0.114	0.424**	0.56**	1	
Number of transaction contracts	**6.673**	**5.093**	**0.227***	**0.495****	**0.625****	**0.913****	**1**
	17.158	27.161	0.126	0.468**	0.515**	0.914**	1

*< 05,**<.01. *Top: Transformed variable data, bottom: variable data before transformation (original data).

4   Discussion and Future Topics

4.1   Discussion

Let us first focus on the relationships among the "number of accesses", "number of entries", and "number of transaction contracts" in Model B. The path coefficients from the "number of accesses" to the "number of entries" and from the "number of entries" to the "number of transaction contracts" showed that an increase of accesses would increase entries, and an increase of entries would increase transaction contracts. Particularly the number of entries has a strong effect on the number of transaction contracts. This signifies that increased information distribution promotes the use of the system. As there was an indirect effect of the "number of accesses" on the "number of transaction contracts", it was shown effective to increase accesses. The hypothesized process was proved, in which active accesses by members increase information distribution and transaction contracts.

On the other hand, there was indirect effect of "metadata" on the "number of transaction contracts". The path coefficient from "metadata" to the "number of transaction contracts" was as small as .077, showing almost no direct effect, but the indirect effect was as large as .479. "Metadata" also affected the "number of accesses" and "number of entries", and was shown to increase them.

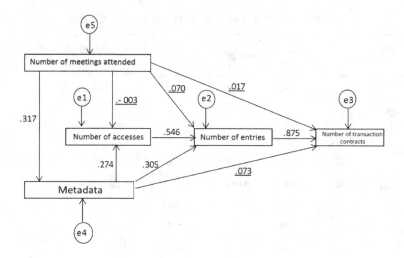

Fig. 3. Results of Model A (hypothetical model). *Underlined paths were not significant. The other paths were significant (significance level: 1%).

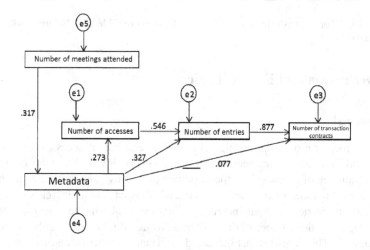

Fig. 4. Results of Model B (adopted model). *Underlined path was significant (significance level: 10%). The other paths were significant (significance level: 1%).

Attendance to meetings was initially hypothesized to increase metadata sharing and the numbers of accesses, entries and transaction contracts. However, attending meetings had direct effect only on spread of metadata (therefore, Model B was adopted). This suggests that attending a meeting can be effective only when metadata is collected via information exchange. This agrees with the importance of personal exchange meetings mentioned in previous studies[8,11].

4.2   Conclusions

In this study, key factors that promote use of a web-mediated service were investigated by performing a questionnaire survey on Japan Local Network System, which has been a successful web-mediated load matching service and has encouraged members to exchange and distribute information via face-to-face communication, and a positive analysis of the results. A covariance structure analysis revealed that increasing information distribution is a factor that promotes the intermediary service and that sharing of metadata has contributed to transaction increase. The metadata was also found to be acquired via face-to-face communication. In this study, we showed that a system of providing occasions for human communication for members to share metadata has been effective.

In Japan, the majority of load matching systems have been suspended or closed. However, the Japan Local Network System has been successful. Unlike the former, which were designed aiming to increase the number of entries, the latter was shown to have been designed so as to promote metadata sharing via human communication. This study proved the effects of face-to-face communication on use of a web-mediated intermediary service.

4.3   Future Topics

In this study, a model was presented to show the effect of information sharing via human communication on system use. The model is the first step for building a theory on intermediary service system design that incorporates human communication. The following three topics need to be studied to build such a theory.

The first topic is constructing a model that can further explain the roles of on-site human communication. According to the model developed in this study, the role of human communication was promoting metadata sharing. Other means for acquiring metadata should also be investigated and compared, including social networking services. Such a study will also help the generation procedure of metadata to be explained in relation to media and thus provide detailed knowledge on system designing.

Secondly, this study could not show the effects of attending meetings on the numbers of accesses, entries and/or transaction contracts. It was not shown that building relationship of mutual trust and unity among members promoted transactions. The effects of face-to-face communication should be further investigated by establishing variables for and modeling a relationship of mutual trust and unity.

Thirdly, the model developed in this study is in the framework of the management information system. It just modeled the effect of metadata possessed by a corporation on promoting the use of the service and does not reproduce information exchange with another. Because a transaction is made by not only spreading its own information but also acquiring information of the other, a model should be constructed to show the effects of information exchange.

Acknowledgements. We thank Chairman Sadao Aoyama and Secretary Yoko Tsujioka of Kanto Branch of the Japan Local Network System, Ichiro and Kiyoe Mitsuhashi of Mitsuhashi Unyu Kompo Service, and members of Hokkaido, Tohoku and Kanto Branches of the Japan Local Network System for their cooperation in this study.

And, This work was supported by Hokkai-Gakuen Kenkyu Josei(Sogo)2014,2015.

References

1. Hayashi, K., Yukawa, K., Tagawa, Y.: Evolving Networking. NTT Shuppan, Tokyo (2006). (in Japanese)
2. Website of NC Network: http://www.nc-net.or.jp/ (December 1, 2014)
3. Nakata, S.: Appearance of New Distribution Industry. In: Ito, M. (ed.) Shin Ryutsu Sangyo. NTT Shuppan, Tokyo (2005). (in Japanese)
4. Seki, N.: Load Matching System in Truck Business Cooperative Association. Journal of Japan Society for the Study of Office Automation **26**, 81–89 (2006). (in Japanese)
5. Website of Japan Local Network System: http://www.jln.or.jp/ (December 1, 2014)
6. Delanty, G.: Community. Routledge, London (2003)
7. Yoshida, J.: Sociology of Internet Space. Sekaishisosha, Tokyo (2000). (in Japanese)
8. Ohe, H.: A study on information acquisition effect by human network – focusing on substitution and complementation of forum activities and the internet. In: Proceedings of Study Meeting of the Information Processing Society of Japan. IPSJ Special Interest Group on Information Systems, (53), pp. 29-36 (2005)
9. Daft, R.L., Lengel, R.H.: Organizational Information Requirements, Media Richness and Structural Design. Management Science 32(5) 1984
10. Futamura, T. (ed.): Current Micro-Organization Theory. Yuhikaku, Tokyo (2004). (in Japanese)
11. Toyama, A.: Study on Present Management Information System. JUSE Press Ltd, Tokyo (1998). (in Japanese)
12. Ishikawa, H.: Information Sharing and Use. Chuo Keizaisha, Tokyo (2001). (in Japanese)
13. Toyoda, H.: Covariance Structure Analysis [Amos]. Tokyo Tosho, Tokyo (2007). (in Japanese)
14. Yoshida, K.: Statistics Oriented Management. Nikkei BP, Tokyo (2010). (in Japanese)
15. Saito, M. (ed.): Work Adaptation Engineering. Nippon Shuppan Service, Tokyo (1998). (in Japanese)

Evaluating Online Peer Assessment as an Educational Tool for Promoting Self-Regulated Learning

Pei-Lin Hsu[1](✉) and Kuei-Hsiang Huang[2]

[1] General Education Center, Oriental Institute of Technology, No.58, Sec. 2, Sih-Chuan Rd., Ban-Ciao District, New Taipei City 22061, Taiwan, R.O.C.
peiling@mail.oit.edu.tw
[2] Department of Marketing and Logistics Management, Chihlee Institute of Technology, 313, Sec. 1, Wunhua Rd., Banciao District, New Taipei City 22050, Taiwan, R.O.C.

Abstract. A study was conducted to investigate how grade evaluations for a university course assignment derived from peer assessment, compare to grade evaluations derived from self-assessment and teacher assessment. One hundred thirty-six undergraduate students in a university course were asked to submit an assignment to Moodle Workshop in June, 2014. The assignments were assessed separately by the students themselves (self-evaluation), a random group of their classroom peers (peer evaluation), and the course teacher (teacher evaluation). Students' opinions on various aspects of the peer assessment process were also collected and analyzed. Results indicate that grade evaluations from peer assessments are inclined to be lower than evaluations from self-assessments. Grade evaluations from peer assessments and teacher assessments were more similarly matched, with a significant positive correlation of r=0.75. Many students (84%) agreed that doing peer assessment on other students' assignments led them to reflect on how they personally performed their own assignment. The benefits of peer assessment as a tool for promoting self-regulated learning are explored. Implications of the results and potential future research on this topic are discussed.

Keywords: Peer assessment · Peer feedback · College student

1 Introduction

"Assessment is a goal-oriented process [1]". Assessments not only emphasize grading, but are also a way to help students learn. By comparing educational performance with educational purposes and expectations, instructors can modify and improve the program goals to be taught by the teachers and learned by the students, to foster better teaching and learning overall.

Traditionally, students' finishing of quizzes and tests means the end of their study responsibility. After teachers check the tests and quizzes, another learning journey begins for the students. Can assessments be used to develop a new and better way to start the next learning journey? The fast growth of technology makes it possible. Learning can happen quickly among peers in the Digital Age. Online assessment has

© Springer-Verlag Berlin Heidelberg 2015
L. Wang et al. (Eds): MISNC 2015, CCIS 540, pp. 161–173, 2015.
DOI: 10.1007/978-3-662-48319-0_13

been popular since the invention of internet technology and change the process of assessment [2][3]. Peer assessment can be both a form of formative and summative assessment, by helping students learn and also checking if students have attained what they need to learn.

Many researchers are interested in studying peer assessment [4][2][5][6][7]. How can peer assessment help students learn? What is the effect of peer assessment on learning? What do peer assessers and student assessees think of the peer assessment process?

Therefore, in this paper, three specific research objectives are examined:

1. Are peer assessment grades consistent with self-assessment and teacher assessment grades?

2. Are the student participants aware of using the benefits of peer assessment to promote learning?

3. What are the student participants' opinions on the peer assessment process?

2   Literature Review

2.1   Self-Regulated Learning

Self-regulated learning is a process by which students manage their thoughts, emotions, and behaviors toward attaining personal goals.

Self-regulation was once viewed from an individual-focused perspective. However, the new model of regulated learning has taken a more social perspective. Social context is now regarded as an element in the process of self-regulation [8]. Indeed, self-regulated learning behaviors include self-initiated forms of social learning, such as seeking help from peers, coaches, and teachers. These behaviors are especially useful because peer learning is an efficient way to transmit knowledge [9].

Social context is important for students' self-regulation [10]. Social support from teachers and peers plays an important role in being self-regulative. Nicol, et. al. [11] argued that formative assessment and feedback should be used to empower students to be self-regulated learners. It is much more difficult to achieve self-regulated learning without motivation [12][13]. It has been found that effective feedback from others can promote students' motivation [14]. A study of 5th grade students' math achievement found that providing the students with more accurate teacher feedback improved their math scores [15]. Peer feedback itself can be also beneficial for learning. Students can learn meta-process such as reflecting on and justifying what they have done [16].

As such, one way to help students become self-regulated and life-long learners is for teachers to provide the students with a supportive social learning environment that incorporates feedback techniques such as peer assessment.

2.2   Assessment, Peer Assessment and Learning

Assessment of learning and assessment for learning are the keys to classroom assessment. Both these aspects of assessment are important. Before teaching, the instructors should decide what is to be learned, determine how to evaluate the learning at the end, and assess how the students' progress in their learning. Traditionally, assessments occur at the end of learning, where the objective is to measure, record, and report what the students achieve with regards to a specific goal. Teachers use assessment to check if the students attain the criteria. Assessments happen periodically after learning takes place. Ultimately after the final examination, teachers need to give the students report card grades. This is known as summative assessment – to evaluate the teaching effect and learning outcome. However, assessment of learning (summative assessment) is different from assessment for learning (formative assessment). Assessment of learning shows the results and symbolizes how well students are learning. Assessment for learning emphasizes the formative assessment and happens during learning. In assessment for learning, students understand what to learn, what is expected of them to learn, and are always given feedback in order to improve. It empowers the students, improve their learning. It involves helping students to understand the learning goal, engage in self-assessment, and see and feel in their continuous growth, and discussing the expected growth and learning for the next stage [17][18].

Regarding assessments as a learning process allows students to make use of the opportunity to improve their learning. Topping [4] encouraged effective peer assessment to be part of evaluation. Self-assessment and peer assessments let students reflect on their own learning. They offer students a chance to think about their own learning and begin to set goals for it. Modern technology provides many opportunities for assessment. Moodle is a recommended online course management system [19]. Moodle (Modular Object-Oriented Dynamic Learning Environment) is an open source learning platform that is also popular as a learning management system and virtual learning environment. As a learning platform, it can enhance the learning environment. Moodle also provides a web-based platform geared towards facilitating peer assessment. Use of web-based peer assessment can be a way to develop students' cognitive schema, construct their knowledge, and promote students' engagement and collaborative learning. The "workshop" activity is the most complex tool in Moodle. It provides both the instructor and peer feedback on open-ended assignment, such as essays and research paper. Users can upload assignment, perform self-assessment and peer reviews of other students' papers.

Peer assessment happens between an "assessee" who receives feedback, and an "assessor" who gives feedback. The assessee and the assessor interact cognitively. Lu and Law[2] found that students benefit more as assessors than as assesses. Peer grading seems to be less effective than peer feedback for assessors. The researchers suggested students should be encouraged to give thoughtful and meaningful comment instead of grading.

Items that can be assessed include writing, presentations, portfolios, test performance, skill behaviors, and reports [5][4]. Sadler and Good [20] integrated the results of several studies and showed that there are four main advantages for self-evaluation and peers' grading the tests   : logistical, pedagogical, metacognitive, and affective. It can improve students' writing and group work ability. It can also save teachers' time. Feedback from peers can be more immediate and more individualized to the assessee than feedback from the teacher. This is important because low executive thinkers have been found to better be able to revise their work after receiving specific feedback, compared to holistic feedback[21].

Peer assessment has been practiced and studied at all types of schools including elementary school, high school, and university. Such research has mostly confirmed the advantages of peer assessment, though the findings sometimes differ by target school due to school-level differences in implementation. Therefore, in implementing peer assessment it is important to follow some guidelines and training in order for it to have effective, consistent, and productive outcomes [4][5].

We also concern about the correlation between rater and ratee scores. It was found that a high correlation between raters' scores (peer or teacher assessment) and ratees' scores (self-assessed) can be fostered by training and experience [5][22][23][24].

2.3   Peer Feedback

Peer feedback means to give comments on peers' work or performance. Much research confirms the benefits of peer feedback. Peer feedback has been shown to be more important and effective than peer grading [2]. External feedback influences how students feel about themselves, and what and how they learn [25]. A student who accepts confirmatory and suggestive feedback can develop and practice self-regulatory skill. Peer assessment includes several functions, such as confirming existing information, adding new information, identifying errors, correcting errors, providing descriptive feedback on classmates' presentations, and evaluating a task performance [4][26]. Good feedback can strengthen students' ability to self-regulate their own performance. Nicol and Macfarlane-Dick[11] strongly recommend seven principles of good feedback practice: (1) help clarify what good performance is, (2) facilitate the development of self-assessment in learning, (3) deliver high quality information to students about their learning, (4) encourage teacher and peer dialogue around learning, (5) encourage positive motivational beliefs and self-esteem, (6) provide opportunities to close the gap between current and desired performance, (7) provide information to teachers that can be used to help shape teaching.

For lifelong learning, students must be taught to develop and regulate their own learning. Making use of good feedback from peer assessment makes that possible.

3   Methods

3.1   Context and Participants

Learning and Problem Solving is a required two-credit course for freshmen at the Oriental Institute of Technology in Taiwan. One hundred thirty-six students (92 males, 44 females) participated in the study. They were asked to finish an assignment called Helping Behavior in April 2014. After submitting their assignments, the students were asked to grade the works online. They were to self-grade their own assignment, as well as peer-grade the assignments of their course peers and give descriptive feedback.

After finding out the results of the peer and teacher assessments, the students answered a questionnaire and open-ended questions designed to gauge their opinion on the peer assessment process.

3.2   Measure

The students submitted the assignments in the 'Workshop' module for self- and peer assessment in Moodle. After online self-assessment and peer assessment, the researchers received the grades from both self- and peer assessments, as well as the grades from the teacher (teacher assessment). Grading of the assignments and grade criteria are described in the procedure section below. The researchers also created a questionnaire and open-ended questions to understand students' opinions about the peer assessment system.

There are three open-ended questions. 'According to the experience of peer assessment, please analyze the advantages and disadvantages. Do you approve or disapprove of it?' 'For the peer rating of "Helping Behavior", who is the most objective rater? Why?' 'Who is the most reliable rater? Why?'

3.3   Procedure

Students submitted assignments to Moodle Workshop. The Moodle Workshop module is a peer assessment activity with many options. The course coordinator distributes the assessments. Then each student receives two marks – one for their own work, and another one for their peer assessments of other students' work.

'Helping Behavior' is an assignment in which students are asked to perform three actions of assisting three people in six weeks. After submitting their works, students were asked to rate their own work and to rate five other students' works which were assigned randomly. They were asked to give marks out of 100. The grading instruction was, 'Please grade yourself and the other five works. There are three dimensions with a reference for evaluation. The helping behavior happens in different place. The helper assists deeply. The helpers do the assignment seriously, describe the helping behavior in detail and reflect on what they did. Finally, grade it and give the feedback.'

Students' writing the reflection after doing 'Helping Behavior' were assessed including grading and giving feedbacks.

There were one hundred thirty-six students who submitted the 'Helping Behavior'. Five students were not cancelled because they submit late and late for the rating time. The researchers split the data into high and low categories, set to contain upper 1/3 and lower 1/3 of the teacher-assessed grades, in order to compare the groups and analyze them.

3.4   Analysis

The collected data were analyzed using SPSS Statistics 20.0. Descriptive statistical analyses were conducted to illustrate the demographic characteristics of the participants. Correlations were examined to evaluate the consistency among the grades of self, peer, and teacher. Content analysis was performed on peer feedback.

4   Results

4.1   Consistency of Student-Grading and Teacher- Grading

The assignment, Helping Behavior, was assessed by the teacher, peers, and students themselves. Grades included the teacher grading, an average of the random five peers' grading, and self-grading. The grades assigned by students (self and peers) were compared to the grades assigned by the teacher to assess criterion validity.

As Table 1 shows, the teacher assessment grades (M=75.64, SD=16.84, N=86) and peer assessment grades (M=75.99, SD=12.59, N=86) were quite close. The self-assessment grades were the highest (M=79.81, SD=15.52, N=86). The t-test statistic was checked among the three raters' grades. The teacher assessment grades were significantly lower than the self-assessment grades (t=-2.005, p<.05). The peer assessment grades were also significantly lower than self-assessment grades (t=-2.433, p<.01). There was no significant difference between teacher assessment grades and peer assessment grades (t=-.290, p>.05).

Table 1. T-tests on the grades of different raters

Grading source	N	M	SD	t-value
Teacher assessment	86	75.64	16.84	-2.005[*]
Self-assessment	86	79.81	15.52	
Teacher assessment	86	75.64	16.84	-.290
Peer assessment	86	75.99	12.59	
Self-assessment	86	79.81	15.52	2.433[**]
Peer assessment	86	75.99	12.59	

[*]P<.05   [**]P<.01.

As Table 2 shows, the Pearson's product-moment correlation coefficient between the teacher assessment grades and peer assessment grades was high (r=.75, P<.01). The correlation between the teacher assessment grades and self-assessment grades was low (r=.291, P<.01). The correlation between peer assessment grades and self-assessment grades was moderate (r=.478, P<.01).

Table 2. Correlation among different raters (N=86)

Grading source	teacher	self	peer
Teacher	-	-	-
Self	.291[**]	-	-
Peer	.750[**]	.478[**]	-

[**]P<.01.

4.2 Advantage and Disadvantage for Peer Assessment

Students expressed their ideas for peer assessment shown them as Table3.

Table 3. Students ideas about advantage and disadvantage of peer assessment

advantage	disadvantage
1. When we assess peer work, it provides me a reference to improve myself.	1. a bias in favor of someone familiar
2. Peer work as a model to be my benchmark.	2. Knowing someone makes me irrational.
3. When finding peer error in the work, it reminds me not to make it again.	3. Giving close friends higher ranks is unfair.
4. Compare with my work. I know how to improve it.	4. Affection influence judgement.
5. It's worthy to evaluate peer work. I approve it.	5. Give the teammates higher score, of course.
6. Feedback from peers makes me grow. I'm happy to receive peers' compliments. I also make progress when I get negative remark.	6. Students forgot the guideline when they assess.
7. Peer feedback is practical.	7. We are students. We can't assess peer work with objectively.
8. I learn how to comment peer work.	8.We have to keep the relationship with classmates, so we can't be objective to assess peer work.
9. Grading is a learning way. More graders make it possible to be objective.	9. Assessing more peer work might lose patience because of being tired.
10. Students are trained to express their ideas and make a judgement.	10. I found someone is lazy to assess because he didn't access the file.
11. I learn to be reasonable and objective.	11. Some students assess peer work lazily and impatiently.
12. It's a new way for freshmen. It's worthy doing it.	12. The grade and the feedback are not matched.

The students also provide some solution for the peer assessment. '*The amount of peer work is needed to be controlled.*' ' *Anonymity for raters and ratees is necessary. It avoids being subjectivity.*' '*Rank with grading peer work is easier than grading out of 100.*'

4.3 Participants' Opinions about the Peer Assessment

Regarding of the pros and cons of the peer assessment, fifty-nine percent of students approve of the peer assessment contributing their learning. And twenty-nine percent of students didn't express their ideas on it. As the experience of peer assessment, forty-nine percent of students expressed that they don't have the experience. As shown on Table 4, eighty-six percent of students agreed that they were serious about grading their peers' works. Eighty-four percent of students agreed that when grading their peers' works, they reflected on themselves and their own performance, finding the mistakes they made in their work. Regarding the objectivity of peer assessments, seventy-one percent of students agreed that peer assessment is objective. Sixty-seven percent of students agreed that they learned a lot from peer assessments. But only forty-five percent of students agreed that peers grade more highly than the teacher does.

Table 4. Opinions about peer assessment

Degree[*]/ Item	I assess peer work serious-ly	when grading, I reflect myself	It's objective for peer assessment	Peer grade is higher than the teacher grade	I learn a lot from peer assessment
1	0	0	2 (1.5%)	1 (0.7%)	3 (2.2%)
2	1 (0.7%)	3 (2.2%)	4 (2.9%)	5 (3.7%)	6 (4.4%)
3	18 (13.2%)	19 (14%)	34 (25.0%)	69 (50.7%)	36 (26.5%)
4	78 (57.4%)	82 (60.3%)	66 (48.5%)	49 (36.0%)	72 (52.9%)
5	39 (28.7%)	32 (23.5%)	30 (22.1%)	12 (8.8%)	19 (14.0%)
Total	136 (100%)	136 (100%)	136 (100%)	136 (100%)	136 (100%)

Note[*] 1 extremely disagree   2 don't agree   3 partly agree and partly disagree
        4 agree   5 extremely agree

5   Discussion

5.1   Peer Assessment Source of Self-Regulation

The analyses showed that the correlation between the teacher-grade and peer-grade was high (r= .75) on the assignment, Helping Behavior. This was considered an indicator of the high criterion validity of peer assessment, because the teacher rating was perceived as the criterion. These findings are similar in part to other related research [20] [22] [23]. Another finding was that the scores of self-grader are the highest of all graders. The self-grade scores are significantly higher than the peer-grade scores and the teacher-grade scores. This can be explained by the common phenomenon of self-expansion. Self-graders might care about their own grades. Lower score ratees graded by teachers tended to inflate their own low performance by giving themselves a higher score than their actual performance deserved. When rating others, they tended to give good performers a lower score than the teacher did. Some students rate themselves leniently, and rate peers strictly.

There were a lot of merits to the peer assessment process that students responded to. When they reviewed others' work, it provided them with a lot of models to learn from. Students learned ways to improve their work in the next stage. They also collected a lot of descriptive feedback. The more ideas they got from peers, the more their work can progress in the future. Another positive point mentioned by students was learning how to give constructive feedback on a piece of work to the ratee. Some of the students expressed that more peer raters would grade more objectively and accurately because of averaging the peer raters' scores.

As Nicol et. al., [11] argued, formative assessment and feedback can help students regulate their own learning. Therefore, peer assessment can be one source of self-regulation. Observing peers' work makes raters think about their own work. Seeing how their peers write and edit their papers can give students a lot of inspiration. They can think about the feedback and correct their own future work. It can also give them a chance to transfer the ideas they learned to another field. Peers' feedback can also facilitate students' rapid advancement by stimulating their cognitive growth. Additionally, social growth happens during the interactions of peer assessment. It might open students' minds and make them practice metacognition. Students are learning in a social context and getting support from faculty and peers. When they know how to modify their work, they start to control their own learning and regulate themselves.

5.2   Provision for Grading and Feedback

It was found that students graded themselves leniently even though the teacher had given guidelines for checking the work. Eighty-four percent participants showed they reflected the assignments when they rated. No students answered that they "extremely disagree"[ed] with the idea that grading a peer made them reflect on their own performance. This was also the case with agreeing that they were serious about assessing their peers' work. It is clear that most of the students took the peer assessment seriously and used it as an opportunity to reflect on their own assignment performance. This means the students had a positive attitude toward the peer assessment process. Sixty-seven percent of students agreed that they learned a lot from peer assessment. This finding is similar to one from a study by Wen [29]. Forty-five percent of students agreed that peers' ratings will be higher than the teacher's rating. It means fifty percent of students think they will be lower. Students seemed to have a conservative view about peers' ratings. It could be that in their past experience, students have experienced peers being strict, unprofessional, or distrustful.

Before starting peer assessment, the teacher should know the students' motivations and expectations for the grading system. Raters, ratees, the preparation, and training in grading are all factors that impact peer assessment. Ratees can have different learning goals. Some students only want to pass, while other students want to get higher grades to apply for scholarships. Chasing different goals means that they have different motivations. It is for these reasons that researchers like Topping et. al. [4] and Nicol [11] suggested that certain procedures need to be followed when implementing peer assessment. Only when students actually know the nature and the well-preparation for assessment, would it be beneficial to them.

In peer assessment, students give and receive ideas from a peer. To give number grades is easier than to give evaluative feedback. And evaluative feedback is more important than grades, because feedback is more effective for students' learning than grading.

Students' provision of feedback for the purpose of peer assessments needs to be guided. The guide or checklist can be tools to remind students to offer positive feedback, suggestion, and ideas. Good feedback empowers students. But feedback messages can be complex and hard to decode. Planning different training stages was needed [4]. Factors affecting the objectivity of assessments are familiarity, the length of time to assess, and ratees' mature manner. It was found that the older the raters are, the more mature they are to rate work. Establishing the mechanism of peer assessment requires steps such as announcing the meaning of peer assessment, clarifying the way to provide feedback, exchanging ideas about observing work, and finally rating the work. The students, answering the open-ended question in the study, suggested grades be assigned with rankings instead out of 100. It is suggested that averaging the grades of more graders would lower personal bias.

It is not easy to make broad statements or recommend standards in a social science field because generally there is no standard answer. The field is diverse. One way to help with this is if students can engage in the discussion to build the assessing standard. That would really be student-centered learning. It may interest the students to join the criteria discussion and promote interaction between teachers with students. Interaction between teachers and students plays an important role in the 21st century. Besides, given an objective comment, rubrics could be taken into consideration.

5.3   Drawbacks of Peer Assessment

Some feedback is positive, while others are not. Negative feedback might be not satisfied with some students. They engaged in denial, finding excuses to protect themselves. And some raters are not responsible or serious, which contributes to the controversy about grades and peer feedback.

Bias was felt by some students. Because peer assessment in this study was not anonymous, raters could have been influenced by other factors such as an unfriendly relationship, close friendship, familiarity, etc. Anonymity was recommended to improve the next peer assessment. It will promote the objectivity of the system. Also some students in the study expressed that the raters are irresponsible and impatient. If so, it would lower the quality of individual peer assessments.

The students suggested that the number of works to be rated should be controlled. One rater suggested that they should not rate more than five pieces of work, or they would lose patience.

Drawbacks of peer assessment were mentioned by the students. The most challenging drawback was assessment objectivity. It was found that the error score is bigger for the teacher grade and self-grade, vs. the teacher grade and peer grade. Different graders might evaluate from different standards. Therefore, assessment criteria are needed to develop uniformity. This would let the students feel a sense of ownership and decrease any anxiety from perceived inequality. Training and education are needed, as well as following suggested best practice procedures [4][11].

In order to keep the work load reasonable for the students' peer rating, this study asked each rater to assess five pieces of work. Blind reviewing or double blind reviewing might decrease the amount of the raters' subjectivity. But there were some clues in the assignment that could not be covered in this study. The students' names on the assignments could not be covered in this case, and even if they could, handwriting could still be recognized by some students. To avoid familiarity between raters and ratees, the assignments should be considered and designed in advance.

Online peer assessment is more convenient. However, decoding online peer feedback is sometimes difficult because providing accurate, precise feedback via text alone is not easy. Written feedback can be misunderstood. When providing feedback face-to-face, body language and vocal expression and accent can provide additional cues to better explain meaning.

To develop an effective peer assessment, researchers need to be concerned about the target work to be assessed. It is necessary to prepare it well and make preparations step by step. If the teachers do not prepare well, the process is slowed down. An effective assessment depends on the faculty's philosophy and wisdom of teaching.

6   Conclusion

Three conclusions are drawn: (1) Among the self-assessment, peer assessment, and teacher assessment, there is a significantly difference. Self-grading scores seem to be over-estimated. Peer rating scores seem to be lower than self-rating ones. The grading by the teacher and peers is more consistent. (2)According to the students' expression for the merits of peer assessment, it seems that they are aware of it. Over half of the students approved the peer assessment and confirm it for their learning. (3) Attitudes expressed by the students about online peer assessment include both pro and against.

Based on the students' awareness and experience of peer assessment, it opens another learning journey for them. But the present results are not yet to be generalized to other student populations. Therefore, it is suggested that there be a further experiment with control and experimental groups to explore the effects of peer feedback on self-regulation.

In the future we would like to continue this research with a larger and more varied sample of students (many colleges/universities participating) and types of assignments so that conclusions can be extended to greater levels.

References

1. Banta, T.W., Lund, J.P., Black, K.E., Oblander, W.: Assessment in Practice: putting principles to work on college, p. 17. Jossey-Bass Publishers, San Francisco (1996)
2. Lu, J., Law, N.: Online peer assessment: effects of cognitive and affective feedback. Instr. Sci. **40**, 257–275 (2012)

3. Tseng, S.C., Tsai, C.C.: On-line peer assessment and the role of the peer feedback: A study of high school computer course. Comput. & Educ. **49**(4), 1161–1174 (2007)
4. Topping, J.: Peer assessment. Theor. into Pract. **48**, 20–27 (2009)
5. van Zundert, M., Sluijsmans, D., van Merrienboer, J.: Effective peer assessment processes: Research findings and future directions. Learn. and Instr. **20**, 270–279 (2010)
6. Kollar, I., Fisher, F.: Peer assessment as collaborative learning : A cognitive perspective. Learn. and Instr. **20**, 344–348 (2010)
7. Lu, R., Bol, L.: A comparison of anonymous versus identifiable e-peer review on college student writing performance and the extent of critical feedback. J. of Interac. Online Learn. **6**(2), 100–115 (2007)
8. Schunk, D.H., Zimmerman, B.J.: Social origins of self-regulatory competence. Educ. Psychol. **32**, 195–208 (1997)
9. Guimette, J.H.: The power of peer learning : networks and development cooperation. Academic Foundation, New Delhi (2007)
10. Hadwin, A.F., Jarvela, S., Miller, M.: Self-regulated, co-regulated, and socially shared regulation of learning. In: Zimmerman, B., Schunk, D.H. (eds.) Handbook of self-regulation of learning and performance, pp. 65–84. Routledge, New York (2011)
11. Nicol, D.J., Macfarlane-Dick, D.: Formative assessment and self-regulated learning : a model and seven principles of good feedback practice. Stud. in High. Educ. **31**(2), 199–218 (2006)
12. Zimmerman, B., Schunk, D.H.: Self-regulated learning and performance- an introduction and overview. In: Zimmerman, B., Schunk, D.H. (eds.) Handbook of Self-Regulation of Learning and Performance, pp. 1–12. Rouledge, New York (2011)
13. Zumbrunnm, S., Tadlock, J., Roberts, E.D.: Encouraging self-regulated learning in the classroom: a review of the literature. MERC. Virginia Commonwealth University (2011)
14. Wigfield, A., Klauda, S.L., Cambria, J.: Influences on the development of academic self-regulatory process. In: Zimmerman, B., Schunk, D.H. (eds.) Handbook of Self-Regulation of Learning and Performance, pp. 33–48. Rouledge, New York (2011)
15. Labuhn, A.S., Zimmerman, B.J., Hasselhorn, M.: Enhancing students' self-regulation and mathematics performance: The influence of feedback and self-evaluative standards. Metacog. and Learn. **5**(2), 173–194 (2010)
16. Liu, N.F., Carless, D.: Peer feedback: the learning element of peer assessment. Teach. in High. Educ. **11**(3), 279–290 (2006)
17. Stiggins, R., Chappuis, J.: An introduction to student-involved assessment for learning, pp. 29–30. Pearson, Boston (2012)
18. Andrade, H., Valtcheva, A.: Promoting learning and achievement through self-assessment. Theo. Into Prac. **48**, 12–19 (2009)
19. Shen, C.H., Huang, X.Y.: The application of Moodle for Web-Based Peer Assessment. J. of Educ. Media & Libr. Sci. **43**(3), 267–284 (2006)
20. Sadler, P.M., Good, E.: The impact of self-and peer-grading on student learning. Educa. Assess. **11**(1), 1–31 (2006)
21. Lin, S.S.J., Liu, E.Z.F., Yuan, S.M.: Web-based peer assessment: feedback for students with various thinking-styles. J. of Compu. Assis. Learn. **17**, 420–432 (2001)
22. Smith, R.A.: Are peer ratings of student debates valid? Teach. of Psycho. **17**, 188–189 (1990)
23. Hughes, I.E., Large, B.J.: Staff and peer-group assessment of oral communication skills. Stud. in High. Educ. **18**, 379–385 (1993)
24. Magin, D.: Reciprocity as a source of bias in multiple peer assessment of group work. Stud. in High. Educ. **26**, 53–63 (2001)

25. Dweck, C.: Self-theories: Their Role in Motivation, Personality and Development. Psychology Press, Philadelphia (1999)
26. Winne, P.H.: Self-regulation is ubiquitous but its forms vary with knowledge. Educa. Psycho. **30**, 223–228 (1995)
27. Strang, K.D.: Exploring summative peer assessment during a hybrid undergraduate supply chain course using Moodle. In: Carter, H., Gosper, M., Hedberg, J. (eds.) Electric Dreams. Proceedings ascilite, Sydney, pp. 840-853 (2013)
28. Zimmerman, B.J., Kitsantas, A.: Acquiring writing revision and self-regulatory skill through observation and emulation. J. of Educ. Psych. **94**, 660–668 (2002)
29. Wen, M.L., Tsai, C.C.: University students' perception of and attitudes toward (online) peer assessment. High. Educ. **51**, 27–44 (2006)

Network Analysis of Comorbidities: Case Study of HIV/AIDS in Taiwan

Yi-Horng Lai[✉]

Department of Health Care Administration,
Oriental Institute of Technology, New Taipei City, Taiwan
FL006@mail.oit.edu.tw

Abstract. Comorbidities are the presence of one or more additional disorders or diseases co-occurring with a primary disease or disorder. The purpose of this study is to identify diseases that co-occur with HIV/AIDS and analyze the gender differences. Data was collected from 536 HIV/AIDS admission medical records out of 1,377,469 admission medical records from 1997 to 2010 in Taiwan. In this study, the comorbidity relationships are presented in the phenotypic disease network (PDN), and φ-correlation is used to measure the distance between two diseases on the network. The results show that there is a high correlation in the following pairs/triad of diseases: human immunodeficiency virus infection with specified conditions (042) and pneumocystosis pneumonia (1363), human immunodeficiency virus infection with specified malignant neoplasms (0422) and kaposi's sarcoma of other specified sites (1768), human immunodeficiency virus acquired immunodeficiency syndrome, and unspecified (0429) and progressive multifocal leukoencephalopathy (0463), and lastly, human immunodeficiency virus infection with specified infections (0420), meningoencephalitis due to toxoplasmosis (1300), and human immunodeficiency virus infection specified infections causing other specified infections (0421).

Keywords: Phenotypic disease network (PDN) · HIV/AIDS · Network analysis

1 Introduction

The human immunodeficiency virus infection and acquired immune deficiency syndrome (HIV/AIDS) epidemic is one of the most important and crucial public health risks facing governments and civil societies in the world. Adolescents were at the center of the pandemic in terms of transmission, impact, and potential for changing the attitudes and behaviors that underlie this disease. Therefore, HIV/AIDS prevention has become a priority all over the world.

The first HIV/AIDS case in Taiwan was reported in 1984. As of the end of 2013, the total number of HIV/AIDS cases had been accumulated to 26,475. Faced with this serious situation, Taiwan's Centers for Disease Control worked with other departments and dedicated a tremendous amount of effort and resources to introduce harm reduction programs. Total reported cases dropped in 2006, which was the first trend

© Springer-Verlag Berlin Heidelberg 2015
L. Wang et al. (Eds): MISNC 2015, CCIS 540, pp. 174–186, 2015.
DOI: 10.1007/978-3-662-48319-0_14

reversal since 1984. In 2008 and thereafter, the epidemic took a turn; infections mainly occurred through sexual encounter [1].

There are no clear boundaries between many diseases, as diseases can have multiple causes and can be related in several dimensions. From a genetic perspective, a pair of diseases can be related because they have both been associated with the same gene, whereas from a proteomic perspective, diseases can be related because disease-associated proteins act on the same pathway [2].

During the past half-decade, several resources have been constructed to help understand the entangled origins of many diseases. Many of these resources have been presented as networks in which interactions between disease-associated genes, proteins, and expression patterns have been summarized. Goh, Cusick, Valle, Barton, Vidal, and Barabási created a network of Mendelian gene-disease associations by connecting diseases that have been associated with the same genes [3]. Besides, more and more studies have applied the network approach in diseases, such as neurodegenerative diseases [4], infertility etiologies [5], SARS, and HIV/AIDS [6].

A comorbidity relationship exists between two diseases whenever they affect the same individual substantially more than chance alone. In the past, comorbidities have been used extensively to construct synthetic scales for mortality prediction [7, 8], yet their utility exceed their current use. Studying the structure defined by entire sets of comorbidities might help the understanding of many biological and medical questions from a perspective that is complementary to other approaches. For example, a recent study built a comorbidity network in an attempt to elucidate neurological diseases' common genetic origins [9]. Heretofore, however, neither this data nor the data necessary to explore relationships between all diseases is currently available to the research community.

This present study decided to provide this data in the form of a phenotypic disease network (PDN) that includes all diseases recorded in the medical claims. Additionally, this study illustrates how a PDN can be used to study disease progression from a network perspective by interpreting the PDN as the landscape where disease progression occurs and shows how the network can be used to study phenotypic differences between patients of different demographic backgrounds. This study indicates the directionality of disease progression, as observed in our dataset, and finds out that more central disease in the PDN are more likely to occur after other diseases and that more peripheral diseases tend to precede other illnesses.

In order to guide HIV/AIDS-related diseases prevention program, this study conducted the PDN of HIV/AIDS to explore the relationship between HIV/AIDS and other diseases. The objective of this study is to identify diseases that are highly correlated with HIV/AIDS, and discuss gender differences.

2 Materials and Method

2.1 Data Source and Sample

The National Health Insurance (NHI) program was initiated in Taiwan in 1995 and covers nearly all residents. In 1999, the Bureau of NHI began to release all claims data in electronic form to the public under the National Health Insurance Research Database (NHIRD) project. The structure of the claim files is described in detail on the NHIRD website and in other publications [10].

Table 1. Data characteristics of admission medical records in this study

Variable		N	%
Gender	Female	701528	50.93
	Male	673985	48.93
	Unknown	1956	.14
Age	~19	219762	15.95
	20~21	180822	13.13
	30~39	182862	13.28
	40~49	169302	12.29
	50~59	171601	12.46
	60~69	168399	12.23
	71~	284721	20.67
Year	1997	71925	5.22
	1998	74422	5.40
	1999	79087	5.74
	2000	84089	6.10
	2001	89181	6.47
	2002	95232	6.91
	2003	91465	6.64
	2004	108586	7.88
	2005	117954	8.56
	2006	114446	8.31
	2007	113092	8.21
	2008	111843	8.12
	2009	113112	8.21
	2010	113035	8.21
HIV/AIDS	Yes	536	.04
	No	1376933	99.96
		1377469	100.00

NHIRD offers reliable, systematic, and complete data for disease detection. The datasets contained only the visit files, including dates, medical care facilities and specialties, patients' genders, dates of birth, and the four major diagnoses coded in the International Classification of Disease, 9th Revision, Clinical Modification (ICD-9-CM) format [10, 11]. In total, the ICD-9-CM classification consists of 657 different categories at the 3 digit level and 16,459 categories at 5 digits. Human immunodeficiency virus infection and acquired immune deficiency syndrome (HIV/AIDS) is coded 042 as Human immunodeficiency virus (HIV) infection disease, 0420 as human immunodeficiency virus infection with specified infections, 0421 as specified infections causing other specified infections, 0422 as human immunodeficiency virus infection with specified malignant neoplasms, and 0429 as acquired immunodeficiency syndrome, and unspecified. To protect privacy, the data on patient identities and institutions had been scrambled cryptographically.

The visit files in this study represented 1,377,469 admission activities within the NHI from 1997 to 2010. Demographically, the data set consists of 1,377,469 admission medical records from 1,000,000 patients. Of all these patients, 50.93% were females, 48.93 were males, 20.67% were over 71 years of age, and 536 persons were diagnosed with HIV/AIDS (Table 1).

Table 2. Data characteristics statistics of HIV/AIDS admission medical records in this study.

Variable		N	%
Gender	Female	45	8.40
	Male	491	91.60
Age	~19	12	2.24
	20~21	118	22.01
	30~39	168	31.34
	40~49	149	27.80
	50~59	68	12.69
	60~69	9	1.68
	71~	12	2.24
Year	1997	3	0.56
	1998	4	0.75
	1999	10	1.87
	2000	6	1.12
	2001	15	2.80
	2002	22	4.10
	2003	16	2.99
	2004	40	7.46
	2005	69	12.87
	2006	78	14.55
	2007	56	10.45
	2008	62	11.57
	2009	70	13.06
	2010	85	15.86
HIV/AIDS	042	461	86.01
	0420	21	3.92
	0421	3	.56
	0422	4	.75
	0429	47	8.77
		536	100.00

These HIV/AIDS admission medical records included 45 females (8.40%) and 491 males (91.60%). Of all the 536 HIV/AIDS records, 461 (86.01%) had major diagnoses coded 042, 21 (3.92%) had major diagnoses coded 0420, 3 (.56%) had primary diagnoses coded 0421, and 7 (8.77%) had major diagnoses coded 0429 (Table 2).

2.2   Data Limitations

There are several limitations to the current study. First, although data gathered from NHIRD is comprehensive and reliable, there are still some mistakes that the system couldn't find, such as code entry errors. These errors may be carried into data pre-processing, and it is beyond the control of this study. Second, the data was not up-to-date. Although future researchers are still recommended to apply the method of this study to analyze the characteristics of patients for the purpose of disease prevention, changes of medical treatments and other factors should be considered. Third, this study does not have a global sample, so there might be limits to replicate the findings of this study in all the other countries. It might, however, be generalized to other ethnic Chinese population due to the similarity in genes and physiology.

2.3   Measure of the Strength of Comorbidity Relationships

To measure the comorbidity relationships, it is necessary to quantify the strength of comorbidities by introducing a notion of distance between two diseases. A difficulty of this approach is that different statistical distance measures have biases that over- or under-estimate the relationships between rare or prevalent diseases. These biases are important given that the number of times a particular disease is diagnosed, such as its prevalence, follows a heavy tailed distribution [2], meaning that while most diseases are rarely diagnosed, a few diseases have been diagnosed in a large fraction of the population.

In this study, the φ-correlation is used to quantify the distance between two diseases. The φ-correlation, which is Pearson's correlation for binary variables, can be expressed mathematically as [2, 12]:

$$\phi_{ij} = \frac{C_{ij}N - P_iP_j}{\sqrt{P_iP_j(N - P_i)(N - P_j)}} \tag{1}$$

where C_{ij} is the number of patients affected by both diseases, N is the total number of patients in the population and P_i and P_j are the prevalence of diseases i and j. The distribution of φ values representing all disease pairs where $C_{ij}>0$ is presented in Fig 1, 3 and 5.

2.4   Network Approach

This research utilizes data from NHIRD to obtain the four major diagnoses codes of all patients. This study calculates φ-correlation with Equation 1. Pajek 4.03 program was used to compute the compute the degree of centrality and betweenness of each node and the path value (φ-correlation). This study is focused on the path between each disease as network and the correlation as the value of line (path weight), and they could be affected by different populations, which indicates differences in gender for each population.

3   Results

It can be summarized the set of all comorbidity associations between all diseases expressed in the study population by constructing a Phenotypic Disease Network (PDN). In the PDN, nodes are disease phenotypes identified by unique ICD9 codes, and links phenotypes that show significant comorbidity according to the measures introduced above.

In principle, the number of disease-disease associations in the PDN is proportional to the square of the number of phenotypes, yet many of these associations are either not strong or are not statistically significant [2]. The structure of the PDN can be explored by focusing on the strongest and the most significant of these associations. The PDN can be seen as a network of the phenotypic space. This network allows people to understand the relationship between illnesses.

3.1   The PDNs of HIV/AIDS

The distribution of φ-values representing all disease pairs is presented in Figure 1. Most of them are between .000 and .005. A discussion on the confidence interval and statistical significance of these measures can be found in Hidalgo, Blumm, Barabási, and Christakis's study, and φ-correlation > .06 is statistically significant [2].

In Figure 2, nodes are diseases; links are correlations. Node color identifies the ICD9 category; node size is proportional to disease prevalence. Link color indicates correlation strength. The PDN built using φ-correlation. All statistically significant links where φ-correlation>.06 are shown here.

Human immunodeficiency virus infection with specified conditions (042) and pneumocystosis pneumonia (1363), cryptococcal meningitis (3210), candidiasis of mouth (1120), unspecified secondary syphilis (0919), kaposi's sarcoma of unspecified (1769), cryptococcosis (1175), kaposi's sarcoma of palate (1762), neurosyphilis, unspecified (0949), kaposi's sarcoma of lung (1764), kaposi's sarcoma of lymph nodes (1765), and late syphilis, latent (096) are highly correlated. Those φ-correlations are all above .06.

The relationship between human immunodeficiency virus infection with specified conditions (042) and pneumocystosis pneumonia (1363) is the strongest. Pneumocystis pneumonia is a form of pneumonia, caused by the yeast-like fungus Pneumocystis jirovecii. Pneumocystis pneumonia is not commonly found in the lungs of healthy people, but, being a source of opportunistic infection, it can cause a lung infection in people with a weak immune system [13].

Human immunodeficiency virus infection with specified infections (0420) and human immunodeficiency virus infection specified infections causing other specified infections (0421), kaschin-beck disease (7160), pneumocystosis (1363), meningoencephalitis due to toxoplasmosis (1300), falciparum malaria (0840), kaposi's sarcoma of other specified sites (1768), with specified malignant neoplasms (0422) are highly correlated. Those φ-correlations are all above .06.

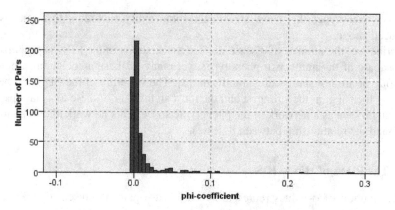

Fig. 1. Distribution of the φ-correlation between all disease pairs.

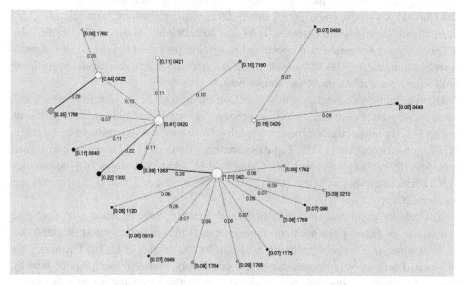

Fig. 2. The PDNs of HIV/AIDS, φ-correlations > .06. [ ] is the summary of φ-correlations with other node (diseases) of this node (disease), and the size of the node is dependent on the value. Such as the summary of φ-correlations with other diseases of 042 is 1.01. Node color is based on the ICD9 Category.

The φ-correlation of human immunodeficiency virus infection with specified infections (0420) and meningoencephalitis due to toxoplasmosis (1300) is highest. Toxoplasmosis is a parasitic disease caused by the protozoan Toxoplasma gondii. The parasite infects most genera of warm-blooded animals, including humans, but the primary host is the felid family. Infection occurs by eating infected meat, particularly swine products. By ingesting water, soil, or food that has come into contact with infected animals' fecal matter [14]. Besides, human immunodeficiency virus infection with specified infections (0420) and Specified infections causing other specified infections (0421) is highly correlated, and the φ-correlation is .11.

3.2   The PDNs of HIV/AIDS for Females

Human immunodeficiency virus infection with specified malignant neoplasms (0422) and kaposi's sarcoma of skin (1760), human immunodeficiency virus infection with specified infections (0420), and kaposi's sarcoma of other specified sites (1768) are highly correlated. Those φ-correlations are all above .06.

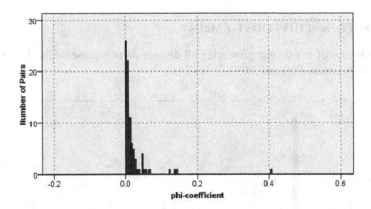

Fig. 3. Distribution of theφ-correlation between all disease pairs for female.

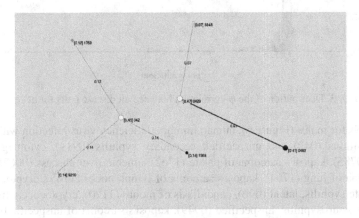

Fig. 4. The PDNs of HIV/AIDS for female, φ-correlations > .06.

The φ-correlation of Human immunodeficiency virus infection with specified malignant neoplasms (0422) and kaposi's sarcoma of other specified sites (1768) is the highest.

Human immunodeficiency virus acquired immunodeficiency syndrome, and unspecified (0429) and progressive multifocal leukoencephalopathy (0463) is highly correlated. Those φ-correlations are all above .06.

The distribution of w values representing all disease pairs is presented in Figure 3. Most of them are between .000 and .005.

In PDNs for females (Figure 4), human immunodeficiency virus infection with specified conditions (042) and cryptococcal meningitis (3210), kaposi's sarcoma of unspecified (1769), and pneumocystosis (1363) are highly correlated. Human immunodeficiency virus acquired immunodeficiency syndrome, unspecified (0429) and other cerebellar ataxia (3343), and progressive multifocal leukoencephalopathy (0463) are highly correlated. Those φ-correlations are all above .06.

3.3   The PDNs of HIV/AIDS for Males

The distribution of w values representing all disease pairs is presented as Fig. 5. Most of them are between .000 and .005.

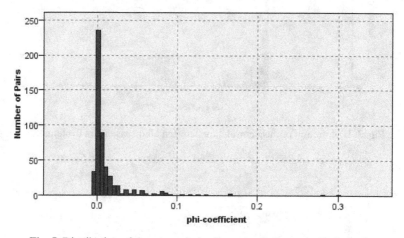

Fig. 5. Distribution of the φ-correlation between all disease pairs for male.

In PDNs for males (Figure 6), human immunodeficiency virus infection with specified conditions (042) and unspecified secondary syphilis (0919), cytomegaloviral disease (0785), kaposi's sarcoma of palate (1762), amebic liver abscess (0063), kaposi's sarcoma of lung (1764), kaposi's sarcoma of lymph nodes (1765), cryptococcosis (1175), late syphilis, latent (096), candidiasis of mouth (1120), cryptococcal meningitis (3210), neurosyphilis, unspecified (0949), kaposi's sarcoma of unspecified (1769), and pneumocystosis (1363) are highly correlated. Those φ-correlations are all above .06. The φ-correlation of human immunodeficiency virus infection with specified conditions (042) with pneumocystosis (1363) is highest.

Human immunodeficiency virus infection with specified infections (0420) and specified infections causing other specified infections (0421), meningoencephalitis due to toxoplasmosis (1300), pneumocystosis (1363), kaschin-beck disease (7160), kaposi's sarcoma of other specified sites (1768), with specified malignant neoplasms (0422), and falciparum malaria (0840) are highly correlation. Those φ-correlations are all above .06. The φ-correlation of specified infections causing other specified infections (0421) with human immunodeficiency virus infection with specified infections (0420) is highest.

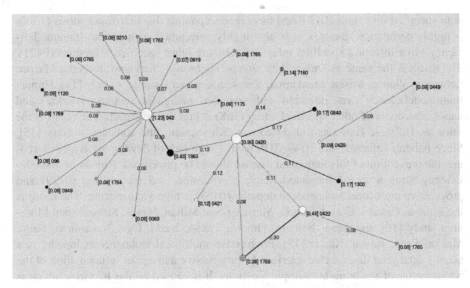

Fig. 6. The PDNs of HIV/AIDS for male, φ-correlations > .06.

Human immunodeficiency virus infection with specified malignant neoplasms (0422) and kaposi's sarcoma of skin (1760), human immunodeficiency virus infection with specified infections (0420), and kaposi's sarcoma of other specified sites (1768) are highly correlation. The φ-correlation of Human immunodeficiency virus infection with specified malignant neoplasms (0422) and kaposi's sarcoma of other specified sites (1768) is highest one.

Human immunodeficiency virus acquired immunodeficiency syndrome, unspecified (0429) and HIV infection, unspecified (0449) is highly correlation, and the φ-correlation is .09.

4   Conclusions

Through the PDN, this paper has identified the diseases that are associated with HIV/AIDS. It could be showed that the PDN has a complex structure where some diseases are highly connected while others are barely connected at all. While not conclusive, these observations can explain the observation that more connected diseases are seen to be more lethal, as patients developing highly connected diseases are more likely those at an advanced stage of disease, which can be reached through multiple paths in the PDN.

The findings suggest that human immunodeficiency virus infection with specified conditions (042) and pneumocystosis pneumonia is highly correlated (1363). This result is consistent with Aliouat-Denis, Chabé, Demanche, Aliouat el, Viscogliosi, Guillot, Delhaes, and Dei-Cas's study [13]. Pneumocystis pneumonia is especially seen in people with cancer undergoing chemotherapy, HIV/AIDS, and the use of medications that suppress the immune system. Human immunodeficiency virus infection with specified infections (0420) and meningoencephalitis due to toxoplasmosis (1300) is highly correlation. Besides, it is also highly correlation with human immunodeficiency virus infection specified infections causing other specified infections (0421). The result is the same as Dubey, Hill, Jones, Hightower, Kirkland, Roberts, Marcet, Lehmann, Vianna, Miska, Sreekumar, Kwok, Shen, and Gamble''s study [14]. Human immunodeficiency virus infection with specified malignant neoplasms (0422) and kaposi's sarcoma of other specified sites (1768) is highly correlation. The result is the same as Holmes, Hawson, Liu, Friedman, Khiabanian, and Rabadan's study [15]. Since patients infected with HIV/AIDS have a high risk of developing Kaposi sarcoma, the prevention of this malignant disease should be prioritized. Human immunodeficiency virus acquired immunodeficiency syndrome, and unspecified (0429) and progressive multifocal leukoencephalopathy (0463) is highly correlation. The result is the same as Casado, Corral, García, Martinez-San Millán, Navas, Moreno, and Moreno's study [16] and Sano, Nakano, Omoto, Takao, Ikeda, Oga, Nakamichi, Saijo, Maoka, Sano, Kawai, Kanda [17]. Progressive multifocal leukoencephalopathy is a usually fatal viral disease characterized by progressive damage or inflammation of the white matter of the brain at multiple locations. It is caused by the JC virus, which is normally present and kept under control by the immune system. JC virus is harmless except in cases of weakened immune systems. Progressive multifocal leukoencephalopathy occurs almost exclusively in patients with severe immune deficiency, most commonly among patients with acquired immune deficiency syndrome, but people on chronic immunosuppressive medications including chemotherapy are also at increased risk of progressive multifocal leukoencephalopathy [16, 17].

For females, human immunodeficiency virus infection with specified conditions (042) and cryptococcal meningitis (3210), kaposi's sarcoma of unspecified (1769), and pneumocystosis (1363) are highly correlation. Human immunodeficiency virus acquired immunodeficiency syndrome, unspecified (0429) and other cerebellar ataxia (3343), and progressive multifocal leukoencephalopathy (0463) are highly correlation. For males, human immunodeficiency virus infection with specified conditions (042) and pneumocystosis (1363) is highly correlation. Human immunodeficiency virus infection with specified infections (0420) and meningoencephalitis due to toxoplasmosis (1300), and falciparum malaria (0840) are highly correlation. Human immunodeficiency virus infection with specified malignant neoplasms (0422) and kaposi's sarcoma of other specified sites (1768) is highly correlation. Human immunodeficiency virus acquired immunodeficiency syndrome, unspecified (0429) and HIV infection, unspecified (0449) is highly correlation.

Exploring comorbidities from a network perspective could help determine whether differences in the comorbidity patterns expressed in different populations indicate differences in races, country, or socioeconomic status for each population. Here this study show as an initially stage that there are differences in the strength of comorbidities measured for patients of different gender. The PDN could be the starting point of studies exploring these and related questions.

Acknowledgements. This study is based in part on data from the National Health Insurance Research Database provided by the Bureau of National Health Insurance, Department of Health and managed by National Health Research Institutes (NHRI). The interpretation and conclusions contained herein do not represent those of Bureau of National Health Insurance, Department of Health or National Health Research Institutes.

References

1. Taiwan Centers for Disease Control: Communicable Diseases & Prevention-HIV/AIDS, Health topics. http://www.cdc.gov.tw (retrieved January 20, 2015)
2. Hidalgo, C.A., Blumm, N., Barabási, A.L., Christakis, N.A.: A Dynamic Network Approach for the Study of Human Phenotypes. PLoS Computational Biology **5**(4), 1–11 (2009)
3. Goh, K.I., Cusick, M.E., Valle, D., Barton, C., Vidal, M., Barabási, A.L.: The Human Disease Network. Proceedings of the National Academy of Sciences of the United States of America **104**(21), 8685–8690 (2007)
4. Santiago, J.A., Potashkin, J.A.: A Network Approach To Clinical Intervention In Neurodegenerative Diseases. Trends in Molecular Medicine **20**(12), 694–703 (2014)
5. Tarín, J.J., García-Pérez, M.A. Hamatani, T., Cano, A.: Infertility Etiologies Are Genetically and Clinically Linked With Other Diseases in Single Meta-Diseases. Reproductive Biology and Endocrinology **13** (2015). http://www.rbej.com/content/13/1/31 (retrieved April 20, 2015)
6. Moni, M.A., Liò, P.: Network-Based Analysis of Comorbidities Risk During an Infection: SARS and HIV Case Studies. BMC Bioinformatics **15**, 1471–2105 (2014)
7. Iezzoni, L.I., Heeren, T., Foley, S.M., Daley, J., Hughes, J., Coffman, G.A.: Chronic Conditions and Risk of In-Hospital Death. Health Services Research **29**(4), 435–460 (1994)
8. Schneeweiss, S., Wang, P.S., Avorn, J., Glynn, R.J.: Improved Comorbidity Adjustment for Predicting Mortality in Medicare Populations. Health Services Research **38**(4), 1103–1120 (2003)
9. Rzhetsky, A., Wajngurt, D., Park, N., Zheng, T.: Probing Genetic Overlap among Complex Human Phenotypes. Proceedings of the National Academy of Sciences of the United States of America **104**(28), 11694–11699 (2007)
10. National Health Research Institutes: National Health Insurance Research Database (NHIRD). http://w3.nhri.org.tw/nhird/ (retrieved December 1, 2014)
11. Centers for Disease Control and Prevention. International Classification of Diseases, Ninth Revision, Clinical Modification (ICD-9-CM). http://www.cdc.gov/nchs/icd/icd9cm.htm (retrieved December 1, 2014)
12. Ekström, J.: The Phi-coefficient, the Tetrachoric Correlation Coefficient, and the Pearson-Yule Debate. http://statistics.ucla.edu/preprints/uclastat-preprint-2008:40 (retrieved February 20, 2015)
13. Aliouat-Denis, C.M., Chabé, M., Demanche, C., el Aliouat, M., Viscogliosi, E., Guillot, J., Delhaes, L., Dei-Cas, E.: Pneumocystis Species, Co-Evolution and Pathogenic Power. Infection, Genetics and Evolution **8**(5), 708–726 (2008)

14. Dubey, J.P., Hill, D.E., Jones, J.L., Hightower, A.W., Kirkland, E., Roberts, J.M., Marcet, P.L., Lehmann, T., Vianna, M.C., Miska, K., Sreekumar, C., Kwok, O.C., Shen, S.K., Gamble, H.R.: Prevalence Of Viable Toxoplasma Gondii in Beef, Chicken, and Pork from Retail Meat Stores in The United States: Risk Assessment to Consumers. The Journal of Parasitology **91**(5), 1082–1093 (2005)
15. Holmes, A.B., Hawson, A., Liu, F., Friedman, C., Khiabanian, H., Rabadan, R.: Discovering Disease Associations by Integrating Electronic Clinical Data and Medical Literature. PLoS ONE **6**(6), 1–11 (2011)
16. Casado, J.L., Corral, I., García, J.: Martinez-San Millán, J., Navas, E., Moreno, A., & Moreno, S.: Continued Declining Incidence and Improved Survival of Progressive Multifocal Leukoencephalopathy in HIV/AIDS Patients in The Current Era. European Journal of Clinical Microbiology & Infectious Diseases **33**(2), 179–187 (2014)
17. Sano, Y., Nakano, Y., Omoto, M., Takao, M., Ikeda, E., Oga, A., Nakamichi, K., Saijo, M., Maoka, M., Sano, H., Kawai, M., Kanda, T.: Rituximab-associated Progressive Multifocal Leukoencephalopathy Derived from Non-Hodgkin Lymphoma: Neuropathological Findings and Results of Mefloquine Treatment. Internal Medicine **54**(8), 965–970 (2015)

Extending [K_1, K_2] Anonymization of Shortest Paths for Social Networks

Yu-Chuan Tsai[1], Shyue-Liang Wang[2]([⊠]), Tzung-Pei Hong[3], and Hung-Yu Kao[4]

[1] Library and Information Center, National University of Kaohsiung,
Kaohsiung 81148, Taiwan
yjtsai@nuk.edu.tw
[2] Department of Information Management,
National University of Kaohsiung, Kaohsiung 81148, Taiwan
slwang@nuk.edu.tw
[3] Department of Computer Science and Information Engineering,
National University of Kaohsiung, Kaohsiung 81148, Taiwan
tphong@nuk.edu.tw
[4] Department of Computer Science and Information Engineering,
National Cheng Kung University, Tainan 70101, Taiwan
hykao@mail.ncku.edu.tw

Abstract. Privacy is a great concern when information are published and shared. Privacy-preserving social network data publishing has been studied extensively in recent years. Early works had concentrated on protecting sensitive nodes and links information to prevent privacy breaches. Recent studies start to focus on preserving sensitive edge weight information such as shortest paths. Two types of privacy on sensitive shortest paths have been proposed. One type of privacy tried to add random noise edge weights to the graph but still maintain the same shortest path. The other privacy, *k-shortest path privacy*, minimally perturbed edge weights, so that there exists at least *k* shortest paths. However, there might be insufficient paths that can be modified to the same path length. In this work, we extend previously proposed *[k_1, k_2]-shortest path privacy*, $k_1 \leq k \leq k_2$, to not only anonymizing different number of shortest paths for different source and destination vertex pair, but also modifying different types of edges, such as partially visited edges. Numerical experiments showing the characteristics of the proposed algorithm is given. The proposed algorithm is more efficient in running time than the previous work with similar perturbed ratios of edges.

Keywords: Privacy preservation · Shortest path · Anonymity · *k-shortest path privacy* · *[k_1, k_2]-shortest path privacy*

1   Introduction

On-line social networking not only shares information with extreme ease, but also exposes personal identifiable information such as full name, email address, phone numbers, hobbies, and interests to public without notices. Privacy is a great concern

© Springer-Verlag Berlin Heidelberg 2015
L. Wang et al. (Eds): MISNC 2015, CCIS 540, pp. 187–199, 2015.
DOI: 10.1007/978-3-662-48319-0_15

when information are published and shared. In order to protect personal data from public breaches, national laws on freedom of information and on-line privacy protection acts have been proposed and implemented in many countries [4]. However, while these laws can deter or punish the offenders, but they cannot completely stop the infraction of privacy from publicly available information. Privacy-preserving social network publishing is intended to preserve privacy of individual information on on-line social networking websites and is developed to complement the laws on freedom of information.

A social network is a graph structure made up of entities, the connections between entities and the strength of connections. The entities represent the individuals and the connections between entities represent the relationships or interactions between entities. The edge weights represent the strength of the linked relationships [10], in which the positive edge weights express the trust relation, and the negative edge weights express the bad relation. [14] showed a sensitive shortest path protection example on a weighted social network, which representing an automotive business network between Japanese corporations and American suppliers in North America [9]. The lowest cost path (shortest path) might be desired to be preserved by the Japanese corporation as well, to receive most competitive suppliers. How to protect the sensitive business paths is a great concern in the competitive business environment these days.

In this work, we study the problem of anonymizing the sensitive shortest paths between given pairs of vertices on the weighted directed graph with positive-only edges and positive/negative edge weights. The previous works [19][20], *k-shortest path privacy* and *[k_1, k_2]-shortest path privacy*, made minimal perturbation of edge weights, so that there exists at least k or $k_1 \leq k \leq k_2$ shortest paths for all pairs of source and destination vertices respectively. However, it is quite possible that certain sensitive vertex pairs do not have enough paths (or edge weights) to achieve exactly *k-shortest paths privacy* or *[k_1, k_2]-shortest path privacy*. Current work further extends the *[k_1, k_2]-shortest path privacy*, $k_1 \leq k \leq k_2$, to modify edges of different types. Based on the greedy approach, we present an algorithm to modify two types of edges, namely None-Visited (*NV*), and partially-visited (*PV*) edges to achieve the *[k_1, k_2]-shortest path privacy*, $k_1 \leq k \leq k_2$.

The major contributions are summarized as follows:

- We extend the *[k_1, k_2]-shortest path privacy*, $k_1 \leq k \leq k_2$, and propose an algorithm to modify different types of edges, namely none-visited (*NV*) and partially-visited (*PV*), on two types of weighted graph, positive-only and positive/negative edge weight graphs.
- Based on two metrics: running time, and ratio of modified edges, we examined the performance on two types of weighted graphs, positive-only and positive/negative edge weight graphs. Comparison with previous work shows our proposed algorithm is more efficient in running time with similar perturbed ratios of edges.

The rest of the paper is organized as follows. Section 2 describes the related works. Section 3 gives the problem description. Section 4 presents the proposed algorithm. Section 5 reports the numerical experiments. Section 6 concludes the paper.

2   Related Works

Privacy preserving on social networks has concentrated on dealing with re-identification attacks of nodes and links [3][5][6][8][11][12][13][14][24][25]. In order to protect the nodes and link relationships before data is published, most of the techniques modify the graph by adding/deleting nodes/edges. The attackers basically use the background knowledge to attack the published networks, such as passive and active attack [1][2]. The passive attack doesn't change the structure of graph, and the active attack changes the structure of graph.

Weighted graphs can be used for analyzing the formation of communities within the network, business transaction networks, viral and targeted marketing and advertising [6][13][14][18][19][22]. Depending on the applications, the edge weights could be used to represent "degree of friendship", "ratios of opinion", and "business transaction", etc. In addition, edges in a social network could be directed with positive and negative weights that show complex linked relationships [10], with negative links representing the relationships as 'dislike", "distrust", "foes", and so on.

In order to protect the privacy of sensitive edges, four types of works have been proposed on weighted graphs. The first type of works tries to protect the shortest path characteristic between pairs of source and destination vertices. The shortest path remains to be the shortest path after all edge weights are minimally modified [6][14]. The second type of works tries to preserve the privacy of the weights of edges emitting from a given vertex within a predefined parameter, called *k-anonymous weight privacy* [13]. The third type of works studies the shortest distance computing in the cloud which aims at preventing outsourced graphs from neighborhood attacks [7]. The adversary in an outsourced graph cannot calculate the shortest path or shortest distance between neighboring nodes. The fourth type of works studied the *k-shortest path privacy* anonymization on social networks [19][20]. It extends *k-anonymity* concept on relational data to graph data and minimally perturbed edge weights so that there exists at least *k* shortest paths. There are other works on weighted graphs [15][21] that preserve the node identities. Liu et al. [15] proposed a generalization based anonymization approach to achieve *k-possible anonymity*, which used the edges generalization to achieving generalized anonymization groups, on weighted social networks. Yuan et al. [21] proposed a *k-weighted-degree anonymous* model and it prevented the attacks using the node's degree and weight information on the edges adjacent to the nodes as the background knowledge.

For the first two types of works that preserve the shortest paths between pairs of vertices, Gaussian randomization perturbation and greedy perturbation techniques that minimally modify the edge weights without adding or deleting any vertices and edges have been proposed [14]. A linear programming abstract model that can preserve linear properties of edge weights (including shortest paths) after anonymization is presented in [6]. These works do not change the property of the selected shortest path on the anonymized graph. Our work is similar to fourth type of works, but is an extension and more flexible. In our work, we achieve the shortest path privacy for sensitive vertex pairs by modifying two different types of edges simultaneously.

3   Problem Description

In this work, we study the problem of how to flexibly achieve anonymization of sensitive shortest paths between specified source and destination vertices on directed weighted graphs. An information network is represented as a graph $G=(V, E, W)$, where $V=\{v_1, v_2, ..., v_n\}$ represents a set of entities, and $E=\{e_{1i}, e_{2i}, ..., e_{nn}\}$ represents relationship between entities and $W=\{w_{1i}, w_{2i}, ..., w_{nn}\}$ represents strength of the relationships, which could be positive or negative.

For given target source and destination vertex pairs in H, the edges on the shortest paths may overlap with each other. Three types of edges can be classified according to their involvement in the shortest paths on a weighted graph [13][19]: *None-Visited* (*NV*) edges, *Partially-Visited* (*PV*) edges, and *All-Visited* (*AV*) edges. An edge $e_{i,j}$ is a *None-Visited* (*NV*) edge if the edge $e_{i,j}$ does not belong to any shortest path to be preserved. An edge $e_{i,j}$ is a *Partially-Visited* (*PV*) edge if some shortest paths (including those modified), but not all shortest paths, pass through the edge. An edge $e_{i,j}$ is an *All-Visited* (*AV*) edge if all shortest paths (including those modified) from all pairs of source and destination vertices pass through the edge. In this work, we use the definition in [19] and consider two types of edges, *NV* and *PV* edges, and each edge can be anonymized only once.

Figure 1 shows a weighted graph with seven vertices. There are two sensitive shortest paths, SP_1 and SP_2, for two specified sensitive vertex pairs in $H = \{(v_3, v_5), (v_2, v_6)\}$ respectively. The first shortest path, SP_1, is between v_3 and v_5, $\{e_{3,5}, e_{2,5}\}$. The second shortest path, SP_2, is between v_2 and v_6, $\{e_{2,5}, e_{5,6}\}$. The edges $e_{1,2}, e_{1,3}, e_{2,7}, e_{3,5}, e_{4,5}, e_{4,6}$ are *NV* edges in the initial state, in which they are not passed by any of the shortest paths SP_1 and SP_2 from target node pairs in H. The *PV* edges are those edges that are passed through by only one of the shortest paths SP_1 or SP_2, but not both, such as $e_{3,2}, e_{5,6}$, in the initial state. In this example, there is only one *AV* edge, $e_{2,5}$, as it is on both shortest paths SP_1 and SP_2. In practices, the *NV* edges could be the majority edge on a graph. Intuitively, *AV* edges could be rare in a graph. There are more *NV* edges than *PV* edges and more *PV* edges than *AV* edges. In addition, modifying the weights of *NV* edges only change the length of specific path, and modifying the weights of *PV* edges have side effects on other already anonymized shortest paths and perhaps new paths to be anonymized. In addition, the modified edges of each path might be overlapped and required the cyclic check process [18].

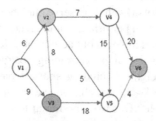

Fig. 1. The Original Network G

For previously proposed k-shortest path privacy [19] and with $k=4$ on Figure 1, there are only three paths each for both of two sensitive node pairs and there is no available *NV* edges to be modified for second sensitive vertex pair on its second shortest path. In such case, *k-shortest path privacy* cannot be achieved, unless artificial edges or vertices are added so that more paths become available for modification. For previous work [20] of $[k_1, k_2]$-*shortest path privacy*, when $2 \leq k \leq 4$ on Figure 1, both of these two nodes pairs have two paths respectively. However, there are no *NV* edges available to be modified.

As such, in this work, we propose to extend the $[k_1, k_2]$-*shortest path privacy*, in previous work, so that, when modifying edge weights, *NV* edges and *PV* edges can be considered simultaneously. The $[k_1, k_2]$-*shortest path privacy* and its privacy value are given as follows.

Definition 1. ($[k_1, k_2]$-*shortest path privacy*)
Given a graph G, a set of source and destination vertices H, a privacy level k, the graph G* satisfies $[k_1, k_2]$-shortest path privacy if there exists k_i shortest paths, $k_1 \leq k_i \leq k_2$, for i-th vertex pair specified in H.

According to the definition of $[k_1, k_2]$-*shortest path privacy*, $k_1 \leq k \leq k_2$, there might exist different numbers of anonymized paths for different source and destination vertex pair, so that we use the privacy value to evaluate the level of privacy as follows.

Definition 2. (*Privacy value of $[k_1, k_2]$-shortest path privacy*)
Given a graph G, a set of source and destination vertices H, a privacy level k, the privacy value of an anonymized graph G* that satisfies $[k_1, k_2]$-shortest path privacy is defined as:

$$privacy \quad value = \frac{1}{n} \sum_i \frac{1}{k_i}$$

The $[k_1, k_2]$-*shortest path privacy*, $k_1 \leq k \leq k_2$, can be applied to multiple sensitive vertex pairs on both weighted un-directed/directed graphs, and prevent the adversary to infer the true sensitive shortest path relationship between any vertex pair. In other words, we try to hide the true sensitive information by cloaking with other shortest paths, depending on how many paths exist between the given vertex pair.

According to the definition, the higher k value will result in lower privacy value, which implies more private and secure. If there is less than k shortest paths for certain vertex pair, compared to *k-shortest path privacy*, $[k_1, k_2]$-*shortest path privacy* can be anonymized, but *k-shortest path privacy* could not be anonymized by its definition.

4 Proposed Algorithm

This section presents an algorithm, *EKMP*, to achieve the $[k_1, k_2]$-*shortest path privacy*, $k_1 \leq k \leq k_2$, by considering and modifying two types of edges, *NV* and *PV* at the same time. The proposed greedy-based algorithm try to modify the *NV* and *PV* edge weights from the top k, $k_1 \leq k \leq k_2$, shortest paths so that they all possess the same path length after modification. We use the weighed-proportional-based strategy to modify the edge weights. For a path to be anonymized, reducing the edge weights of *NV* edges make the path length equal to the length of the shortest path, the path is anonymized. If it does not have *NV* edges, then will also consider *PV* edges. In addition, the modified edges of already anonymized paths might be overlapped and required the cyclic check process [18]. Due to some paths weights cannot be further reduced, the next longer path will be considered. The algorithm will return the k anonymized paths, $k_1 \leq k \leq k_2$. The algorithm will not try to modify the structure of the graph.

As an example, for $k_1 = 2$ and $k_2 = 4$, Figure 2 shows the anonymized result from Figure 1. There are two sensitive shortest paths, SP_1 and SP_2, for the two target vertex pairs in $H = \{(v_3, v_5), (v_2, v_6)\}$ respectively in Figure 1. The first shortest path, SP_1, is between v_3 and v_5, $\{e_{3,2}, e_{2,5}\}$. The second shortest path, SP_2, is between v_2 and v_6, $\{e_{2,5}, e_{5,6}\}$. Their shortest path lengths are 13 and 9, respectively. For first target vertex pair v_3 and v_5, the second shortest path is $P_{12} = \{e_{3,5}\}$ and the path length is 18. The difference between SP_1 and P_{12} is 5. The *NV* edges, $e_{3,5}$ is modified to 13. The third shortest path is $P_{13} = \{e_{3,2}, e_{2,4}, e_{4,5}\}$ and the path length is 30. The difference between SP_1 and P_{13} is 17. The *NV* edges are $e_{2,4}$ and $e_{4,5}$, and they are modified to 1.6 and 3.4, respectively. In first vertex pair, it only has two other paths. For second target vertex pair v_2 and v_6, the second shortest path is $P_{22} = \{e_{2,4}, e_{4,5}, e_{5,6}\}$ and the updated path lengths is 9. The difference between SP_2 and P_{22} is 0. So, this path has been done. The third shortest path is $P_{23} = \{e_{2,4}, e_{4,6}\}$ and the updated path length is 21.6. The difference between SP_2 and P_{23} is 12.6. The *NV* edge is $e_{4,6}$ and is modified to 7.4. The second vertex pair also has two other paths to be anonymized. For this example, according to our definition, the privacy value is 1/3.

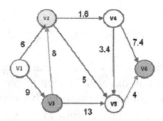

Fig. 2. The [2,4]-shortest Path Privacy Network

The proposed algorithm includes the following major steps.

1. In the initial state, for every vertex pair in H, it finds all paths between the vertices and arranges them in increasing order.
2. Select one target vertex pair and use the weighted-proportional strategy to modify the *NV* edge weights first so that the modified path has the same length as the sensitive target shortest path.
3. If the target path does not find have enough *NV* edges to modify, then we include the *PV* edges in the target path. Double check and update the length of all anonymized paths.
4. Repeat step two to four until all vertex pairs are processed.

Algorithm. [k_1, k_2]-Multiple Paths Anonymization Algorithm (*EKMP*)

Input: W, weighted adjacency matrix of a given graph G,

      H, the set of source and destination vertices for which the shortest paths are to be anonymized,

      k_1, the minimum number of shortest path between each pair of source and destination vertices,

      k_2, the maximum number of shortest path between each pair of source and destination vertices,

Output: anonymized weighted adjacency matrix W^*,

for (each distinct start vertices v_i of H)

//find the shortest path and top-k' shortest paths, where $k_1 \leqq k' \leqq k_2$

{ $SPL := SPL$ + shortest path $p_{i,j}$; //SPL: shortest path list for (v_i, v_j) in H

  $DL := DL + d_{i,j}$;   //DL: shortest paths length list for (v_i, v_j) in H

  If ($k_1 \leqq$ |top-k' shortest paths| $\leqq k_2$)

  { $APL := APL$ + top-k' shortest paths $tp_{i,j}$; //APL: top-k' shortest paths list for (v_i, v_j) in H order by path length increasing

    $PDL := PDL$ + top-k' shortest path lengths $td_{i,j}$;   //PDL: top-k_2 shortest path lengths list for each pair (v_i, v_j) in H order by path length increasing

    $k_p :=$ |top-k' shortest paths|;

  }

  Else { //|top-k' shortest paths| > k_2 or |top-k' shortest paths| < k_1

    If (|top-k' shortest paths| > k_2)

    { $APL := APL$ + top-k_2 shortest paths $tp_{i,j}$; //APL: top-k' shortest paths list for (v_i, v_j) in H order by path length increasing, where $k' = k_2$

      $PDL := PDL$ + top-k_2 shortest path lengths $td_{i,j}$;

      //PDL: top-k' shortest path lengths list for each pair (v_i, v_j) in H order by path length increasing, where $k' = k_2$

      $k_p :=$ |top-k' shortest paths|;

    }

    Else { continue; }

}//end of for, finding the shortest paths and top-k' shortest paths in H, where $k_1 \leqq k' \leqq k_2$

while ($H \neq \emptyset$) {

  $d_{r,s} := \min_H d_{i,j}$; //minimum of all shortest paths

  $H := H - (v_r, v_s)$;

  while (|$TSPL$| < k_p) {   //$TSPL$: there are at most k paths for current vertex pair

    $d'_{r,s} :=$ pop up first of $PDL_{r,s}$;

    p'r,s := pop up first of $APL_{r,s}$;

    If ($d'_{r,s} = d_{r,s}$) {   //same length

      $TSPL := TSPL + p'_{r,s}$; // add to anonymized list

      continue; }   //to find next shortest path

    Else {   // different length

      If ($d'_{r,s}$ and $d_{r,s}$ satisfy cyclic check) then continue;

        // $d'_{r,s}$ cannot be anonymized

      let $diff := d'_{r,s} - d_{r,s}$;   //the weight to be reduced

      $modified\_Process(p'_{r,s}, diff)$;

      //call a modifying procedure to do the weighted-proportional-process

    };   //end of else, different length

  };   // end of while (|$TSPL$|< k)

  $SPL := SPL + TSPL$;

};   // end of while ($H \neq \phi$)

$modified\_Process(p'_{r,s}, diff)$
$ML = NVL := p'_{r,s} - \{$edges in SPL and $TSPL\}$;   //NVL: the NV edges list
If $(ML \neq \phi)$ {   //exist NV edges to be modified
    for (each edge $(eML)_{i,j}$ on the ML) {   //reduce proportionally

$$w'_{i,j} = w_{i,j} - \frac{w_{i,j}}{\sum w_{i,j}} \times diff \; ;$$

    update the adjacency matrix;
      $TSPL := TSPL + p'_{r,s}$;   //save the modified path
    };   // end of for each edge $(eML)_{i,j}$
  } ;   // end of if, check NV edges to be modified
Else { //Use the PV edges to be modified
      $PV\_Process()$; // end of check PV edges to be modified
};   //end of if , check NV edges to be modified

$PV\_Process (p'_{r,s}, diff)$
$PVL :=$ edges in $SPL -$ edges in $TSPL$;   //find the PV edges that are not used;
$PML := PVL \cap p'_{r,s}$;   //consider the PV edges that are not modified in the selected path
If $(PML \neq \phi)$ {   // Use PV edges to be modified
    for (each edge $(eML)_{i,j}$ on the PML) {   // modified proportionally

$$w'_{i,j} = w_{i,j} - \frac{w_{i,j}}{\sum w_{i,j}} \times diff \; ;$$

    update the adjacency matrix;
      $TSPL := TSPL + p'_{r,s}$;   //save the modified path
    }; // end of for each edge $(eML)_{i,j}$
Update the path lengths in $TSPL$ and new length $uad_{i,j}$;
Update the path lengths in PDL and DL;
for (each path $AP_{i,j}$ in $(TSPL \cup SPL)$)
{ // updated the path lengths which had be reduced
  let $PVdiff = d_{i,j} - uad_{i,j}$;
  If $(PVdiff > 0)$ {   // re-modified the NV edges
    $RML :=$ edge $(eRML)_{i,j}$ on the path $AP_{i,j}$ in $(TSPL+SPL-H)$;
    for (each edge $(eRML)_{i,j}$ on the RML) {

$$w''_{i,j} = w'_{i,j} - \frac{w'_{i,j}}{\sum w_{i,j}} \times PVdiff \; ;$$

    update the adjacency matrix;
      };   //end of for each edge $(eRML)_{i,j}$
    }; //end of if, check the $PVdiff$
  };   //end of for each $AP_{i,j}$
};   // check PV edges to be modified

In lines 1 to 16 of the $EKMP$ algorithm, it first finds all the shortest paths and top-k' shortest paths for all source and destination vertices in H, where $k_1 \leq k' \leq k_2$. In lines 17 to 33, it performs the anonymization process using the first path of APL in increasing order of the path lengths for the selected vertices pair (v_r, v_s). In the *modified_Process()* procedure, it applies the weighted-proportional strategy to modify the edges weights for NV and PV edges. In the *modified_Process()* procedure, it applied the NV edges to modify the edges weights firstly. If certain path does not have any NV edges, then it called *PV_Process()* procedure to process this situation. In the *PV_Process()*, it applied the PV edges firstly in line 1 to 10, and after modifying PV

edges, due to the side effect, it only modifies the *NV* edges in lines 11 to 21. The *PV* edges are only used one time.

5   Numerical Experiments

To evaluate the performance of the proposed algorithm, we run simulations on a real world dataset, *hep-th* [16]. In the experiments, we randomly generate four target vertex pairs and construct sets of vertex pairs *H* for |*H*|=2, 3, respectively. All experiments reported in here were carried out on Intel core i7 CPU, 2.67GHz machine with 4GB RAM, running Microsoft Windows 7 operating system. The algorithm was implemented in Microsoft Visual Studio 2005. We fixed the k_1, k_1=2, and varied the k_2 values of anonymity in the range of 3 and 20.

The following experimental results demonstrate the performance of proposed algorithm. We evaluated the running times, and the ratios of modified edges of our proposed algorithm on two types of graphs: positive/negative edge weight graph and positive-only edge weight graph. We also compare with previous work in [20], namely *K1K2MPN* on both positive and negative edges and, *K1K2MPP* on positive-only edges, with two metrics: running time and the ratio of modified edges.

A.   Dataset

To demonstrate the characteristics and evaluate the performance of proposed algorithm, we run simulations on one real world dataset, *hep-th*. The *hep-th* dataset was a weighted network that describes a co-authorship network of high energy physics scientists, which was compiled by M. Newman in 2001 [16]. The hep-th dataset contains 7,610 scientists (nodes) and 15,751 co-authorships (edges), and each edge is assigned with a real value weight.

B.   Performance Analysis

We examine the performance of our proposed algorithm with two metrics: running times, ratios of modified edges. The running time indicates the computation efficiency of the algorithm. The ratios of modified edges indicate the percentage of edges affected and modified by the algorithm. It is defined as the number of modified edges over the total number of edges on the *k'* shortest paths, $k_1 \leq k \leq k_2$, for vertices specified in *H*. In the experiments, our algorithm run simulation on two types of graphs, the first type allows both positive and negative edge weights, namely *EKMPN*, and the second type allows positive-only edges, namely *EKMPP*.

C.   Discussion

To compare the performance of *EKMPN* and *EKMPP*, Figure 3 shows the results of running times on different types of graphs for different sizes of *H*. On positive-edge-weight-only graphs, there exist paths that cannot be modified to the same length as shortest path. As such, more paths need to be checked and modified and therefore take longer running time.

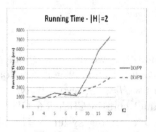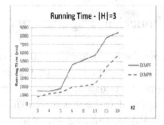

Fig. 3. Comparison of running times with different sizes of H.

Figure 4 shows the ratios of modified edges on different types of graphs with different sizes of H on varies k_2 values. Both types of graphs have similar ratios of modified edges, the *EKMPN* modified less number of edges than *EKMPP*.

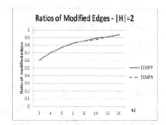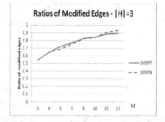

Fig. 4. Comparison of the ratios of modified edges with different sizes of H.

As shown in Figure 5, the proposed algorithms, *EKMPN* and *EKMPP*, have faster running time than previous work, *K1K2MPN* and *K1K2MPP* (modify *NV* edges only) with two different size of *H*. *EKMPN* has better performance then *EKMPP* on running time with two different size of *H*. In Figure 6, *EKMPN* and *EKMPP* have similar ratios of modified edges compared to *K1K2MPN* and *K1K2MPP*. Hence, our proposed algorithms have better performance than the previous work.

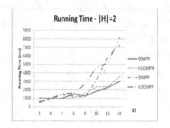

Fig. 5. Comparison of running time with different sizes of H for different methods.

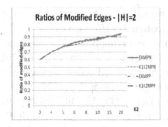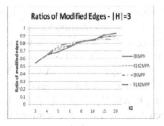

Fig. 6. Comparison of ratios of modified edges with different sizes of H for different methods.

In summary, the proposed *EKMPN* on positive/negative edge weight graphs has better performance than *EKMPP* on positive-only edge weight graphs. In addition, *EKMPN* and *EKMPP* have faster running times than *K1K2MPN* and K1K2MPP, but with similar ratios of modified edges.

6   Conclusion

In this work, we have studied the problem of anonymizing sensitive shortest paths on information networks. We extended the $[k_1, k_2]$-*shortest path privacy*, $k_1 \leq k \leq k_2$, and presented an algorithm based on the greedy approach to modify two different types of edges, namely *NV* and *PV* edges at the same time. We used the privacy value (the equation) to quantify the privacy level. Numerical experiments examining the characteristics of the proposed algorithms were given. The results demonstrated that the proposed algorithm is feasible to achieve the $[k_1, k_2]$-*shortest path privacy*, $k_1 \leq k \leq k_2$, with different performances on both types of graphs and more efficient than previous approach.

In the future, we will consider how to increase the data utility of anonymized graph. In addition, different approaches of modification should be considered, such as adding/deleting nodes/edges and preserving other types of sensitive characteristics such as k-degree for each node on the shortest paths.

Acknowledgment. This work was supported in part by the National Science Council, Taiwan, under grant NSC-101-2221-E-390-030-028-MY3.

References

1. Backstrom, L., Huttenlocher, D.P., Kleinberg, J.M., Lan, X.: Group formation in large so-cial networks: membership, growth, and evolution. In: Proceedings of KDD, pp. 44–54 (2006)
2. Backstrom, L., Dwork, C., Kleinberg, J.M.: Wherefore art thou r3579x?: anonymized so-cial networks, hidden patterns. and structural steganography. In: Proceedings of World Wide Web, pp. 181–190 (2007)
3. Bhagat, S., Cormode, G., Krishnamurthy, B., Srivastava, D.: Class-based graph anonymi-zation for social network data. In: Proceedings of Very Large Data Bases, pp. 766–777 (2009)

4. Banisar, D.: Freedom of information around the world 2006 – A global survey of access to government information Laws. www.privacyinternational.org/foi/survey
5. Cheng, J., Fu, A., Liu, J.: K-isomorphism: privacy preserving network publication against structural attacks. In: Proceedings of ACM SIGMOD Conference, pp. 459–470 (2010)
6. Das, S., Egecioglu, O., Abbad, A.E.: Anonymizing weighted social network graphs. In: Proceedings of International Conference on Data Engineering, pp. 904–907 (2010)
7. Gao, J., Xu, Y., Jin, R.M., Zhou, J.S., Wang, T.J., Yang, D.Q.: Neighborhood-privacy protected shortest distance computing in cloud. In: Proceedings of ACM SIGMOD Conference, pp. 409–420 (2011)
8. Hay, M., Miklau, G., Jensen, D., Towsley, D.F., Weis, P.: Resisting structural re-identification in anonymized social networks. Proceedings of the VLDB Endowment 1(1), 102–114 (2008)
9. Inkpen, A.: The Japanese corporate network transferred to North America: implications of North American firms. The International Executive 36(4), 411–433 (1994)
10. Leskovec, J., Huttenlocher, D., Kleingerg, J.: Signed networks in social media. In: Proceedings of CHI Conference on Human Factors in Computing Systems, pp. 1361–1370 (2010)
11. Li, Y., Shenm, H.: On Identity Disclosure in Weighted Graphs. In: Proceedings of the 11th International Conference on Parallel and Distributed Computing, Applications and Technologies, pp.166–174 (2010)
12. Liu, K., Terzi, E.: Towards identity anonymization on graphs. In: Proceedings of ACM SIGMOD Conference, pp. 93–106 (2008)
13. Liu, L., Liu, J., Zhang, J.: Privacy preservation of affinities in social networks. In: Proceedings of International Conference on Information Systems (2010)
14. Liu, L., Wang, J., Liu, J., Zhang, J.: Privacy preservation in social networks with sensitive edge weights. In: Proceedings of the SIAM International Conference on Data Mining, pp. 954–965 (2009)
15. Liu, X., Yang, X.: A Generalization Based Approach for Anonymizing Weighted Social Network Graphs. In: Wang, H., Li, S., Oyama, S., Hu, X., Qian, T. (eds.) WAIM 2011. LNCS, vol. 6897, pp. 118–130. Springer, Heidelberg (2011)
16. Newman, M.E.J.: The structure of scientific collaboration networks. In: Proceedings of Natl. Acad. Sci. 98, pp. 404409 (2001)
17. Nobari, S., Karras, P., Pang, H., Bressan, S.: L-opacity: linkage-aware graph anonymization. In: Proceedings of 17th International Conference on Extending Database Technology, pp. 583–594 (2014)
18. Wang, S.-L., Tsai, Z.-Z., Hong, T.-P., Ting, I.-H.: Anonymizing Shortest Paths on Social Network Graphs. In: Nguyen, N.T., Kim, C.-G., Janiak, A. (eds.) ACIIDS 2011, Part I. LNCS, vol. 6591, pp. 129–136. Springer, Heidelberg (2011)
19. Tsai, Y.C., Wang, S.L., Kao, H.Y., Hong, T.P.: Edge types vs privacy in K-anonymization of shortest paths. In Applied Soft Computing 31, 348–359 (2015)
20. Tsai, Y.C., Wang, S.L., Kao, H.Y., Hong, T.P.: [k₁, k₂]-anonymization of Shortest Paths, In: Proceedings of 4th International Workshop on Intelligent Data Analysis and Management (2014)
21. Yuan, M., Chen, L., Yu, P.S.: Personalized privacy protection in social networks. In: Proceedings of the 36rd International Conference on Very Large Data Bases, pp.141–150 (2010)
22. Yuan, M., Chen, L.: Node Protection in Weighted Social Networks. In: Yu, J.X., Kim, M.H., Unland, R. (eds.) DASFAA 2011, Part I. LNCS, vol. 6587, pp. 123–137. Springer, Heidelberg (2011)

23. Yen, J.Y.: A shortest path algorithm, Ph.D. dissertation, University of California, Berkeley (1970)
24. Zhou, B., Pei, J.: Preserving privacy in social networks against neighborhood attacks. In: Proceedings of the 24th International Conference on Data Engineering, pp. 506–515 (2008)
25. Zou, L., Chen, L., Ozsu, M.T.: K-automorphism: A general framework for privacy preserving network publication. In: Proceedings of the 35rd International Conference on Very Large Data Bases, pp. 946–957 (2009)

E-Tail Quality and Brand Loyalty
in Thai Social Commerce

Chawalit Jeenanunta[⊠], Narongak Pongathornwiwat, Kwanchanok Chumnumporn,
Atchara Parsont, Kanisorn Lunsai, and Ramida Piyapaneekul

School of Management Technology, Sirindhorn International Institute of Technology,
Thammasat University, 131 Moo 5, Tiwanon Road, Bangkadi, Pathum-Thani, Thailand
chawalit@siit.tu.ac.th

Abstract. Social Networks are becoming one of powerful channel of transaction. Customers are comfortable to shopping online and easy to switch their mind to other competitors as well. Hence, sellers have to find the suitable techniques to get competitive advantage to maintain customers' loyalty. E-tail qualities, including information and graphic designs, website reputation and service reliability was used to measure diverse aspects of online transaction base on consumers' online experience which help to evaluation customer behavior and find out the right technique to fulfill customers' need to become loyalty to the brand. The objective of this study is the first to empirically test whether E-tail qualities are positively impacted on brand loyalty in social commerce Thailand. The statistical analysis was analyzed using factor analysis and linear regression. The results have shown that E-tail qualities components are positively significance on brand loyalty.

Keywords: E-tail quality · Social marketing · Customer behavior · Brand loyalty

1 Introduction

As the numbers of internet users are still increasing in the last ten years, this gains the opportunity of merchandisers to increase the sale channels such as social-media. There are 1.5 billion internet users, and still ongoing by five-percent year-on-year [13]. The social channels are turned advantage for small businesses by communicating directly to customers and could take market share from the large business [14]. Therefore, it is necessary that company has to become brand loyalty to gain competitive advantage. The number of the purchasing through the social media is popular especially the skin care products. The issue of brand loyalty for E-commerce has been intensively studying in various literatures such in the South Korea [5]. Online customer base their base their repurchase on previous experience through full service include online transaction and offline fulfillment which determine as E-tail quality of the full service offer [3,9].The E-tail quality consists of the information and graphic design, the website reputation and reliability and also how service the provided to the desired customer. The behavioral of customers are influenced by the

© Springer-Verlag Berlin Heidelberg 2015
L. Wang et al. (Eds): MISNC 2015, CCIS 540, pp. 200–209, 2015.
DOI: 10.1007/978-3-662-48319-0_16

process of retailing based on the internet channel as E-tail quality, strongly influence the customer decision whether they are willing to stay with the same brand or not [11]. As guided intensively in literature; however, there is lack of empirical study testing such E-tail quality on brand loyalty. The purpose of this research is therefore to explore and examine the factors that affect online loyalty in context of Thailand. The research questions are (1) what are the main factors influencing the successful online loyalty and (2) what are the impacts of such factors on the online loyalty.

This paper was organized as follows; theory background and hypotheses is in the section 2, followed by the research methodology and data analysis is in section 3 and 4 respectively. The last section is the conclusions and suggestions for further investigation.

2   Theory Background and Hypothesis

2.1   Brand Loyalty in E-Business

A definition of Brand loyalty is defined as integrating this multidimensional construct has been given by Oliver (1999) as: a deeply held commitment to repurchase a preferred product/service consistently in the future, thereby causing repetitive same-brand purchasing [8]. Brand loyalty helps to increasing in market share, higher profits and better goodwill among consumers which is direct impact on the revenue and profitability of a company.

Brand loyalty in E-business is widely defined as customer's favorable attitude and commitment towards the online retailer that result in repeat purchase behavior [10]. As an important measure of online loyalty, a level of brand loyalty causes the action of customer to support their loyalty brand activities as action loyalty. Customer satisfaction and trust in the product or service create e-loyalty which effect customer action. For satisfaction and trust, this two facet effect by E-tail quality from previous study in online loyalty [6,4]. Furthermore in this research is consist of three main point study of brand loyalty which are repurchase, word-of-mouth, and support the brand activities and new product which affect from the overall satisfaction and trust to customer loyalty brand.

2.2   E-Tail Quality

E-tail quality is arguably the first psychometrically robust instrument that focuses specifically on online quality, and is proved to generalize across different product categories and cultures [1]. E-tail quality composes of four dimensions: fulfillment/reliability, customer service, website design, and privacy/security. In this study E-tail quality defines as a measured "from the beginning to the end of transaction, including information search, website navigation, ordering, interactions, delivery and satisfaction with the ordered product" [11]. We hypothesize that E-tail quality will have effect to Brand loyalty on online merchandise. For four dimension of E-tail quality we determine website design as graphic/information design that online retailer provide for customer; privacy/security as website/social network credibility of brand and online retailer; Customer service and Online communication channel are the service of online transaction include transaction process, deliver product, after service, and the connection through social media such as Facebook, Line, and Twitter etc.; and Traditional communication channel is the connection through e-mail, face-to-face, message, and calling.

Hypothesis: There is a positive relationship between Graphic/information, Website/social network credibility, Fulfillment/Reliability, and Customer service on Brand loyalty.

Hypothesis 1: Graphic/information design is positively on Brand loyalty
Hypothesis 2: Website/social network credibility is positively on Brand loyalty
Hypothesis 3: Customer service and online communication channel on Brand loyalty
Hypothesis 4: Traditional communication channel is positively on Brand loyalty

3   Research Methodology

3.1   Samples and Data Collection

The sample size (n) for estimation quality data of the unknown population with the error below 5 percent and the reliability of 95 percent can evaluate by the following equation:

$$n = \frac{Z^2}{4E^2}$$

Hence, $Z_{i \bullet \alpha/2} = Z_{.975} = 1.96$, E =0.05

$$n = \frac{(1.96)^2}{4(0.05)^2} = 384.16$$

Therefore, the sample size should be 385 samples [12].

The survey method in this study is based on a stratified sampling method by two ways approaches including online and face-to-face survey. The target respondent is those who had an online purchase experience on skin care/cosmetic product. The target group trends to have higher percentage of female than male and higher percentage of age range 18-40 years old than less than 18 and more than 40 years old, as the female have a strong interest in a skin care and a beauty more than male. The online survey is posted to Facebook in the cosmetic fan page and follower of skin care/cosmetic selling page. Face-to-face survey data is collected at public facilities in Bangkok and out skirting city of Bangkok including five university locations, a public library, and ten local shopping malls. There are 80 responses from online and 327 responses from face-to-face survey in total there are 407 responses. This study only interested in respondents who had online transaction on skin care/cosmetic product which are 120 valid responses for data analysis.

3.2   Measures Goodness of Measure

All of constructs and measurement items were adopted form the intensively review of the literature. The survey questions apply from the study of Kim et al (2007), Caruana and Ewing (2006), and the review of Thai online shopping (skin care/cosmetic product) [6, 1]. There were 4 components to measure E-tail quality as follows, 4 items for Graphic/information design, 5 items for Website/ social network credibility, 7 items for Customer service and Online Communication channel, 4 items for Tradition Communication channel, and lastly, Brand loyalty consists of 4 items. Each construct was measured by a five-point Likert type scale rang in from 1 (Strongly Disagree) to 5 (Strongly Agree). In this study, the data analyze in three steps. First step, we checked reliability of all measurement items, then used factor analysis with VARIMAX to ensure the measurement items in each component (E-tail quality and Brand loyalty) can group by using Kaiser-Meger-Olkin (KMO) technique, the component are inter correlation if KMO value is nearly to one, should be greater than0.60, and the factor loading of all items to their latent variable are over 0.50 threshold [2]. Then, we used reliability analysis to check internal consistency of components by Cronbach's alpha coefficient, which is a good method for Likert scale data [12]. Cronbach's alpha recommended threshold over 0.70 [7]. Second step, we use Levene's test (test statistics) which aimed at the homogeneity of variance to test that the two samples have the similar variance and it mean that these two samples groups are equal. The two groups equally when the Sig. value of Levene's test is less than the significant level that we limit. Last step, we conducted a simple regression analysis to test hypothesis, one independent to one dependent each test. The relationship of two components represent by F-test and t-test. The beta (β) number show level of relationship of each item in the component and the adjust R^2 number show how items can explained the different of items and that component [12].

4   Result and Data Analysis

4.1   Data Description

Data description of 120 samples show in Table 1. The majority of respondents were female with 84.17% (male 15.83%). The respondents consisted large number of studying age people (31.67% of 18-23 years old), and working age people (27.50% of 24-29 years old, and 22.50% of 30-40 years old), which trend to have well-education (60.83% had a bachelor's degree). The time of shopping online was generally 1-5 times per month of 86.67%.The online shopping channel that popular are Facebook, IG (Instagram), Line, Other website/social network, Google+, YouTube, Twitter, and lastly WhatsApp.

Table 1. Demographic of the respondents

		Frequency	%
Gender	Male	19	15.83
	Female	101	84.17
Age	< 18	6	5.00
	18-23	38	31.67
	24-29	33	27.50
	30-40	27	22.50
	40+	16	13.33
Education	High school graduate	15	12.50
	Associate degree, occupational	8	6.67
	Bachelor's degree	73	60.83
	Higher than Bachelor's degree	24	20.00
Personal income per month	0-10,000 bath	34	28.33
	10,001-20,000 bath	35	29.17
	20,001-30,000 bath	23	19.17
	> 30,000 bath	28	23.33
Time of shopping online	1–5 times	104	86.67
	6–10 times	13	10.83
	11–20 times	2	1.67
	21–30 times	0	0.00
	>30 times	1	0.83
Shopping channel	Facebook	75	62.50
	Line	60	50.00
	IG (Instragram)	53	44.17
	Other	9	7.50
	Google+	11	9.17
	YouTube	6	5.00
	Twitter	2	1.67
	WhatsApp	2	1.67

Table 2 show the detail, E-tail quality compose of four groups, the first group is Graphic/information design has mean from 3.42 to 3.66. Second group, Website/social network credibility has mean from 3.17 to 3.38. Third group, Customer service and Online Communication channel has high mean from 3.84 to 4.07. And the last group, Tradition Communication channel has mean from 2.90 to 3.31.

Table 2. Reliability Test and Factor Analysis

Constructor	Mean	Std.	Factor Loading	KMO	α
E-tail quality:				0.884	0.924
Graphic/ information design:					
Photos of product	3.66	0.942	0.780		
Video clips description of product	3.59	0.915	0.749		
Current topic/ interesting topic from seller	3.59	0.973	0.574		
Compare photo (before vs. after)	3.42	0.934	0.673		
Website/ social network credibility:					
Profile and brand activity	3.38	0.995	0.769		
The reviews of product from customers	3.37	1.010	0.728		
Recommended brand from famous people	3.31	0.958	0.661		
Brand owner write the beauty block/page	3.29	1.005	0.623		
Recommended brand from friend	3.17	1.027	0.834		
Customer service and Online Communication channel:					
Seller renew product from shipping damage	4.07	0.979	0.761		
The product delivered correctly as preview	4.03	1.008	0.856		
Quick response from the brand	4.00	1.002	0.842		
The product delivered on the time	3.99	0.966	0.827		
After service	3.98	1.047	0.833		
Describe for product information	3.98	0.970	0.729		
Customer service via social network (i.e. Facebook, IG (Instagram), Line etc.)	3.84	1.030	0.729		
Tradition Communication channel:					
E-mail	3.31	1.104	0.695		
Calling	3.21	1.197	0.811		
SMS	3.13	1.128	0.760		
Face-to-face	2.90	1.314	0.764		
Brand Loyalty:					
Customers choose their brand loyalty before others	3.53	0.989	0.872	0.789	0.851
Customers are willing to try brand loyalty's new product	3.49	0.989	0.776		
Customers support the offerings from their brand loyalty	3.40	0.941	0.885		
Activities to get reward from the brand	3.38	1.010	0.795		

For Brand loyalty, first, we do factor analysis for all activities to get reward from the brand then the mean of Brand loyalty ranks from 3.38 to 3.53. The activities to get reward from brand include like brand's photo/page on website/social network, visit brand's website, watching a video product trial, referring friends, share/mention brand's website/social network, and checking in to a location via social network (i.e. Facebook, IG (Instagram)).

4.2   Factor Analysis, Reliability Test and Homogeneity Test

The reliability of each measurement items was computed whether they suitable for factor analysis by using KMO and Cronbach's alpha (α).The four components of E-tail quality have Cronbach's alpha of 0.942. Brand loyalty has Cronbach's alpha of 0.851. As Cronbach's alpha of E-tail quality and Brand loyalty were over 0.70, the measurement are reliable.

Factor analysis was considered. The four components of E-tail quality have KMO of 0.884, and Brand loyalty has KMO of 0.789. The KMO values of both items are more than 0.60 and factor loading values are more than 0.50, thus the components can be grouped as show in the table 2.

Table 3 show the homogeneity test of variances. This is a test that determines if the two sample groups (online and face to face survey) have about the same or different amounts of variability between scores. You will see columns labeled Sig. Look in the Sig. column. They are greater than significant value of 0.01 that we limited. A value greater than 0.01 means that the variability in our two sample groups are about the same. That the scores in one condition do not vary too much more than the scores in our second condition. Put scientifically, it means that the variability in the two conditions is not significantly different. This is a good thing so we can combine two sample groups together for our analyzed.

Table 3. Test of Homogeneity of Variances

Variable	Levene Statistic	Sig.
E-tail Quality:		
Graphic/ information design	0.284	0.595
Website/ social network credibility	2.972	0.086
Customer service and Online Communication channel	4.507	0.035
Tradition Communication channel	0.081	0.776
Brand loyalty	4.004	0.047

4.3   Test of Hypothesis

In this section, the main hypothesis and four sub-hypotheses are tested using the sample data sets

Table 4. Result for linear regression analysis on hypothesized

Independent variable	Dependent variable: Brand loyalty				
	F-test	t-test	Beta	Adjust R^2	Conclusion
H1: Graphic/ information design	4.035**	2.009**	0.182**	0.025**	Support
H2: Website/ social network credibility	42.287*	6.503*	0.514*	0.258*	Support
H3: Customer service and Online Communication channel	16.734*	6.503*	0.352*	0.117*	Support
H4: Tradition Communication channel	3.390**	1.841**	0.167**	0.020**	Support

Note: *p=0.01, **p=0.05

The hypothesis model is tested by simple linear regression method result is showed in table 4. Simple linear regression show as follow:

Hypothesis 1: Brand loyalty = a_1 + b_1 (Graphic/ information design)
Hypothesis 2: Brand loyalty = a_2 + b_2 (Website/ social network credibility)
Hypothesis 3: Brand loyalty = a_3 + b_3 (Customer service and Online Communication channel)
Hypothesis 4: Brand loyalty = a_4 + b_4 (Tradition Communication channel)

By using simple linear regression to test hypothesizes we used F-test and t-test to examine the relationship between the independent variable and the dependent variable. At the level of significance of p = 0.01, the value of F-test should over 6.85 and the value of t-test should over 2.33, and at significance level of p = 0.05, the hypothesis acceptable if F-test over the threshold of 3.92 and the value of t-test over 1.64 threshold, the independent variable has relationship with the dependent variable and it mean that the hypothesis is acceptable to show that the independent variable has relationship with the dependent variable and it mean that the hypothesis is acceptable [12]. Hypothesis 1, Graphic/ information design interaction on Brand loyalty was significant (b=0.182, F=4.035, t=2.009, p=0.05, sig. =0.047) which means Graphic/ information design has positive effect on Brand loyalty. Hypothesis 2, Website/ social network credibility interaction on Brand loyalty is significant (b=0.514, F=42.287, t=6.503, p=0.01, sig. =0.000) which means Website/ social network credibility has positive effect on Brand loyalty. Hypothesis 3, Customer service and Online Communication channel on Brand loyalty is significant (b=0.352, F=16.734, t=4.091, p=0.01, sig. =0.000) which means Customer service and Online Communication channel has positive effect on Brand loyalty. Hypothesis 4, Tradition Communication channel interaction on Brand loyalty is significant (b=0.167, F=3.390, t=1.841, p=0.05, sig. =0.068) which means Tradition Communication channel do not has enough evident to support on Brand loyalty. Therefore, we can imply that Graphic/ information design, Website/ social network credibility, and Customer service and Online Communication channel are the factors to make Brand loyalty, supporting the notion of Kim et al (2009) and Caruana and Ewing's study (2006) [6, 1].

5   Conclusion

In this research, we studied the E-tail quality influences on brand loyalty in Thailand. We found that E-tail quality has positive effect to customer on Brand loyalty. This mean to maintain the customer brand loyalty online, the online merchants must carefully consider the E-tail quality which includes a Graphic/ information design, Website/ social network credibility, Customer service and Online Communication channel, and Tradition Communication channel. For Graphic/ information design, product detail should be illustrated clearly. Photo and video clips to introduce product could help customers to make decision. Online retailer should build more credibility in their Website/ social network and online retailer also need to keep promise on deliver product on time. The product delivered should appear correctly as show in photo. Customers prefer to have connection with seller through social network channel (i.e. Facebook, Line, Twitter, etc.) and get good responsive service both before and after the transaction. Tradition Communication channel such as E-mail, Calling, SMS and face-to-face are not imperative on online transaction. Follow these factors from the study, sellers can maintain customers' loyalty to the brand easily.

Since this study provides a significant contribution in social marketing literature, there are some limitations needing for further investigation. There are many factors such as brand identity, the distinctive and relatively enduring characteristics of a focal brand (or company), that influence brand loyalty. Therefore, further study should be investigated the effects of such factor on brand loyalty. Lastly, the further study should be utilized the structural equation modeling to analysis in order to provide an in-depth understanding.

References

1. Caruana, A., Ewing, M.T.: The psychometric properties of eTail quality. International Marketing Review **23**(4), 353–370 (2006)
2. Fornell, C., Larcker, D.F.: Evaluating structural equation models with unobservable variables and measurement error. J. Mark. Res. **18**(1), 39–50 (1981)
3. Gronroos, C., Heinonen, F., Isoniemi, K., Lindholm, M.: The NetOffer model: a case example from the virtual marketspace. Management Decision **38**(3/4), 243–252 (2000)
4. He, H., Li, Y., Harris, L.: Social identity perspective on brand loyalty. Journal of Business Research **65**, 648–657 (2012)
5. Koo, D.M.: The fundamental reasons of e-consumers' loyalty to an online store. Electronic Commerce Research and Applications **5**, 117–130 (2006)
6. Kim, J., Fiore, A.M., Lee, H.-H.: Influences of online store perception, shopping enjoyment, and shopping involvement on consumer patronage behavior towards an online retailer. Journal of Retailing and Consumer Services **14**, 95–107 (2007)
7. Mouakket, S., Al-hawari, M.A.: Examining the antecedents of e-loyalty intention in an online reservation environment. Journal of High Technology Management Research **23**, 46–57 (2012)
8. Oliver, R.L.: Whence consumer loyalty? Journal of Marketing **63**, 33–34 (1999)
9. Porter, M.E.: Strategy and the internet. Harvard Business Review **79**(3), 63–78 (2001)

10. Srinivasan, S.S., Anderson, R., Ponnavolu, K.: Customer loyalty in e- commerce: an exploration of its antecedents and consequences. Journal of Retailing **78**(1), 41–50 (2002)
11. Wolfinbarger, M., Gilly, M.C.: E-tailQ: dimensionalizing, measuring and predicting etail quality. Journal of Retailing **79**(3), 193–198 (2003)
12. Vanichbuncha, K.: Statistic for research. Chulalongkorn University Bookshop, Bangkok (2012)
13. Bennett, C.: Social Media's Role Will Soon Shift From Driving Awareness to Creating Revenue (November 19, 2014). http://www.entrepreneur.com/article/239764 (accessed November 27, 2014)
14. Kemp S.: Social, Digital & Mobile Worldwide in 2014 (January 23, 2014). http://wearesocial.net/blog/2014/01/social-digital-mobile-worldwide-2014/ (accessed October 3, 2014)

Network-Based Analysis of Comorbidities: Case Study of Diabetes Mellitus

Yi-Horng Lai[1(✉)], Tse-Yao Wang[2], and Hsieh-Hua Yang[1,]

[1] Department of Health Care Administration, Oriental Institute of Technology,
New Taipei City, Taiwan
FL006@mail.oit.edu.tw, yansnow@gmail.com
[2] Taipei Veterans General Hospital, Taipei City, Taiwan
tseyao85@gmail.com

Abstract. Comorbidity is the presence of one or more additional disorders or diseases co-occurring with a primary disease or disorder. The purpose of this study is to identify diseases that co-occur with diabetes mellitus and analyze the gender differences. Data was collected from 154,434 diabetes mellitus admission medical records out of 1,377,469 admission medical records from 1997 to 2010 in Taiwan. In this study, the comorbidity relationships are presented in the phenotypic disease network (PDN), and φ-correlation is used to measure the distance between two diseases on the network. The results show that there is a high correlation in the following pairs/triad of diseases: diabetes mellitus (250) and essential hypertension (401), diseases of the circulatory system (390~459) are highly correlated. Diabetes mellitus without mention of complication (2500) and essential hypertension, unspecified (4019), and diseases of the circulatory system (390~459) are highly correlated. Diabetes with renal manifestations (2504) and nephritis and nephropathy, not specified as acute or chronic, with other specified pathological lesion in kidney (5838), and diseases of the genitourinary system (580~629) are highly correlated. Diabetes with ophthalmic manifestations (2505) and diabetic retinopathy (3620), and diseases of the nervous system and diseases of the sense organs (320~389) are highly correlated. Diabetes with neurological manifestations (2506) and polyneuropathy in diabetes (3572), and diseases of the nervous system and diseases of the sense organs (320~389) are highly correlated. Diabetes with peripheral circulatory disorders (2507) and gangrene (7854), and symptoms, signs, and ill-defined conditions (780~799) are highly correlated. Diabetes with other specified manifestations (2508) and ulcer of lower limbs, except decubitus (7071), and diseases of the skin and subcutaneous tissue (680~709) are highly correlated.

Keywords: Phenotypic Disease Network (PDN) · Diabetes mellitus · Network analysis

1   Introduction

Diabetes mellitus is a group of metabolic diseases mainly characterized by hyperglycemia resulting from an array of dysfunctions of insulin. Persons with diabetes are at risk of many complications, including retinopathy, nephropathy, neuropathy, cardiovascular disease and other miscellaneous complications. There is mounting evidence of other diabetes

© Springer-Verlag Berlin Heidelberg 2015
L. Wang et al. (Eds): MISNC 2015, CCIS 540, pp. 210–222, 2015.
DOI: 10.1007/978-3-662-48319-0_17

mellitus related complications in different stages in these years [1, 2]. These chronic complications occurred when prolonged hyperglycemia or as time goes by. Diabetes mellitus, a systemic disease, may be presented as various symptoms and multiple complications in different stages.

Prevention of the long-term complications is one of the important goals of diabetes mellitus care. Besides, appropriate diagnosis and treatment of some established complications may improve quality of life and delay their progression [1, 2]. Additionally, some diseases coexist with diabetes mellitus will increase the risk of complications. For example, the hyperlipidemia and hypertension accompanied with diabetes mellitus may increase the risk of cardiovascular event and even mortality [1].

On the other hands, it was showed that the glucose intolerance and other conditions share some similar underlying causes, such as obesity and physical inactivity. The metabolic syndrome is constellation of abdominal obesity, hypertension, glucose intolerance, and an atherogenic lipid profile [3]. The diabetes mellitus could coexist with other specified diseased due to the similar common causes.

Thus, the understandings of various complications and comorbidities of diabetes mellitus are fundamental. To picture the relationship between diabetes mellitus and other diseases helps us to target the complications and comorbidities and establish appropriate screening examinations and interventions [1, 2].

There are no clear boundaries between many diseases, as diseases can have multiple causes and can be related in several dimensions. From a genetic perspective, a pair of diseases can be related because they have both been associated with the same gene, whereas from a proteomic perspective, diseases can be related because disease-associated proteins act on the same pathway [4].

During the past half-decade, several resources have been constructed to help understand the entangled origins of many diseases. Many of these resources have been presented as networks in which interactions between disease-associated genes, proteins, and expression patterns have been summarized. Goh, Cusick, Valle, Barton, Vidal, and Barabási created a network of Mendelian gene-disease associations by connecting diseases that have been associated with the same genes [5]. Besides, more and more studies have applied the network approach in diseases, such as neurodegenerative diseases [6], infertility etiologies [7], SARS, and HIV/AIDS [8].

A comorbidity relationship exists between two diseases whenever they affect the same individual substantially more than chance alone. In the past, comorbidities have been used extensively to construct synthetic scales for mortality prediction [9, 10], yet their utility exceed their current use. Studying the structure defined by entire sets of comorbidities might help the understanding of many biological and medical questions from a perspective that is complementary to other approaches. For example, a recent study built a comorbidity network in an attempt to elucidate neurological diseases' common genetic origins [11]. Heretofore, however, neither this data nor the data necessary to explore relationships between all diseases is currently available to the research community.

This present study decided to provide this data in the form of a phenotypic disease network (PDN) that includes all diseases recorded in the medical claims. Additionally, this study illustrates how a PDN can be used to study disease progression from a network perspective by interpreting the PDN as the landscape where disease progression occurs and shows how the network can be used to study phenotypic differences between patients of different demographic backgrounds. This study indicates the directionality of disease progression, as observed in our dataset, and finds out that more central diseases in the PDN are more likely to occur after other diseases and that more peripheral diseases tend to precede other illnesses.

In order to guide diabetes mellitus -related diseases prevention program, this study conducted the PDN of diabetes mellitus to explore the relationship between diabetes mellitus and other diseases. The objective of this study is to identify diseases that are highly correlated with diabetes mellitus, and discuss gender differences.

2   Materials and Method

2.1   Source Data and Study Population

The National Health Insurance (NHI) program was initiated in Taiwan in 1995 and covers nearly all residents. In 1999, the Bureau of NHI began to release all claims data in electronic form to the public under the National Health Insurance Research Database (NHIRD) project. The structure of the claim files is described in detail on the NHIRD website and in other publications [12].

NHIRD offer reliable, systematic, and complete data for disease detection. The datasets contained only the visit files, including dates, medical care facilities and specialties, patients' genders, dates of birth, and the four major diagnoses coded in the International Classification of Disease, 9th Revision, Clinical Modification (ICD-9-CM) format [12]. In total, the ICD-9-CM classification consists of 17 different categories at the 3 digit level and 16459 categories at 5 digits. The 17 different include infectious and parasitic diseases (001~139), neoplasms (140~239), endocrine, nutritional and metabolic diseases, and immunity disorders (240~279), diseases of the blood and blood-forming organs (280~289), mental disorders (290~319), diseases of the nervous system and diseases of the sense organs (320~389), diseases of the circulatory system (390~459), diseases of the respiratory system (460~519), diseases of the digestive system (520~579), diseases of the genitourinary system (580~629), complications of pregnancy, childbirth, and the puerperium (630~679), diseases of the skin and subcutaneous tissue (680~709), diseases of the musculoskeletal system and connective tissue (710~739), congenital anomalies (740~759), certain conditions originating in the perinatal period (760~779), symptoms, signs, and ill-defined conditions (780~799), and injury and poisoning (800~999) [13].

Diabetes mellitus be coded with 042 as diabetes mellitus, 2500 as diabetes mellitus without mention of complication, 2501 as diabetes with ketoacidosis, 2502 as diabetes with hyperosmolar coma, 2503 as diabetes with other coma, 2504 as diabetes with renal manifestations, 2505 as diabetes with ophthalmic manifestations, 2506 as diabetes with neurological manifestations, 2507 as diabetes with peripheral circulatory disorders, 2508 as diabetes with other specified manifestations, and 2509 as diabetes with unspecified complication.

Table 1. Data characteristics statistics.

Variable		N	%
Gender	Female	701528	50.93
	Male	673985	48.93
	Unknown	1956	.14
Age	~19	219762	15.95
	20~21	180822	13.13
	30~39	182862	13.28
	40~49	169302	12.29
	50~59	171601	12.46
	60~69	168399	12.23
	71~	284721	20.67
Year	1997	71925	5.22
	1998	74422	5.40
	1999	79087	5.74
	2000	84089	6.10
	2001	89181	6.47
	2002	95232	6.91
	2003	91465	6.64
	2004	108586	7.88
	2005	117954	8.56
	2006	114446	8.31
	2007	113092	8.21
	2008	111843	8.12
	2009	113112	8.21
	2010	113035	8.21
Diabetes mellitus	Yes	154434	11.21
	No	1223035	88.79
		1377469	100.00

To protect privacy, the data on patient identities and institutions had been scrambled cryptographically. These visit files in this study represented 1377469 admission activities within the NHI from 1997 to 2010. Demographically, the data set consists of 1,377,469 admission medical records come from 1,000,000 patients and is composed 50.93% of female, 48.93% of male, 20.67% of over 71 years old, with 154434 diabetes mellitus admission medical records (Table 1).

These diabetes mellitus admission medical records include 76579 female (49.59%) and 77849 male (50.40%).

177 diabetes mellitus admission medical records (.11%) are diabetes mellitus (250), 107912 diabetes mellitus admission medical records (69.88%) are diabetes mellitus without mention of complication (2500), 2015 diabetes mellitus admission medical records (1.30%) are diabetes with hyperosmolar coma (2501), 1990 diabetes mellitus admission medical records (1.29%) are diabetes with hyperosmolar coma (2502),593 diabetes mellitus admission medical records (.38%) are diabetes with other coma (2503), 12446 diabetes mellitus admission medical records (8.06%) are

Table 2. Data characteristics statistics of diabetes mellitus

Variable		N	%
Gender	Female	76579	49.59
	Male	77849	50.40
	Unknown	6	.01
Age	~19	713	.46
	20~21	1390	.90
	30~39	4554	2.95
	40~49	14192	9.19
	50~59	30251	19.59
	60~69	39737	25.73
	71~	63597	41.18
Year	1997	3431	2.22
	1998	4466	2.89
	1999	5184	3.36
	2000	5850	3.79
	2001	6894	4.46
	2002	8556	5.54
	2003	9356	6.06
	2004	12259	7.94
	2005	14680	9.51
	2006	14893	9.64
	2007	16026	10.38
	2008	16814	10.89
	2009	17460	11.31
	2010	18565	12.02
Diabetes Mellitus	250	177	.11
	2500	107912	69.88
	2501	2015	1.30
	2502	1990	1.29
	2503	593	.38
	2504	12446	8.06
	2505	3455	2.24
	2506	3522	2.28
	2507	2485	1.61
	2508	6181	4.00
		154434	100.00

diabetes with renal manifestations (2504), 3455 diabetes mellitus admission medical records (2.24%) are diabetes with ophthalmic manifestations (2505), 3522 diabetes mellitus admission medical records (2.28%) are diabetes with neurological manifestations (2506), 2485 diabetes mellitus admission medical records (1.61%) are diabetes with peripheral circulatory disorders (2507), and 6181 diabetes mellitus admission medical records (4.00%) are diabetes with other specified manifestations (2508) (Table 2).

2.2   Data Limitations

There are several limitations to the current study. First, although data gathered from NHIRD is comprehensive and reliable, there are still some mistakes that the system couldn't find, such as code entry errors. These errors may be carried into data pre-processing, and it is beyond the control of this study. Second, the data was not up-to-date. Although future researchers are still recommended to apply the method of this study to analyze the characteristics of patients for the purpose of disease prevention, changes of medical treatments and other factors should be considered. Third, this study does not have a global sample, so there might be limits to replicate the findings of this study in all the other countries. It might, however, be generalized to other ethnic Chinese population due to the similarity in genes and physiology.

2.3   Quantifying the Strength of Comorbidity Relationships

To measure the comorbidity relationship, it is necessary to quantify the strength of comorbidities by introducing a notion of distance between two diseases. A difficulty of this approach is that different statistical distance measures have biases that over- or under-estimate the relationships between rare or prevalent diseases. These biases are important given that the number of times a particular disease is diagnosed, such as its prevalence, follows a heavy tailed distribution [4], meaning that while most diseases are rarely diagnosed, a few diseases have been diagnosed in a large fraction of the population. In this study, the φ -correlation is used to quantify the distance between two diseases. The φ-coefficient relates to the 2×2 table [2, 14]:

		Y		
		Pos.	Neg.	
X	Pos.	p_a	p_b	p_X
	Neg.	p_c	p_d	$1-p_X$
		p_Y	$1-p_Y$	

The φ-coefficient is given by

$$r_{phi} = \frac{p_a - p_X p_Y}{(p_X p_Y (1 - p_X)(1 - p_Y))^{1/2}} \tag{1}$$

if $0 < p_X, p_Y < 1$, and $r_{phi} = 0$ otherwise.
The distribution of r_{phi} values representing all disease pairs is presented in Fig. 1.

2.4   Network Approach

This research utilizes data from NHIRD to obtain the four major diagnoses codes of all patients. This study calculates φ-correlation with Equation 1. Pajek 4.03 program was used to compute the compute the degree of centrality and betweenness of each node and the path value (φ-correlation). This study is focused on the path between each disease as network and the correlation as the value of line (path weight), and they could be affected by different populations, which indicates differences in gender for each population.

3   Results

It can be summarized the set of all comorbidity associations between all diseases expressed in the study population by constructing a Phenotypic Disease Network (PDN). In the PDN, nodes are disease phenotypes identified by unique ICD9 codes, and links phenotypes that show significant comorbidity according to the measures introduced above.

In principle, the number of disease-disease associations in the PDN is proportional to the square of the number of phenotypes, yet many of these associations are either not strong or are not statistically significant [4]. The structure of the PDN can be explored by focusing on the strongest and the most significant of these associations. The PDN can be seen as a network of the phenotypic space. This network allows people to understand the relationship between illnesses.

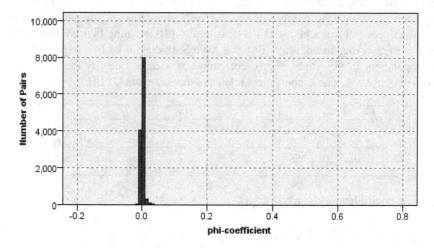

Fig. 1. Distribution of the φ-correlation between all disease pairs.

The distribution of w values representing all disease pairs is presented in Fig. 1. Most of them are between .000 and .005. A discussion on the confidence interval and statistical significance of these measures can be found in Hidalgo, Blumm, Barabási, and Christakis's study, and φ-correlation > .06 is statistically significant [4].

In Fig. 2, nodes are diseases; links are correlations. Node color identifies the ICD9 category; node size is proportional to disease prevalence. Link color indicates correlation strength. The PDN built using φ-correlation. Here all statistically significant links where φ-correlation>.06 are shown in Fig. 2.

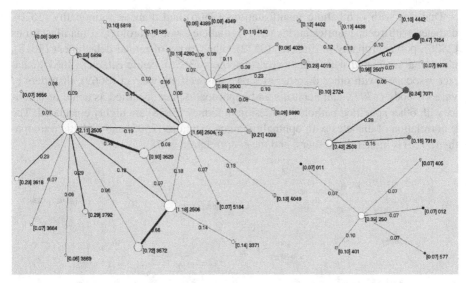

Fig. 2. The PDNs of diabetes mellitus, φ-correlations > .06. [ ] is the summary of φ-correlations with other node (diseases) of this node (disease), and the size of the node is dependent on the value. Such as the summary of φ-correlations with other diseases of 042 is 1.01. Node color is based on the ICD9 Category.

Diabetes mellitus (250) and pulmonary tuberculosis (011), other respiratory tuberculosis (012), essential hypertension (401), secondary hypertension (405), and diseases of pancreas (577) are highly correlated. Those φ-correlations are all above .06. The φ-correlation of diabetes mellitus (250) and essential hypertension (401) is higher than others, and the φ-correlation is .10 (Fig. 2).

Diabetes mellitus without mention of complication (2500) and other and unspecified hyperlipidemia (2724), essential hypertension, unspecified (4019), hypertensive heart disease, unspecified (4029), hypertensive renal disease, unspecified (4039), coronary atherosclerosis (4140), congestive heart failure (4280), unspecified cerebral artery occlusion (4349), unspecified late effects of cerebrovascular disease (4389), and urinary tract infection, site not specified (5990) are highly correlated. The φ-correlation of diabetes mellitus without mention of complication (2500) and essential hypertension, unspecified (4019) is higher than others, and the φ-correlation is .23.

Diabetes with renal manifestations (2504) and diabetes with ophthalmic manifestations (2505), diabetes with neurological manifestations (2506), diabetic retinopathy (3620), hypertensive renal disease, unspecified (4039), hypertensive heart and renal disease, unspecified (4049), congestive heart failure (4280), acute edema of lung, unspecified (5184), nephrotic syndrome, with other specified pathological lesion in kidney (5818), nephritis and nephropathy, not specified as acute or chronic, with other specified pathological lesion in kidney (5838), and chronic renal failure (585) are highly correlated. The φ-correlation of diabetes with renal manifestations (2504) and nephritis and nephropathy, not specified as acute or chronic, with other specified pathological lesion in kidney (5838) is higher than others, and the φ-correlation is .49.

Diabetes with ophthalmic manifestations (2505) and diabetic retinopathy (3620), diabetes with renal manifestations (2504), diabetes with neurological manifestations (2506), polyneuropathy in diabetes (3572), other forms of retinal detachment (3618), glaucoma associated with other ocular disorders (3656), senile cataract (3661), cataract associated with other disorders (3664), unspecified cataract (3669), disorders of vitreous body (3792), and nephritis and nephropathy, not specified as acute or chronic, with other specified pathological lesion in kidney (5838) are highly correlated. The φ-correlation of diabetes with ophthalmic manifestations (2505) and diabetic retinopathy (3620) is higher than others, and the φ-correlation is .78.

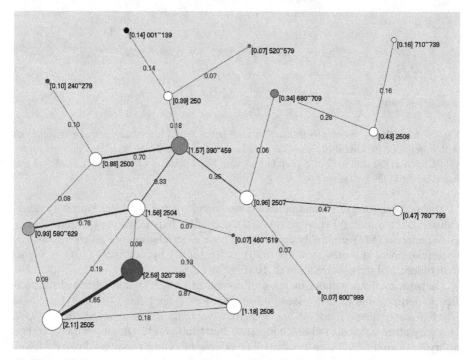

Fig. 3. The PDNs of diabetes mellitus with categories, φ-correlations > .06. [ ] is the summary of φ-correlations with other node (diseases) of this node (disease), and the size of the node is dependent on the value. Such as the summary of φ-correlations with other diseases of 042 is 1.01. Node color is based on the ICD9 Category.

Diabetes with neurological manifestations (2506) and diabetes with renal manifestations (2504), diabetes with ophthalmic manifestations (2505), peripheral autonomic neuropathy in disorders classified elsewhere (3371), polyneuropathy in diabetes (3572), and diabetic retinopathy (3620) are highly correlated. The φ-correlation of diabetes with neurological manifestations (2506) and polyneuropathy in diabetes (3572) is higher than others, and the φ-correlation is .66.

Diabetes with peripheral circulatory disorders (2507) and atherosclerosis of native arteries of the extremities (4402), other specified peripheral vascular diseases (4438), arterial embolism and thrombosis of arteries of the extremities (4442), ulcer of lower limbs, except decubitus (7071), gangrene (7854), and late amputation stump complication (9976) are highly correlated. The φ-correlation of diabetes with peripheral circulatory disorders (2507) and gangrene (7854) is higher than others, and the φ-correlation is .47.

Diabetes with other specified manifestations (2508) and ulcer of lower limbs, except decubitus (7071), and other bone involvement in diseases classified elsewhere (7318) are highly correlated. The φ-correlation of diabetes with other specified manifestations (2508) and ulcer of lower limbs, except decubitus (7071) is higher than others, and the φ-correlation is .28.

The PDNs of diabetes mellitus with categories is as Fig.3. Diabetes mellitus (250) and infectious and parasitic diseases (001~139), diseases of the digestive system (520~579), and diseases of the circulatory system (390~459) are highly correlated. The φ-correlation of diabetes mellitus (250) and diseases of the circulatory system (390~459) is higher than others, and the φ-correlation is .18.

Diabetes mellitus without mention of complication (2500) and endocrine, nutritional and metabolic diseases, and immunity disorders (240~279), diseases of the genitourinary system (580~629), and diseases of the circulatory system (390~459) are highly correlated. The φ-correlation of diabetes mellitus without mention of complication (2500) and diseases of the circulatory system (390~459) is higher than others, and the φ-correlation is .70.

Diabetes with renal manifestations (2504) and diseases of the circulatory system (390~459), diseases of the genitourinary system (580~629), diseases of the nervous system and diseases of the sense organs (320~389), diseases of the respiratory system (460~519), diabetes with ophthalmic manifestations (2505), and diabetes with neurological manifestations (2506) are highly correlated. The φ-correlation of diabetes with renal manifestations (2504) and diseases of the genitourinary system (580~629) is higher than others, and the φ-correlation is .76.

Diabetes with ophthalmic manifestations (2505) and diseases of the genitourinary system (580~629), diseases of the nervous system and diseases of the sense organs (320~389), diabetes with renal manifestations (2504), and diabetes with neurological manifestations (2506) are highly correlated. The φ-correlation of diabetes with diabetes with ophthalmic manifestations (2505) and diseases of the nervous system and diseases of the sense organs (320~389) is higher than others, and the φ-correlation is 1.61.

Diabetes with neurological manifestations (2506) and diseases of the nervous system and diseases of the sense organs (320~389), Diabetes with renal manifestations (2504), and diabetes with ophthalmic manifestations (2505) are highly correlated. The φ-correlation of diabetes with neurological manifestations (2506) and diseases of the nervous system and diseases of the sense organs (320~389) is higher than others, and the φ-correlation is .87.

Diabetes with peripheral circulatory disorders (2507) and diseases of the circulatory system (390~459), diseases of the skin and subcutaneous tissue (680~709), symptoms, signs, and ill-defined conditions (780~799), and injury and poisoning (800~999) are highly correlated. The φ-correlation of diabetes with peripheral circulatory disorders (2507) and symptoms, signs, and ill-defined conditions (780~799) is higher than others, and the φ-correlation is .47.

Diabetes with other specified manifestations (2508) and diseases of the skin and subcutaneous tissue (680~709), and diseases of the musculoskeletal system and connective tissue (710~739) are highly correlated. The φ-correlation of diabetes with other specified manifestations (2508) and diseases of the skin and subcutaneous tissue (680~709) is higher than other one, and the φ-correlation is .28.

4   Conclusions

The results had shown a suggestive proof that patients develop diseases close in the PDN to those already affecting them. The PDN we found has a complex structure that some diseases are highly connected while others are barely connected at all.

With the results, we pictured the structure of diabetes mellitus and its central associated diseases. While the PDN is blind to the pathophysiology underlying the observed comorbidities, the most of the results could be explained by previous studies [1]. For example, some central associated diseases are readily proved complications of diabetes mellitus, such as nephropathy, retinopathy, neuropathy, gastropathy, macrovascular disease and peripheral vascular disease. Some associated diseases accompanied with diabetes mellitus belong to metabolic syndrome, such as hypertension and hyperlipidemia. Some associated diseases attribute to the adverse effects of medication, such as injury and poisoning. However, the relationships of some associated diseases are still questionable, such as diseases of the genitourinary system. The diseases of genitourinary system may be contributed partially by the urinary tract infection which is one of other common admission causes. On the other hand, the causality of the close relationship between disease of genitourinary system and diabetes mellitus among inpatients, as well as other strong or vague relationships, still remains further studies.

As this study collects the data from admission medical records, it could only appear the disease structure among inpatients. The figure among patients in community may be different. Although we did not do the subgroup analysis, we could reasonably assume our study imply the hospitalization risk of those patient with diabetes mellitus. Besides, while not conclusive, these observations can also explain that more connected diseases are seen to be more severe, as patients developing highly connected diseases are more equivalent to those who at an advanced stage of disease, which can be reached through multiple paths in the PDN.

We cannot directly tell the mechanism from the network perspective study. Yet, we could sketch the disease progression, as patients tent to develop complications in the vicinity of PDNs [4]. The further inpatients network perspective study and subgroup analysis may help to capture the specific admission causes which could be the secondary target in the diabetes mellitus care plans. Therefore, for reducing hospitalization and increasing quality of life, we hope this initial stage inpatients network perspective study may help to strengthen and adjust the diabetes mellitus care plans among different subgroups and in different disease stages.

As diabetes mellitus is a group of diseases, and the nature courses of subtypes are not totally the same, we could separate the subtypes and other factors in further studies.

Exploring comorbidities from a network perspective helps determine whether differences in the comorbidity patterns expressed in different populations indicate differences in races, country, or socioeconomic status for each population. Here this study show as an initially stage that there are differences in the strength of co-morbidities measured for patients of different gender. The PDN could be the starting point of studies exploring these and related questions.

Acknowledgements. This study is based in part on data from the National Health Insurance Research Database provided by the Bureau of National Health Insurance, Department of Health and managed by National Health Research Institutes (NHRI). The interpretation and conclusions contained herein do not represent those of Bureau of National Health Insurance, Department of Health or National Health Research Institutes.

References

1. American Diabetes Association: Standards of Medical Care in Diabete-2014. Diabetes Care. **37**, S14–S80 (2014)
2. Handelsman, Y., Grunberger, G., Zimmerman, R.S., Blonde, L., Cohen, A.J., Davidson, J.A., Ganda, O.P., Henry, R.R., Hurley, D.L., Lebovitz, H.E., McGill, J.B., Mestman, J.H., Orzeck, E.A., Rosenblit, P.D., Wyne, K.: American Association of Clinical Endocrinologists and American College of Endocrinology – Clinical Practice Guidelines for Developing a Diabetes Mellitus Comprehensive Care Plan - 2015. AACE/ACE Guidelines http://pic1.cmt.com.cn/newspic/files/dm-guidelines-ccp.pdf (retrieved April 30, 2015)
3. Lakka, H., Laaksonen, D.E., Lakka, T.A., Niskanen, L.K., Kumpusalo, E., Tuomilehto, J., Salonen, J.T.: The Metabolic Syndrome and Total and Cardiovascular Disease Mortality in Middle-aged Men. JAMA **288**(21), 2709–2716 (2002)
4. Hidalgo, C.A., Blumm, N., Barabási, A.L., Christakis, N.A.: A Dynamic Network Approach for the Study of Human Phenotypes. PLoS Computational Biology **5**(4), 1–11 (2009)
5. Goh, K.I., Cusick, M.E., Valle, D., Barton, C., Vidal, M., Barabási, A.L.: The Human Disease Network. Proceedings of the National Academy of Sciences of the United States of America **104**(21), 8685–8690 (2007)
6. Santiago, J.A., Potashkin, J.A.: A Network Approach To Clinical Intervention In Neurodegenerative Diseases. Trends in Molecular Medicine **20**(12), 694–703 (2014)
7. Tarín, J.J., García-Pérez, M.A., Hamatani, T., Cano, A.: Infertility Etiologies Are Genetically and Clinically Linked With Other Diseases in Single Meta-Diseases. Reproductive Biology and Endocrinology 13 (2015). http://www.rbej.com/content/13/1/31 (retrieved April 20, 2015)
8. Moni, M.A., Liò, P.: Network-Based Analysis of Comorbidities Risk During an Infection: SARS and HIV Case Studies. BMC Bioinformatics **15**, 1471–2105 (2014)
9. Iezzoni, L.I., Heeren, T., Foley, S.M., Daley, J., Hughes, J., Coffman, G.A.: Chronic Conditions and Risk of In-Hospital Death. Health Services Research **29**(4), 435–460 (1994)

10. Schneeweiss, S., Wang, P.S., Avorn, J., Glynn, R.J.: Improved Comorbidity Adjustment for Predicting Mortality in Medicare Populations. Health Services Research **38**(4), 1103–1120 (2003)
11. Rzhetsky, A., Wajngurt, D., Park, N., Zheng, T.: Probing Genetic Overlap among Complex Human Phenotypes. Proceedings of the National Academy of Sciences of the United States of America **104**(28), 11694–11699 (2007)
12. National Health Research Institutes: National Health Insurance Research Database (NHIRD) http://w3.nhri.org.tw/nhird/ (retrieved December 1, 2014)
13. Centers for Disease Control and Prevention. International Classification of Diseases, Ninth Revision, Clinical Modification (ICD-9-CM) http://www.cdc.gov/nchs/icd/icd9cm.htm (retrieved December 1, 2014)
14. Ekström, J.: The Phi-coefficient, the Tetrachoric Correlation Coefficient, and the Pearson-Yule Debate http://statistics.ucla.edu/preprints/uclastat-preprint-2008:40 (retrieved February 20, 2015)

Product Innovation and ICT Use in Firms of Four ASEAN Economies

Hiroki Idota[1(✉)], Teruyuki Bunno[2], Yasushi Ueki[3], Somrote Komolavanij[4], Chawalit Jeenanunta[5], and Masatsugu Tsuji[6]

[1] Faculty of Economics, Kinki University, Osaka, Japan
idota@kindai.ac.jp
[2] Faculty of Business Administration, Kinki University, Osaka, Japan
tbunno@bus.kindai.ac.jp
[3] Institute of Development Economies/JETRO, Chiba, Japan
Yasushi_ueki@ide.go.jp
[4] Faculty of Management Sciences, Panyapiwat Institute of Management, Bangkok, Thailand
somrote@gmail.com
[5] Sirindhorn International Institute of Technology, Thammasat University, Bangkok, Thailand
chawalitj@gmail.com
[6] Graduate School of Applied Informatics, University of Hyogo, Hyogo, Japan
tsuji@ai.u-hyogo.ac.jp

Abstract. In order to achieve successful innovation, firms have to elevate their capability for innovation. ICT becomes one of essential tools of business management. The aim of this papers to examine the causal relationship between ICT use and innovation capability. Based on survey data from four ASEAN economies of Indonesia, the Philippines, Thailand, and Vietnam, this study examines what kind of ICT use such as SCM, CAD/CAM, SNS and so on, has been introduced by firms, and how it enhances internal innovation capability and contributes to achieve innovation. In so doing, to make the concept of internal innovation capability more tractable for empirical analysis, AHP (Analytical Hierarchy Process) was employed to construct an index of internal innovation capability. By using instrumental variable probit estimation (or two Stage Least Square: 2SLS), the following results are obtained: (1) ICT significantly influences to enhance internal innovation capability; and (2) internal innovation capability promotes innovation.

Keywords: ICT · Innovation · Innovative capability · AHP · MNCs

1 Introduction

Economic development in East Asia has continued in the midst of a global recession after the economic crisis. Further agglomeration has been transforming the area from a simple production base to a knowledge-based economy. In order to achieve this, the further empowerment of regional firms to enhance innovation is required. This is a difficult and time-consuming. A large number of factors in the innovation process

© Springer-Verlag Berlin Heidelberg 2015
L. Wang et al. (Eds): MISNC 2015, CCIS 540, pp. 223–235, 2015.
DOI: 10.1007/978-3-662-48319-0_18

inside the firm in regions and in economies is required, as endogenous economic growth theory emphasizes, i.e. capital, labor and technology. In reality, it is difficult to raise these factors and promote economic development in the entire economy, but it is more difficult to improve the power of innovation within an individual firm. This paper attempts to identify factors behind the innovation process in an individual firm. Such power is referred to as internal innovation capability, or internal capability, for short. Typical factors are technology, managerial organization, and human resources. An important issue is how to promote these factors which are required for innovation. There are two measures for individual firms to promote these factors: (1) investment in technology and in human resources; (2) utilizing factors outside the firm to promote internal capability.

This paper defines internal capability as an integrated ability of a firm to create innovation consisting of all resources, core competence, or competitiveness. In addition, ICT use has been recognized as an important factor for innovation. ICT stimulates innovation activities [1, 2, 3, 4, 5, 6, 7, 8]. Some previous studies have also showed that ICT was viewed as an effective tool for innovation [9, 10, 11, 12, 13, 14, 15, 16, 19]. If firms obtain information and knowledge from outside the firm by using ICT, firms can transform them to innovation by using their internal capability.

In what follows, this paper examines therefore research questions:

RQ 1: How innovative capability promotes innovation in firms of four ASEAN economies

RQ 2: Whether ICT significantly influence enhanced innovative capability.

2 Hypothesis

According to the above discussions, this study postulates the following hypotheses to test:

Hypothesis I: Internal innovation capability affects (product) innovation.
Hypothesis II: ICT use enhances internal innovation capability.

The aim of Hypothesis I is that firms perceive new information on technologies or the shifts of consumer's taste, assimilate them with existing knowledge and resources inside the firm, and transform new information or knowledge into innovation. In this process, internal capability plays an essential role, which is defined as an integrated ability of a firm to create innovation.

In addition, ICT has been regarded as a tool not only that improves the productivity of firms but also that enhances innovation. ICT improves greatly the power of obtain and sharing information. It comes to enable to sort new findings from so-called big data on the real time basis, and information thus obtained can be shared instantly by various related people.

In addition to ICT use inside the firm, ICT also supports collaboration with entities outside the firm, in particular, cooperation with other related firms, universities, and local research institutions by creating a virtual environment in which ICT transmit much larger volume of information at much faster speed than face-to-face communication. In the open innovation process, a strategy for sharing information and resources with other firms, from suppliers to customers, is required. In this sense, ICT enhances internal innovation capability, which is the objective of Hypothesis II.

3   Summary of Survey

3.1   Data

The summary of data obtained from the surveys is presented here. This study is based on mail surveys and phone interviews which were conducted with 1,232 companies in the Hanoi area and 1,000 in the Ho Chi Minh City area, Vietnam, 239 in the Batangas and other areas in the Philippines, 437 in the Jabodetbek area, Indonesia, and 878 in Greater Bangkok, Thailand. The surveys were conducted from 2012 to 2013 by ERIA. The numbers of valid responses were 149 from the Hanoi area (13.16%), 171 from Ho Chi Minh City (17.10%), 157 from Indonesia (35.93%), 237 from the Philippines (99.16%) and 284 from Thailand (32.35%).

3.2   Indices of the Internal Innovation Capability and ICT Use

The construction of an index of the internal innovation capability of firms and ICT index are presented.

(1) Internal innovation capability index.
This study follows the same fundamentals and methodology as used in the previous study, that is, the AHP approach [17, 18], which explains how the internal capability of firms to create innovation is defined according to the questionnaire.

It postulates the following three factors that contribute to innovation: (i) technology, (ii) managerial organization, and (iii) human resources. The technological factor is clearly the basis of innovation. These three constitute the "first layer" and are referred to as first-layer factors. Moreover, each of these factors has its own detailed sub-factors, which form the "second layer." These sub-factors are called second-layer factors. Let us take the example of (i) the technological factor, which includes the following two second-layer factors: (a) "ratio of R&D expenditure to sales at present," and (b) "established OEM (Original Equipment Manufacturer)," asked about high technical ability. (ii) Managerial organization indicates whether the managerial organization is designed and functioning to encourage exchange and a sharing of information among employees. This first-layer factor consists of the following three second-layer factors: (c) "adopted ISO9000/14000;" (d) "cross-functional team;" and (e) "practicing QC circle." Finally, the first-layer factor of human resources is an important factor for engaging in innovation activities as well as for design and managing R&D. This consists of the following three second-layer factors: (f) "career of top management;" (g) "career of factory manager;" and (h) "training/HRD program." Figure 1 shows the tree structure of the index and related questions in the questionnaire. The obtained weights of factors of the first and second layers are shown in Table 1, which are the same as the previous study since it is less possible that the weights which are reflections of evaluators of whom we asked the rating, such as specialists, academics, and top management of firms does change in the short run.

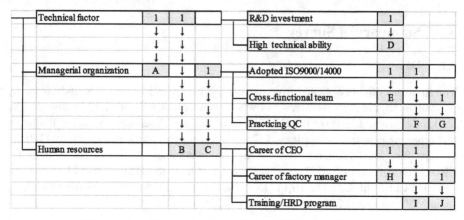

Fig. 1. Internal Innovation Capability AHP Hierarchy Diagram

Source: Authors.

Table 1. Weights of Factors by AHP

Technical factor	0.490	R&D investment	0.240
		High technical ability	0.760
Managerial organization	0.164	Adopted ISO9000/14000	0.104
		Cross-functional team	0.623
		Practicing QC	0.273
Human resources	0.346	Career of CEO	0.675
		Career of factory manager	0.153
		Training/HRD program	0.172

Source: Authors.

(2) Indices of ICT use for inside and outside the firm

Indices of ICT use are constructed for empirical analysis. This study constructs the indices in a simple way. ICT use is categorized into two groups; inside and outside the firm. ICT use for inside the firm consists of (i) ERP (Enterprise Resource Planning) packages, (ii) groupware, (iii) CAD/CAM and (iv) Intra-Social Networking Services (SNSs). The latter consist of (i) B2B e-commerce, (ii) B2C e-commerce, (iii) EDI, (iv) SCM, and (v) Public SNSs.

If firms utilize these systems, they replied "yes." The number of positive replies is then calculated. To match the scale of two variables, the number of ICT use for inside the firm is multiplied by 1.2. Thus the number of systems used takes a value between 0-5, which is referred to as a "score." Accordingly, the total score is a maximum of three, which is termed the index of ICT use inside the firm, abbreviated to "ICT index (internal)."

The "ICT index (external)" is also constructed in the same manner, consisting of e-commerce, EDI, SCM and public SNS. The number of positive replies to the use of these systems becomes the score, and then the index. The score thus takes a value of 0-5.

Finally, ICT index of all categories is calculated by ICT index (internal) plus ICT index (external), abbreviated to "ICT index (all)." The summary statistics of ICT indices is shown in Table 2.

It should be added that values of C.I.s of Technical capability, Organizational learning, and Human resource management are, 0.000, 0.037, and 0.007, respectively. Since overall C.I. is 0.001, it is less than 0.1 and accordingly it is statistically fitted.

Table 2. The summary statistics of ICT index

	ICT index (internal)	ICT index (external)	ICT index (all)
Mean	1.193	1.375	2.556
Median	1.250	1.000	2.000
Max.	5	5	10
Min.	0	0	0

Source: Authors.

4   Analyses and Results

4.1   Methodology

The models for estimations on innovations, capability, and ICT use are formulated as follows:

$$Capability_i = \beta_0 + \beta_1 ICTindex_i + \beta_2 X_i + v_i \tag{1}$$

$$Innovation_i = \alpha_0 + \alpha_1 Capability_i + \alpha_2 X_i + u_i \tag{2}$$

where $Innovation_i$ denotes whether the i-th firm achieves a particular type of product innovation which will be explained later. It takes 0 if it replied "not tried yet," while it takes 1 if it replied "achieved." The product innovations are represented by four types: Innovation I (redesigning packaging), II (new products by new technologies), III (new products by existing technologies), and IV (improvement of existing products). $Capability_i$ implies its capability index which continuously takes between 0 and 1, and ICT_i indicates the indices the i-th firm has, which are represented between 0 and 5, except ICT index (all) which takes between 0 and 10.

X_i is the i-th firm's characteristics, including "whether firms have overseas affiliates (dummy, 1: Yes, 0: No)," "year of establishment," "local firms or not (dummy)," "number of full-time employees (categorical data)," "amount of total assets (categorical data)," and "main business activities (dummy)." X also includes dummy variables such as Indonesia, Thailand, the Philippines, and Ho Chi Minh City in Vietnam, while Hanoi in Vietnam is excluded as a reference area.

In the models for estimation where product innovations are taken as dependent variables (equation 2), since estimated coefficients contain the endogenous bias due to the correlation between explanatory variables and the error term using simple probit analysis, estimation should be based on so-called "models with dichotomous dependent variables and endogenous regressors."

Table 3. Summary Statistics

Variable	N	Mean	S.D.	Min.	Max.
Innovation					
Redesigning packaging or significantly changing appearance design	851	0.320	0.467	0	1
Significantly improving existing products	851	0.353	0.478	0	1
New product based on the existing technologies	851	0.261	0.439	0	1
New product based on new technologies	851	0.152	0.359	0	1
Internal Capability					
Capability index	851	0.399	0.236	0	0.909
ICT use					
ICT index (internal)	850	1.181	1.276	0	5
ICT index (external)	850	1.375	1.353	0	5
ICT index (all)	850	2.556	2.210	0	10
Firms' Characteristics					
Own overseas affiliates	851	0.329	0.470	0	1
Year of establishment	793	1995.391	14.355	1891	2013
100% locally-owned	840	0.548	0.498	0	1
No. of full-time employees (category)	846	5.053	2.795	1	11
Total assets (category)	758	7.384	2.218	1	10
Food, beverages, tobacco	843	0.076	0.265	0	1
Textiles	843	0.057	0.232	0	1
Footwear	843	0.012	0.108	0	1
Chemicals, chemical products	843	0.051	0.220	0	1
Plastic, rubber products	843	0.094	0.292	0	1
Iron, steel	843	0.030	0.170	0	1
Metal products	843	0.074	0.261	0	1
Other electronics & components	843	0.127	0.333	0	1
Precision instruments	843	0.007	0.084	0	1
Automobile, auto parts	843	0.079	0.271	0	1
Other business activity	843	0.115	0.319	0	1
Hanoi (dummy)	851	0.175	0.380	0	1
Ho Chi Minh (dummy)	851	0.201	0.401	0	1
The Philippines (dummy)	851	0.168	0.374	0	1
Thailand (dummy)	851	0.334	0.472	0	1
Indonesia (dummy)	851	0.122	0.328	0	1

Source: Authors.

In equation 2, for instance, estimator α_1 contains possibly endogenous bias. In order to avoid this, instrumental variables are introduced to probit estimation. Similarly, since other models such as equation 1 also require instrumental regression models, two-stage least squares (2SLS) for estimation is used. Variables for estimation as well as their summary statistics are shown in Table 3.

Based on the above discussions, the procedures of estimations as well as the hypotheses in this study are summarized in Figure 2.

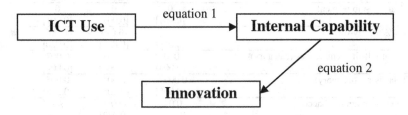

Fig. 2. Relationship of Variables and Procedure of Estimations

Source: Authors.

4.2   Causality from ICT Use to Internal Capability: Equation 1

In this section, ICT use effect on internal capability (equation 1) is examined. The estimation method here is based on the instrumental variable probit two-stage least squares (IV2SLS). The result of estimation is summarized in Table 4 in which both Sargan and Basmann test for overidentifying restrictions that show instruments are selected appropriately.

According to the result, all ICT indices are positively significant to internal capability, namely "ICT index (internal)" ($p < 0.01$), "ICT index (external)" ($p < 0.01$), and "ICT index (all)" ($p < 0.01$). The hypothesis related to the equation 1 is clarified for the part of "ICT use to internal capability". Especially, the coefficient of "ICT index (external)" is the highest among them. "ICT index (external)" which is consisted by B2B e-commerce, B2C e-commerce, EDI, SCM and Public SNSs is useful for innovation, in order to gather information and knowledge from outside the firm.

As for the other variables, "whether firms have overseas affiliates," "Number of full-time employees," "Total assets," "Textiles," "Plastic, rubber products," "Metal products," "other electronics & components," "Automobile, auto parts," "Thailand," and "Indonesia" are positive significant.

"Whether firms have overseas affiliates" showed strongly positive effects in raising internal capability, since owning overseas affiliates seems to be a proxy for those firms having a high absorptive capability to learn from outside the firms and that they are reliable to partner with.

Table 4. Relationship between ICT use and internal innovation capability

	Case 1	Case 2	Case 3
ICT index (internal)	**0.063*****		
	[0.021]		
ICT index (external)		**0.116*****	
		[0.033]	
ICT index (all)			**0.066*****
			[0.017]
Own overseas affiliates	0.058***	0.043**	0.047**
	[0.020]	[0.021]	[0.021]
Year of establishment	0	0	0
	[0.001]	[0.001]	[0.001]
100% locally-owned	-0.008	-0.013	-0.007
	[0.019]	[0.020]	[0.020]
No. of full-time employees (category)	0.010**	0.009**	0.007
	[0.004]	[0.004]	[0.004]
Total assets (category)	0.009**	0.005	0.005
	[0.005]	[0.005]	[0.005]
Food, beverages, tobacco	0.042	-0.009	0.026
	[0.040]	[0.042]	[0.040]
Textiles	0.072*	0.047	0.071*
	[0.042]	[0.043]	[0.041]
Footwear	0.036	-0.078	-0.002
	[0.071]	[0.075]	[0.070]
Chemicals, chemical products	0.043	0.013	0.031
	[0.043]	[0.045]	[0.043]
Plastic, rubber products	0.090**	0.096**	0.101***
	[0.037]	[0.038]	[0.037]
Iron, steel	0.05	0.021	0.03
	[0.053]	[0.056]	[0.054]
Metal products	0.070*	0.101**	0.077**
	[0.038]	[0.039]	[0.038]
Other electronics & components	0.151***	0.107***	0.135***
	[0.035]	[0.039]	[0.037]
Precision instruments	0.007	0.06	0.037
	[0.088]	[0.092]	[0.088]
Automobile, auto parts	0.130***	0.124***	0.120***
	[0.041]	[0.042]	[0.042]
Ho Chi Minh (dummy)	-0.011	-0.169**	-0.156**
	[0.040]	[0.076]	[0.068]
The Philippines (dummy)	-0.002	0.019	0.001
	[0.030]	[0.031]	[0.029]
Thailand (dummy)	0.143***	0.117***	0.129***
	[0.030]	[0.032]	[0.031]
Indonesia (dummy)	0.110***	0.099***	0.100***
	[0.031]	[0.033]	[0.031]
Constant	0.107	0.673	0.627
	[1.277]	[1.367]	[1.322]
Observations	710	709	703
R-squared	0.251	0.187	0.254
Sargan chi-squared (overidentifying restrictions)	7.051	5.950	7.231
Prob > chi-squared	0.424	0.203	0.204
Basmann chi-squared (overidentifying restrictions)	6.750	5.712	6.943
Prob > chi-squared	0.455	0.222	0.225

Note 1: Standard errors in brackets *** $p<0.01$, ** $p<0.05$, * $p<0.1$,

Note 2: Case 1: ICT index (internal), Case 2: ICT index (external), Case 3: ICT index (all)

Source: Authors.

4.3   Causality from Internal Capability to Innovation: Equation 2

This section examines whether internal capability promote product innovation which is summarized in Table 5.

In Table 5, the capability index is included as explanatory variables. According to the obtained results, instrument variables are properly chosen, since Amemiya-Lee-Newey minimum statistics for overidentifying restriction are not significant at the 10% level in all of the estimation models.

Regarding the estimated coefficients, all of the coefficients of the internal capability index are significant ($p< 0.01\sim0.1$). The hypothesis related to the equation (2) is thus proved for the part of "internal capability to product innovations." In other words, internal capability positively raises the probability of achieving product innovation. Equivalently, causality from internal capability to produce innovation is demonstrated.

As for the other variables, such as firms' characteristics (including size of firms in terms of the number of employees and total assets), they showed positively significant for Innovation I (redesigning packaging), only total assets for Innovation III (new products by existing technologies), and only employees for Innovation IV (new products by new technologies). In addition, firms with shorter operating histories are innovative in Innovation II (improvement of existing products), since the year of establishment is negatively significant. The probability of achieving Innovation I (redesigning packaging) is higher in firms whose business activities are "foods," and "precision instruments." Similarly, the probability of achieving Innovation II (improvement of existing products) is higher in "footwear," "chemicals," and "plastic, rubber products." Only firms of "plastic, rubber products" have probability of Innovation IV (new products by new technologies).

Table 5. Relationship between internal innovation capability and innovation

	Case1	Case2	Case3	Case4
Capability index (by AHP)	**2.266*****	**1.717***	**1.490***	**1.905***
	[0.798]	[0.927]	[0.765]	[1.056]
Own overseas affiliates	-0.08	0.009	-0.051	0.081
	[0.160]	[0.147]	[0.150]	[0.197]
Year of establishment	-0.004	**-0.013*****	-0.006	0.001
	[0.005]	[0.004]	[0.004]	[0.006]
100% locally-owned	-0.18	0.056	0.029	-0.09
	[0.147]	[0.133]	[0.140]	[0.178]
Q4.1. No. of full-time employees (category)	**0.060***	0.035	0.036	**0.098****
	[0.033]	[0.030]	[0.030]	[0.040]
Q4.2. Total assets (category)	**0.084****	0.057	**0.100*****	-0.036
	[0.038]	[0.035]	[0.037]	[0.045]
Food, beverages, tobacco	**0.591***	0.254	0.405	0.187
	[0.346]	[0.292]	[0.292]	[0.375]
Textiles	-0.053	0.18	-0.201	-0.111
	[0.355]	[0.296]	[0.317]	[0.412]
Footwear	0.595	**0.955***	-0.126	0.07
	[0.558]	[0.502]	[0.526]	[0.655]

Table 5. (*continued*)

Chemicals, chemical products	0.32	**0.530***	0.229	0.579
	[0.373]	[0.300]	[0.316]	[0.387]
Plastic, rubber products	0.297	**0.494***	0.373	**0.669****
	[0.316]	[0.268]	[0.275]	[0.325]
Iron, steel	-0.6	-0.232	-0.442	-0.013
	[0.601]	[0.410]	[0.467]	[0.582]
Metal products	0.22	0.394	0.236	0.475
	[0.318]	[0.277]	[0.278]	[0.329]
Other electronics & components	0.087	-0.085	0.199	-0.281
	[0.313]	[0.289]	[0.279]	[0.360]
Precision instruments	**1.784****	-0.024	0.122	1.019
	[0.659]	[0.581]	[0.675]	[0.702]
Automobile, auto parts	-0.063	-0.283	-0.16	-0.45
	[0.351]	[0.324]	[0.318]	[0.430]
Ho Chi Minh (dummy)	0.297	-0.171	**0.788****	**0.671****
	[0.186]	[0.179]	[0.202]	[0.295]
The Philippines (dummy)	**0.555****	**-0.357***	**0.977****	**1.547****
	[0.217]	[0.196]	[0.230]	[0.316]
Thailand (dummy)	-0.118	**-0.771****	0.423	**0.875****
	[0.338]	[0.249]	[0.260]	[0.444]
Indonesia (dummy)	**-1.195****	**-1.657****	-0.002	**0.804****
	[0.306]	[0.271]	[0.269]	[0.351]
Constant	5.68	**23.985****	9.775	-4.929
	[10.109]	[8.969]	[8.892]	[11.507]
Observations	611	708	703	611
Amemiya-Lee-Newey minimum chi-squared (overidentifying restriction)	5.17	2.69	2.83	2.45
Prob > chi-squared	0.522	0.442	0.726	0.785

Note 1: Standard errors in brackets and *** $p<0.01$, ** $p<0.05$, * $p<0.1$
Note 2: Case 1: Redesigning packaging or significantly changing appearance design, Case 2: Significantly improving existing products, Case 3: New product based on the existing technologies, Case 4: New product based on new technologies.
Source: Authors.

5   Conclusion

Let us summarize the results of the analysis in this paper (Figure 3).

(1) ICT significantly has significant influence on enhancing internal capability.
(2) Internal capability significantly promotes product innovation.

According to the above results, this study verifies the existence of a cumulative process between internal capability and ICT use. Internal capability itself directly enhances product innovation, while ICT use promote product innovation via an enhancement of internal capability. In this sense, internal capability is the core of innovation, but it is not necessarily promoted without supplementary external information. The following processes related to how information is transferred to firms and assimilated inside the firm: (1) information required for innovation is transmitted along the distribution networks such as SCM and SNSs; and (2) knowledge creation for innovation inside the firm is prompted by using groupware and inter-SNSs. ICT use is thus demonstrated to enhance innovative capability through these processes.

Although this study thus demonstrated interesting results, it has some limitations. The directions of further research to overcome these are discussed here. First, there are other approaches to obtain internal innovation capability such as factor analysis or principal component analysis. Although AHP is based on rigorous methodology, its validity cannot be tested by statistical analysis. Second, among the causal relationships, this study assumes that ICT is one of the fundamentals of innovation, that is, as shown Figure 5, ICT causes to raise internal capability, on the other hand, some of studies postulated on the other way around, namely internal capability is a cause of ICT use. Firms own a larger internal capability index tend to introduce ICT more easily[1]. Which way is more realistic must be analyzed by using ASEAN economies data[2]. In this analysis, less concern is paid to the relationship between ICT use and external linkages such as universities, local research organizations, and MNCs (multinational corporations). How firms in the developing economies utilize ICT for obtaining new information, and through what channel they obtain information are important to decide strategic policy to promote innovation in these areas. Lastly, this study does not successfully identifies the mechanism inside the firm regarding how ICT use enhances internal capability or innovation, and the reality is much more complicated than the hypothesis shown in Figure 2 and 3. In cope with these issues, questions must be designed more properly.

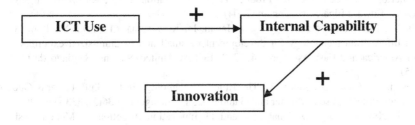

Fig. 3. Summary of Results

Source: Authors.

---

[1] Our previous research clarified organizational learning which is one element of internal capability promotes ICT use in ASEAN economies [19].

[2] In our previous research, the relationship between ICT use and innovation in Japanese SMEs was analyzed. As the results of the analyses, innovation capability is activated by ICT use, and thus innovation is promoted. On the other hand, the model that "ICT use is improved by innovation capability, and then innovation is promoted," was less suitable [16].

Acknowledgements. This paper is based on research outcomes of an ERIA (Economic Research Institute for ASEAN and East Asia) Supporting Study Project "Innovation Between and Within Supply Chain: Empirical Study of Tracing Local and Global Production-Knowledge Network in East Asia." The financial support is gratefully acknowledged.

This paper is revised "ICT Use in the Innovation Process of Firms in ASEAN Economies," which is presented in poster session of TPRC 41: The 41st Research Conference on Communication, Information and Internet Policy.

References

1. Thomke, S.H.: Managing Experimentation in the Design of New Products. Management Science **44**(6), 743–762 (1998)
2. Thomke, S.H.: Simulation, Learning and R&D Performance: Evidence from Automotive Development. Research Policy **27**(2), 55–74 (1998)
3. Thomke, S.H.: Experimentation matters. Harvard Business School Press, Boston (2003)
4. Henderson, K.: On-line and on-paper: Visual Representations, Visual Culture, and Computer Graphics in Design Engineering. MIT Press, Cambridge (1999)
5. Schrage, M.: Serious Play: How the World's Best Companies Simulate to Innovate. Harvard Business School Press, Boston (1999)
6. von Hippel, E.: User Toolkits for Innovation. Journal of Product Innovation Management **18**, 247–257 (2001)
7. Debackere, K., Van Looy, B.: Managing integrated design capabilities in new product design and development. In: Dankbaar, B. (ed.) Innovation Management in the Knowledge Economy, pp. 213–234. Imperial College Press, London (2003)
8. D'Adderio, L.: Inside the Virtual Product: How Organizations Create Knowledge through Software. Edward Elgar, Cheltenham (2004)
9. Tsuji, M.: Information technology use in Japan. In: Kuwayama, M., Ueki, Y., Tsuji, M. (eds.) Information technology for development of small and medium-sized exporters in Latin America and East Asia, pp. 345–373. ECLAC/United Nations, Santiago de Chili (2005)
10. Dodgson, M., Gann, D., Salter, A.: The Role of Technology in the Shift Towards Open Innovation. In: the Case of Procter and Gamble. R&D Management **36**(3), 333–346 (2006)
11. Lee, G., Xia, W.: Organizational Size and IT Innovation Adoption: A Meta-analysis. Information & Management **43**, 979–985 (2006)
12. Tsuji, M., Miyahara, S.: A comparative analysis of organizational innovation in Japanese SMEs generated by information communication technology. In: Kuchiki, A., Tsuji, M. (eds.) From Agglomeration to Innovation: Upgrading Industrial Clusters in Emerging Economies, pp. 231–269. Palgrave Macmillan, Basingstoke (2010)
13. Spiezia, V.: Are ICT users more innovation? An analysis of ICT-enabled innovation in OECD firms. OECD Journal: Economic Studies, 99–119 (2011)
14. Idota, H., Ogawa, M., Bunno, T., Tsuji, M.: An empirical analysis of organizational innovation generated by ICT in Japanese SMEs. In: Allegrezza, S., Dubrocard, A. (eds.) Internet Econometrics, pp. 259–287. Macmillan, Hampshire, UK (2012)
15. Idota, H., Bunno, T., Tsuji, M.: An empirical analysis of innovation success factors due to ICT use in Japanese firms. In: Kargidis, T., Katsaros, P. (eds.) Approaches and processes for managing the economics of information systems, pp. 324–347. IGI, Hershey (2013)

16. Idota, H., Bunno, T., Tsuji, M.: Impact of ICT on innovation: the case of Japanese SMEs. In: Kaur, S. (ed.) Handbook of Research on Cultural and Economic Impacts of the Information Society. IGI, Hershey (2015, in print)
17. Saaty, T.L.: The Analytic Hierarchy Process: Planning, Priority Setting, Resource Allocation. McGraw-Hill, New York (1980)
18. Saaty, T.L.: Absolute and Relative Measurement with the AHP: The Most Livable Cities in the United States. Socio-Economic Planning Sciences **20**(6), 327–331 (1986)
19. Idota, H., Ueki, Y., Bunno, T., Shinohara, S., Tsuji, M.: Role of ICT in the innovation process based on firm-level evidence from four ASEAN economies: an SEM approach. In: proceedings of 23th European Regional ITS Conference in Brussels, pp. 1–29 (2014). http://econstor.eu/bitstream/10419/101397/1/795228937.pdf

How Social Media Enhances Product Innovation in Japanese Firms

Hiroki Idota[1(✉)], Teruyuki Bunno[2], and Masatsugu Tsuji[3]

[1] Faculty of Economics, Kinki University, Osaka, Japan
idota@kindai.ac.jp
[2] Faculty of Business Administration, Kinki University, Osaka, Japan
tbunno@bus.kindai.ac.jp
[3] Graduate School of Applied Informatics, University of Hyogo, Hyogo, Japan
tsuji@ai.u-hyogo.ac.jp

Abstract. A large number of firms use social media as new communication tools for obtaining information on consumer's needs and market for developing new goods and services. Authors' previous research found social media use promotes product innovation based on questionnaire data of Japanese firms. However, it was not clear whether social media effected directly on product innovation. This paper introduces a new variable constructed by achievement of product innovation and social media use and utilizes this as a dependent variable. The estimation methods conducted in the previous research are employed for probit analysis. This study finds all estimations have the same results. In addition, this paper also examines what kind of effect of social media is important for product innovation. The effects of social media such as obtaining trends and customer's needs, and opening the FAQ site are found to be useful for achieving product innovation.

Keywords: Social media · Product innovation · Effect of social media · R&D orientations

1 Introduction

Social media including SNS (Social Network Service/Social Network Site), Twitter, and the blogs (a short form of 'web log') has been spreading all over the world. In these circumstances, a number of firms come to recognize social media as new communication tools for obtaining information on consumer's needs and market for developing new goods and services, and for promoting marketing. In particular, industries such as automobile, PC, mobile phones, transportation and finance came to utilize widely social media. In spite of increasing its popularity in the real society, academic research on whether or how social media contributes to product innovation remains few [1, 2, 3]. One reason is that it is less intuitive to recognize how social media leads to innovation. Our previous research attempted to analyze empirically how social media use enhances product innovation based on the survey data of Japanese firms using statistical method such as ordered probit analysis [2]. In this estimation, it was not clear, however, whether

© Springer-Verlag Berlin Heidelberg 2015
L. Wang et al. (Eds): MISNC 2015, CCIS 540, pp. 236–248, 2015.
DOI: 10.1007/978-3-662-48319-0_19

or not social media directly effects on product innovation due to small sample size. In order to answer this question, this paper attempts to analyze this by using the cross term constructed by two questions; "Product innovation was achieved" multiplied by "Social media is thought to be useful to enhance product innovation." The reason is that this cross term examines whether the firms which consider social media important to product innovation achieve actually product innovation. This variable is a proxy for connecting innovation and social media use. By taking this as an independent variable, the same estimation procedure as the previous paper, probit analysis, is employed. Thus this paper analyzes what kind of effect of social media is important for product innovation.

2   Social Media and Product Innovation

Social media has recently been introduced and utilized by a large number of firms. In such firms, blogs are used to transmit and share business-related information. Managers, for instance, send messages to their subordinates via social media for the purpose of a bulletin board posting an administrative report, in-house rules, news of employees, and so on. A working team needs to share specific information such as on the daily report, projects, customers, market, and technology through social media. Social media thus can activate in-house communications. It is reported that even the simple use of social media among employees promotes innovation [1, 2]. Moreover, social media can be used for obtaining useful consumer's needs for product innovation and enables collaboration with entities outside the firm. As mentioned above, a large number of studies examining the roles of social media in business focus mainly on marketing and sales force; they analyze how social media is utilized for advertising their products, sampling consumer's needs, reputation of products, and so on [3, 4, 5, 6, 7, 8, 9, 10, 11]. Useful contents for product improvement and a new marketing strategy are contained in information that users exchange through social media [3]. Moreover, firms can contact with opinion leaders by identifying from information exchanged in social media. These are especially necessary for the product innovation of goods and services for consumers.

3   Methodology

3.1   Research Questions

As explained in the previous section, social media is a meaningful tool to communicate with not only consumers but also colleagues of teams or firm as a whole. The objectives of this study are thus to examine whether and how social media promotes product innovation. In so doing, this paper postulates the following research questions:

RQ1: How social media promotes innovation, particularly what type of social media enhances innovation.
RQ2: How such firms implement social media for innovation.
RQ3: What kind of R&D activities are conducted.

3.2   Questionnaire and Data

The analysis employs the data obtained by authors' mail survey of product innovation and social media use conducted in February 2014; the questionnaire was sent to 2,000 firms throughout Japan. The firms were selected from industries which have possibility to use social media such as automobile, apparel, electric machinery, consumer electronics, medicine, cosmetics, soap, detergent, beverage, food, telecommunications, software, restaurant, department store, supermarket, other retailing, trading company, wholesale, travel agent, transportation (e.g., railway and airline), construction, bank, brokerage, insurance, and other industries. Among these industries, the number of firms which are successfully implementing social media is not expected to be so large, and they were selected from previous surveys on social media and marketing, innovation, and so on. The data contains listed and non-listed firms, and regional quota was not considered.

Under these setting, 70 (3.5%) valid responses were received, and the summary of replies to specific questions is shown in Table 1. It should be noted that the question of "Achievement of product innovation in 2011-2013," which will be used in estimation in what follows, takes "2," if firms achieved entirely new goods and services; it takes "1," if they achieved the improvement of existing commodities; and it takes "0,", if there was no such innovation. The question about "Importance of social media to product innovation" takes "2," "1," and "0," if they replied to "yes," "not decided," and "no," respectively. Regarding "Orientations toward social media," the values of "Max" and "Min" were obtained by factor analysis, which will be shown later.

Table 1. Descriptive statistics of replies to selected questions

	Variables	Obs.	Mean	Std. Dev.	Min	Max
Achievement of product innovation in 2011-2013		70	1.671	0.717	0	2
Importance of social media to product innovation		67	1.209	0.664	0	2
Social media use (Yes or No)	Owing official accounts	68	0.574	0.498	0	1
	Owning community sites	69	0.159	0.369	0	1
	Owning in-house use of social media	69	0.188	0.394	0	1
Years of operation		65	63.554	49.239	3	271
Capital (ln) (unit: Japanese Yen)		65	3.253	1.099	1	5.7741
Industry	Manufacturing	70	0.429	0.498	0	1
	Service	70	0.386	0.490	0	1
	Other industries	70	0.186	0.392	0	1
Orientations toward social media	Consumer orientation	59	0.000	0.967	-2.477	1.8351
	Prototype orientation	65	0.000	0.881	-0.708	2.8592
	Initiative orientation	65	0.000	0.996	-1.486	0.8354
	Collaboration orientation	65	0.000	0.832	-0.715	3.4163
Importance of items in collaborating with consumers (five scales)	Developing the concept of goods and services	63	3.492	1.014	1	5
	Evaluating the concept of goods and services	63	3.683	1.148	1	5
	Decision on function and content of goods and services	63	3.270	1.003	1	5
	Decision on design package	63	3.048	1.099	1	5
	Decision on brand name	63	2.746	1.077	1	5
	Decision on sales price	62	3.161	1.011	1	5
	Proposal of how to use	63	3.476	1.030	1	5
	Proposal of easy maintenance in daily use	62	3.065	1.186	1	5
	Proposal of the abandonment method	62	2.629	1.090	1	5
	Evaluating PR projects and events	62	3.306	1.080	1	5
	Evaluation of characters and talents of advertisement or commercial	62	2.871	1.166	1	5

Table 1. (*continued*)

	Ideas of product development are discovered by questionnaire and word of mouth	65	0.585	0.497	0	1
	Ideas of new goods and services are obtained by exchanging with opinion leaders.	65	0.277	0.451	0	1
	Appointing persons in charge of product development who knead concepts of products based on discovered ideas.	65	0.692	0.465	0	1
	Software, equipment, tools, etc. are lent to consumers so that they can make a prototype by themselves.	65	0.031	0.174	0	1
Actions taken for new product development (No=0, Yes = 1)	Presenting trial plans until a prototype is made, and consumers' reactions are collected.	65	0.154	0.364	0	1
	Trial parties or trail usage of a prototype are offered to consumers.	65	0.200	0.403	0	1
	Pricing, advertising expressions, and packages are presented to consumers to obtain their opinions, and their opinions are consolidated.	65	0.138	0.348	0	1
	Consolidating results of consumer opinions is opened to the public via the web.	65	0.046	0.211	0	1
	To rouse consumers' interest in products, firms urge them to recommend it to others.	65	0.154	0.364	0	1
	For consumers to solicit to friend, events and campaigns are held.	65	0.246	0.434	0	1
	Opportunities such as discussing problems in use and exchanging opinions for improvement after purchased are offered.	65	0.123	0.331	0	1
	Asking to consumers about ideas improving products	65	0.154	0.364	0	1
	Understanding of the trend of consumer awareness	61	0.393	0.493	0	1
	Understanding of the reputation of existing products	62	0.516	0.504	0	1
	Recognition of customers' needs	63	0.397	0.493	0	1
	Recognition of potential customers' needs	62	0.323	0.471	0	1
Effect of social media(No=0, Yes = 1)	Providing the prototypes or the samples and gathering evaluation information of them	62	0.145	0.355	0	1
	Specific of opinion leaders	63	0.127	0.336	0	1
	Joint development of new products and services with opinion leaders	63	0.016	0.126	0	1
	Improvement degree of completion of new products and services	63	0.254	0.439	0	1
	Obtaining claims information	63	0.286	0.455	0	1
	Opening FAQ such as how to use the goods	63	0.190	0.396	0	1

Source: Idota et al. (2014, p.8)

3.3 Methodology and Construction of Variables

This study employs ordered probit analysis to examine the relationship between innovation and social media, and some of questions in the questionnaire as well as new variables which are constructed from questions are used. It is necessary to specify potential variables which causes innovation, which are referred to as "factors" in what follows. For this purpose, factor analysis and the maximum likelihood method (Varimax Rotation) are employed. Table 2 indicates the result of factor analysis which aims to extract a variable named "Collaboration with consumers." The related questions were asked to the five-point Likert scales and one factor is identified as collaborating with consumers, which are related to questions such as "Decision on function and content of goods and services," "Developing the concept of goods and services," "Evaluating the concept of goods and services," "Proposal of easy maintenance in daily use," "Decision on design package," "Decision on brand name," and so on. As a result, one factor converged. Accordingly we denoted this factor as "Consumer orientation."

Table 2. Factor analysis of collaboration with consumers

	Consumer orientation
Decision on function and content of goods and services	.851
Developing the concept of goods and services	.798
Evaluating the concept of goods and services	.791
Proposal of easy maintenance in daily use	.788
Decision on design package	.766
Decision on brand name	.765
Proposal of how to use	.701
Proposal of the abandonment method	.700
Decision on sales price	.677
Evaluating PR projects and events	.649
Evaluation of characters and talents of advertisement or commercial	.561
Sums of squares of loadings after extraction	54.161

Source: Idota et al. (2014, p.9)

In addition, three factors are identified for R&D orientations. The first factor of R&D orientations consists of following questions: "Presenting trial plans until a prototype is made," and "Consumer' reactions are collected," "Pricing, advertising expressions," "Packages are presented to consumers to obtain their opinions, and their opinions are consolidated," "For consumers to solicit to friend, events and campaigns are held," "Trial parties or trail usage of a prototype are offered to consumers," "Opportunities such as discussing problems in use and exchanging opinions for improvement after purchased." Accordingly we denoted this first factor as "Prototype orientation." As for the second factor, questions such as "Appointing persons in charge of product development who knead concepts of products based on discovered ideas" and "Ideas of product development are discovered by questionnaire and word of mouth are extracted" are singled out, which is referred to as "Initiative orientation." The third factor consists of "Ideas of new goods and services are obtained by exchanging with opinion leaders," "Software, equipment, tools, etc. are lent to consumers so that they can make a prototype by themselves," and "Consolidating results of consumer opinions is opened to the public via the web," and it is named as "Collaboration orientation (Table 3)."

Table 3. Factor analysis of R&D orientations

	Prototype orientation	Initiative orientation	Collaboration orientation
Presenting trial plans until a prototype is made, and consumers' reactions are collected.	.780	-.078	.075
Pricing, advertising expressions, and packages are presented to consumers to obtain their opinions, and their opinions are consolidated.	.589	.074	.033
For consumers to solicit to friend, events and campaigns are held.	.544	.299	-.230
Trial parties or trail usage of a prototype are offered to consumers.	.470	-.028	.134
Opportunities such as discussing problems in use and exchanging opinions for improvement after purchased are offered.	.444	.034	.216
Appointing persons in charge of product development who knead concepts of products based on discovered ideas.	-.040	.972	.133
Ideas of product development are discovered by questionnaire and word of mouth	.195	.402	-.204
Ideas of new goods and services are obtained by exchanging with opinion leaders.	-.097	.196	.684
Software, equipment, tools, etc. are lent to consumers so that they can make a prototype by themselves.	.213	-.137	.519
Consolidating results of consumer opinions is opened to the public via the web.	.114	-.076	.515
Factors Correction Matrix			
1	1.000	.344	.111
2	.344	1.000	.242
3	.111	.242	1.000

Source: Idota et al. (2014, p.10)

3.4　Estimation Result I

(1) Innovation and Social Media in General

This study employs ordered probit analysis which enables to clarify the relationships between innovation and social media. In the first estimation, the dependent variable is "Achievement of product innovation in the recent three years," while the explanatory variables are selected from questions which consist of the followings: (i) "Importance of social media to product innovation"; and (ii) firm characteristics such as years of operation, industry, and size of firm. In other words, this estimation examines whether firms which recognize the importance social media tend to achieve more innovation. The estimation result is shown on Table 4, and "Importance of social media to product innovation" is found to be significant for product innovation (0.57, P=0.07).

Table 4. Social media and Innovation

	Product innovation Coeff. / t value
Importance of social media to product innovation	0.57* (1.81)
Operation years	-0.003 (-0.83)
Capital (ln)	0.033 (0.16)
Manufacturing	0.182 (0.43)
Other industries	5.139 (0.01)
Cut1/Constant	-0.205 (-0.26)
Cut2/Constant	-0.053 (-0.07)
log-likelihood	-30.132
Pseudo R^2	0.155
N	61

Note: * $p<0.1$, ** $p<0.05$,*** $p<0.001$

Source: Idota et al. (2014, p.11)

(2) Innovation and Types of Social Media Use

Next, let us examine how the firms which achieve product innovation and consider social media important to product innovation utilize social media. In so doing, taking the cross term constructed by "Achievement of product innovation in the recent three years" multiplied by "Importance of social media to product innovation" as an explained variable, while important explanatory variables included are the following three social media uses: (i) "Owing official accounts"; (ii) "Owing community sites"; and (iii) "Owing in-house use of social media." Other control variables are selected from firm characteristics such as years of operation, industry, and size of firm (Table 5). In Table 5, the first column indicated by "All" shows the model in which above three variables are included, while other three cases contain one of three variables.

Regarding the results of estimations, in case of "All," "Owing official accounts" (0.962, p=0.00) is significant. Regarding other three individual cases, the following variables are found to be significant for achievement of product innovation and importance of social media about product innovation. (i) "Owing official accounts" (1.04, p=0.00); (ii) "Owing community sites" (0.865, p=0.06); and (iii) "Owing in-house use of social media" (0.716, p=0.09). Among other controlled variables in case of "All," the following three are significant, namely "Operation year" (-0.007, p=0.05); "Manufacturing" (0.704, p=0.04); and "Other industries" (0.893, p=0.07).

Table 5. Social media use

| Social media use | Innovation * Importance of social media | | | |
| | Coeff./ z value | | | |
	All	1	2	3
1. Owing official accounts	0.962*** (2.97)	1.04*** (3.26)		
2. Owning community sites	0.708 (1.47)		0.865* (1.90)	
3. Owing in-house use of social media	0.393 (0.89)			0.716* (1.72)
Operation years	-0.007* (-1.96)	-0.005 (-1.54)	-0.008** (-2.39)	-0.005 (-1.64)
Capital (ln)	-0.07 (-0.43)	-0.088 (-0.55)	-0.082 (-0.51)	-0.094 (-0.59)
Manufacturing	0.704** (2.01)	0.821** (2.41)	0.493 (1.48)	0.592* (1.80)
Other industries	0.893* (1.81)	0.978** (2.01)	0.568 (1.26)	0.67 (1.49)
Cut1/Constant	-0.33	-0.303	-1.075	-0.89
Cut2/Constant	-0.268	-0.245	-1.016	-0.833
Cut3/Constant	0.997	0.959	0.14	0.307
log-likelihood	-58.355	-60.157	-64.695	-65.025
Pseudo R^2	0.154	0.128	0.075	0.07
N	60	60	61	61

Note: Standard errors in brackets, and *** $p<0.01$, ** $p<0.05$, * $p<0.1$
Source: Authors

(3) Orientations Toward Social Media

Next, we examine what kind of attitude toward social media use the firms which achieve product innovation and think social media being important to product innovation tend to have. In this case, the dependent variable is the cross-term variable as previous estimation, while the explanatory variables are constructed by the following four orientations: (i) "Consumer orientation"; (ii) "Prototype orientation"; (iii) "Initiative orientation"; and (iv) "Collaboration orientation." Besides these, firm characteristics such as years of operation, industry, and size of firm are included (Table 6). Again in estimation, the first column indicated "All" shows the model in which above four orientations are included, while in other four cases indicated, one of four variables is included for estimation.

Regarding the results of estimations, if four orientations are taken together in the estimation equation, that is, in case of "All," "Consumer orientations" (0.404, p=0.05), "Initiative orientations" (0.478, p=0.03) and "Collaboration orientation" (0.818, p=0.01) are significant. In the separated cases, the following variables are significant, namely achievement of product innovation and significant for importance of social media to product innovation: (i) "Consumer orientation" (0.465, p=0.00); (ii) "Prototype orientation" (0.328, p=0.08); (iii) "Initiative orientation" (0.495, p=0.00); and (iv) "Collaboration orientation" (0.816, p=0.00). In addition to these, in case of "All," "Capital (ln)" (-0.433, p=0.04) and "Other industries" (0.620, p=0.00) are also significant.

Table 6. R&D orientations

R&D orientations	Innovation * Importance of social media				
	All	1	2	3	4
1. Consumer orientation	0.404* (1.94)	0.465*** (2.62)			
2. Prototype orientation	0.016 (0.06)		0.328* (1.73)		
3. Initiative orientation	0.478** (2.14)			0.495*** (2.98)	
4.Collabora-tion orientation	0.818** (2.46)				0.816*** (3.15)
Operation years	-0.005 (-1.43)	-0.005*** (-1.47)	-0.005 (-1.60)	-0.006* (-1.77)	-0.006* (-1.81)
Capital (ln)	-0.433** (-2.11)	-0.247*** (-1.38)	-0.139 (-0.82)	-0.097 (-0.57)	-0.173 (-1.00)
Manufacturing	0.694 (1.48)	0.705* (1.93)	0.363 (1.00)	0.223 (0.61)	0.750* (2.13)
Other industries	1.784*** (2.88)	1.203 (2.34)	0.707 (1.53)	0.404 (0.87)	1.022** (2.08)
Cut1/Constant	-2.137	-1.323	-1.176	-1.264	-1.267
Cut2/Constant	-2.031	-1.252	-1.115	-1.196	-1.201
Cut3/Constant	-0.292	0.068	-0.017	-0.033	-0.011
Log-likelihood	-42.188	-53.318	-62.299	-59.268	-57.518
Pseudo R2	0.323	0.144	0.070	0.115	0.141
N	54	54	58	58	58

Note: Standard errors in brackets, and *** p<0.01, ** p<0.05, * p<0.1
Source: Authors

3.5   Estimation Results II

This section examines in more detail what kinds of effect of social media are engaged in to achieve product innovation. Again the previous cross-term variable is taken as the dependent variable, while the explanatory variables are listed in Table 7.

Table 7. Effect of social media

Effect of social media	Innovation * Importance of social media				
	1	2	3	4	5
1.Understanding of the trend of consumer awareness	1.012*** (2.93)				
2.Understanding of the reputation of existing products		0.893*** (2.69)			
3.Recognition of customers' needs			0.954*** (2.75)		
4.Recognition of potential customers' needs				0.701** (2.08)	
5.Providing the prototypes or the samples and gathering evaluation information of them					0.667 (1.48)
Operation years	-0.008** (-2.37)	-0.006** (-2.01)	-0.006* (-1.74)	-0.007** (-2.02)	-0.006* (-1.86)
Capital (ln)	-0.162 (-0.97)	-0.095 (-0.57)	-0.226 (-1.31)	-0.116 (-0.71)	-0.096 (-0.59)
Manufacturing	0.787** (2.26)	0.462 (1.31)	0.41 (1.19)	0.637* (1.86)	0.562* (1.69)
Other industries	1.074** (1.98)	1.004* (1.88)	0.926* (1.72)	0.968* (1.82)	1.13** (2.10)
Cut1/Constant	-0.955	-0.699	-1.172	-0.952	-0.899
Cut2/Constant	-0.881	0.465	-1.104	-0.88	-0.835
Cut3/Constant	0.324		0.007	0.221	0.215
Log-likelihood	-52.94	-51.71	-56.90	-56.60	-59.14
Pseudo R2	0.149	0.133	0.12	0.105	0.071
N	54	55	56	55	55

Note: Standard errors in brackets, and *** $p<0.01$, ** $p<0.05$, * $p<0.1$

Source: Authors

Table 7. *(continued)*

Effect of social media	Innovation * Importance of social media				
	6	7	8	9	10
6.Specific of opinion leaders	0.032 (0.07)				
7.Joint development of new products and services with opinion leaders		0.242 (0.22)			
8.Improvement degree of completion of new products and services			1.088*** (2.78)		
9.Obtaining claims information				0.804** (2.26)	
10. Opening FAQ such as how to use the goods					1.209*** (2.69)
Operation years	-0.006* (-1.75)	-0.006* (-1.76)	-0.007** (-2.06)	-0.007** (-2.04)	-0.008** (-2.40)
Capital (ln)	-0.095 (-0.6)	-0.101 (-0.62)	-0.176 (-1.06)	-0.141 (-0.87)	-0.089 (-0.54)
Manufacturing	0.602* (1.83)	0.614* (1.84)	0.744** (2.18)	0.632* (1.88)	0.626* (1.85)
Other industries	1.012* (1.91)	1.026* (1.92)	0.866 (1.6)	1.067** (2.02)	0.864 (1.64)
Cut1/Constant	-0.966	-0.976	-1.049	-0.991	-0.95
Cut2/Constant	-0.904	-0.914	-0.981	-0.922	-0.883
Cut3/Constant	0.111	0.101	0.132	0.158	0.233
Log-likelihood	-60.81	-60.78	-56.75	-58.18	-56.81
Pseudo R2	0.059	0.059	0.122	0.1	0.121
N	56	56	56	56	56

Note: Standard errors in brackets, and *** $p<0.01$, ** $p<0.05$, * $p<0.1$
Source: Authors

Those are as follows: (1) "Understanding of the trend of consumer awareness"; (2) "Understanding of the reputation of existing products"; (3) "Recognition of customer's needs"; (4) "Recognition of potential customers' needs"; (5) "Providing the prototypes or the samples and gathering evaluation information of them"; (6) "Specific of opinion leaders"; (7) "Joint development of new products and services with opinion leaders"; (8) "Improvement degree of completion of new products and services"; (9) "Obtaining claims information"; and (10) "Opening the FAQ site such as how to use the goods." In estimation, only one of the above variables is included in an equation. Other variables of firm characteristics such as "Operation years," "Capital (ln)," "Manufacturing," and "Other industries" are controlled.

The results of estimation are shown in Table 7. As a result, the above variables (1) "Understanding of the trend of consumer awareness"; (2) "Understanding of the reputation of existing products"; (3) "Recognition of customer's needs"; (4) "Recognition of potential customer's needs"; (8) "Improvement degree of completion of new products and services"; (9) "Obtaining claims information"; and (10) "Opening the FAQ site such as how to use the goods" are found to be significant.

4   Discussion and Conclusion

From the previous analyses, social media is found to be useful for the innovation. These results of this study are same with authors' previous research [2]. The firms which satisfy the above innovation have official accounts and original community sites to communicate with consumers. Similarly, in-house social media is used for the promotion of information sharing and knowledge creation. These firms understand the advantages of social media for communicating with outside entities as well as with those inside the firm. Moreover, they understand the value of social media as collaborating with consumers in the process of product development and upgrading. They also aim to collaborate with consumers closely for making the concept of new goods, deciding of package design and brand name, and proposing how to use products and how to dispose them. These are answers to RQ 1 which was raised at the beginning of this paper.

The results show that the importance of social media will be increasing more to the types of business which provide goods and services for final consumers in the future. However, the firms which make product innovation successful by using social media positively are conducting product development not from consumers' initiation, but from their initiatives to obtain the idea of the product development through questionnaires, word of mouth, and exchanging with the opinion leaders. Employees in charge of development knead those ideas according to their own technology and knowhow, and they make prototypes, that is, they make final commodities by taking consumer opinion as a source of ideas and by confirming the difference between their images and actual needs through consumer evaluation of prototypes. Consumers evaluate those prototypes. In such firms, employees in charge of development can create the concept of commodities by taking their own ideas as a professional of product development. Social media therefore is important for the firms to establish such process. Moreover, obtaining market trends, customer needs, and opening the FAQ site are useful effects of social media for product innovation. In contrast, the effects of social media, such as specifying opinion leader and collaborating with them are not significant for product innovation. The firms utilize them for a different manner. They seem to achieve product innovation by distinguishing social media and traditional method. These are answers to RQ 2.

Regarding RQ 3, R&D activities which are found significant are (i) Consumer orientation, (ii) Prototype orientation, (iii) Initiative orientation, and (iv) Collaboration orientation. In more detail, Consumer orientation includes "Decision on function and content of goods and services." Prototype orientation contains "Presenting trial plans until a prototype is made, and consumers' reactions are collected." Initiative orientation includes "Appointing persons in charge of product development who knead concepts of products based on discovered ideas," Finally, Collaboration orientation includes "Ideas of new goods and services are obtained by exchanging with opinion leaders." These are identified R&D activities which utilize social media.

According to the above statements, this paper has some limitations which are due the number of sample firms, that is, the number of firms uses social media for product innovation is still too small to analyze how social media is used for successful product innovation. The further case studies explaining their success will be required and then success factors will be identified.

Acknowledgements. This paper is partly supported by JSPS grants (c-24530435). Financial support is gratefully acknowledged.

References

1. Idota, H., Bunno, T., Tsuji, M.: Empirical analysis of internal social media and product innovation: focusing on SNS and social capital. In: Proceedings of 22nd European regional ITS conference, pp. 1–20 (2011)
2. Idota, H., Bunno, T., Tsuji, M.: Empirical study on how social media promotes product innovation. In: Proceedings of International Telecommunications Society 2014 Biennial Conference, Rio de Janeiro, Brazil, pp. 1–24 (2014)
3. Haavisto, P.: Social media discussion forums and product innovation - The way forward? First Monday 17 (2012). http://firstmonday.org/ojs/index.php/fm/article/view/3984/3332 (last check, October 20, 2014)
4. Noone, B.M., McGuire, K.A., Rohlfs, K.V.: Social media meets hotel revenue management: Opportunities, issues and unanswered questions. Journal of Revenue & Pricing Management. **10**(4), 293–305 (2011)
5. Agnihotri, R., Kothandaraman, P., Kashyap, R., Singh, R.: Bringing 'social' into sales: the impact of salespeople's social media use on service behaviors and value creation. Journal of Personal Selling & Sales Management **32**(3), 333–348 (2012)
6. Groza, M., Peterson, R., Sullivan, U.Y., Krishnan, V.: Social media and the sales force: the importance of intra-organizational cooperation and training on performance. The Marketing Management Journal **22**(2), 118–130 (2012)
7. Hausmann, A.: Creating 'buzz': opportunities and limitations of social media for arts institutions and their viral marketing. International Journal of Nonprofit & Voluntary Sector Marketing **17**(3), 173–182 (2012)
8. Kate, V., Pavan, S.: Social Media and Revenue Management; Where Should the Two Meet? Journal of Technology Management for Growing Economies **3**(1), 33–46 (2012)
9. Rodriguez, M., Peterson, R.M., Vijaykumar, K.: Social Media's Influence on Business-To-Business Sales Performance. Journal of Personal Selling & Sales Management **32**(3), 365–378 (2012)
10. Schultz, R.J., Schwepker, C.H., Good, D.J.: An exploratory study of social media in business-to-business selling: salesperson characteristics, activities and performance. The Marketing Management Journal **22**(2), 76–89 (2012)
11. Luo, X., Zhang, J.: How Do Consumer Buzz and Traffic in Social Media Marketing Predict the Value of the Firm? Journal of Management Information Systems **30**(2), 213–238 (2013)

Sociomateriality Implications of Software as a Service Adoption on IT-Workers' Roles and Changes in Organizational Routines of IT Systems Support

Freddie Mbuba[✉], William Yu Chung Wang, and Karin Olesen

Faculty of Business and Law, Auckland University of Technology, Auckland, New Zealand
{fmbuba,wiwang,kolesen}@aut.ac.nz

Abstract. This paper aims to deepen our understanding on how sociomateriality practices influence IT workers' roles and skill set requirements and changes to the organizational routines of IT systems support, when an organization migrates an on-premise IT system to a software as a service (SaaS) model. This conceptual paper is part of an ongoing study investigating organizations that migrated on-premise IT email systems to SaaS business models, such as Google Apps for Education (GAE) and Microsoft Office 365 systems, in New Zealand tertiary institutions. We present initial findings from interpretive case studies. The findings are, firstly, technological artifacts are entangled in sociomaterial practices, which change the way humans respond to the performative aspects of the organizational routines. Human and material agencies are interwoven in ways that reinforce or change existing routines. Secondly, materiality, virtual realm and spirit of the technology provide elementary levels at which human and material agencies entangle. Lastly, the elementary levels at which human and material entangle depends on the capabilities or skills set of an individual.

Keywords: Cloud computing · Sociomateriality · IT workers · Materiality · Organizational change · Organizational routines · On-premise IT system · Software as a service (SaaS)

1 Introduction

Software as a service (SaaS) adoption has the potential to change IT workers' roles and skill set requirements as well as organizational routines of IT systems support. The conceptualization of the relationship between the social and the technological has been a recurring concern in information systems (IS) research, not only in how human interact with the material, but also the mutually constitutive relationship between social practices and materiality [1, 2]. In this paper, we combine the sociomateriality perspective [3] and the organizational routines theory [4] to understand how sociomateriality practices influence IT workers' roles and skill set requirements and changes in the organizational routines of IT systems support. Where the term materiality refers to "the arrangement of an artifact's physical and/or digital materials into particular forms that endure across differences in place and time" [2]. Similarly, Leonardi

© Springer-Verlag Berlin Heidelberg 2015
L. Wang et al. (Eds): MISNC 2015, CCIS 540, pp. 249–263, 2015.
DOI: 10.1007/978-3-662-48319-0_20

defines sociomateriality as the "enactment of a particular set of activities that meld materiality with institutions, norms, discourses, and all other phenomena we typically define as social" [5]. Therefore, the sociomateriality perspective examines the role of materiality in social life [5] and its implications on organizational routines [6]. Feldman and Pentland define an organizational routine as a "repetitive, recognizable pattern of interdependent actions, involving multiple actors" [4]. An organizational routine is the primary means by which organizations accomplish what they do. The sociomateriality perspective is useful in explaining how organizational and technological changes are interwoven [7] and how the social and material are entangled in practice [3]. The sociomateriality perspective has not sufficiently examined and linked changes in organizational routines with changes in the materiality of work practices in SaaS context [3].

The *SaaS business model* refers to a hosted software application delivered online via the Internet and accessed through a web browser [8]. The SaaS business model is proven useful because it has the potential to change IT workers' roles and skill set requirements, and the organizational routines of IT systems support [9]. More specifically, by using the sociomateriality perspective [3], this conceptual paper examines how materiality inscribes aspects of social structures [7] on IT workers and on the organizational routines [4] of IT systems support, when an organization moves an on-premise IT system to SaaS business model. An on-premise IT system is a software service model in which customers purchase permanent licenses of the commercially available software, and the organizations IT workers maintain the application and the infrastructure associated with it. We present initial findings of interpretive case studies from New Zealand tertiary institutions, of the migration of on-premise emailing system to SaaS business model, such as Google Apps for Education and Microsoft Office 365.

1.1   Research Context: SaaS Business Models in Tertiary Institutions

Many organizations, such as tertiary institutions, have embraced the use of SaaS business model, such as Google Apps for Education and Microsoft Office 365 [10]. This is because in these institutions, migrating on-premise IT systems to SaaS business model provide improved real-time collaboration and research capabilities, at reduced costs while providing better levels of computing services to faculty members [10].

Exaggerated by the need to cut overhead costs at a time when public and private institutions are coping with significant budget shortfalls, the massification of higher education, globalization effects, the advent of the knowledge society, and IT capabilities [11]. On massification, governments demand for expansion of enrolment that has taken place worldwide, while tertiary institutions are coupled with a financial struggle to meet the need for an enlarged infrastructure and a larger teaching staff [11]. Globalization has shaped the world through factors such as an integrated world economy, new IT innovations, the emergence of an international knowledge network, and the rise of English as the universal language of scientific communication. These factors of globalization have increased student mobility worldwide [11] and consequently have affected the tertiary education sector globally. Institutions have implemented a variety of policies and programs to respond to them. These include, setting up branch campuses overseas, internationalizing curricula, engaging in international partnerships, and the use of latest IT innovations-to enhance real-time collaboration and research capabilities, as well better computing powers at reduced costs [11].

The SaaS business model, therefore, allows institutions not only to reduce IT infrastructural costs and tap into latest IT innovations from commercial cloud service providers via the Internet, but as well many of these resources are available to them either for free or at reduced costs [10].

With the SaaS business model, students and faculty staff take advantage of the ability to undertake research and collaborate with international researchers from anywhere and on any device using SaaS applications. Other benefits SaaS provides to customers include: no upfront investment cost required, elasticity of computing resources, vendor support and upgrades, agile response to markets, usage metered as a utility, resource pooling and the ability to add computing resources as needed [12]. As a result, organizations adopt the SaaS business model based on these potential benefits. However, these benefits may have some implications on the IT workers and on the organizational routines of IT systems support. These implications might include: changing IT workers' roles and skill set requirements, widespread of layoffs of hardware IT workers, and IT department lose control of IT-Servers, and focusing more on data security, vendor management, as IT systems support moves to cloud service providers [8]. The migration of on-premise IT system to the SaaS business model therefore, is proven to be useful in understanding how technological changes influence changes in organizational routines of IS work practices [3]. To control the diversity of various types of SaaS systems, cases were selected based on being on-premise emailing systems that had migrated to similar SaaS emailing systems across tertiary institutions in New Zealand.

1.2   Research Purpose

There is a lack of research conducted on the implications of the SaaS business model on organizational routines and related human resource management [13], as well as the linked changes in organizational routines with changes in the materiality of IS work practices [3]. Drawing on organizational routines theory [4], and the sociomateriality perspective [2], this paper therefore, seeks to fill this gap in the literature by examining how SaaS adoption, may trigger variations in performative actions in organizational routines of IT workers during IT systems support [14].These variations in performative actions may be institutionalized resulting in changes in organizational routines. This ongoing study extends earlier work on the implications of artifacts on organizational routines [13], and demonstrates the performative to ostensive change which accounts for the effects such as creation, maintenance or modifications of the organizational routines [4]. In addition, using the sociomateriality lens, this study examines the entanglement between human and material agencies in an on-premise IT system and in the SaaS environment [15].

By studying how technological changes from on-premise IT system to SaaS business model occurs, the inscribing features of social structures on IT workers' roles and skill set requirements, and on the organizational routines of IT systems support, our aim is to answer the following research questions:

1. At what materiality level does human and material agencies entangle?
2. What social outcomes do these entanglements produce in on-premise IT systems and SaaS contexts?
3. How can technological artifact entangled in sociomaterial practices, influence the changes on IT workers' roles and skills set requirements, and on the organizational routines of IT systems support?

The paper is structured as follows: First, we review SaaS business model. Thereafter, a theoretical review on the sociomateriality and organizational routines theory. Subsequently, we discuss research methodology. The paper concludes with a discussion of contributions and limitations of our paper, and future research directions.

2   SaaS Business Model

SaaS is one of the cloud computing services levels. Here cloud computing refers to applications delivered as services over the Internet and have infrastructure in data centers that enable these services [16]. Other cloud services levels include, hardware as a Service (HaaS), Infrastructure as a Service (IaaS) and Platform as a Service (PaaS) [12]. Extant IS literature discuss these services in much depth [17, 18]. This study however, focuses on the SaaS business model, not only because of the elasticity, multitenancy configurability, and ubiquitous features that SaaS offers to adopting organizations, but also, it represents a huge transformation in the way organizations use and support IT services [8, 9].

3   Theoretical Review

We briefly review theoretical lenses used in this study, sociomateriality and the organizational routines theory, followed by research methodology.

3.1   Sociomateriality Perspective

Technology presents a conceptual challenge when faced with complex materiality of everyday IS- mediated work practices [3, 5, 19]. IS researchers, have encountered this challenge in studying the relationship between *materiality* and human, or *technology* and social life, and how this contributes to organizational change [1, 3, 5]. In this endeavor therefore, IS scholars have coined a term sociomateriality, that describes the constitutive entangling of the material and social in work practice [3].

Sociomateriality thinking roots from sociotechnical systems thinking [20], actor network theory [21] and practice theory [22]. Two ontologies guide the sociomateriality thinking- *relational ontology* [1, 19] and *substantiality ontology* [5]. *Relational ontology* of sociomateriality assumes that "social and material are inherently inseparable" [1]. It challenges the separation of material from human or technology from social life. This represents a shift from understanding people and technologies, each characterized by specific properties and boundaries that interact and mutually impact each other in the performative nature of practices when are enacted and re-enacted [22].

Substantiality ontology on the other hand, takes the notion of "substances of various kinds (things, beings, essences)" as self-subsistent entities, which come preformed and involve themselves in dynamic relations [23]. According to substantiality views, materiality is inherent to technology, independent of its use and the context in which technology is used [15]. Leonardi argues that once technology is built its materiality is fixed, unless some subsequent redesign is undertaken [5]. When implemented in an organizational context, a technology's materiality becomes important, because users react to the technology's materiality—a materiality they perceived as bounded and stable—when translating it from the realm of the artifactual into the realm of the social [2].

This paper takes the substantiality ontological stance, in understanding the intertwining of human and material agencies and how this bring changes in the organizational routines [2]. According to Leonardi, "routines and technologies are the infrastructure that the imbrication of human and material agencies produce" [7]. Here "imbrication" describes the ways human and material agencies intermingle to create or change routines, or alter technologies [7]. Imbrication is useful in two ways: First, it suggests that human and material agencies are effectual at producing outcomes (e.g. routines or technologies), only when are joined together, but their interdependences does not alter their specificity and distinct characters [7]. In other words, "people have agency and technologies have agency, but ultimately, people decide how they will respond to a technology" [7]. Secondly imbrication reminds us that all interactions between human and material agencies produce "organizational residue", such as routines or technologies and these figurations persist absent of their creators [7].

3.2 Organizational Routines Theory

Organizational routines are an essential aspect of organized work, a source of inertia, inflexibility and mindlessness [4]. In addition, organizational rules and routines have been seen as a source of accountability, political protection, and a source of stagnation. Routines enable bureaucracies to organize expertise and exercise power efficiency [4]. For example, in order to effectively support IT users, the IT-department has designed an organizational routine - the IT systems support to cater for user support. IT systems support allows any IT user to log a job to the IT-helpdesk, for any IT related request or a faulty piece of hardware to be fixed. The jobs are executed by IT workers in hierarchical manner from IT technicians at the helpdesk (known as first tier IT support level), to systems engineers and IT-experts at the second tier, which could be escalated to the third and fourth tier IT-support levels [24].

The organizational routine comprises two intertwined aspects (shown on Figure 1) known as performative and ostensive [4]. Performative (or agency) aspect describes the actual actions of the routine by specific people, at specific times, in specific places [4, 25]. The role of performative aspects bring subjectivity and power into organizational routines and involves the ability to remember the past, imagine the future, and respond to present circumstances [4]. Thus, the performance of routines involve adapting to contexts that require either idiosyncratic or ongoing changes and reflecting on the meaning of actions for future realities [4].

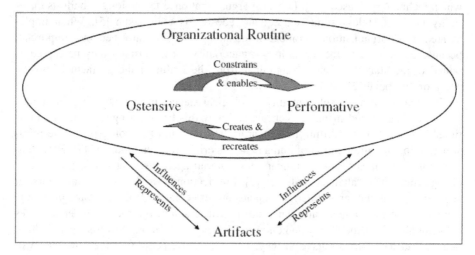

Fig. 1. Organizational routines are generative systems [6].

Ostensive defines the structure or abstract idea of routines [4]. It describes abstract regularities and expectations that enable participants to guide, account for, and refer to specific performances of a routine [6]. Ostensive aspects are not written rules or procedures, but rather consists of the understandings embodied and cognition of the actors [6]. Artifacts associated with the organizational routines may attribute variations in performative aspects, [4, 14]. Artifacts take forms like written rules, procedures, or IT systems. In an organizational routine, artifacts are used to ensure the reproduction of particular patterns of action [6].

For example, an IT-department has setup a web based IT system support to help fix IT users' problems in an effective, efficient and accountable manner. However, this technology can been seen as enabling and constraining action [26]. The ostensive and performative aspects, therefore, are produced and reproduced through actions taken by actors and the actions taken are constrained and enabled by structure [25]. Hence, these aspects of organizational routines are viewed as recursively related, with the performances creating and recreating the ostensive aspect and the ostensive aspect constraining and enabling the performances [4].

The relationship between ostensive and performative aspects of routines therefore, create an on–going opportunity for variation, selection and retention of new practices and patterns of actions within routines and allows routines to generate a wider range of outcomes from apparent stability to considerable change [27].

4   Research Methodology

In this ongoing study, the empirical data was obtained through semi-structured interviews with IT workers of their experiences and events, when migrating on-premise IT–system to SaaS business model. We conducted a multi-site, interpretive case study [28] in four tertiary institutions in New Zealand to investigate how the technology changed IT workers roles and skill set requirements and functions of the organizational routines of the IT systems support, during the adoption of SaaS business model [4].

We used both primary and secondary sources of empirical evidence. Semi structured interviews and field notes are among the primary sources of data, while the secondary sources of data came from official documentation and published information.

The semi-structured interviews assisted us in uncovering understandings of series of events, stories, and experiences [29] from IT workers and IT managers involved in migrating on-premise IT system to a SaaS business model. In total 17 interviews were conducted, of which four of the interviewees were IT managers, seven systems engineers and six IT technicians. All of those interviewed were involved in the on-premise IT systems migration project and supporting the SaaS business model.

The principles of data coding and hermeneutic circles were used for data analysis within and cross-cases [28], to identify situations in which human and materiality entanglements were seen to be significant in maintaining IT systems support. These human and material entanglements were analyzed in relationship with the five notions of sociomateriality [30].

5   Analysis and Discussions

Our discussions aim is to answer the following research questions:

1. At what materiality level does human and material agencies entangle?
2. What social outcomes do these entanglement produce in on-premise IT systems and SaaS contexts?
3. How can technological artifact entangled in sociomaterial practices, influence the changes on IT workers' roles and skills set requirements, and on the organizational routines of IT systems support?

First we discuss the imbrication of human and material agencies, followed by how technological artifacts entangled in sociomaterial practices, influence the changes of IT workers' roles and skill set requirements and performative and ostensive aspects of organizational routines.

5.1   Imbrication of Human and Material Agencies

Following the principles of chemistry, matter exists in three basic forms: solid, liquid, and gaseous forms [31, 32]. Principally, it's at these forms human may imbricate with the material artifact [31]. Human and material agencies therefore imbricate: physical-ly, at the virtual realm and the spirit levels. For example, a car driver imbricates with a car at least physically, while a non-driver, for instance a car enthusiast, imbricates with cars virtually, whereas a car designer imbricate with cars at the spirit level (the essence of the car to meet the users' requirements, affordances, look and feel), as well as physically and in a virtual realm. In the same vein, this paper, argues that human and material agencies imbricate at three different elementary levels: materiality, vir-tually and/or at spirit (essence) level of sociomateriality. Our initial findings consis-tently show how IT technicians, systems engineers and IT managers imbricate diffe-rently with technology at *materiality*, *virtual* and *spirit* of the technology.

5.1.1   Materiality of Technology

Materiality is not only inherent to technology, independent of its use and the context in which technology is used [15], but it is also a point of imbrication of human and material agencies. In other words, materiality provides affordances and constrains to accord imbrication of human and material agencies to happen [33]. For "materiality" to exist as a concept separate from "sociomateriality", implies that some materials are not simultaneously social [2]. Our study shows IT workers perform well within the materiality of IT, that is the hardware and software of the IS, in terms of IT systems' maintenance and support.

5.1.2   Virtual Realm of Technology

The virtual realm of technology [34], relates to technological features that are mallea-ble and changes per system use. Unlike, materiality of technology, virtual features are inherent of technology, but depends on context and use of technology. For example, unlike on-premise IT systems built on a client-server architecture, SaaS business model has characteristics of utility features, such as: configurability, multitenancy, elasticity, scalability and ubiquitous, and it's a web-based application accessible via a browser [8]. This does not mean that traces of them did not necessarily exist in client-server architecture of the on-premise IT systems, but the richness, malleability and ease integrations through application programming interfaces (APIs) with other IT-systems created a different abstraction layer of technology, hence SaaS demands dif-ferent IT workers' skills to maintain and support. Our study consistently shows how IT technicians and systems engineers imbricate differently at the virtual realm of technology.

5.1.3   Spirit of Technology

Spirit of technology refers to the essence or strategic goals of the technology [35]. Spirit can be identified by analyzing the philosophy of the technology based on the design metaphor underlying the system (e.g., SaaS is ubiquitous-anywhere and on any device access via Internet browsers). The spirit of technology can also be identified by the features it incorporates and how they are named and presented (e.g., elasticity, scalability, configurability and multitenancy of SaaS applications). Spirit can also be seen through the nature of the user interface (e.g., SaaS is web-based) and training materials and online guidance facilities (e.g., dashboards and online tutorials). Lastly spirit of technology can be seen through the systems support and maintenance (e.g., vendor maintains and supports SaaS applications) and integrations with other systems (e.g., use of web-services and APIs) [35].

Our initial findings show IT managers imbricate with technology at the spirit level, by enforcing organizational routines to meet the organization's strategic goals of the technology.

For Example.
Our findings indicate that IT technicians are more inclined to support and maintenance of the IT system-both hardware and software components. IT technicians therefore, tend to follow IT system support routines on "how to" fix the system, and thus imbricating with the materiality of the technology. Systems engineers, however, tend to think of the design architecture of the IT system and "why" it is behaving as it is. Systems engineers, therefore have an ability to change the system's code, to align with the organizational needs. The coding skills enable systems engineers to manage system changes from on-premise IT system to SaaS model more efficiently, than the IT technicians. Our empirical data shows that IT managers imbricate with the technology at spirit level-by analyzing the purpose of the system. IT managers use organizational routines to enforce rules and guidelines, and to ensure the system meets its strategic goals.

Following Orlikowski and Scott [1], Jones discusses five main notions of sociomateriality: materiality, inseparability, relationality performativity and practice [30]. Materiality describes physical, digital and nonphysical features of material artifacts that does not change over time and space [2]. Inseparability explains whether human and material agencies exist as separate or inseparable entities [1]. Relationality describes whether imbrications of human and material agencies have any "analytical boundaries" [1]. Performativity explains the ability of human and material agencies to achieve social outcomes [30]; and practice discusses the forms of bodily, and mental activities, things and their use in form of understanding, states of emotion and motivational knowledge [30].

To gain deeper understanding of the social outcomes, this study examines further, on the imbrication of human and material agencies at these elementary levels (materiality, virtual and spirit) with respect to five notions of sociomateriality [30]. Our discussions in the context of the implications of SaaS business model on IT workers' roles and skill set requirements and changes in organizational routines of IT-system support are summarized in Table 1. In building Table 1, we ask what it means with respect to five notions of sociomateriality, when human and material agencies imbricate at materiality, virtual and/or spirit levels of technology, in the context of both on-premise IT systems and SaaS business models?

5.2   Changes in Organizational Routines

5.2.1   Selective Variations (Performative to Ostensive Changes)
The performative aspect of routines is critical for creation, maintenance and modifications of the ostensive aspects. Actors might alter the performative aspects (actions or

performances) that the ostensive aspect of the routine guides or accounts for [4]. Thus, performative aspects enact ostensive aspect of the routine as an unintended effect of action [25, 26]. The variations in performative actions therefore, might be institutionalized into and alter ostensive aspects of the organizational routine and hence bring organizational change [6].

For example, IT workers might change the way IT systems support is performed. For instance, when an IT user logs a job at the IT technician help-desk, IT technicians then advises an IT user to call a second tier support systems engineer directly instead. Then the systems engineer solves a problem and closes the job. Subsequent IT users' requests bypass IT helpdesk and calls second tier support directly. The trend might continue until it becomes a norm in solving such a problem. These performative actions might be institutionalized into ostensive aspects of the organizational routine and bring effects such as creation, maintenance and modification of organizational routines, and bring changes to a larger organizational context as part of relevant actions that IT workers take in organizational routines .

In this example, the motivation of IT workers performing user support is not to create, maintain or modify the ostensive aspect of the routine, but rather aim to fix a computer problem logged by a user. However, some outcomes of engaging in actions has some effect on the structure (ostensive) that constrain and enable further actions [4].

5.2.2 Selective Retention (Ostensive to Performative Changes)

Actors can use the ostensive aspect of routines as a guide on how actions ought to be taken, or for accounting actions already taken [4]. The use of the ostensive aspect in relation to the performative aspect of routine is referred to as "*guiding, accounting and referring*" [4]. Employers, and authorities use guiding, accounting and referring to exert power over their own people's performances to signify that some performances are part of a recognizable routine and to legitimate some performances as appropriate to that routine [4].

Guiding serves as a template for behavior or a goal [4]. For example, IT managers enforce the use of IT system support structures to accomplish any task. *Accounting* allows actors to explain what they are doing in ways that make sense of their activities [4]. It provides ready-made justification for actions that seem unusual and provides reasonable accounts when called to explain [4]. *Referring* ostensive aspect of routines, therefore, helps to refer to pattern of activity that would otherwise be incomprehensible. For example, IT workers referring to systems manuals and user guides to accomplish certain tasks.

With the introduction of the SaaS business model changes occurred in the ways human and material agencies imbricate at the materiality, virtual realm and spirit level of technology. These imbrications between human and material agencies produce "organizational residue" –organizational routines or technologies and these figurations persist absent of their creators [7]. Similarly, the recursive nature of performative and ostensive aspects of organizational routines brings organizational change (see Figure 2). Both changes in organizational routines and organizational residues might be institutionalized to form a new sociomateriality assemblage, as illustrated in figure 2.

Table 1. Human and Material Agencies Imbrications Levels

Human/material Agencies Imbrication Levels	What does it mean with respect to five notions of sociomateriality, when human and material agencies imbricate at these levels? Any changes when moving to SaaS?				
	Notions of Sociomateriality [30]				
	Materiality	Inseparability	Relationality	Performativity	Practice
Materiality	Technician level-supporting hardware and software of IT. This level requires experience and know 'how' skills and capabilities. Human/material imbricated at material level, the materiality of technology remains unchanged over differences in time and places. **Example:** In SaaS business model, the hardware relocates to data centers ("cloud"), and software accessed via a web-based virtualized abstraction layer	Human/material agencies remain independent entities from one another when imbricating at materiality level. **Example:** IT technician may decide to restart the IT-system, otherwise will remain 'on' unless some physical and electronic fault occurs.	There is a clear boundary between human/material agencies at materiality level. Human and material agencies emerge in their imbrication in specific practices [2].**Example:** Human agency or an IT-system's faulty calling for human intervention might initiate social activities.	Human/material agencies have capabilities to achieve "social outcomes" [2]. Activities are interpreted physically, digitally and by non-physical means. **Example:** Human agency might initiate software application to run or be installed. The software runs to produce desired outcomes.	Ability to initiate activities human/material were different during the on-premise and SaaS systems. **Example:** Our findings show IT technicians and systems engineers have different capabilities to engage with SaaS system. IT technicians altered certain organizational routines to circumvent direct support on SaaS
Virtually	When human and material agencies imbricate at virtual level, the materiality of technology is seen to be malleable. At this level, IT workers need to understand 'why' the system behaves that way and demands deeper IT skills, such as coding skills. Our findings show imbrication at this level is predominantly common in SaaS than in an on-premise IT system. **Example:** IT technician failing to troubleshoot SaaS systems and escalates these issues to systems engineers, showing most IT technicians were unable to imbricate at virtual realm of IT.	Imbricating at virtual realm of technology, human/material agencies remain separable and independent entities across space and time [2]. **Example:** Multitenancy, elasticity and scalability features of SaaS systems remain independent from IT workers' interventions, and material performances as these features follow prescribed IT settings.	Analytical boundaries exists between human/material agencies imbricated at virtual realm of technology [2]. **Example:** Though SaaS has multitenancy, elasticity and scalability features, these exist independent of the designer following virtualization characteristics of IT.	Human/material agencies have capabilities to achieve social outcomes in non-physical and digital means. **Example:** The scaling up or down of IT resources by SaaS system achieves social outcomes as a utility to IT users.	Human/material agencies imbricated at virtual realm-tend to operate at coding layer. Systems engineers creates APIs to integrate with other organizational systems
Spirit	Human/material agencies imbricated at spirit level calls for IT managers to understand the ability of materiality of technology to meet the organization's strategic goals. **Example:** IT manager opted for Serverless (SaaS) system to cut down operational and human resource costs as well as tap into latest cloud based technologies.	At the spirit level, human/material agencies imbrication becomes inseparable. As at this level, "social and material becomes inherently inseparable" [1]. In other words, managers' "quotidian interaction" [2] with technology becomes a subconscious norm of everyday life. **Example:** IT managers concerns of on-premise IT system's inability to meet users' expectations were a major concern in their social work life. As it was for Steve Jobs and is for Bill Gates imbricated at spirit levels to Apple and Microsoft products respectively. These men and their products are inseparable.	When human/material agencies imbricate at spirit level, the assemblage "dissolves analytical boundaries between human and technology" [1]. Human/material agencies exist in an intricate assemblage of the human's mind. **Example:** IT managers shared their on-premise IT systems' problems with colleagues, family members and outsiders/vendors to solicit insights for a comprehensive solution.	Human/material agencies imbricated at spirit level, forms a human/material assemblage recursively intertwined, having a resultant social outcome. **Example:** Our study shows that IT managers embodied with troubled technological issues, have a strong decisive desire to make changes to the system.	With relation to practice, human/material imbrication at spirit level remains an enthusiastic stance and differs among individuals. **Example:** Our study consistently shows that IT managers were enthusiastic in migrating on-premise IT-system to SaaS model.

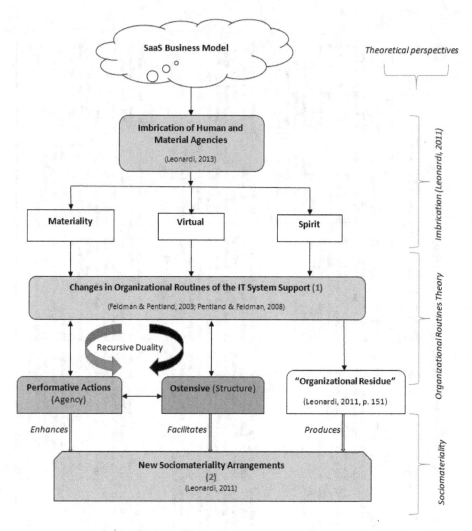

Fig. 2. Implications of SaaS Business Model

6   Conclusion

Our study examined theories of organizational routines and the sociomateriality perspective of the SaaS business model that remain an under-explored topic. Past IS research on implications of cloud computing on human resources, have explored general implications without paying particular attention on specific service levels [18]. The exception is study that examined the implications of SaaS adoption on organization's operations and business [9]. His study however, did not address how organizations were affected due to new social structures of the SaaS business model. Further, he did not account for the events and processes that led to organizational changes. To answer our research questions, this paper has demonstrated that (1) human and material agencies imbricates/entangles at materiality, virtual realm and/or the spirit of technology. The imbrications though, depend on the capabilities or skill sets of an individual; (2) The imbrications produce varied social outcomes, such as routines or new technologies as organizational residues; (3) The influences of sociomaterial practices varies, depending on individual's capabilities when imbricates with artifact at either materiality, virtual realm and/or the spirit of technology.

6.1   Contributions

Our study makes three contributions. Firstly, we show the way technological artifacts, which are entangled in sociomaterial practices can change the way human responds to the performative aspects of the organizational routines. Leonardi supports this that "human and material agencies interweave in ways that create or change routines…"[7]. Secondly, we demonstrate how materiality, virtual realm and spirit of the technology provides elementary levels at which the human entangles with material. Thirdly, we show that the elementary levels at which human and material entangle depends on the "capability" [5] or skill set of an individual.

6.2   Limitations

The study has several limitations. First, it is a multi-case study in need of further in-depth longitudinal studies. Second, the unique nature of SaaS business model needs to be further examined associated with other technologies. Third, this study accounts for individual experiences of IT workers, and does not account for a wider range at an organizational level. This paper, however, examined IT workers experiences at the project level.

6.3   Future Research

We suggest several opportunities for future research. First, to extend the organizational routines theory we need to examine the changes in IT systems support with other IT technologies. Second, there possibly other organizational routines changes on IT systems management. Third, we suggest deeper examination of other micro-level routines that might inform how performative and ostensive aspects of routines changes, due to the introduction of new technologies. Finally, further examination of changes in organization routines as a result of changes in sociomateriality of technology in different contexts could be explored.

References

1. Orlikowski, W.J., Scott, S.V.: Sociomateriality: challenging the separation of technology, work and organization. Annals **2**, 433–474 (2008)
2. Leonardi, P.M.: Theoretical foundations for the study of sociomateriality. Information and Organization **23**(2), 59–76 (2013)
3. Orlikowski, W.J.: Sociomaterial practices: Exploring technology at work. Organization Studies **28**(9), 1435–1448 (2007)

4. Feldman, M.S., Pentland, B.T.: Reconceptualizing Organizational Routines as a Source of Flexibility and Change. Administrative Science Quarterly **48**(1), 94–118 (2003)
5. Leonardi, P.M.: Materiality, sociomateriality, and socio-technical systems: what do these terms mean? how are they related? do we need them? In: Leonardi, P.M., Nardi, B.A., Kallinikos, J. (eds.) Materiality and Organizing: Social Interaction in a Technological World, pp. 25–48. Oxford University Press, Oxford (2012)
6. Pentland, B.T., Feldman, M.S.: Designing routines: on the folly of designing artifacts, while hoping for patterns of action. Information and Organization **18**(4), 235–250 (2008)
7. Leonardi, P.M.: When flexible routines meet flexible technologies: Affordance, constraint, and the imbrication of human and material agencies. MIS Quarterly: Management Information Systems **35**(1), 147–167 (2011)
8. Carraro, G. Chong, F.: Software as a service (SaaS): An enterprise perspective. Microsoft Corporation. (2006) [cited 2012 July 14]. http://msdn.microsoft.com/en-us/library/aa905332.aspx
9. Wu, W.-W.: Mining significant factors affecting the adoption of SaaS using the rough set approach. The Journal of Systems & Software **84**(3), 435–441 (2011)
10. Wyld, D.C., Juban, R.L.: Education in the clouds: how colleges and universities are leveraging cloud computing. In: Elleithy, K., et al. (eds.) Technological Developments in Networking, Education and Automation, pp. 1–6. Springer, Netherlands (2010)
11. Altbach, P.G., Reisberg, L., Rumbley, L.E.: Tracking a global academic revolution. Change: The Magazine of Higher Learning **42**(2), 30–39 (2010)
12. Mell, P., Grance, T.: The NIST definition of cloud computing (draft). NIST special Publication **800**(145), 1–7 (2011)
13. Lyytinen, K., Rose, G., Yoo, Y.: Learning routines and disruptive technological change: Hyper-learning in seven software development organizations during internet adoption. Information Technology and People **23**(2), 165–192 (2010)
14. Feldman, M.S.: Organizational Routines as a Source of Continuous Change. Organization Science **11**(6), 611–629 (2000)
15. Barley, S.R., Leonardi, P.M.: Materiality and change: challenges to building better theory about technology and organizing. Information and Organization **18**(3), 159–176 (2008)
16. Wyld, D.C.: Risk in the clouds?: security issues facing government use of cloud computing. In: Innovations in Computing Sciences and Software Engineering, pp. 7–12. Springer (2010)
17. Benlian, A., Hess, T., Buxmann, P.: Drivers of SaaS-Adoption - An Empirical Study of Different Application Types. Business & Information Systems Engineering **1**(5), 357 (2009)
18. Cusumano, M.: Cloud computing and SaaS as new computing platforms. Communications of the ACM **53**(4), 27 (2010)
19. Barad, K.M.: Meeting the universe halfway: quantum physics and the entanglement of matter and meaning. Duke University Press, Durham (2007)
20. Mumford, E.: The story of socio-technical design: reflections on its successes, failures and potential. Information Systems Journal **16**(4), 317–342 (2006)
21. Latour, B.: Reassembling the social: an introduction to actor-network-theory. Oxford University Press, Oxford (2005)
22. Cecez-Kecmanovic, D., et al.: The Sociomateriality of Information Systems: Current Status. Future Directions. MIS Quarterly **38**(3), 809–830 (2014)
23. Emirbayer, M.: Manifesto for a Relational Sociology 1. American Journal of Sociology **103**(2), 281–317 (1997)

24. Gao, F., Qiu, X., Meng, L.: SLA based business-driven adaptive QoS maintenance mechanism for multi-tier service in virtualized IT environment. In: GLOBECOM Workshops (GC Wkshps). IEEE (2010)
25. Giddens, A.: The constitution of society: outline of the theory of structuration. Polity Press, Cambridge (1984)
26. Orlikowski, W.J.: Using technology and constituting structures: A practice lens for studying technology in organizations. Organization Science **11**(4), 404–428 (2000)
27. Pentland, B.T., Rueter, H.H.: Organizational Routines as Grammars of Action. Administrative Science Quarterly **39**(3), 484–510 (1994)
28. Myers, M.D.: Qualitative research in business & management. SAGE, London (2011)
29. Walsham, G.: Interpretive case studies in IS research: nature and method. European Journal of Information Systems **4**(2), 74–81 (1995)
30. Jones, M.: A matter of life and death: exploring conceptualizations of sociomateriality in the context of critical care. Management Information Systems **38**(3), 895–925 (2014)
31. Blin-Stoyle, R.J.: Eureka!: physics of particles, matter and the universe. CRC Press (1997)
32. Ursul, A.: Informational and cosmological foundations of the cybernetic systems security phenomenon. Scientific and Technical Information Processing **35**(2), 87–98 (2008)
33. Markus, M.L., Silver, M.S.: A Foundation for the Study of IT Effects: A New Look at DeSanctis and Poole's Concepts of Structural Features and Spirit. Journal of the Association for Information Systems **9**(10), 609–632 (2008)
34. Wetmore, A.: The Poetics of Pattern Recognition: William Gibson's Shifting Technological Subject. Bulletin of Science, Technology & Society **27**(1), 71–80 (2007)
35. DeSanctis, G., Poole, M.S.: Capturing the complexity in advanced technology use: Adaptive structuration theory. Organization Science, 121–147 (1994)

Corporate Usage of Social Media and Social Networking Services in the USA

William A. Sodeman[1(✉)] and Lindsey A. Gibson[2]

[1] D.W. Johnston School of Business, Martin Methodist College,
433 W Madison Street, Pulaski, TN 38478, USA
wsodeman@martinmethodist.edu
[2] College of Business, Hawaii Pacific University,
1 Aloha Tower Drive, Honolulu, HI 96813, USA
lgibson@hpu.edu

Abstract. Corporations and organizations in the United States have used social media and social networking services (SNS) in a variety of ways since the 1970s. The Internet, mobile communications, and smartphones have led to widespread adoption of social media for marketing, public relations, commerce, stakeholder management, and other corporate uses. Examples of best practice and poor performance are examined.

Keywords: Advertising · BBS · Compuserve · Corporate social performance · Facebook · Management · Marketing · Point of sale · POS · Retail · Social media · Social networking services · Stakeholder · Twitter · USA · Yelp

1 Introduction

Businesses in the United States have used various forms of social networking services (SNS), including social media and bulletin board services (BBSes), to communicate with customers, suppliers, current and potential employees, and many other organizational stakeholders. The stakeholder concept is especially relevant to business SNS usage, as stakeholder groups can form very easily during online discussions. Managers who communicate regularly with stakeholder groups can craft mutually acceptable goals, and are in a better position to identify unforeseen opportunities while avoid conflicts. [1] From the earliest years of the personal computer, the commercial use of SNS has varied widely – from a truly disruptive business strategy [2] to a less effective tactic. [3] This paper examines the history of SNS in the United States, with a specific focus on their use in business. Important predecessors to Internet-based SNS are examined, as well as the most important SNS currently in use in the United States. Several examples of business use are discussed, with a specific focus on marketing, reputation, and overall strategy.

1.1 Evolution and History of Social Networking Services in the United States

Early attempts at online services were described as online networks, computer mediated communication, and virtual communities. [4] Personal computing was an

© Springer-Verlag Berlin Heidelberg 2015
L. Wang et al. (Eds): MISNC 2015, CCIS 540, pp. 264–278, 2015.
DOI: 10.1007/978-3-662-48319-0_21

emerging concept during the 1970s. Apple and IBM created popular hardware and software standards, but the wide variety of computer networking standards made communication between individual computers costly and difficult.

CompuServe. Before the Internet was commercialized, CompuServe was one of the most popular computer networks used by United States businesses. The service was established in 1969 as a way to access minicomputers by using a computer terminal connected by modem to a standard telephone line. [4] CompuServe offered businesses an early solution for electronic commerce by providing a payment network that billed charges to the customer's CompuServe account. By 1987, CompuServe had 380,000 subscribers, and held a 43% market share. [5] In 1989, CompuServe introduced an Internet gateway that let CompuServe users exchange e-mail messages with Internet users. By early 1997, CompuServe had grown to 2.6 million customers. A few companies ran their own forums on CompuServe, but the vast majority of forums were based around specific interests and hobbies. [6]

Forums and BBSes. Local and regional BBSes emerged in the 1980s as a popular way to share information online through personal computer servers connected to telephone modem banks. These systems offered forums where users could exchange information. Colleges and universities would provide local dial-up numbers for students, staff, and faculty to access campus information networks from remote locations [7] Many large software companies provided customer support activities through toll-free telephone numbers and centralized call centers, but some companies ran their own BBSes.

America Online. During the 1990s, a new competitor emerged in America Online or AOL. AOL quickly surpassed CompuServe as the dominant national provider of dial-up online and Internet gateway services. In 1997, AOL had 9 million subscribers, representing about half of the US market. [8] AOL purchased CompuServe's consumer business, [9] and acquired Netscape, the first widely successful Web browser. When AOL merged with Time Warner in 2000, many local and regional companies were offering their own Internet gateway services. By 2002, Microsoft Internet Explorer held a 97 percent market share worldwide for web browsing software. [8]

1.2   The Dot-com Era

Between 1997 and 2001, corporations raced to install Internet connections to provide employees and managers with email and web access, while adding servers for internal and public use. This period was known as the dot-com era, when venture capitalists and corporate interests pumped billions of dollars into a host of new media companies. One of the first new industries that was created by the surge in Internet usage was online advertising. The addition of banner ads to web pages gave media companies a new revenue stream, and provided businesses with new ways to engage with consumers, suppliers, and other stakeholders. There are many reasons that the dot-com era ended, including an investment bubble fed by lax government regulation, ambitious commercial banks, as well institutional and small investors who ignored the signs of economic overexpansion. [10,11] One example is MySpace, founded in 2003. MySpace users could create highly individualized web pages that listed their favorite music groups, television shows, and other interests. The service had problems generating enough revenue to scale its operations. Service quality declined as MySpace management struggled to expand the company beyond the United States. By 2008, millions of MySpace users had moved to other SNS. [12,13]

2   Major Social Networking Service Providers in the United States

Much of the SNS traffic in the United States is concentrated among a small number of SNS. The earliest companies in this group rose to prominence in 2007, when Apple introduced the first generation iPhone. Smartphones equipped with cameras and GPS technology were an excellent platform for SNS; users could read and post content whenever they had spare time to do so, instead of working at a laptop or desktop computer. By 2010, Apple had become the dominant mobile phone brand in the United States, and a number of mobile phone manufacturers had licensed Google's Android operating system to create competing products.

Facebook. Mark Zuckerberg took a different approach with Facebook. The service started in 2004 as a social networking site for college students at a small number of elite United States colleges and universities. The company gradually expanded by first opening up the service to anyone with a .edu email account, commonly used by faculty, staff, and students of United States colleges and universities. The user base was further expanded when Facebook opened the service to the general public in 2006. In 2007, Facebook had over 100,000 business pages. In 2008, Facebook surpassed MySpace as the largest SNS on the Internet. [14]

Gehl [13] presented a novel argument: MySpace failed because the service allowed its users too much control over their profile pages. MySpace users could use CSS to radically change the appearance of the page and block advertising banners. MySpace users often assumed fantasy identities that did not mention their real name or present an actual photograph of the user. Facebook, in comparison, provided layers of software abstraction for its users by rigidly controlling user customization options, and aggregated user-generated content into newsfeeds and groups. Gehl [13] explored the idea that Facebook also provided a better social environment, by encouraging each user to post under a single real online identity. Facebook users could post group photos, and tag or identify each user in the photo so Facebook could show the same photo in relevant user accounts.

Facebook also allowed business to run brand pages on the site. Brand pages could have multiple administrators, but the brand page could not be presented as an individual user. A common mistake made by novice Facebook marketers is to create an individual account to represent a business or organization. Facebook prohibits identity switching, as part of the service's efforts to ensure that individual Facebook users have only one Facebook account. Facebook also loses opportunities to earn advertising and placement revenue when an individual account is used for business purposes.

Twitter. Started as a SNS for text messaging users in 2006, Twitter was designed to let individual users broadcast text messages simultaneously to other users who became followers of an individual Twitter user account. The service soon added a web site and an application programming interface (API) that allowed software developers to let their apps send and receive Twitter messages. The API was a major advantage for Twitter, as most other SNS offerings required users to use a desktop computer or a specific smartphone app to interact with their services. Twitter limited messages, commonly called tweets, to a maximum of 140 characters. Developers quickly realized that URLs could be included in tweets to lead users to web pages, images, and other digital content. There was no need to embed digital content directly within Twitter messages; Twitter became a universal platform for sending short text messages and delivering digital content. Twitter has updated its service to include geolocation information, direct messages, analytics, and other features. [15]

Twitter users can be individuals or businesses. Twitter operates a program to validate celebrity and corporate accounts, but allows users to have multiple accounts. Businesses often use Twitter to broadcast marketing information, updates, and URLs of press releases. The direct messaging feature users to have private conversations; companies such as Delta Airlines and Comcast use this feature to provide customer support on Twitter, or to guide customers to specific employees who can provide further assistance.

LinkedIn. Designed as an SNS for businesspeople, LinkedIn struggled for several years to gain traction with users. By 2014, LinkedIn had established itself as the dominant SNS for businesspeople, by recruiting influential decision leaders and executives to write content for the service. LinkedIn earned the majority of its revenue through paid user accounts. A basic individual user account was free of charge, but several levels of paid membership were offered. Sales professionals and corporate recruiters placed a high value on LinkedIn's broad membership of business professionals and potential employees.

Facebook responded in late 2014 by announcing an enterprise-level service called Facebook At Work. While the Facebook At Work SNS uses the familiar Facebook newsfeeds and groups, each business user's account is kept separate from their personal Facebook account. [16] Facebook At Work uses a separate smartphone app, and was put into pilot testing with a small group of companies in January 2015. [17]

Instagram. Launched in late 2010 as an iOS app that let users take and post photos from their iPhone to a user account, Instagram became one of the leading SNSes for photo and video sharing. Some SNS users came to prefer Instagram because they can post as much text as possible in an image to help users evade spam and profanity filters. Other SNS users, especially young adults, preferred Instagram because their parents were using more web-based services such as Facebook. Instagram was purchased by Facebook in 2012 for US$1 billion. [8,18] According to one report, 91 percent of retail brands in the United States had a presence on Instagram in 2014. Instagram surpassed Twitter in user numbers in 2014. [19] One drawback of Instagram is that the service does not provide a direct method for buying a product or service from a post. [20] This issue is partially mitigated by the close integration of Facebook and Instagram user accounts.

FourSquare. Smartphone apps and companion websites that allow consumers to post their location and reviews of businesses have gained in popularity in the United States since 2010, with an app that let mobile users check in to a specific location. The Foursquare service has gamification elements; a user can earn badges by checking into multiple locations of a specific type of restaurant, or by checking into several different locations over a period of time. [21]

Yelp. The Yelp service provides high quality reviews and local venue information to help consumers find and compare restaurants, bars, and other retail establishments. [22] Yelp recruits local users to as Yelp ambassadors, setting up social events to reward elite users while attracting new users. There have been legal enquiries regarding the validity of Yelp reviews; in a small number of cases, businesses were found to have paid third parties to post false, positive reviews on Yelp and other sites. [23] Yelp has implemented filtering systems to identify biased reviews and fake reviews, but businesses are still vulnerable to other types of harassment by third parties, including extortion over false, negative reviews that may hurt sales revenues if left untouched on Yelp. [24,25] A few public health departments in the United States have used Yelp reviews to identify potential violations of food safety regulations by restaurants and bars. [26]

3   Examples of Social Networking Service Usage by Business in the United States

A 2011 conceptual framework for corporate SNS usage [27] presented four options that help organize the following discussion. The *predictive practitioner* focuses on a specific area, such as customer service. *Creative experimenters* employ small-scale tests with employees before trying larger projects. *Social media champions* coordinate activities across business functions and units to optimize SNS campaign effectiveness. *Social media transformers* develop branded platforms that graft SNS features such as wells, newsfeeds, groups, and video onto an extranet environment.

3.1   Opportunities Linked to Traditional and Digital Media Usage

The Nielsen television ratings service started monitoring SNS usage several years ago, when it became apparent that SNS users were watching television shows and having online discussions about the program. Twitter sponsored a 2013 Nielsen study that indicated up to 95 percent of the live online conversation during television programs took place on Twitter. [28] Dual screen viewers watch a television program on one device while sending SNS messages on a second device. [29] Some advertisers believe dual screen viewing may encourage young adults who have grown up with DVRs and on demand viewing to watch live television, by providing real-time interaction with an audience of peers. [30] One global example of dual screen viewing is the Olympic Games. Because the schedule of events is so broad, and multiple time zones are involved, some SNS users have turned to many-screen experience in order to follow the events while interacting with other SNS users. [31]

Recording artists and their distribution companies have used SNS to promote new songs, albums, and videos. For this industry, YouTube and Vevo have used social media transformer strategies for hosting and promoting music video. Businesses and artists use Facebook and Twitter to post URLs that link to the artist's recordings. [32]

3.2   Opportunities at the Point of Sale (POS)

Corporations have made effective use of SNS to drive customers towards traditional retail outlets. L'Oreal, a marketer of beauty products, reexamined its digital strategies in 2010 and adopted a social media champion model in its use of Facebook and YouTube as its major SNS platforms. L'Oreal focused its SNS campaigns on its B2C marketing efforts, and found that it received a better ROI from SNS than from traditional media campaigns. L'Oreal was invested in technology to tie its SNS campaigns to POS data. [33]

Location-Based Technologies in Retail and Entertainment Environments. Before the deployment of near field communications (NFC) technology and Bluetooth Low Energy in smartphones and networks, retailers and technology providers relied on wifi to track individual smartphones. [34,35] Mobile digital wallets require precise location sensing technology in order to work effectively. [36] Apple developed iBeacon to solve the problem of tracking individual users within an indoor environment such as an Apple Store, and the technology was implemented in all US Apple Stores in 2014. [37] iPhone users must install the Apple Store app and turn on Bluetooth to use iBeacon; when enabled, the Apple Store app provides shopping suggestions as the user moves through the Apple Store, alerts users when their Genius Bar appointment is available, and offers digital payment options for purchases. [38]

Major League Baseball employed a social media transformer strategy by installing iBeacons in 20 MLB stadiums in the United States for the 2014 regular season. iPhone users who have installed the MLB At the Park app received notifications about events and promotions as they moved through each stadium during a game. [39] A typical iBeacon can cover an 84,000 square foot building, and a set of 3 beacons costs as little as US$99. General Electric offers an iBeacon integrated in an LED light bulb; this solution is being used by Wal-Mart in its US stores. [37] As US retailers upgrade their point-of-sale transaction systems for a planned 2015 cutover to RFID chip-based credit and debit cards, some retailers have deployed POS systems that include NFC readers that work with smartphone-based payment systems like Apple Pay and Google Wallet. American Eagle Outfitters has used NFC payment systems since 2011, but added beacons in 2014 to find more selling opportunities within each retail store. [40] Macy's has also deployed iBeacons in a few stores on a trial basis. [41] Apple is developing enhanced indoor positioning technology that would estimate the wait times at retail stores, airports and restaurants. [42]

In 2014, the electronic payments company, Square, added a customer feedback feature to its point of sale system for smartphones and tablets. The new feature allowed merchants to ask for feedback on the transaction. For several years, online ratings companies such as Yelp and Foursquare had offered consumers opportunities to post public reviews and comments about businesses. The new Square feature sent consumer feedback directly to the business owner, who could then use Square's system to send the customer an SNS or email message to address the issue in private. This can create risks and opportunities for the business, depending upon how employees and management respond to the challenge of delivering consistently high service levels. [43] A social media champion strategy may be effective here.

3.3   Opportunities on Mobile Platforms

Mobile technology can turn almost any location into a point of sale. In a 2014 research study that used mobile phone data gathered from urban train stations in the northeast United States, consumers in public environments were sent targeted mobile advertisements via text messages; consumers in crowded environments were more likely to make purchases. One possible explanation is that people in crowded environments tend to turn inwards; a targeted mobile advertisement with a sales offer is actually a diversion from interacting with other commuters. [44]

SNS Usage in High Service and Luxury Markets. This heightened level of communication can be very helpful in industries where reputation and high levels of service are highly valued by customers and businesses. Examples of such an industry include hotels and lodging. One research study indicates that high-reputation hotels use a higher percentage of informal SNS contacts with their customers than do lower-reputation hotels. Informal contacts through SNS services such as Facebook and Twitter usually come during the sales process. It is likely that high-reputation hotels have dedicated more staff and information technology resources towards attracting and retaining customers. [45] Because privacy is often a concern for luxury customers, managers must be careful in their selection and implementation of social media strategies. For high-reputation firms and businesses that offer luxury products and services, SNS usage offers the benefits of a virtual community of brand advocates and admirers; the same SNS can easily be used by individuals and businesses who want to focus unwelcome attention on specific brands. Before SNS usage became popular, individual consumers would register domain names that portrayed target companies in unflattering ways. Twitter and Facebook have made this process even easier, by allowing individual SNS users to start groups and accounts with names that resemble or mock specific businesses. [46,47]

Advertising and Hashtags. Companies should also be aware of how they engage SNS users. An Irish study of consumer responses to SNS advertising indicated that aggressive push marketing strategies may disrupt the marketing message and repel potential customers in some markets. One example is placing advertisements within the personal profile pages of SNS users; some SNS users regard this as a commoditization of their online identity. Opt-in campaigns that allow the consumer to choose when and where advertisements appear may help pull campaigns. [48]

It has become much more common to see corporate social media accounts and recommended hashtags in print, broadcast, and digital advertising. A hashtag is a short text label, usually proceeded with a number sign (#), that serves as a clickable keyword or search term. Clothing brands such as Coach and Burberry sponsored Instagram campaigns that encouraged customers to post selfies with specific hashtags while wearing branded products. [49] Television networks advertise specific hashtags tied to specific shows, episodes, or story arcs. The hashtags also help marketers and analysts track aggregate users during live programming, as well as viewer interest in recorded programming available on DVRs and on demand services such as Netflix, Hulu, Amazon Prime, as well as dedicated apps and sites provided by distribution companies and broadcast networks.

3.4   Opportunities Within and Across Businesses

Some US businesses have invested in enterprise SNS to facilitate communication among employees. When employees are encouraged by management to exchange relevant information about work processes, organizational knowledge and work group performance can be enhanced. This type of SNS usage requires champion or transformer strategies, as well as significant corporate investments in training and implementation, to be effective. [50]

Employee Usage of SNS. Changes in business SNS usage may affect B2B relationships and corporate business functions. A 2015 study of research on social network usage in business environments indicated that the nature of the business sales force itself is changing, as new hires who have used SNS as young adults and students take jobs with companies. Younger employees may be more accepting of new motivational techniques, such as dashboards and online gamification of the sales process. Training on new sales methods is more likely to be delivered in with online content or through informal methods. Sales personnel have far more access to accurate, up-to-date information about current and potential clients, and supervisors can track employees with SNS usage, geolocation data, and video calls. Geographic sales territories may be less relevant in some industries, as travel and face-to-face selling are replaced by less expensive contact methods like Skype and Facebook. [51]

When Dell shifted from public to private ownership in 2013, the company's SNS team was in the midst of realigning its operations. The need for an SNS team had diminished as Dell employees become more adept at business SNS usage. Dell focused on B2B computer hardware and consulting services. A redesigned SNS consulting unit helped Dell leverage its expertise in a variety of areas. [52]

Engagement with Customers and Stakeholders. Maersk Line, a multinational logistics company based in Denmark that has significant operations in the United States, launched an integrated B2B social media champion campaign called Maersk Line Social across multiple SNS in 2011. The offering attracted over 1 million Facebook users; many more Facebook users shared Maersk Line Social posts with other users than the Maersk Line Social staff had forecast. Maersk management encountered significant internal problems in getting the marketing department to use Maersk Line Social in traditional campaigns. LinkedIn was used to share business-related news with Maersk customers through posts and groups. Instagram became a hub for hobbyists called container spotters who enjoyed taking photos of ships and posted them online. [53] While the majority of Maersk Line Social users were Maersk employees, between 15 and 20 percent of the user base were associated with Maersk B2B customers. The campaign used the company's digital photo archives, which were not being utilized by the marketing department, as an initial source of compelling content. During the Egyptian uprising of 2013, a Maersk vessel was in the Suez Canal. Maersk Line Social staff were able to verify that the ship's location with canal authorities. Staff members then posted updates on the ship's location and status on Twitter. Before Maersk Line Social, many more employees would have been involved in a less efficient effort to contact Maersk customers. [54]

3.5   Opportunities Related to Reputation

As US businesses worked to integrate SNS usage into their strategy and operations, many examples of best practice have emerged. Corporate reputation is an especially valuable asset for businesses. Reputational capital is difficult to build, but can be easily lost through inattention to the core values of a business. [55] SNS users are mobile, so employees who are recorded when they perform poorly may become the focus of online discussions and viral memes. [56]

Airlines. When 130,000 JetBlue passengers were trapped by winter storms and unable to leave airports and airplanes during a 2008 blizzard, some passengers turned to SNS to voice their frustration. The company's CEO became involved, working with employees to address passenger concerns during the storm. The resulting flurry of media attention led to policy changes within JetBlue, including a customer bill of rights that the airline released on YouTube, Twitter, and Facebook. [57] Active support of business SNS usage, especially by the CEO, is an important element in a social media champion strategy. In some companies, the CEO is the most visible user of SNS to external stakeholders. [58,59]

Electric Utilities. In July 2011, Commonwealth Edison, the electric utility company for much of the Chicago area, experienced massive service outages during a series of storms. The company's SNS team used Facebook and Twitter to update customers on service restoration, safety issues, and other relevant items. During the following week, question and answer (Q&A) sessions with the company's senior vice president of customer operations were held on Twitter and Facebook. While the company's SNS usage did not deflect all of the public criticism of the company's performance during the storm, the company considered the SNS team's activities successful because more customers had used SNS to engage with the company than ever before. [60]

Disseminating Information about Corporate Social Performance. Businesses have also used SNS to disseminate information about corporate social performance (CSP). Researchers have studied the financial linkages between CSP and financial performance, and found evidence to support a relationship between the two concepts. [61,62,63,64] Popular ratings of corporate social performance, such as the rankings provided by Fortune magazine, may have a variety of biases. [65] A review of business Twitter and Facebook activities of companies indicated that green companies tend to use SNS more often than companies that lag in natural environmental responsibility. Businesses that excel in CSP may need to use SNS to reach their target audiences and stakeholders. SNS usage by socially responsible businesses may also reinforce positive aspects of corporate culture, which aids recruiting and retention of employees who support that culture. Companies that exhibit poor CSP may resort to greenwashing and obfuscation to avoid public discussion of these issues. [66]

Legal Rights and Privacy. Some US businesses also have concerns regarding legal liability. In 2014, General Mills amended the company's privacy policy such that individual users give up their right to sue the company when they interacted with the company's web sites and online forums. Facebook pages and Twitter accounts were exempted from the policy. [67]

Employee use of SNS may be protected by US law, especially when the employee discusses workplace conditions. For example, employees who use their own smartphones and computers outside of work time to discuss an organizational performance issue on Facebook cannot be disciplined by management, unless the business has posted specific rules about off-hours SNS usage by employees. While some businesses may choose to reduce legal risks by banning personal SNS usage in the workplace, this policy may harm corporate culture and employee retention. A policy that encourages employees to be professional in their personal SNS usage may be easier to implement and more acceptable to employees and management. [68]

4   Discussion and Conclusions

Postman [69] (2003) remarked that the telegraph reduced public conversation to a language of headlines. The same might be said regarding SNS; however, there are more than a billion individual SNS users. Far fewer businesses are using SNS in the United States. This section will examine how businesses can better utilize SNS in their operations and strategy.

Researchers of business SNS usage suggest that careful planning is essential. Recruiting highly skilled employees to manage corporate SNS activities is an essential step. These individuals tend to be millennials; a recent estimate indicated that 96% of millennials belong to at least one SNS. [70,71] Managers should pay attention to work-life balance issues while providing opportunities for collaboration and career development. [72] The SNS must understand business and industry policies, regulations, and stakeholder expectations, especially in regards to accuracy. [73] There is an inherent tension between the activities of marketing departments, especially their attempts to impose definitions upon brands, and the fundamental individuality of the general SNS user. Some individual SNS users regard businesses as an unwanted participant in SNS activity. SNS such as Facebook and Twitter make it very easy for similarly minded consumers to find each other. The pace of SNS discussions may be too rapid for some businesses to address in a timely manner. [73]

Paniagua and Sapena [75] studied the Facebook and Twitter activity terms of 9 NASDAQ businesses. Their results suggest that these SNS user actions such as likes and follows can be a positive influence on a company's stock price, but only after a critical mass of SNS followers is obtained. The analysis also suggested that Twitter activity has a more positive effect on business performance than does Facebook activity. Weinberg and Pehlivan [76] suggest that businesses should develop a comprehensive plan for SNS activity to convert users into brand evangelists.

Madrigal [8] posits that SNS that support anonymity will become popular. Smartphone apps such as YikYak and Snapchat have gathered significant user bases in the United States by offering anonymous accounts and self-deleting user content. Employees may be drawn towards anonymous SNS, especially when presented by management with restrictive rules on personal SNS usage. Carefully designed SNSes like Maersk Line Social can help employees voice their support for their organization, as well as contribute to organizational knowledge. It is important that businesses establish a proper organizational context for employee SNS usage; employee SNS usage tends to go unnoticed unless and until problems occur. [77]

Businesses encounter many challenges and opportunities when dealing with SNS usage. Consistent environmental scanning by staff and managers is an essential task in SNS usage. Changes in technology and SNS policies may require swift reassessment and adjustments by corporate SNS managers and users.

Acknowledgment. This work was supported by JSPS KAKENHI (Grant-in-Aid for Scientific Research C) Grant Number 25380491v.

References

1. Phillips, R.: Stakeholder Theory and Organizational Ethics. Berrett-Koehler, San Francisco (2003)
2. Christensen, C.: The Innovator's Dilemma: When New Technologies Cause Great Firms to Fail. Harvard Business School Press, Boston (1997)
3. Rafii, F., Kampas, P.: How to identify your enemies before they destroy you. Harvard Bus. Rev. https://hbr.org/2002/11/how-to-identify-your-enemies-before-they-destroy-you/
4. Lapachet, J.A.: Virtual communities: The 90's mind altering drug or facilitator of human interaction. University of California, Berkeley (1994). http://besser.tsoa.nyu.edu/impact/s94/students/jaye/jaye_asis.html
5. Pollack, A.: Ruling may not aid videotex. New York Times. http://www.nytimes.com/1987/09/15/business/ruling-may-not-aid-videotex.html

6. Solberg, R.: Networking in the '90s via Computer. Public Relations Tactics **2**, 21 (1995)
7. Quinn, M.: Ethics for the information age, 2nd edn. Pearson/Addison-Wesley, Boston (2006)
8. Madrigal, A. The Fall of Facebook. The Atlantic. http://www.theatlantic.com/magazine/archive/2014/12/the-fall-of-facebook/382247
9. Mermigas, D.: CompuServe Deal: Both AOL. WorldCom Win. Electronic Media **16**, 30 (1997)
10. Cassidy, J.: Dot.con: The Greatest Story Ever Sold. HarperCollins, New York (2002)
11. Lowenstein, R.: Origins of the Crash: The Great Bubble and Its Undoing. Penguin, New York (2004)
12. Gillette, F. The Rise and Inglorious Fall of Myspace. http://www.bloomberg.com/bw/magazine/content/11_27/b4235053917570.htm
13. Gehl, R.: Real (Software) Abstractions: On the Rise of Facebook and the Fall of MySpace. Social Text **30**, 99–119 (2012)
14. Arrington, M.: Facebook No Longer the Second Largest Social Network. http://techcrunch.com/2008/06/12/facebook-no-longer-the-second-largest-social-network
15. Carlson, N.: The Real History of Twitter. http://www.businessinsider.com/how-twitter-was-founded-2011-4
16. Abbruzzese, J.: Facebook is making 'Facebook at Work,' so you can Facebook at work. http://mashable.com/2014/11/16/facebook-at-work-2
17. Abbruzzese, J.: Facebook at Work is here — but only for a select few. http://mashable.com/2015/01/14/facebook-at-work-soft-launch
18. Price, E.: Facebook buys Instagram for $1 billion. http://mashable.com/2012/04/09/facebook-instagram-buy
19. Griffith, E.: Twitter co-founder Evan Williams: 'I don't give a shit' if Instagram has more users. http://fortune.com/2014/12/11/twitter-evan-williams-instagram
20. Dishman, L.: Instagram is shaping up to be the world's most powerful selling tool. http://www.forbes.com/sites/lydiadishman/2014/02/13/instagram-is-shaping-up-to-be-the-worlds-most-powerful-selling-tool
21. Frith, J.: Communicating through Location: The Understood Meaning of the Foursquare Check-In. J. Computer-Mediated Comm. **19**, 890–905 (2014)
22. Lim, Y.S., Van Der Heide, B.: Evaluating the Wisdom of Strangers: The Perceived Credibility of Online Consumer Reviews on Yelp. J. Computer-Mediated Comm. **20**, 67–82 (2014)
23. Streitfeld, D.: Give Yourself 5 Stars? Online, It Might Cost You. http://www.nytimes.com/2013/09/23/technology/give-yourself-4-stars-online-it-might-cost-you.html
24. Kugler, L.: Keeping Online Reviews Honest. Comm. ACM **57**(11), 20–23 (2014)
25. Rahman, M., Carbunar, B., Ballesteros, J., Burri, G., Chau, D.H.P.: Turning the tide: curbing deceptive yelp behaviors. In: Proceedings of SIAM Data Mining Conference (2014)
26. Booth, D.: Yelp Partners with Health Departments to Improve Food Safety. J. Environmental Health **76**, 52 (2014)
27. Wilson, H., Guinan, P., Parise, S., Weinberg, B.: What's Your Social Media Strategy? Harvard Bus. Review. https://hbr.org/2011/07/whats-your-social-media-strategy
28. Koetsier, J. Nielsen: Tweets Drive Higher Broadcast TV Ratings for 48% of Shows. http://venturebeat.com/2013/08/06/nielsen-tweets-drive-higher-broadcast-tv-ratings-for-48-of-shows
29. Wang, J.: TV, Digital, and Social: A Debate. Media Industries **1** (2015)
30. Cassino, K.: Media's Millennial Myth: 'They'll Grow into It'. http://adage.com/article/guest-columnists/media-s-millennial-myth-grow/291730

31. Anstead, E., Benford, S., Houghton, R.J.: Many-screen viewing: evaluating an olympics companion application. In: Proceedings of the 2014 ACM International Conference on Interactive Experiences for TV and Online Video (2014)
32. Kaplan, A.M., Haenlein, M.: The Britney Spears Universe: Social Media and Viral Marketing at Its Best. Bus. Horizons **55**, 27–31 (2012)
33. Neff, J.: L'Oreal's Digital Spending Boost is Paying Off Big Time. http://adage.com/article/cmo-interviews/l-al-digital-spending-boost-paying-big-time/237204/
34. Danova, T.: Apple's New iBeacon May Spell The End Of NFC Technology. http://www.businessinsider.com/how-apples-new-ibeacon-may-be-the-end-of-nfc-2013-9
35. Brustein, J.: Apple Cuts Off a Way to Secretly Track Shoppers. http://www.bloomberg.com/bw/articles/2014-06-10/apple-cuts-off-a-way-to-secretly-track-shoppers
36. Husson, T.: The Future of Mobile Wallets Lies Beyond Payments. Forrester Research (2015)
37. Gilpin, L.: 10 Ways iBeacon is Changing the Future of Shopping. http://www.techrepublic.com/article/10-ways-ibeacon-is-changing-the-future-of-shopping
38. Hein, B.: iBeacons turn Apple Store into seamless spam machine. http://www.cultofmac.com/257218/ibeacons-in-action-at-the-apple-store
39. Calimlim, A.: MLB.com At The Ballpark 3.0 Features iBeacon Support And iOS 7 Redesign. http://appadvice.com/appnn/2014/03/mlb-com-at-the-ballpark-3-0-features-ibeacon-support-and-ios-7-redesign
40. Grobart, S.: Apple's Location-Tracking iBeacon Poised for Retail Sales Use. http://www.bloomberg.com/bw/articles/2013-10-24/apples-location-tracking-ibeacon-poised-for-retail-sales-use
41. Cole, S.: Macy's begins pilot test of Apple's iBeacon in flagship New York, San Francisco stores. http://appleinsider.com/articles/13/11/20/macys-begins-pilot-test-of-apples-ibeacon-in-flagship-new-york-san-francisco-stores
42. Hughes, N.: Apple's 'indoor traffic' concept would estimate wait times at the airport, grocery store, DMV, more. http://appleinsider.com/articles/15/02/05/apples-indoor-traffic-concept-would-estimate-wait-times-at-the-airport-grocery-store-dmv-more
43. Wohlsen, M.: Square Turns the Lowly Receipt into a Giant Opportunity. http://www.wired.com/2014/05/square-feedback-digital-receipts/
44. Andrews, M., Luo, X., Fang, Z., Ghose, A.: Mobile Crowdsensing. Marketing Science forthcoming. Fox School of Business Research, 15–040 (2014)
45. Floreddu, P.B., Cabiddu, F., Evaristo, R.: Inside your Social Media Ring: How to Optimize Online Corporate Reputation. Bus. Horizons **57**, 737–745 (2014)
46. Krishnamurthy, S., Kucuk, S.U.: Anti-Branding on the Internet. J. Bus. Rsrch **62**, 1119–1126 (2009)
47. Jin, S.A.A.: The Potential of Social Media for Luxury Brand Management. Marketing Intelligence & Planning **30**, 687–699 (2012)
48. Diffley, S., Kearns, J., Bennett, W., Kawalek, P.: Consumer Behaviour in Social Networking Sites: Implications for Marketers. Irish J. of Mgmt. **30**(2), 47–65 (2011)
49. Dishman, L.: Instagram is shaping up to be the world's most powerful selling tool. http://www.forbes.com/sites/lydiadishman/2014/02/13/instagram-is-shaping-up-to-be-the-worlds-most-powerful-selling-tool/
50. Beck, R., Pahlke, I., Seebach, C.: Knowledge Exchange and Symbolic Action in Social Media-Enabled Electronic Networks of Practice: A Multilevel Perspective on Knowledge Seekers and Contributors. Mgmt. Info. Sys. Q. **38**, 1245–1270 (2014)
51. Moncrief, W.C., Marshall, G.W., Rudd, J.M.: Social media and related technology: Drivers of change in managing the contemporary sales force. Bus. Horizons **58**, 45–55 (2015)

52. Deshpandé, R., Norris, M.: Building a Social Media Culture at Dell. Harvard Business School Case 514-096 (2014)
53. Katona, Z., Sarvary, M.: Maersk Line: B2B social media-"It's Communication, not Marketing". California Mgmt. R. **56**(3), 142–156 (2014)
54. Wichmann, J. Edelman, D.: Being B2B social: A Conversation with Maersk Line's Head of Social Media. http://www.mckinsey.com/insights/marketing_sales/being_b2b_social_a_conversation_with_maersk_lines_head_of_social_media
55. Fombrun, C.: Reputation: Realizing value from the corporate image. Harvard Business School Press, Boston (1996)
56. Brownstein, M.: Keeping Your Reputation out of the Jaws of Social Media. http://adage.com/article/small-agency-diary/keeping-reputation-jaws-social-media/243716
57. Hanna, J.: JetBlue's Valentine's Day Crisis. http://hbswk.hbs.edu/item/5880.html
58. Gaines-Ross, L.: Get social: A Mandate for New CEOs. Sloan Management R. (2013). http://sloanreview.mit.edu/article/get-social-a-mandate-for-new-ceos
59. Rokka, J., Karlsson, K., Tienari, J.: Balancing Acts: Managing Employees and Reputation in Social Media. J. Marketing Management **30**, 802–827 (2014)
60. Diermeier, D.: Commonwealth Edison: The Use of Social Media in Disaster Response (2012). https://cb.hbsp.harvard.edu/cbmp/product/KEL734-PDF-ENG
61. Aupperle, K.E., Carroll, A.B., Hatfield, J.D.: An Empirical Examination of the Relationship between Corporate Social Responsibility and Profitability. Acad. Mgmt. J. **28**, 446–463 (1985)
62. Sodeman, W.A.: Social Investing: The Role of Corporate Social Performance in Investment Decisions. Bus. & Society **33**(2), 222–223 (1994)
63. Carroll, A.B.: Social Issues in Management Research Experts' Views, Analysis, and Commentary. Bus. & Society **33**, 5–29 (1994)
64. Wood, D.J.: Measuring Corporate Social Performance: A Review. Int. J. Mgmt. Reviews **12**, 50–84 (2010)
65. Sodeman, W.A.: Commentary: Advantages and Disadvantages of Using the Brown and Perry database. Bus. & Society **34**, 216–221 (1995)
66. Reilly, A.H., Hynan, K.A.: Corporate Communication, Sustainability, and Social Media: It's Not Easy (Really) Being Green. Bus. Horizons **57**, 747–758 (2014)
67. Strom, S.: General Mills Amends New Legal Policies. http://www.nytimes.com/2014/04/18/business/general-mills-amends-new-legal-policies.html
68. Jennings, S.E., Blount, J.R., Weatherly, M.G.: Social media—A Virtual Pandora's Box Prevalence, Possible Legal Liabilities, and Policies. Bus. Prof. Comm. Q. **77**, 96–113 (2014)
69. Postman, N.: Amusing Ourselves To Death: Public Discourse In The Age Of Show Business (20th Anniversary Edition). Penguin Books, New York (2006)
70. Childs, R.D., Gingrich, G., Piller, M.: The future workforce: Gen Y has arrived. Pub. Manager **38**(4), 21 (2009)
71. Gibson, L.A., Sodeman, W.A.: Millennials and Technology: Addressing the Communication Gap in Education and Practice. Org. Dev. J. **32**(4), 63–75 (2014)
72. DiMauro, V., Zawel, A.: Social Media for Strategy Focused Organizations. Balanced Scorecard Report (2012). https://hbr.org/product/social-media-for-strategy-focused-organizations/an/B1201C-PDF-ENG
73. Ray, A.: Ethics in Social Media Marketing: Responding to the Boston Tragedy. http://www.socialmediatoday.com/content/ethics-social-media-marketing-responding-boston-tragedy

74. Fournier, S., Avery, J.: The Uninvited Brand. Bus. Horizons **54**, 193–207 (2011)
75. Paniagua, J., Sapena, J.: Business Performance and Social Media: Love or Hate? Bus. Horizons **57**, 719–728 (2014)
76. Weinberg, B., Pehlivan, E.: Social Spending: Managing the Social Media Mix. Bus. Horizons **54**, 275–282 (2011)
77. Miles, S.J., Mangold, W.G.: Employee voice: Untapped resource or social media time bomb? Bus. Horizons **57**, 401–411 (2014)

The Impact of Knowledge Property and Social Capital on Creation Performance: Moderation of Goal-Driven Strategy

Shu-Chen Kao[1(✉)], Chien-Hsing Wu[2], Jhe-Yu Syu[1], and Ru-Jin Sie[1]

[1] Department of Information Management, Kun Shan University, Tainan, Taiwan
kaosc@mail.ksu.edu.tw
[2] Department of Information Management,
National University of Kaohsiung, Kaohsiung, Taiwan
chwu@nuk.edu.tw

Abstract. In this paper, effects of knowledge property (tacitness and complexity) and social capital cultivated by social networks (structure, relationship, and cognition) on knowledge creation performance are examined. A saliency of this research is the moderating effect of creation strategy. Based on 142 valid samples from the manufacturing industry and the service industry, the research findings were obtained, showing that (1) the knowledge creation performance depends likely on the knowledge property that companies consume and social capital generated from social networks existed in companies, (2) knowledge creation strategy plays a moderating role in the relationship between precedent variables and creation performance, (3) complex knowledge and social capital derived from social team cognition have more influences on the knowledge creation performance while adapting goal-driven creation mode in knowledge creation. Discussion and implications are also addressed.

Keywords: Knowledge property · Social capital · Creation mode · Creation performance

1 Introduction

As global competition increases, the challenges that companies confront are short product life cycle and rapid change of product to meet customer's needs. In such competitive environments, the importance of traditional capitals, such as factories, machinery or facilities, has gradually decreased. Instead of, intangible capitals such as knowledge resource or creation capabilities steadily become the key resources determining companies survive and grow. Knowledge, regarded as valuable capitals and accumulated by team collaboration, is a driver for forming creation capability and obtaining competitive advantages [1]. Accompanied with those had established knowledge, the climate that encourages employees creating and forming consensus of creation vision are also indispensable for ensuring valuable knowledge be continually generated. Innovation or creation, consisting of generating new ideas and their

© Springer-Verlag Berlin Heidelberg 2015
L. Wang et al. (Eds): MISNC 2015, CCIS 540, pp. 279–291, 2015.
DOI: 10.1007/978-3-662-48319-0_22

implementation into new products, manufacturing process or service by infusing new elements into original thinking context, can lead to the dynamic growth of profit for the innovative business enterprise [2,3]. From resource point of view, resource-based theory further argues that the outstanding capabilities to utilize or to transform those original resources into resources beneficial to gaining competitive position is importance and such capability itself is a kind of competitive resource. This "core capability" not only can keep company at a competitive position in the market, but also can form an obstacle keeping other competitors from approaching [4,5]. Furthermore, knowledge-based theory is proposed based on the viewpoint which illustrated that knowledge can be viewed as a kind of important resources in creation and has significant effect on the creation outcome [6,7,8,9]. Accordingly, understanding how to possess creation competences by transforming, utilizing and leveraging knowledge, and to embed the created knowledge inside companies as a source of competition have become the key issue for companies.

Although the importance of knowledge creation has been addressed in related literature, the question about the roles that knowledge characteristics play in creation is still unknown. Nonaka (1994) viewed knowledge with two dimensions: explicit and tacit according to knowledge contents easy to codify or not [10]. The explicit knowledge is articulated, codified, and communicated using symbols. And the tacit one is based on experience, thinking, and feelings in a specific context, and is comprised of both cognitive and technical components [11]. Knowledge with different characteristics, such as complexity and tacitness, could cope with different creation context and have different contribution on the final creation outcomes. For example, incremental innovation mostly is done by recombining explicit knowledge, and whereas radical innovation usually is succeeded by utilizing tacit knowledge [12,13]. Furthermore, knowledge can have another classification: simple and complex knowledge based on the numbers of conceptual elements make up and the relationship exists among those elements [12]. In general, complex knowledge is composed of multiple independent and novel conceptual elements so that it is hard to be realized through existing knowledge mapping and it probably require lots of extra information to comprehend such knowledge. Although complex knowledge is hard to understand compared with simple knowledge, more imaginations or novel ideas can be stimulated while figuring out those complicated concepts and their relationships [14].

In additions, the resource generated from social networks in a team is another consideration in knowledge creation. Social capital theory, argues that the actual and potential resources embedded within are available through and derived from the networks of relationships by actors to complete collective tasks [12,15,16,17]. Based on social psychology and organization theories, social capital can explain some phenomenon existing in team work, such as why reliable information to be volunteered, reduce malfeasance, cause agreements to be honored, enable employees to share tacit information, and place negotiators on the same wave-length [18]. According to related research, social capital can highlight three dimensions: structural dimension, relational dimension, and cognitive dimension [12,15,19]. The first one refers to overall pattern of connections between actors, that is, who you reach, how you reach them, and how often you reach them. The second one describes the personal relationships that people develops with each other through interactions. The third one illustrates the common view that all the collective members develop for innovation vision and collective objective. Thus, through connectivity, densely interaction and common consensus among team members, abundant social capital can be created and embedded in enterprise social networks, which is benefit for achieve the innovation goal [12,17,19].

Creation strategies, a planning used to call creation attention on the thinking space and direct team member's creating behavior, are importance factors to make creation task successful [2,20]. The creation mode, one kind of creation strategies, is generally divided into goal-driven mode and goal-free mode characterized by defined goal and degree of freedom [21]. In Kao et al. (2011), different creation mode usually represents different level of thinking limitation and has influence on thinking direction, finally lead to different creation results [21]. Although been viewed important, the role of creation mode plays in the creation is not received much attention and discussed in the literature. In addition to the effects of knowledge property (in terms of tacitness and complexity) and of social capital (in terms of structure, relation, cognition) on creation performance, the moderating effect of creation goal on the above relationships will be discussed in this research.

2   Research Hypotheses

2.1   Creation Performance

Creation or innovation refers to the generation of new idea and its implementation into new products, processes, or services, leading to dynamic growth of the profit for innovative business [2]. Urabe (1988) further indicated that through creation process, the new idea can be developed and commercialized into a new marketable product or a new process with attendant cost reduction and increased productivity. However, because of the difficulty of directly describing the connection within knowledge creation and the value it creates, the creation performance can be used to evaluate it. In general, the benefit of knowledge creation can be evaluated by the following dimensions: (1) product creation performance, (2) manufacturing creation performance, (3) management creation performance, (4) organization creation performance, and (5) strategy creation performance [21,22]. In this research, the first three dimensions will be adopted to measure creation performance for reducing the complexity of evaluation [21].

2.2   Knowledge Property

In most of resource-based related research, knowledge is thought as an intangible capital which is critical for companies indulging in varieties of creation activities. However, knowledge with different characteristics, such as tacitness vs. explicitness, simplicity vs. complexity, will contribute to the innovation process differently and lead to different impact on the creation outcomes [12,14,23]. Tacit knowledge, representing individual intuition, cognition, wisdom, and insight derived from past work, can denote individual deep viewpoint and mental state for surrounding events. Owing to tacit knowledge can help people think outside of the logical box, more ideas or solutions can be developed differently from the present, finally resulting in better creation performance [10,12,13].

Complexity, another property affecting knowledge creation, can be defined as the number of knowledge components contained in knowledge and the difficulties to realize the relationships among knowledge components [24]. Gopalakrishnan et al. (1999) illustrated three characteristics about complex knowledge: its difficulty, its intellectual sophistication, and its originality [25]. When knowledge be transferred is more complex, more novel idea could be generated because of the uncertainty derived from the process of understanding its originality. That novelty, if applied to product, could give assistance on radical innovation [12,25]. By the way, Chang et al. (2011) demonstrated that knowledge with more complexity usually means this kind of knowledge is likely to be helpful in resolving the complicated problems or the problems never met [26]. In transferring or utilizing complex knowledge, novel solution for product, manufacturing process or management sometimes could also be generated.

According to the description above, the hypothesis H1 is proposed as:

H1: Knowledge property is significantly related to creation performance
H1a: The tacitness of knowledge has positive effect on creation performance
H1b: The complexity of knowledge has positive effect on creation performance

2.3 Social Capital

Social capital is the aggregation of actual and potential resources within a specific social network, where is composed of relationship of mutual acquaintance and mutual recognition [17]. Moreover, social capital is the resource embedded within social network can be mobilized by the actors (individual or enterprise) to increase the rate of success for specific actions [16]. Generally, research on social capital highlights three dimensions: structural dimension, relational dimension, and cognitive dimension [15]. The first one refers to the connection pattern among the members, which can be measured by centrality, density, connectivity, hierarchy, and size [12,17,27,28]. Based on the connectivity, the structural pattern of social networks can be roughly divided into two types: central and distributed [28]. Because of their different connectivity, it not only leads to diverse interactivities among members but also results in different influences on the resource creation in a team [12,29].

The relationship exists in the social networks not only can reflect the degree of familiar each other, but also the foundation of bringing up trust each other [12,15]. While considering the question of how social capital is generated in specific social networks, relationship quality with respect with trust and cohesion are thought as the critical factors because trusted and close relationship is liable to inspire members to exchange their own resource voluntarily [12]. Refer to relational dimension, social capital is generally measured by interaction frequency, relationship duration, emotional intensity, closeness of a bond [30]. Once a cohesive and trust relationship is form in a team, individual member is able to access the produced social capital from social networks directly or indirectly to supplement personal sufficiency toward the success of knowledge creation in a team [31].

Cognitive form of social capitals, referring to norms, values, attitudes, and beliefs affecting interdependence in a team, is manifested as shared vision and shared language since the resources provides shared representation and interpretation [15,19]. Shared vision helps members unite together and target some specific common goals to create relevant resource, such as knowledge, information, specialty for goal approaching. On the other hand, shared language, which could beyond a language itself and be the acronyms, subtleties, and underlying assumptions that are the staples of day-to-day interactions [32]. Shared language can not only provide an avenue in which participants can understand each other and build common vocabulary in their domain, but also is the base to stimulate more novel thinking [19]. Thus, shared vision and shared language can help team members aggressively complete the assigned creation mission based on the similar cognitive background and obligation. Accordingly, hypothesis H2 is proposed as below:

H2: Social capital is significantly related to creation performance
H2a: The structural dimension of social capital has positive effect on creation performance
H2b: The relational dimension of social capital has positive effect on creation performance
H2c: The cognitive dimension of social capital has positive effect on creation performance

2.4   Goal-Driven Creation Mode

Knowledge creation is not only a process of stimulating knowledge, but also a model for deriving behavior that results from existing knowledge [33]. In knowledge creation, different creation modes usually lead to different creation processes and to obtain diversity of outcomes because creation mode can inspires different creation intention with respect to the thinking behavior and thinking space. Goal-driven creation mode usually focus on the goal defined in advance and is helpful in allocating relevant resources efficiently with the guide of this defined goal. The goal-driven mode has four features: (1) the goal should be recognized and is regarded as an important index which should be satisfactory, (2) processes of goal achievement can be ignored, (3) part of goal achievement can be allowed, and (4) the goal is fixed [34,35]. The goal-driven creation mode gives enterprise or individual a clear direction of resource allocation so that the goal achievement can be approached efficiently in top-down way [36]. In goal-driven mode, how to allocate and utilize resource to achieve predefined goal is considered more than possible solutions searching [37].

Creation mode, the driver for enterprise innovation, always brings different impacts on the final creation performance in different contexts [33]. For example, goal-driven creation mode possibly make intrinsic knowledge be neglect in knowledge creation since it is not efficient to transfer such uncodified knowledge under the limitation of predefined goal. Oppositely, explicit knowledge is thought valuable in goal-driven creation process because of well-organized knowledge diffusion which can directly forward the setting goal. In such situation, goal-driven mode can help team members transmit and learn the required or complex background knowledge in rapid and efficient way, and could indirectly find the better ways to reach the goal. But, the limitation of goal-driven and explicit knowledge used also could lead to few novel ideas breaking through the rigid frame developed to bring the huge improvement for creation performance.

In additions, goal-driven creation mode is also influential on the final creation performance resulting from different social capital contexts. For example, goal-driven creation will suitable for core-periphery structure because knowledge can diffused efficiently by way of some core members located at the central position of the social networks. In this circumstance, peripheral members are tent to develop cohesive relationship with the core member more than other peripheral members based on the existing norms and in a passive manner which is harmful for creation. Although the social capital produced by goal-driven creation mode satisfy the predefined goal more than deriving the novel ideas beyond the expectation, the predefined goal can give all the team members a clear vision for creation so that a common language and conscious can be easily formed. Thus, from structural, relational, and cognitive perspective, creation mode could have different impacts on the relationship between social capital and the creation outcomes. Accordingly, hypothesis 3 and 4 are illustrated as follow:

H3: Goal-driven creation mode has moderating effects on the relationship of knowledge property and creation performance

H3a: Goal-driven creation mode has negative moderating effects on the relationship of intrinsic knowledge and creation performance

H3b: Goal-driven creation mode has positive moderating effects on the relationship of complex knowledge and creation performance

H4: Goal-driven creation mode has moderating effects on the relationship of social capital and creation performance

H4a: Goal-driven creation mode has negative moderating effects on the relationship of structural social capital and creation performance

H4b: Goal-driven creation mode has negative moderating effects on the relationship of relational social capital and creation performance

H4c: Goal-driven creation mode has positive moderating effects on the relationship of cognitive social capital and creation performance

3   Method

According to the hypothesis described above, the research model is demonstrated in Fig. 1. There are two independent variables: knowledge property and social capital, dependent variable: creation performance, and moderating variable: goal-driven creation mode.

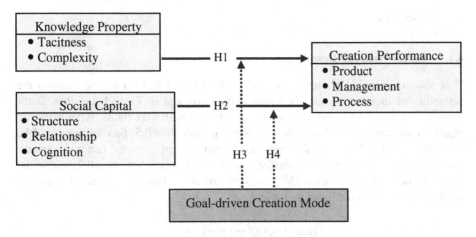

Fig. 1. Research Model

3.1   Measurement

The questionnaire was designed with four major parts: knowledge property, social capital, creation mode, and creation performance. The constructs are based on the related literature and measure by using Liker five-point scale for each question, where 1 represented "strongly disagree" while 5 represented "strongly agree". Before the questionnaire was sent out to the target, a pilot test was carried out to evaluate the validity, readability and reliability of the questionnaire.

3.2   Sample Analysis

For achieving the research goal, this research targeted the population of the industries who are demonstrating more activities in knowledge creation, in comparison with other industries. Thus, the manufacturing industry and the service industry are selected for considering their relevant experiences on the creation of product, processes and management. Of the 200 questionnaires mailed to company's administrative office at 2014/11/1, a total of 142 were returned after four weeks, revealing 71% valid returned rate.

For the survey samples, most of companies come from information manufacturing industry (58.5%), next is service industry (40.1%). In additions, 40.1% companies with less than 100 employees, 21.1% companies with employees 100 to 500, and 19.7% companies with over 1500 employees. Moreover, companies whose capital is less than 100,000 accounts for 50%, companies with capitals with less than one billion is 18.3%, and companies with capitals over 10 billions is 17.6%.

4   Results

4.1   Factor Analysis Results

This research adopted Kaiser-Meyer-Olkin (KMO) and Bartlett test to confirm the suitability of factor analysis. The result is demonstrated in Table 1 and then factor analysis is derived by principal component analysis with max rotate. After selecting Eigen value greater than 1 and factor loading greater than 0.5, four factors could be extracted as Table 2. Meanwhile, most the Cronbach α of extracted factor are greater than 0.7 except "complexity knowledge" factor which is little less than 0.7, so that the questionnaire can be accepted [38]. As to the result of Pearson's correlation coefficient of these factors are listed in Table 3.

Table 1. KMO and Bartlett's Test

Factor Name	KMO	Significance
Knowledge property	0.765	.000
Social capital	0.840	.000
Goal-driven Creation mode	0.642	.000
Knowledge Creation Performance	0.891	.000

Table 2. Factor Analysis Results

Factor Name		Questionnaire Item	Total explained variance	Cronbach α
Knowledge property	Tacitness	1-1,1-2,1-4,1-5,1-8	39.56%	0.81
	Complexity	1-6,1-7	60.52%	0.65
Social capital	Structure	2-2,2-3,2-10	21.16%	0.70
	Relationship	2-1,2-4,2-5,2-6	43.21%	0.77
	Cognition	2-7,2-8,2-9	64.11%	0.71
Goal-Driven Creation mode		3-1,3-2,3-3,3-4,3-5,3-6	45.62%	0.74
Creation Performance		4-1,4-2,4-3,4-4,5-1, 5-2,5-3,6-1,6-2,6-3	53.66%	0.90

Table 3. Correlation Coefficient Analysis

	KP	SC	CM	KCPe
Knowledge Property (KP)	0.762			
Social capital (SC)	0.229**	0.715		
Creation mode (CM)	0.399**	0.514**	0.773	
Creation Performance (KCPe)	0.248**	0.534**	0.514**	0.821

**.p<0.001

Diagonal line: AVE square value.

4.2   Analysis Results

(1) The relationship of knowledge property, social capital, goal-driven creation mode
     and knowledge creation performance

This study conducts regression analysis to realize the effects of knowledge property
and social capital on knowledge creation performance. First, Table 4 shows the relationships between two main precedent variables and the dependent variables. Model 1
illustrated that knowledge creation performance is affected positively and significantly by knowledge property (β=0.421, p= 0.000) and hypothesis H1 is supported. Model
3 also indicates that social capital has significantly positive effects on knowledge
creation performance (β=0.586, p= 0.000) so that hypothesis H2 is supported. With
respect to the moderating effect of creation mode, Model 2 shows that creation mode
has significant moderating effect on the relationship of knowledge property and
knowledge creation performance (β=0.455, p= 0.001). Similarly, Model 4 also reveals
that creation mode plays moderating role on the relationship of social capital and
knowledge creation performance (β=0.35, p= 0.003). According to the results of
Model 2 and Model 4, hypothesis H3 and hypothesis H4 are proven to be supported.

Table 4. Regression result of knowledge creation performance – Main dimensions

Dependent	Knowledge creation performance			
Independent	Model 1	Model 2	Model 3	Model 4
Constant	2.048***	7.553***	1.395***	5.746**
Knowledge property *(KP)*	0.421***	-1.493**		
Social capital *(SC)*			0.586***	-0.862*
Creation mode *(CM)*		-1.257*		-1.025*
Interactive effect *KP × CM*		0.455**		
SC × CM				0.35**

*p<0.1; **p<0.01; ***p<0.001

(2) The relationship of knowledge property, social capital, creation mode and know-
     ledge creation performance - Sub dimensions

To verify the hypothesis mentioned above in details, the relationship existing between
sub dimensions of precedent variables and dependent variables would be tested in
depth. In Model 5, it discloses that sub-dimensions of knowledge property: tacitness
and comlexity both have positively and significantly effects on knowledge creation
performance (β=0.262, p= 0.001; β=0.175, p= 0.016), so that hypothesis H1a and
hypothesis H1b are supported. Model 7 also indicates that three sub-dimensions of
social capital: structure, relationship, and cognition all have significantly positive
effects on knowledge creation performance (β=0.185, p= 0.013; β=0.232, p= 0.011;
β=0.171, p= 0.052) and hypothesis H2a, H2b, and H2c can be supported. Moreover,
Model 6 indicates that the moderating effects of goal-driven creation mode on the
relationship of tacit knowledge and creation performance is not significant (β=0.077,
p=0.607) so that hypothesis H3a is not fully supported. On the other hand, because

goal-driven creation mode has significantly positive moderating effect on the relationship of complex knowledge and knowledge performance (β=0.385, p= 0.003), hypothesis H3b can be supported here. Moreover, Model 8 reveals that moderating effect of goal-driven creation mode on the relationship of structural social capital and creation performance is not positive significantly (β=0.007, p=0.956). The similar result is seen on the relationship of relational social capital and creation performance (β=0.083, p=0.642). However, goal-driven creation mode indeed plays significant positive moderating role on the relationship of social cognition and knowledge performance (β=0.271, p= 0.086). Consequently, both hypothesis H4a and H4b are not supported, but hypothesis H4c is supported here.

Table 5. Regression of knowledge creation performance – Sub-dimensions

| Dependent | | Knowledge creation performance | | | |
Independent		Model 5	Model 6	Model 7	Model 8
Constant		1.977***	7.769***	1.388***	5.912**
Knowledge property	Tacitness *(TC)*	0.262**	-0.177		
	Complexity(**CX**)	0.175*	-1.325**		
Social capital	Structure *(ST)*			0.185*	0.012
	Relationship *(RS)*			0.232*	-0.091
	Cognition *(CG)*			0.171*	-0.825
Creation mode *(CM)*			-1.336*		-1.065*
Interactive effect	*TC × CM*		0.077		
	CX × CM		0.385**		
	ST × CM				0.007
	RS × CM				0.083
	CG × CM				0.271*

*p<0.1; **p<0.01; ***p<0.001.

5   Discussion and Implications

According to the analysis results, some findings are obtained and describes as follows. First, it is found that knowledge property itself including tacitness and complexity are benefit for knowledge creation (Table 4 and Table 5). The result is partially coincide with some prior research, the reasons could be original ideas are easy to be stimulated in the process of transferring tacit knowledge and consequently improve the creation performance. Meanwhile, hard to comprehension of complex knowledge are likely to initiate generating more possibilities for searching problem solving or product development in innovative manner, and finally aid the creation performance. Accordingly, although difficulties in transferring knowledge, tacit or complex knowledge is necessary and can be seen as an accelerator for achieving better creation performance.

Second, social capital itself and representing structural, relational, and cognitive dimensions have significantly positive effects on knowledge creation performance (Table 4 and Table 5). This finding is partially consistent with the suggestions of related literature indicating that social capital influences individual's knowledge sharing in virtual communities. Thus, we suggest that companies should develop their own social capital by way of suitable structure, cohesive relationship, or unity cognition as strategic resources for preparing for any possible creation with respecting to new product, manufacturing/service process, and management toward the creation performance improvement.

Third, from Table 4, it can be found that the effect of knowledge property on creation performance could be strengthened as creation goal is predefined clearly. If exam in details, it can be found that goal-driven creation mode plays extremely significant and positive moderating effect on the casual relationship of complex knowledge and creation performance (Table 5). The reasons could be diffusion of complex knowledge can be achieved efficiently in the scope of predefined goal, which can indirectly enhance final creation performance. Thus, if creation mission is filled with abundant of complex knowledge, we suggest companies are supposed to predefine goal clearly.

Finally, the result of Table 4 shows that moderating effect of goal-driven exists on the relationship of social capital and creation performance. It means that the creation mode in which goal is set clearly in advance will carry out powerful effect of social capital towards final creation performance. It is worth to mention is that the effect of social capital build by cognition on creation performance will be strengthened because of goal-driven consideration (Table 5). The reasons may that predefined goal can aid formation of collaborative vision and help team members to focus on the predefined thinking space, which is beneficial for improving final creation performance. Accordingly, we suggest those companies would like to reinforce the impact of accumulated social capital by way of common vision on creation should consider taking goal-driven creation mode to obtain better creation performance.

6   Conclusion

Nowadays, innovation becomes more important and receives more attention from academy and practice so that knowledge creation is viewed as a critical driver to accelerate companies growing. This research mainly explores those factors affecting knowledge creation performance from the viewpoint of knowledge property, social capital, and creation strategy. In addition to illustrate the effects of characteristics of knowledge and the ways social capital is accumulated on knowledge creation performance, this research also reveals the role of goal-driven creation mode by examining its moderating effect on the original relationship. Especially, it is found that complex knowledge and cognitive social capital can be leveraged by dealing with goal-driven creation activities. The findings of this research are expected to provide suggestions for companies to enhance their knowledge creation performance.

Acknowledgement. The research presented in this paper was funded by research grant No. 103-2410-H-168 -002. The authors would also like to sincerely express their deepest appreciation to the reviewers for their comments and suggestions.

References

1. Davenport, T., Prusak, L.: Working knowledge: How organizations manage what they know. Harvard Business School Press, Boston (1998)
2. Urabe, K.: Innovation and the Japanese management system. In: Urabe, K., Child, J., Kagono, T. (eds.) Innovation and Management International Comparisons. Walter de Gruyter, Berlin (1988)
3. Yang, J.: Innovation capability and corporate growth: An empirical investigation in China. J. Eng. Technol. Manage 29(1), 34–46 (2012)
4. Prahalad, P.K., Hamel, G.: The core competence of the corporation. Harvard Business Review, pp. 79–90, May–June 1990
5. Grant, R.: A resource-based perspective of competitive advantage. California Management Review 33, 114–135 (1991)
6. Drucker, P.F.: Post-capitalist society. Butterworth Henemann, Harper Business, London, Oxford (1993)
7. Grant, R.: The knowledge-based view of the firm: the strategic management of intellectual capital and organizational knowledge, pp. 133–148. Oxford University Press, New York (2002)
8. Wiklund, J., Shepherd, D.: Knowledge-based resources, entrepreneurial orientation, and the performance of small and mediumm-sized businesses. Strategic Management Journal 24, 1307–1314 (2003)
9. Curado, C.: The knowledge based-view of the firm: from theoretical origins theoretical origins to future implications. ISEG – Universidade Técnica de Lisboa Working paper (2006)
10. Nonaka, I.: A dynamic theory of organizational knowledge creation. Organization Science 5(1), 14–37 (1994)
11. Nonaka, I., Takeuchi, H.: The knowledge creating company: How Japanese companies create the dynamics of innovation. Oxford University Press, New York (1995)
12. Ana, P.L., Carmen, C.M., Antonio, C.L., Gloria, C.R.: How social capital and knowledge affect innovation. Journal of Business Research 64(12), 1369–1376 (2011)
13. Brockman, B.K., Morgan, R.M.: The role of knowledge in new product innovativeness and performance. Decision Science 34(2), 385–419 (2003)
14. Gopalakrishnan, S., Bierly, P.: Analyzing innovation adoption using a knowledge-based approach. Journal of Engineer Technology Manage 18(2), 107–130 (2001)
15. Nahapiet, J., Ghoshal, S.: Social capital, intellectual capital, and the organizational advantage. The Academy of Management Review 23(2), 242–266 (1998)
16. Lin, N.: Social capital: A theory of social structure and action. Cambridge University Press (2001)
17. Yu, S.H.: Social capital, absorptive capability, and firm innovation. Technological Forecasting & Social Change 80(7), 1261–1270 (2013)
18. Maskell, P.: Social capital, innovation and competitiveness. In: Baron, S., Field, J., Schuller, T. (eds.) Social capital: Critical perspectives. Oxford Univ. Press, Oxford (2001)
19. Chiu, C.M., Hsu, M.H., Wang, E.T.G.: Understanding knowledge sharing in virtual communities: An integration of social capital and social cognitive theories. Decision Support Systems 42(3), 1872–1888 (2006)
20. Popadiuka, S., Choo, C.W.: Innovation and knowledge creation: How are these concepts related? International Journal of Information Management 26(4), 302–312 (2006)
21. Kao, S.C., Wu, C.H., Su, P.C.: Which mode is better for knowledge creation? Management Decision 49(7), 1037–1060 (2011)

22. Liao, S.H., Wu, C.C.: System perspective of knowledge management, organizational learning, and organization innovation. Expert systems with Applications 37(2), 1096–1103 (2010)

23. Subramaniam, M., Venkatraman, N.: Determinants of transnational new product development capability: testing the influence of transferring and deploying tacit overseas knowledge. Strategy Manage Journal 22(4), 359–378 (2001)

24. McEvily, S.K., Chakravarthy, B.: The persistence of knowledge-based advantage: an empirical test for product performance and technological knowledge. Strategy Manage Journal 23(4), 285–305 (2002)

25. Gopalakrishnan, S., Bierly, P., Kessler, E.H.: A reexamination of product and process innovations using a knowledge-based view. Journal of High Technology Management Research 10(1), 147–166 (1999)

26. Yang, H.L., Tang, J.H.: Team structure and team performance in IS development: a social network perspective. Information & Management 41, 335–349 (2004)

27. Reagans, R., Zuckerman, E., McEvily, B.: How to make the team: Social networks vs. demography as criteria for designing effective teams. Administrative Science Quarterly 49, 101–133 (2004)

28. Borgatti, S., Everett, M.: Models of core-periphery structure. Social Networks 21(4), 375–395 (1999)

29. Janhonen, M., Johanson, J.E.: Role of knowledge conversion and social networks in team performance. International Journal of Information Management 31(3), 217–225 (2011)

30. Collins, C., Clark, K.: Strategic human resource practices, top management team social networks, and firm performance: the role of human resource practices in creating organizational competitive advantage. Academy Manage Journal 46(6), 740–751 (2003)

31. Holmen, E., Pedersen, A.C., Torvatn, T.: Building relationships for technological innovation. Journal of Business Research 8(12), 40–50 (2005)

32. Lesser, E.L., Storck, J.: Communities of practice and organizational performance. IBM Systems Journal 40(4), 831–841 (2001)

33. Sun, H., Wong, S.Y., Zhao, Y., Yam, R.: A systematic model for assessing innovation competence of Hong Kong/China manufacturing companies: A case study. Journal of Engineering and Technology Management 29(4), 546–565 (2012)

34. Scriven, M.: Goal-free evaluation. In: Hamilton, D., MacDonald, B., King, C., Jenkins, D., Parlett, M. (eds.) Beyond the numbers game, pp. 134–138. McCutcheon, Berkeley (1977)

35. Patton, M.Q.: How to use qualitative methods in evaluation. Sage, Newbury Park (1987)

36. Hsia, T.L., Wu, J.H., Li, E.Y.: The e-commerce value matrix and use case model: A goal-driven methodology for eliciting B2C application requirements. Information & Management 45(5), 321–330 (2008)

37. Loraasa, T., Diaz, M.C.: Learning new uses of technology: Situational goal orientation matters. International Journal of Human-Computer Studies 67(1), 50–61 (2009)

38. Hair, J.F., Anderson, R.E., Tatham, R.C., Black, W.C.: Multivariate data analysis. Upper Saddle River. Prentice-Hall, NJ (1998)

The Research on MMR Algorithm
Based on Hadoop Platform

Zhaoyang Qu, Xu-dong Ma[⊠], and Jian-lou Lou

College of Information Engineering, Northeast Dianli University, Jilin 130012, China
qzywww@mail.nedu.edu.cn, {609101577,loujianlou}@qq.com

Abstract. With the application and development of cloud computing technology in the field of mining data, how can a better data mining algorithm and Hadoop platform gather become the hot issue in the field of data mining. But the traditional Apriori algorithm needs to scan the database multiple times when dealing with large-scale data. System memory consumption is also too large to properly run. In this paper, we combined with the Hadoop platform in the MapReduce computing framework by introducing the matrix model. The MMR algorithm of association rules can have high processing capacity design. Test results show that MMR association rules algorithm not only realizes the frequent itemsets optimization, but also can effectively solve the traditional association rules algorithm memory consumption, low efficiency, high processing performance and reliability.

Keywords: Apriori algorithm · Hadoop platform · Matrix · Frequent itemsets

1   Introduction

In recent years, there is rapid development of computer in various fields of communication and network technology. According to statistics, the global data increment reached 1.8ZB (1.8 trillion GB) only in 2011 which is equivalent to each person in the world owing more than 200GB of data. This growth is still accelerating, according to conservative estimates in the next few years. The data will always maintain an annual growth rate of 50% [1,2]. These data are often noisy, huge and complicated, and it is difficult to use them directly. Therefore, how to more rapidly, low costly, high efficiently dig up the valuable and understand knowledge from a large number of database, which can help decision-makers make up better decisions has become a new subject in the field of data mining.

The improvement data mining algorithm based on the Hadoop platform has solved the bottleneck existing in the massive data mining. Hadoop platform is the basic

The work is supported by National Natural Science Foundation of China (No. 51277023).
The work is supported by Hall of Jilin province science and Technology Agency of key science and technology project (No. 20130206085SF).

L. Wang et al. (Eds): MISNC 2015, CCIS 540, pp. 292–302, 2015.
DOI: 10.1007/978-3-662-48319-0_23

architecture of distributed system developed by the Apache foundation. And its core technology is the Hadoop distributed file system HDFS (Hadoop Distributed File System) and MapReduce framework[3]. How to improve the traditional data mining algorithm, and to make it combine with MapReduce programming framework to make the final result solve large-scale data mining problems is a hot issue in data mining.

Association rules algorithm is one of data mining algorithms which is mainly used to reflect the dependency relationships between things and dig relationship things. Its application is extremely common, so the study on association rules of data mining is still a hot issue [4,5].The Apriori algorithm can meet the needs of massive data processing on computational efficiency and memory comsumption.

In order to improve the operation efficiency and reduce system memory consumption, we put forward the MMR association rules algorithm combined with MapReduce computing framework. We calculate the frequency of item sets support by matrix, and optimize and improve the process of generating frequent itemsets. We use association rules mining to improve efficiency and reduce the cost of system memory.

2   Related Work

In the Hadoop platform, HDFS is usually used to save the files and is mainly composed of NameNode and DataNode[6,7]. During processing, HDFS is divided into a plurality of data blocks (the default size is 64M), and each block has multiple copies stored on different machines.

NameNode is the central node, and it saves the original data files, such as the file names, and structures, related attributes corresponding to a file (file editing time, permission). Each file blocks of data statistics belongs to a block of data DataNode and so on. As a part of central server, it is mainly used to hold related file namespace and the clients access files and other information. Through the main node, we can manage each file and its sub files including time, and attributes etc. The main attribute information of all the above information is stored in the local disk. The storage way as the namespace image files and edit log two methods. When processing the original data file, NameNode is used to process the original data file, and the DataNode is responsible for processing file which contains the essence of I/O operation whit something related to file data flow will not through the NameNode, but only queries DataNode which is related to data flow.

DataNode is the data block node which is used to store files on the local system, and including the test result of the data block. DataNode that created from the corresponding NameNode, deleting, moving, or renaming the file, such as commanding, creating files, writing and modifing the contents of the file can not be closed upon completion. After the completion of file creation, writing operation and closing, file contents can not be changed. When the DataNode node data block through the form of documents stored on the local file system, which includes two aspects of the information. One is the data on the other hand is the information of original data, including data block size, checking information of data block, and time of recording. When the DataNode node starts to work, registration will be conducted on the NameNode node.

After it is approved, and a certain time interval all nodes data block information will be reported to the NameNode. This process once every 3 seconds, the result include NameNode to the appropriate instruction DataNode such as data block transplant to another machine, or to give up a block of data. If after 10 minutes have not been processing results of a DataNode node, it proves that the existence of the node.

Over the years, in order to solve the bottleneck of the Apriori algorithm, scholars have done a lot of research research on its algorithm corresponding improvement. One of the typical research results for the proposed FP-growth algorithm, we can prove that the execution of FP - growth algorithm efficiency compared with Apriori algorithm has great progress through the relevant testing with good adaptability. At the present stage, aiming at the existing problems of traditional Apriori algorithm ,the study follows the status of domestic: literature [8,9] by solving node failure, load balancing is not easy to bring the issue to improve the algorithm efficiency and reliability, the literature [10, 11] has improved Apriori parallel algorithm by introducing a statistical grouping, cutting key pair production, therefore reducing the processing time of the algorithm. The literature [12] under the framework of Hadoop to improve the read and write data rate by Hbase. Through the Hbase algorithm is applied to experimental embodied in comparability and necessity. All of the above algorithms and although the realization of parallel Apriori algorithm, improving the computational efficiency, but we did not take into account the memory and communication and other aspects of the cost. The algorithm in the application process still has some unnecessary expenses.

3   MMR Algorithm Based on Hadoop Platform

3.1   The Detailed Description of MMR Algorithm

In order to improve calculation efficiency and reduce the system memory consumption, this paper introduces the concept of matrix, improving and optimizing the process of frequent itemsets. We improve operational efficiency of the algorithm of association rules. Through the matrix we calculate the item set support count, it only need to scan the transaction database two times. Specifically described below:

1.   First, the original data set to be processed is converted into Hadoop platform supports the HDFS file format. The original data set were divided into several blocks, counting for x, then were to be assigned to the node for processing.
2.   Get frequent 1-itemsets by MapReduce model calculations.
3.   In the Map stage, we set each data block owing a unique identifier MapID for introducing the concept of vector. When a data item which appeared also in frequent 1-item focus on the emergence of a project record appears, it did not appear to 0. Finally, every piece of data and frequent 1- itemsets can get comparison vector n, generate   keys for <MapID, n>.
4.   All key-value pairs <MapID, n> reassign nodes perform Reduce operation. All the key-value pairs into a matrix, the corresponding portion of the support column k frequency "and" action items available k.

5. Partial support for the same frequency k data items are combined to get the final frequency of the support, and the minimum support set by the user to compare more than frequent item sets the frequency of the minimum support can be used to generate association rules.

3.2   The Implementation of MMR Algorithm

Combining with these ideas, the process of MMR algorithm is shown in Fig.1:

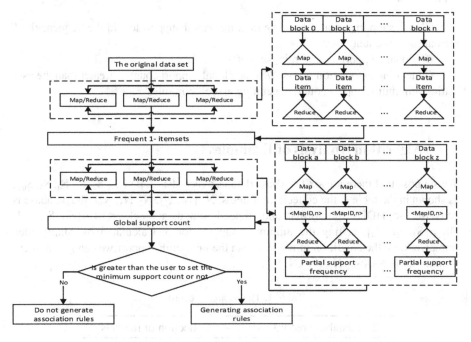

Fig. 1. MMR algorithm implementation process

Through the above statements we can see when the MMR algorithm in scanning the database and other algorithms are very different. Because it improves the traditional algorithm of association rules in database scanning and frequent itemset generation process, to improve the efficiency of the algorithm and only need to consume little memory.

3.3   The Pseudo Code of MMR Algorithm

The pseudo code of MMR algorithm:

```
F1=Frequent_1-itemsets(Dataset);   // The generation of frequent 1-itemsets in F1
Foreach (divide Dataset into x) {    // The original data set level into x data blocks
```

Transform F1 into a row vector n;   // Each data into a row vector n, each data block has a unique identification MapID

WriteOut(<MapID,n>);   // Output value of <MapID, n>

}

Foreach (MapID mapID) {

Matrix L=∅; // Have the same mapID n constitute a matrix L

Partial_support_frequency=[with the operation of "and" of L];   // The partial support frequency

}

Global_support_count=[the same rank for partial support to add the frequency];   // Global support count

if(Global_support_count≥min_support){

Generating association rules;   // If the global support count is greater than the user setting minimum support count, generating association rules

}

4   Algorithm Analysis of Examples

We suppose that there are six DataBase database records: $DB=\{S_1, S_2, S_3, S_4, S_5, S_6\}$, as shown in the table 1, the collection of items: $N=\{N_0, N_1, N_2, N_3, N_4\}$, the database is converted to HDFS file format and divided into data blocks, $DB_1=\{S_1, S_2, S_3\}$, $DB_2=\{S_4, S_5, S_6\}$, setting the minimum support for 3. Calculated by MapReduce framework, on the condition of not less than the minimum support, we can get frequent 1- itemsets:$\{<N_0,4>, <N_1,4>, <N_3,5>\}$.

Table 1. The database record

The database record Sid	Collection of items N
S_1	N_0 N_1 N_3 N_4
S_2	N_0 N_1 N_3
S_3	N_2 N_3
S_4	N_0 N_1 N_4
S_5	N_1 N_2 N_3
S_6	N_0 N_3

In the stage of Map, through frequent 1- itemsets, we can make each data of DB_1 and DB_2 two data blocks turn into row vectors, which contains N_0, N_1 and N_3 three elments. Compared with frequent 1-itemsets and every piece of data, there are results existing in both the corresponding value is 1, otherwise 0. The following table shows two block row vector n. Finally, every piece of data and frequent 1- itemsets getting comparison vector n, generate keys for <MapID, n>.

Table 2. Data block DB_1 vector representation

The database record Sid	N_0	N_1	N_3
S_1	1	1	1
S_2	1	1	1
S_3	0	0	1

Table 3. Data block DB_2 vector representation

The database record Sid	N_0	N_1	N_3
S_4	1	1	0
S_5	0	1	1
S_6	1	0	1

All key-value pairs <MapID, n> key reassigned to the node with the same MapID will be sent to the same node. Reducing processes to generate a matrix, and the matrix transpose to the data block DB_1 example transferring matrix shown in the following Table 4:

Table 4. Matrix transpose of data block DB_1

Collection of items N	S_1	S_2	S_3
N_0	1	1	0
N_1	1	1	0
N_3	1	1	1

As it can be seen from the table 4, the candidate set for the project: $\{N_0, N_1\}$, $\{N_0, N_3\}$, $\{N_1, N_3\}$, $\{N_0, N_1, N_3\}$.Then the candidate set $\{N_0, N_1\}$ frequency of partial support for: 1*1+1*1+0*0=2. In the same way it can be obtained $\{N_0, N_3\}$, $\{N_1, N_3\}$, $\{N_0, N_1, N_3\}$ portions respectively supported frequency: 2, 2, 2. The same calculation process of data block in DB_2 ,$\{N_0, N_1\}$, $\{N_0, N_3\}$, $\{N_1, N_3\}$, $\{N_0, N_1, N_3\}$ frequency of partial support for:1, 1, 1, 1. We sum up the two block of data partially support the frequency, and get the final support count of candidate itemsets. $\{N_0, N_1\}$, $\{N_0, N_3\}$, $\{N_1, N_3\}$, $\{N_0, N_1, N_3\}$ Support for the final frequency: 3, 3, 3, 3.

According to the set of minimum support, we can calculate the final frequent item-sets $\{\{N_0\}, \{N_1\}, \{N_3\}, \{N_0, N_1\}, \{N_0, N_3\}, \{N_1, N_3\}, \{N_0, N_1, N_3\}\}$,these are used to generate association rules.

5   Experimental Results and Analysis

5.1   The Experimental Environment

We used two HP ProLiant DL380 servers of which the processor model for Xeon X5660, RAM 192GB, 8TB hard disk. We install VMware ESXI 5.5 virtual machine software on each server, one of them installing the vCenter 5.0 virtual machine management platform and creating three Ubuntu Linux10.10 operating systems in a virtual machine. Creating two Ubuntu Linux10.10 operating systems on another virtual machine. The configuration works as the Master of the virtual machine system Linux, and at the same time as a Slave, and as a NameNode (virtual machine name node). The rest of the system works only as a Slave, and as a DataNode (data nodes). Three servers form a small LAN through a high performance switch. In addition, the experiment use the JDK1.6 version of JDK, and Eclipse-SDK-4.2.2 programming integrated development environment.

Table 5. The basic settings of each node

Serial number	The server name	The host name	IP address
1		Master	192.168.20.20
2	S_1	Slave1	192.168.20.21
3		Slave2	192.168.20.22
4		Slave3	192.168.20.23
5	S_2	Slave4	192.168.20.24
6		Slave5	192.168.20.25

5.2   The Experimental Process

We choose the online running state data(For a period of two years) in substation transformer of a region as the data sets, which contains the acquisition time, voltage, current, temperature, gas in oil, micro water content as a total of 17 attributes, eventually each file contains more than 500,000 records , each file as a separate data block.Get a different file from a backup, define four data sets Data1 to Data4, Table 6 describes the specific circumstances of each data set.

Table 6. The size of the data sets

Data set	prop-erty	Number of files	The amount of data/strip	Total size
Data1	17	5	2500000	≈300MB
Data2	17	10	5000000	≈650MB
Data3	17	15	7500000	≈900MB
Data4	17	20	10000000	≈1.2G

Here is a representative of the improved algorithm which are compared and analyzed: MR-Apriori algorithm[13] and DPC algorithm[14]. MR-Apriori algorithm using parallel computing project each data set the number of occurrences, thereby generating frequent itemsets, while improving the efficiency of the algorithm. But this method requires scanning the database many times,so it will greatly increase the memory overhead.

Experiment 1: Apriori algorithm, MR-Apriori algorithm and MMR three algorithms under the condition of data sets Data1 to Data4 data size gradually increasing, comparison of three algorithms in the processing time which is shown in Fig. 2.

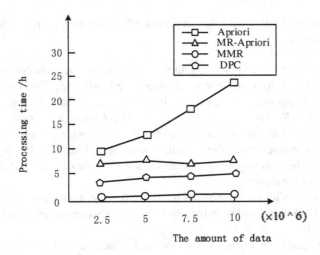

Fig. 2. Comparison of three algorithms in the processing time

As it can be seen from Figure 2, there are obvious advantages of MR-Apriori algorithm and MMR algorithm in processing time. And with the data size gradually increasing, the processing time ratio of them compare with single machine Apriori algorithm gradually has reduced. This shows that the parallel algorithm has a higher speedup ratio, in the small data size, duing to the parallel algorithm needs to open distributed task, consumption of communications so that the ratio is not great.

Experiment 2: With the number of nodes increasing,comparison of MR-Apriori algorithm and MMR algorithm in the processing time when runnig on Data1 data set, as shown in Fig. 3:

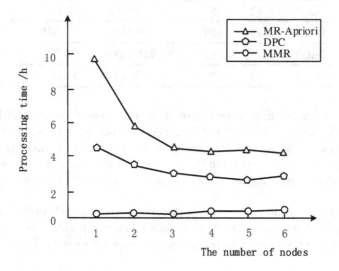

Fig. 3. Comparison of MR-Apriori algorithm and MMR algorithm in the processing time when run on Data1 data set.

As it can be seen from Figure 3, with the nodes increasing, processing time of parallel algorithm tends to be stable starting around the fourth node. This means that from the start of the fourth nodes, all tasks can directly handle without waiting. And we can see that the MMR algorithm is superior to MR-Apriori algorithm, because the MR-Apriori algorithm only needs to scan the database twice during the processing. And it does not produce a lot of intermediate result sets. Therefore, this algorithm can not only reduce the consumption on communication and the memory, but also improve the efficiency of the algorithm.

Experiment 3: With the number of nodes increasing, comparison of MR-Apriori algorithm and MMR algorithm in the processing time when run on Data4 data set, as shown in Fig. 4:

As can be seen from Figure 4, with the nodes increasing, MR-Apriori algorithm processing time is also increasing, in comparison, there are no major fluctuations of MMR algorithm in execution time, once again shows that the MMR algorithm is relatively optimal.

Experiment 4: Comparison of MMR algorithm execution time executed on the four data sets, as shown in Fig. 5:

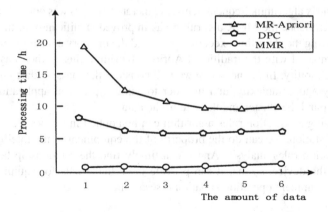

Fig. 4. Comparison of MR-Apriori algorithm and MMR algorithm in the processing time when run on Data4 data set.

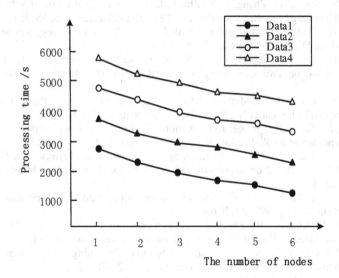

Fig. 5. Comparison of MMR algorithm execution time executed on the four data sets

As it can be seen from Figure 5, when MMR algorithm dealing with four different scale data sets, the time difference between them is very small. And it does not have larger fluctuation under the different scale of data, so it can be seen that the MMR algorithm has better portability and scalability.

6   Conclusion

In this paper, we proposed MMR association rule mining algorithm based on the Hadoop platform, which has improved the traditional Apriori algorithm. The introduction of the concept of matrix algorithm, frequent itemsets generated process was optimized.From the result of experiments, the MMR algorithm has improved significantly in the operation efficiency, saving the algorithm execution time, and the memory consumption is greatly reduced. Compared with the traditional Apriori algorithm, this method has high efficiency and reliability. In the next step we will consider the massive data environment data diversity and complexity, and in order to achieve practical application requirements, we expand the cluster and the size of the data.

MMR mining association rules algorithm can be applied in many fields, such as in the smart substation, we can do the property of the equipment in the intelligent substation association rules analysis. And it can timely find the relationship between the attributes of the device, so that it can provide the reliable data for equipment maintenance and equipment operation situation assessment.

References

1. Qu, Z.-Y., Hou, S.-L., Zhang, Y.-P., Zhang, J.-H., Xin, P.: Realization of Substation Visualization Training Platform. Journal of Northeast Dianli University **3**, 75–79 (2014)
2. Guo, X.-L., Han, X.: Grid Knowledge Collaborative Discovery Strategy Research. Journal Of Northeast Dianli University **1**, 94–98 (2014)
3. Qu, Z.-Y., Chen, S., Yang, F., Yan, J., Xu, S.-Q.: Intelligent Substation Data Classification Method Based on Two-level Analysis. Journal of Northeast Dianli University **2**, 61–65 (2014)
4. Qu, Z.-Y., Sun, P.-F.: A Method of Konwledge Representation in Power Grid Domain Based on OWL. Journal of Northeast Dianli University **4**, 30–34 (2012)
5. Hao, X.-F., Tan, Y.-S., Wang, J.-Y.: Apriori algorithm on Hadoop platform research and implementation of parallelization. Jisuanji Yu Xiandaihua **3**, 1–4+8 (2013)
6. Bai, G., Zhang, L.: A Novel Algorithm of Association Rules Extraction in E-commerce Concept Lattice Based on Improved Apriori. Journal of Convergence Information Technology **8**(5) (2013)
7. Yang, X., Liu, Z., Fu, Y.: MapReduce as a programming model for association rules algorithm on Hadoop, pp. 99–102 (2010)
8. Xiang, L.-H., Chen, Y.-W., Zhang, Y.-L.: High Order Matrix Multiplication by MapReduce Algorithm Based on Hadoop Platform. Computer Science, S1:96–S1:98 (2013)
9. Yu, C.-L., Xiao, Y.-Y., Yin, B.: A parallel algorithm for mining frequent item sets on Hadoop. Journal of Tianjin University of Technology **1**, 25–28+32 (2011)
10. Rong, X., Li, L.-J.: A method for frequent set mining based on MapReduce. Journal of Xi'an University of Posts and Telecommunications **4**, 37–39+43 (2011)
11. Huang, L.-Q., Liu, Y.-H.: Research on improved parallel Apriori with MapReduce. Journal of Fuzhou University (Natural Science Edition) **5**, 680–685 (2011)
12. Sun, Z.-X., Xie, X.-L., Zhou, G.-Q., Ni, J.-S., Hu, X.: Reserch on Apriori algorithm and implementation of Hadoop platform. Journal of Guilin University of Technology **3**, 584–588 (2014)
13. Li, L.: Cloud computing environment based on graphs Apriori algorithm of parallel optimization research. Automation & Instrumentation **7**, 1–4 (2014)
14. Lin, M.-Y., Lee, P.-Y., Hsueh, S.-C.: Apriori-based frequent itemset mining algorithms on MapReduce. In: Proceedings of the 6th International Conference on Ubiquitous Information Management and Communication, ICUIMC 2012 (2012)

Exploring Multidimensional Conceptualization
of Online Learned Capabilities

Cathy S. Lin[✉] and Han-Wei Hsaio

National University of Kaohsiung, 700, Kaohsiung University Rd.,
Nanzih District, Kaohsiung 811, Taiwan, R.O.C.
{cathy,hanwei}@nuk.edu.tw

Abstract. Internet has changed the way of learning, it has become an important part of the adolescents' life, online social networking has become particularly popular with adolescents, yet it is relatively unclear what capabilities the teenagers have learned and developed from the social media. While the most teenagers have their digital life, many parents of the teens have a range of concerns about what do their children do on the Internet. For many parents and teachers, they are anxious about their children's spending much time on the Internet might be risky, wasting time, having effects on their real life and academic performance. Therefore this study tries to close the gap to understand what the teenagers learn from their online activities. The purpose of this study is to propose a multidimensional conceptualization of "Online learned capabilities", and exploring the relationships between positive/negative emotions of teenagers and their online learned capabilities. Based on the empirical study, a field survey was conducted to senior high schools in Taiwan. The paper-and-pencil version questionnaire was distributed to students through teachers after class. A totally 383 valid surveys were collected. The statistical analyses helped validated the proposed second-order "online learned capabilities" and would strengthen explanations of teenagers' multi-dimensional learned from the cyber society.

Keywords: Online learned capabilities · Social network sites · Positive emotion · Negative emotion · Gender

1   Introduction

According to the Taiwan Network Information Center survey on broadband Internet usage in Taiwan (2014), there are 16.23 million (77.66%) Taiwan residents aged 12 and above have Internet experience. The participating of social networking sites (64.31%) is the first ranking among the Internet activities; the second ranking is using instant messaging software (51.45%). According to the "Teens, Social Media & Technology Overview 2015" conducted by Pew Research Center (2015), 92% of teens report going online daily, 71% of teens use more than one social network sites (Lenhart, 2015). For the teenagers who explore the online world, the ever-changing affect no doubt has different effects on each individual. A survey by King Car Education Foundation (2008) on online recreation indicates that 67% of teenagers would

© Springer-Verlag Berlin Heidelberg 2015
L. Wang et al. (Eds): MISNC 2015, CCIS 540, pp. 303–315, 2015.
DOI: 10.1007/978-3-662-48319-0_24

feel bad or insecure when they cannot access the Internet. Another survey by YoungNet (2009) on young people's needs in a digital age shows that 45% of the adolescences believe they are very happy when going online. Internet has become an inseparable part of the young lives; they live in a digital age.

While the most teenagers have their cyber life, many parents of the teens have a range of concerns about what do their children do on the Internet. Many parents are doubt that their children might interact with strangers, the online advertisers are collecting about their children's privacy information. For many parents of younger teens, they are anxious about how their children's online activities might affect their real life, academic performance, and privacy (PewResearchCenter, 2012). Several investigations report that children inclined not to "friend" their parents on social networking sites (Smallwood, 2010).   Obviously, the teens and parents have very different viewpoints about what can the teenagers learn from the Internet.

In the boundless and vast online society, teenagers get to choose the activities they wish to engage in, and they tend to feel more confident when dealing within online, technological activities (National Youth Commission, 2009). YoungNet (2009) also shows young people demonstrate a high degree of digital literacy in terms of how to access electronic media and online communities. Through the Internet, young people communicate with the world, and the abilities they demonstrate in the process eventually become what they learn from the Internet. For example, a survey conducted by the UK National Literacy Trust shows that approximate 60% of teenagers believe their creativity and focus in writing are inspired by the computer (Clark & Dugdale, 2009). Prensky (2002) discovered that teenagers learn to how help each other and work as a team in computer games. The use of instant-messengers promotes the establishment and maintenance of teenagers' friendship and also enhances their ability to express themselves during interpersonal interactions (Lee & Sun, 2008).

1.1 Research Questions

Given the above, the purpose of this study is to explore what the adolescences have learned from online social networking activities. Based on this understanding, the present study answers the following questions:

1.   Propose a multidimensional conceptualization model for understanding teenagers' online learned capabilities.
2.   Exploring the relationship between positive/negative emotions of teenagers and their online learned capabilities.
3.   Does the gender of teenagers make any difference in their emotions and online learned capabilities?
4.   Does the have/have not cyber acquaintance of teenagers make any difference in their emotions and online learned capabilities?
5.   Does the play/not play online game of teenagers make any difference in their emotions and online learned capabilities?

2   Literature Reviews

The literature reviews cover the following research variables, emotion, online learned capabilities, and hypotheses formulations.

2.1   Emotion

Emotion is an important resource to an individual, Ingleton (1999) pointed out that emotion plays an important role in the learning of any subject for an individual of any age. Different emotions have different effects. In the "broaden and build theory," Fredrickson (2001) stated that positive emotions such as joy, interest, satisfaction, confidence and love expand an individual's attention, cognition and behaviours, which in turn establish a person's physical and wisdom and social resources, allowing a better life in the future. In contrast, negative emotions narrow a person's ability to think and act. As pointed out by Seligman (2004), positive emotions encourage teenagers to boldly go where none has gone before, and as these emotions strive towards mastery, an individual continues to have more positive emotions and discover their strengths.

Smahel, Blinka, and Ledabyl (2008) discovered that in MMORPG (massive multiplayer online role-playing game), teenagers have positive and/or negative emotions of their in-game avatars, such as feeling proud or being embarrassed. The positive or negative emotions during online activities play a significant role in adolescence. In other words, the more positive emotions felt by a teenager during an online activity, they may expand the person' psychological, physical, and social resources, such as the ability to work as a team for the greater good and develop social relationships, which in turn provide gains for the person. On the contrary, the more negative emotions such as insult or fear experience by a person, the more likely it is for them to withdraw or give up and in turn narrow their learning. In our study, therefore, we believe the positive/negative emotions experienced by teenagers during online activities would influence their online learning and negative influences.

2.2   Online Learned Capabilities

To teenagers, online activities bring an individual much entertainment, information, and social communication and also allow them to acquire different positive learning. Prensky (2002) believes that playing games helps one to learn the five Ws (how, what, why, where, when/whether); how: how to use a computer; what: what rules to obey; why: why a co-op strategy should be used in a certain situation; where: a person can play games under different cultural settings; when: the person can make the right decision when facing moral decisions. Moreover, in the process of playing games, teenagers may acquire positive learning in online activities when they may not be aware of or do not expect such learning.

Take the learning of team work as an example; Prensky (2004) pointed out that MMORPG helps teenagers understand how team-mates help each other as well as basic computer skills such as quickly sending messages to communicate with other members. In the study on MMORPG, Yee (2006) discovered that by completing quests through team-work and communication, teenagers learn about leadership in the process. Cole and Griffiths (2007) believe that MMORPG provides opportunities for 80.8% of teenagers to interact with their real-life friends/family, and they also believe that in such an environment, team-mates must work together and assign tasks in order to complete challenges, thus teaching teenagers about the importance of team-work and communication.

Regarding the acquisition of creativity, National Literacy Trust (UK) pointed out that teenagers who access blogs or social networks are better in terms of creativity and focus (Clark & Dugdale, 2009). Prensky (2002) also believes computer games help expand teenagers' imagination since there are different elements in a game that help teenagers practice their imagination. Delwiche (2006) used MMORPG as an in-class teaching tool, and students demonstrated a high degree of creativity at the end of the semester.

Further, in terms of learning about interpersonal relationship, past studies on instant messaging (or IM) indicate that the use of IM does not only facilitate the establishment and maintenance of teenager's friendship but also facilitates their interpersonal communication ability (Lee & Sun, 2008; Lin, Sun, Lee, & Wu, 2007; Wang, Chen, Lin, & Wang, 2007).

In the learning of empathy, the study by McCown, Fischer, Page, and Homant (2001) on online chat-rooms indicates that teenagers who have better social skills and stronger verbal skills are more likely to show empathy for others. In the study by Stover (2005), the students were put in a online diplomatic simulation where they got into groups to simulate the release of news and strategies of different nations, and each team had to work together and communicate effectively; at the end, the young respondents' empathy for the countries they each simulated was increased.

Though teenagers acquire positive learning in online activities, emotions still play an important role in the process as Ingleton (1999) pointed out that emotions may develop social relationships and develop and maintain pride. The emotional changes experienced by teenagers in online activities would determine what types of abilities they learn after an activity. In the study by O'Regan (2003) on online learning, a correlation between an individual's emotion and the effectiveness of online learning was found; that is, when an individual has fewer negative emotions and more positive ones, the better the learning result gets. The study by Nummenmaa and Nummenmaa (2008) on "web-based learning environment," or WBLE, indicates that students who have more positive emotions are more likely to log onto WBLE; in contrast, when they have more negative ones, they are less likely to participate in collaborations such as making announcements or joining discussions. Based on the above literature review, we propose two hypotheses:

H1. An individual's perceived positive emotions influence the online learned capabilities.

H2. An individual's perceived negative emotions influence the online learned capabilities.

3   Methods

The literature reviews cover the following research variables, emotion, online learned capabilities, and hypotheses formulations.

3.1   Instruments

The questionnaire used in this study has two major sections, the first part includes demographic information and general Internet use; the second part includes a "positive and negative emotions" scale and a multidimensional scale to access teenagers' "perceived learned capabilities from online activities". Most items came from the previous studies. All items used five-point Likert scales anchored from strongly disagree (=1) to strongly agree (=5).

Positive and Negative Emotions. Positive emotion reflects the affective traits to which a person feels enthusiastic, active, etc., while negative emotion reflects the affective traits to which a person feels anxious, pressured, etc. To assess the teenagers' emotions when they engage in online activities, this study use the PANAS scale (Watson, Clark, & Tellegen, 1988). Sample item for positive emotion include "I feel exciting when I get online" and sample item for negative emotion include "I feel distressed when I am get online.").

Online Learned Capabilities. This study attempts to propose a multidimensional conceptualization model of "learned capabilities from online activities", based on the previous studies, Internet users can learn variety capabilities from the Internet such as team work (Cole & Griffiths, 2007; Yee, 2006), empathy (McCown et al., 2001; Stover, 2005), creativity (Clark & Dugdale, 2009; Delwiche, 2006; Wang et al., 2007), communication (Lee & Sun, 2008; Lin et al., 2007), and computer skills (Prensky, 2002, 2004). We measured this construct with 15 items, which consisted of the five aspects of learned capabilities online.

3.2   Research Subjects

In this study, we have conducted a field survey to senior high schools in Taiwan. The paper-and-pencil version questionnaire was distributed to students through teachers after class. A totally 383 valid surveys were collected. Of the survey respondents,

171(44.6%) were male and 212(55.4%) were female. The details of demographic statistics information are shown in Table 1.

Table 1. Sample Demographics and General Online Activities items (n=383)

Demographic Variables	Sample (N=383) Mean	SD	
Age (years)	16.81	1.30	
Internet experience (years)	6.26	2.25	
Game experience (years)	2.56	2.64	
	Category	Freq.	%
Gender	Male	171	44.6%
	Female	212	55.4%
Have Cyber Acquaintance	Yes	234	61.3%
	No	149	38.7%
Game Playing	Yes	241	63.3%
	No	142	36.7%

3.3   Reliability and Validity

To assess convergent and discriminant validity, exploratory factor analyses were conducted to detect high loadings on factors, factor loadings values fall between 0.626-0.899. All factor loadings are greater than 0.5 and all are statistically significant at $p < 0.01$. This implies that the measures satisfy convergent validity. All eigenvalues associated with the factors were exceeding the level of 1, vary from 1.515 to 4.835. Principal components analysis was used as the extraction method for factor analysis with Varimax rotation. As shown in Table 2, the overall factor structural solution has an appropriate loading pattern and explains 75.90 percent of the variation. Therefore, convergent and discriminant validity are supported.

After the factor analysis, a reliability test was performed for the extracted factors. Table 3 demonstrates the descriptive statistics and reliabilities among the factors. The reliability of each multiple-item measure is estimated using Cronbach's alpha, a commonly used measure of internal consistency. The alpha values fall between 0.76-0.92, these factors provide a reliable dimensions. Consequently, the extracted factor analysis shows that the following factors are justified; for the emotions constructs, they are: (1) positive emotion, and (2) negative emotion; for the learned capabilities online, they are the learned (1) team work, (2) empathy, (3) creativity, (4) communication, and (5) computer skills.

Table 2. EFA for research constructs

| Constructs | No. | Factor Loading | | | | | | | Eigenvalues | Sums of Squared Loadings |
		1	2	3	4	5	6	7		
Negative Emotion	1	0.871	0.044	-0.033	0.012	-0.056	0.063	0.013		
	2	0.845	0.036	-0.108	0.088	0.013	0.020	0.006		
	3	0.818	-0.004	0.014	0.068	-0.151	-0.086	0.069		
	4	0.817	0.032	-0.109	0.175	-0.084	-0.114	0.145	4.835	17.906
	5	0.805	0.037	0.079	-0.112	0.098	0.194	-0.141		
	6	0.805	-0.015	0.015	-0.165	0.037	0.174	-0.115		
	7	0.805	0.019	0.120	-0.064	-0.059	-0.190	0.048		
Positive Emotion	1	0.022	0.895	0.052	0.090	0.047	0.035	0.062		
	2	0.075	0.880	0.081	0.105	0.045	0.062	0.080		
	3	0.062	0.880	0.126	0.149	0.002	0.076	-0.001	3.848	32.157
	4	0.123	0.809	0.043	0.241	-0.013	0.001	0.076		
	5	-0.165	0.765	0.176	-0.041	0.239	0.041	0.029		
Learned Team Work	1	-0.015	0.117	0.815	0.290	0.153	0.084	0.121		
	2	-0.003	0.126	0.804	0.280	0.058	0.121	0.159		
	3	0.005	0.119	0.744	0.263	0.153	0.111	0.163	3.386	44.697
	4	-0.015	0.097	0.714	0.284	0.113	0.214	0.066		
	5	0.004	0.133	0.308	0.792	0.192	0.063	0.135		
Learned Empathy	1	-0.012	0.119	0.359	0.750	-0.011	0.229	0.084		
	2	0.012	0.202	0.380	0.697	0.028	0.230	0.115	3.348	57.096
	3	0.053	0.124	0.206	0.672	0.381	0.089	0.124		
	4	-0.045	0.214	0.343	0.639	0.355	0.144	0.054		
Learned Computer Skills	1	-0.044	0.110	0.142	0.149	0.893	0.099	0.075	1.937	64.272
	2	-0.136	0.089	0.199	0.302	0.784	0.006	0.070		
Learned Communication	1	0.022	0.088	0.237	0.243	0.115	0.782	0.200	1.625	70.290
	2	0.020	0.099	0.279	0.281	0.030	0.745	0.232		
Learned Creativity	1	0.030	0.110	0.173	0.157	0.043	0.257	0.845	1.515	75.900
	2	-0.003	0.121	0.397	0.190	0.168	0.150	0.700		

Table 3. Descriptive statistics and Correlations for research constructs

Constructs	Mean	STD	α	PA	NA	TM	EM	CS	COM	CRE
PA	3.39	0.76	0.91	1						
NA	2.42	0.85	0.92	0.048	1					
LRN-TW	3.42	0.83	0.90	0.369**	0.001	1				
LRN-EM	3.36	0.82	0.89	0.298**	-0.014	0.697**	1			
LRN-CS	3.99	0.88	0.84	0.228**	-0.139**	0.509**	0.404**	1		
LRN-COM	3.12	0.82	0.84	0.230**	0.026	0.529**	0.530**	0.274**	1	
LRN-CRE	3.31	0.87	0.76	0.255**	0.022	0.501**	0.531**	0.294**	0.512**	1

*p<0.05 ; **p<0.01

PA=Positive Affect, NA=Negative Affect, LRN-TW=Learned Team Work, LRN-EM=Learned Empathy, LRN-CS=Learned Computer Skill, LRN-COM=learned Communication, LRN-CRE= Learned Creativity

4   Results

4.1   Validation of the Multidimensional "Learned Capabilities Online"

We employed confirmatory factor analysis (CFA) to examine the assumptions multi-dimensional "learned capabilities online", comprising the four factors of team work, empathy, creativity & communication, and computer skills. The analyses were conducted with LISREL 8.30 (Jöreskog & Sörbom, 1996). All analyses used the covariance matrix and maximum likelihood estimation.

Figure 1 shows that the "learned capabilities online" is an aggregate of four dimensions, they are the learned (1) team work (standardized coefficient=0.93, p<0.01), (2) empathy (standardized coefficient=0.86, p<0.01), (3) creativity (standardized coefficient=0.69, p<0.01) (4) communication (standardized coefficient=0.67, p<0.01), and (5) computer skills (standardized coefficient=0.59, p<0.01). The totally five dimensions can be operationalized as a second-order factor model, in which a latent factor of "online learned   capabilities"   governs   the   correlations   among   the   five sub-dimensions.

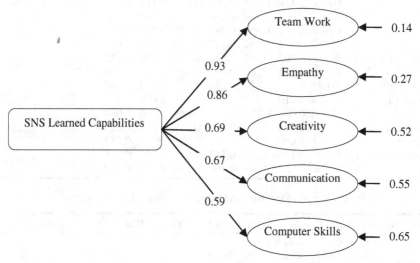

Chi-Square=203.38, df=80, p-value=0.00000, RMSEA=0.064

Fig. 1. The second-order factors of the "online learned capabilities"

4.2   Regression Results

Multiple regression analysis using SPSS 19.0 is conducted to examine the relationship between positive/negative emotions and learned capabilities online. The impact of two predictors on respectively five online learned capabilities was investigated using standard multiple regression (Enter method). Table 4 shows that the five models were significant. For the five online learned capabilities, PA all has a significant positive impact on the learned computer skill, team work, empathy, communication, and creativity. Yet NA has only negative impact on the learned computer skill. The results illustrate that for those teenagers who exhibit positive emotion online usually have dispositions that can be described as interested, active, enthusiastic, and so on; positive emotion therefore boost one's online leaned capabilities.

Table 4. Regression Results

Model	Dependent Variable	Standardized Coefficient (t value)		Adjusted R^2	F value
		Positive Emotion	Negative Emotion		
1	Learned Computer Skill	0.236^{**} (4.764)	-0.135^{**} (-2.725)	0.071^{**}	14.459^{**}
2	Learned Team Work	0.364^{**} (7.603)	-0.014 (-0.291)	0.132^{**}	28.903^{**}
3	Learned Empathy	0.298^{**} (6.080)	-0.028 (-0.568)	0.089^{**}	18.522^{**}
4	Learned Communication	0.228^{**} (4.569)	0.016 (0.325)	0.053^{**}	10.587^{**}
5	Learned Creativity	0.252^{**} (5.074)	0.010 (0.204)	0.064^{**}	12.977^{**}

$*p<0.05$; $**p<0.01$.

4.3　MANOVA Results

4.3.1　Gender Difference in Teenagers' Emotions and Online Learned Capabilities

In order to investigate whether gender makes a difference in teenagers emotions and online learned capabilities, the MANOVA served to examine gender differences in

Table 5. Results of the MANOVA test on gender difference

Constructs	Gender				F value
	Male(n=171)		Female(n=212)		
	Mean	STD	Mean	STD	
Positive Emotion	3.525	0.768	3.282	0.740	9.828^{**}
Negative Emotion	2.571	0.841	2.307	0.841	9.351^{**}
LRN-TW	3.600	0.796	3.267	0.829	15.802^{**}
LRN-EM	3.482	0.855	3.268	0.782	6.563^{**}
LRN-CS	3.921	0.924	4.042	0.843	1.800
LRN-CRE	3.285	0.874	3.335	0.859	0.317
LRN-COM	3.175	0.862	3.083	0.793	1.201

$^*p<0.05$; $^{**}p<0.01$.

each factors. Table 5 show the analytical results reveal that the male students' mean score of positive emotion, negative emotion, learned team work, and learned empathy were higher than their female counterparts ($p<0.05$ and $p<0.01$, respectively). However, no significant difference was found between male and female teenagers in learned creativity and communication and learned computer skills.

4.3.2   Cyber Acquaintance Difference in Teenagers' Emotions and Online Learned Capabilities

This study furthermore analyzed the relationship between teenagers' cyber acquaintance and the research constructs. This study divided the sample students into two groups: (1) have cyber acquaintance (n=234), and (2) don't have cyber acquaintance (n=149). The cyber acquaintance means the teenagers meet those net friends purely on the Internet; they are not a friend in the real life. Table 6 presents the results of the MANOVA, according to our findings, those teenagers who have cyber acquaintance revealed a significant higher mean score in the positive affect, learned team work, learned empathy, and learned computer skills than those who have no cyber acquaintance ($p < 0.05$ and $p < 0.01$, respectively).

Table 6. Results of the MANOVA test on cyber acquaintance difference

	Cyber Acquaintance				
Constructs	Have cyber acquaintance (n=234)		Have No cyber acquaintance (n=149)		F value
	Mean	STD	Mean	STD	
Positive Emotion	3.466	0.816	3.275	0.655	5.760[*]
Negative Emotion	2.487	0.835	2.333	0.867	3.021
LRN-TW	3.542	0.789	3.232	0.840	13.344[**]
LRN-EM	3.488	0.793	3.182	0.813	13.219[**]
LRN-CS	4.137	0.857	3.774	0.844	16.469[**]
LRN-COM	3.188	0.836	3.037	0.785	0.831
LRN-CRE	3.350	0.871	3.268	0.838	3.096

[*]$p<0.05$; [**]$p<0.01$.

4.3.3   Game Playing Difference in Teenagers' Emotions and Online Learned Capabilities

This study also analyzed the relationship between teenagers' game playing and the research constructs. This study divided the sample students into two groups: (1) have game playing (n=241), and (2) don't have game playing (n=142). Table 7 presents the results of the MANOVA, according to our findings, those teenagers who have game playing revealed a significant higher mean score in the positive emotion, learned team work, learned empathy, learned computer skills, and learned creativity than those who have no game playing ($p < 0.05$ and $p < 0.01$, respectively).

Table 7. Results of the MANOVA test on game playing difference

| Constructs | Game Playing Experiences | | | | F value |
| | Game Playing (n=241) | | No Game Playing(n=142) | | |
	Mean	STD	Mean	STD	
Positive Emotion	3.455	0.807	3.286	0.670	4.357*
Negative Emotion	2.481	0.866	2.328	0.814	2.872
LRN-TW	3.527	0.825	3.246	0.790	10.622**
LRN-EM	3.490	0.796	3.166	0.808	14.471**
LRN-CS	4.120	0.848	3.789	0.867	13.277**
LRN-COM	3.199	0.833	3.014	0.784	0.840
LRN-CRE	3.351	0.870	3.268	0.838	4.554*

*$p<0.05$; **$p<0.01$.

5   Discussions and Conclusions

5.1   Discussions

The purpose of this study is to explore what the adolescences have learned from on-line activities. Therefore, a multidimensional conceptualization model of online learned capability is proposed, further, the relationship between emotion and the on-line learned capabilities is examined. In addition, the differences of gender, cyber acquaintance, and game playing are tested.

The results show that emotion is confirmed to have a significant impact on five kinds of online leaned capabilities. Of which, positive emotions have the strongest effects on the learned team work ($\beta=0.364$), followed by empathy ($\beta=0.298$), creativity ($\beta=0.252$), computer skills ($\beta=0.236$), and communication ($\beta=0.228$). However, negative emotion has only negative influence on the learned computer skill ($\beta=-0.135$). The result of this study reveals that even in everyday Internet activities, positive emotions cultivate the adolescence online learned. The result is consistent with those online education studies by O'Regan (2003) and Nummenmaa and Nummenmaa (2008), these studies both indicate that teenagers' emotions influence their learning and performance. Also the "broaden and build theory" proposed by Fredrickson (1998), which states that positive emotions enhance mental capabilities, thinking, and action potentials. For example, the emotion of "interest" arouses our curiosity and encourages us to explore and look for needed information. On the other hand, negative emotions narrow one's thinking and actions and provide no benefits for learning.

5.2   Conclusions

Internet has changed students learning and life, it has become an important part of the adolescents' life, but it is relatively unclear what capabilities the teenagers have learned and developed from the online activities. The purpose of this study is to propose a multidimensional conceptualization of online learned capabilities. Based on the empirical validation, the statistical analyses helped validated the proposed "online learned capabilities" and would strengthen explanations of teenagers' multidimensional learned from the Internet.

Why the teenagers enjoy staying on the Internet? Traditionally, the kids learn social skills at home with their parents, then when they go to school, they learn socialization through peer interactions; and now in the Internet era, the existing of cyber society provide them a true experience to learn and practice different social skills online. Through this study, parents and teachers should try to understand the fact that the teens learn from online activities. Internet is a real cyber environment for adolescents having plenty of opportunities to practice and develop social thought and skills, and social behaviours independently without the supervision of parents and teachers. Examples in this study include the creativity in decorating one's blog, improved teamwork skills in online games, and improved computer skills such as faster typing and keyboard literacy. What also deserve our attention are teenagers' emotional reactions in the Internet life. Positive emotions may encourage teenagers to acquire different computer skills; on the other hand, when teenagers feel nervous or defeated online, they will get less online learned capabilities.

To the youth in Generation N, the Internet is no longer an important medium for gathering information but a life-style that intertwines with their real-life. Therefore, teenagers' Internet life can be improved if we understand the influences of online activities and the roles played by emotions. Our findings indicate the positive emotions in online activities have great benefits on online learned capabilities. It is advised that teenagers need to consider the appropriateness of an online activity and maintain the fun and passion in the process, and avoid online transgression behaviours.

References

Clark, C., Dugdale, G.: Young people's writing: Attitudes, behavior and the role of technology. National Literacy Trust (2009). (retrieved from) http://www.literacytrust.org.uk/research/Writing_survey_2009.pdf

Cole, H., Griffiths, M.D.: Social interactions in massively multiplayer online role-playing gamers. CyberPsychology & Behavior 10(4), 575–583 (2007)

Delwiche, A.: Massively multiplayer online games (MMOs) in the new media classroom. Educational Technology & Society 9(3), 160–172 (2006)

Fredrickson, B.L.: What good are positive emotions? Review of general psychology 2(3), 300 (1998)

Fredrickson, B.L.: The role of positive emotions in positive psychology: The broaden-and-build theory of positive emotions. American psychologist 56(3), 218 (2001)

Ingleton, C.: Emotion in learning: a neglected dynamic. Paper presented at the HERDSA annual international conference, Melbourne (1999)

Jöreskog, K.G., Sörbom, D.: LISREL 8 user's reference guide: Scientific Software International (1996)

KingCarEducationFoundation. The Adolescent Online Leisure Survey (2008). (retrieved from) http://www.kingcar.org.tw/news_txt.asp?NewsID=376&NewsType=1

Lee, Y.-C., Sun, Y.C.: Using instant messaging to enhance the interpersonal relationships of Taiwanese adolescents: evidence from quantile regression analysis. Adolescence **44**(173), 199–208 (2008)

Lenhart, A.: Teens, Social Media & Technology Overview 2015. Pew Research Center Ineternet, Science & Tech. (2015). (retrieved from) http://www.pewinternet.org/2015/04/09/teens-social-media-technology-2015/

Lin, C., Sun, Y., Lee, Y., Wu, S.: How instant messaging affects the satisfaction of virtual interpersonal behavior of Taiwan junior high school students. Adolescence-San Diego- **42**(166), 417 (2007)

McCown, J.A., Fischer, D., Page, R., Homant, M.: Internet relationships: People who meet people. CyberPsychology & Behavior **4**(5), 593–596 (2001)

Nummenmaa, M., Nummenmaa, L.: University students' emotions, interest and activities in a web-based learning environment. British Journal of Educational Psychology **78**(1), 163–178 (2008)

O'Regan, K.: Emotion and e-learning. Journal of Asynchronous learning networks **7**(3), 78–92 (2003)

PewResearchCenter. Parents, Teens, and Online Privacy (2012). (retrieved from) http://www.pewinternet.org/2012/11/20/main-report-10/

Prensky, M.: What kids learn that's positive from playing video games (2002). (retrieved from) http://www.marcprensky.com/writing/default.asp

Prensky, M.: A lesson for parents: how kids learn to cooperate in video games (2004). (retrieved from) http://www.marcprensky.com/writing/default.asp

Seligman, M.E.: Authentic happiness: Using the new positive psychology to realize your potential for lasting fulfillment. Simon and Schuster (2004)

Smahel, D., Blinka, L., Ledabyl, O.: Playing MMORPGs: Connections between addiction and identifying with a character. CyberPsychology & Behavior **11**(6), 715–718 (2008)

Smallwood, J.: Facebook: Should parents 'friend' their children? BBC News Magazine (2010). (retrieved from) http://www.bbc.com/news/magazine-11968954

Stover, W.J.: Teaching and learning empathy: An interactive, online diplomatic simulation of middle east conflict. Journal of Political Science Education **1**(2), 207–219 (2005)

TaiwanNetworkInformationCenter. A Survey on Broadband Internet Usage in Taiwan (2014, 2014/05). from http://www.twnic.net.tw/download/200307/20140820d.pdf

Wang, E., Chen, L., Lin, J., Wang, M.: The relationship between leisure satisfaction and life satisfaction of adolescents concerning online games. Adolescence **43**(169), 177–184 (2007)

Watson, D., Clark, L.A., Tellegen, A.: Development and validation of brief measures of positive and negative affect: the PANAS scales. Journal of personality and social psychology **54**(6), 1063 (1988)

Yee, N.: The demographics, motivations, and derived experiences of users of massively multiuser online graphical environments. Presence **15**(3), 309–329 (2006)

YoungNet. Young people's need in a digital age (2009). (retrieved from) http://www.youthnet.org/mediaandcampaigns/pressreleases/hybrid-lives

Multiple Days Trip Recommendation
Based on Check-in Data

Heng-Ching Liao[1], Yi-Chung Chen[2(✉)], and Chiang Lee[1]

[1] Department of Computer Science and Information Engineering,
National Cheng Kung University, Tainan 701, Taiwan, R.O.C.
{P76014127,leec}@mail.ncku.edu.tw
[2] Department of Information Engineering and Computer Science,
Feng Chia University, Taichung 407, Taiwan, R.O.C.
chenyic@fcu.edu.tw

Abstract. A travel recommender system can generate suggested itineraries for users based on their preferences. However, current systems are not capable of simultaneously considering trip length, distance, user requirements and preferences when making recommendations, being only equipped to consider one or two of these variables at one time. Also, to generate recommendations the system must process all attractions in the database, requiring more data access and longer processing time. We analyzed the check-in records of users and utilized a new concept of time intervals combined with a multiple days trip algorithm to produce itineraries compatible with the interests and needs of users. By applying R-tree to the travel recommender system, we reduced data access times and computation time. Lastly, we propose a trip evaluator equation that can be used to compare the strengths and weaknesses of each algorithm. Experimental results verified the effectiveness of our method.

Keywords: Trip recommendation system · Social network · Check-in data

1   Introduction

Academics have recently been devoting increased attention to travel recommendation systems to provide individuals travelling to a new city with itinerary suggestions according to their travel requirements. In the past, travellers used one of two methods to plan their trips: 1) purchasing travel packages from travel agencies; or 2) collecting information on their destination(s) and organizing an itinerary themselves. Although the first method saves users the time required to research the destination, travel packages are not designed according to individual preferences and may include attractions that do not interest the user. In addition, the length and daily itinerary of a travel package is fixed, leaving no flexibility for users to adjust plans as they please. The second method allows users to select the attractions that interest them and control the length and intensity of the trip; however, gathering information and organizing an itinerary can be time consuming. Travel recommendation systems can solve both of these problems. When a user inputs the destination, length of the trip, and how many

© Springer-Verlag Berlin Heidelberg 2015
L. Wang et al. (Eds): MISNC 2015, CCIS 540, pp. 316–330, 2015.
DOI: DOI: 10.1007/978-3-662-48319-0_25

Table 1. Attraction dataset.

Attraction	Name	Category	Opening Hour	Duration (Hour)	Public Rating
A	Concert hall	Amusement	10:00-22:00	3	9.5
B	Mall	Shopping	10:00-22:00	8	8.7
C	Valley	Landscape	08:00-17:00	2	6.5
D	Night market	Shopping	18:00-23:00	1.5	8.6
E	Mountain	Landscape	08:00-17:00	2	9.4
F	Park	Amusement	10:00-22:00	4	7.5
G	Old street	Shopping	10:00-22:00	2	5.8

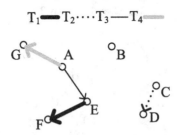

Fig. 1. Attractions on a map and the potential trips.

attractions he/she wishes to visit per day, the system analyzes these preferences and generates a suggested itinerary based on preferences and travel requirements.

A simple example of a travel recommendation system is presented in Figure 1 and Table 1. In Figure 1, each dot represents an attraction on a map and the lines indicate potential trips. Table 1 provides detailed information on each attraction, the attributes of which include Name, Category, Opening hours, Duration, and Public rating. Duration refers to the time that can be spent at the destination and public rating refers to how the destination is evaluated by the general public. Assuming that a user has allocated 10:00 to 15:00 for tourism, then T_1 and T_2 would not be suitable. As shown in Table 1, the finish time for T_1 (E→F) is 10:00 + 2 hours + 4 hours = 16:00, which is past the finish time of 15:00. With T_2 (C→D), the opening hours of D (18:00 – 23:00) do not fit the user's schedule (10:00 - 15:00). The schedules for T_3 (A→E) and T_4 (A→G) both meet the user requirements; however, Table 1 shows that the public rating for T_3 is 18.9, exceeding that of T_4 (15.3). T_3 has the highest score and most suitable itinerary among the four options; therefore, the travel recommendation system would suggest T_3 to the user.

However, these recommendation systems are not always practical, for the following four reasons:

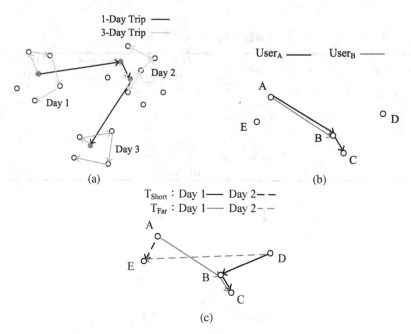

Fig. 2. Problems of existing trip recommendation systems.

- Public ratings do not necessarily reflect the individual preferences of users, each of whom may feel differently about an attraction.
- Each user travels for a different length of time. In Figure 2(a), each dot is a destination, with grey dots those attractions specially recommended by tourism websites. The black line represents a one-day trip and the grey line a three-day trip. On a one-day trip, users only visit the more important attractions (represented by the grey dots). By contrast, on a longer trip, users visit not only the recommended attractions but also neighboring areas. As the length of a trip affects the itinerary, this is a variable that must be considered in travel recommendation systems.
- Users have different preferences regarding the number of attractions they want to visit in one day. Figure 2(b) shows the plans of $User_A$ and $User_B$ in relation to how many attractions they plan to visit in a single day. The black line is the itinerary of $User_A$ while the grey line represents $User_B$. For $User_A$, one day is enough for him/her to visit three attractions, but $User_B$ does not want to be so rushed and chooses to visit only two attractions.
- Distance must also be considered on a trip lasting more than one day. Figure 2(c) shows two possible routes for a two-day trip, with the black line representing T_{Short} and the grey line T_{Far}. The solid line = Day 1 and the dotted line = Day 2. Both T_{Short} and T_{Far} include the same attractions, but T_{Short} covers a shorter distance compared to T_{Far}, because T_{Short} moves between neighboring attractions while T_{Far} travels between further destinations. Most users would choose T_{Short} to save additional transport costs.

No previous study has covered all of these issues. Hsu *et al.* [3] and Huang *et al.* [2] incorporated online information in the development of a personalized recommendation system that suggests only the highest scoring attractions. In that system, only one attraction rather than a continuous itinerary is provided with each response. Using social networks, Bao *et al.* [1], Ye *et al.* [10], Ying *et al.* [11] and Levandoski *et al.* [6] recommended new destinations based on their ratings of previous destinations. However, that method can only be used to recommend attractions, rather than organizing an itinerary. Lu *et al.* [7] first rated attractions and then integrated them into a travel algorithm. Their study is the most similar to the current study; however, they did not consider number of travel days or flexibility. Their algorithm simply uses the brute force method to generate answers and thus requires more time for computation. This study developed a new algorithm was developed to overcome all of these problems.

This study focuses on three issues. (1) How to use check-in records on social networks in order to analyze user preferences for each attraction. We also recommend some formulas that can be used to accomplish this. (2) We propose an entirely new concept, using time intervals, to solve the problems of personalization, user limitations, and opening hours. (3) Our approach includes user preferences, distance, I/O times, and algorithm processing time to rate a trip. An experimental model was used to verify the efficiency and effectiveness of the proposed method.

Chapter 2 introduces the development of trip recommendation systems and their methodologies. Chapter 3 describes our problem definition, and Chapter 4 explains the proposed algorithm in detail. Experiment results are set out in Chapter 5, and conclusions are drawn in Chapter 6.

2 Related Work

This section describes trip recommendation and classifies them into two categories: knowledge-based trip recommendation and collaborative filtering and knowledge-based trip recommendation.

2.1 Knowledge-Based Trip Recommendation

A knowledge-based trip recommendation system recommends a trip based on user's preferences and the characteristics of the attraction [4]. Lee *et al.* [5] proposed an ontological, multi-agent based recommendation system that uses fuzzy logic to select eight attractions as the basis of recommendation. The colony optimization algorithm is then used to select the optimal itinerary. However, this recommendation system is limited with regard to the number of attractions. Lu *et al.* [7] considered check-in records and travel budget when calculating user preference for attractions. Attractions were then arranged in combinations to find the optimal itinerary. Unfortunately, this method cannot be applied to longer trips, because arranging all of the attractions this way is too time consuming. Chiang *et al.* [2] considered user requirements, the number of days in a trip, budgets, and the type of attraction when recommending itineraries, allowing users to amend trip details as desired. However, their method can only recommend attractions based on category, rather than analyzing the previous records of users and making recommendations based on individual preferences.

2.2   Collaborative Filtering and Knowledge-Based Trip Recommendation

This recommendation system usually uses collaborative filtering technique to find the attractions that user may like, and then considers characteristics of the attraction to arrange a trip. Lu *et al.* [9] proposed the Trip-Mine algorithm, which considers trip scores and the time limitations of users when calculating an optimal itinerary. However, as the number of attractions increases, computing time increases exponentially, which makes it impractical to generate an itinerary for a trip lasting several days.

3   Problem Definition

Definition 1. Attraction. $A = \{a_1, a_2, \ldots, a_n\}$, there are n number of attractions in database A and each attraction has six attributes, being longitude, latitude, attraction category (*AC*), opening hours (*hour*), recommended length of visit (*duration*) and total check-in times (*ATC*).   ■

Definition 2. Attraction Score. $AS(u, a)$ indicates how much each user liked the attraction, with u representing the user.   ■

Definition 3. Day. k indicates the number of days as input by the user.   ■

Definition 4. Time Interval. i is the time interval as indicated by the user; this can be used to divide a single day into multiple intervals.   ■

Definition 5. Duration Time Interval. $Dti(a, i)$ indicates the duration interval, which can comprise one or more time intervals.   ■

Definition 6. Trip Table. TT records the k, i, begin time and end time of users.   ■

Definition 7. Trip Score. The trip table lists the average score AS (a) of attractions, which together make up the trip score TS (TT).   ■

Definition 8. Trip Distance. The total distance of recommended attractions is represented by $TD(TT)$.   ■

Problem Formulation: Using the above definitions, if a user inputs k, i, begin time and end time, our system recommends a trip and makes TS as large as possible, TD as short as possible.   ■

The reason our system does not recommend a trip with maximum TS and minimum TD is because the complexity of finding a such trip is O (n!). Assume that we have a database A which contains n attractions. User given k, i, begin time and end time is 10, 6, 8:00, and 18:00 respectively. We pick up enough attractions based on attraction score form high to low and fill up the trip table, we get a set $ca = \{a_1, a_2, \ldots, a_a\}$. In order to make sure that those attractions can be arranged in an appropriate time interval. We must list all kind of combinations to examine this set. If those combinations could not pass the exam of opening hour and duration, we must discard some attractions in ca and find another set. It is easy to understand that the complexity of listing all kind of combinations is O $(a!)$. Similarly, if we want to find a trip with minimum TD, we still must list all kind of combinations to find the shortest distance and the complexity in this case is O $(a!)$.

4   Algorithm

Our trip recommendation system has offline and online phases. During the offline phase, we use an R-tree to storage all attractions in the database. The online phase is divided into two main parts: 1.) Calculating user preference – After users input their parameters, the system analyzes their preferences and calculates attraction scores and user preference for each attraction. 2.) Multiple days trip recommendation algorithm – based on user preferences, this algorithm generates the trip most suited to the preferences and needs of users.

Table 2. User check-in record.

Attraction	Category	Total Check-in times	Attraction	Category	Total Check-in times
a_a	c_1	2	a_f	c_2	1
a_b	c_1	3	a_g	c_2	2
a_c	c_1	5	a_h	c_2	1
a_d	c_1	1	a_i	c_3	1
a_e	c_1	3	a_j	c_3	1

4.1   R-Tree Index Method

In the Off-line phase, the R-tree is used for search purposes. Each category of attractions is arranged into an R-tree (i.e., if there are m types of attractions in the database, then there will be m number of R-trees, R_{c1}, R_{c2}, ...R_{cm}). The latitude (coordinate x), longitude (coordinate y) and check-in times of each attraction form the basis of the tree. The leaf node stores the attractions from the database, and an internal node contains multiple leaf nodes. Each leaf node and internal node includes the x and y coordinates, as well as the maximum and minimum check-in times of the respective attractions. The x and y coordinates are used to mark the area of the nodes. The maximum and minimum check-in values are used during the online search. Note that for ease of explanation, we have indicated the maximum check-in value of node e as max_c(e) and the minimum check-in value as min_c(e).

The R-tree is used effectively in this study to accelerate searches by reducing the I/O of the algorithm. For example, the algorithm may realize that only those attractions rated 80 and over can be included in the itinerary. If the max_c (e) of a node e in R_{ci} = 70, then all the attractions in e will not need to be accessed by the algorithm, as their maximum check-in times falls below 80.

4.2   User Preference

After users enter a search, the first step of the online phase is to analyze their previous check-in records and calculate their preferences for each attraction accordingly. Table 2 shows the check-in records of a user. The first column is all the locations where he/she has checked-in. The second column shows what type of attractions these are, and the third is the times with which this user checks into these attractions. Therefore, user u preference for a certain type of attraction c_m can be calculated as follows:

$$\text{User Preference}(u, c_m) = \frac{\text{User total check-in times in } c_m}{\text{User total check-in times}}, \qquad (1)$$

where the denominator of Equation 1 is the total check-in times of u and the numerator is the total number of times that u checked in at c_m. In Table 2, for example, users checked into c_1 type attractions 14 times (2+3+5+1+3), with their total check-in times being 20. Using Equation 1, we find that user preference for c_1 type attractions is 0.7. Similar calculations show that user preference is 0.2 for attractions in category c_2 and 0.1 for attractions in category c_3.

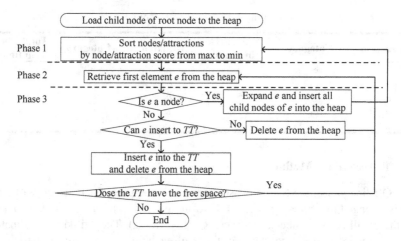

Fig. 3. The flow chat of multiple days trip recommendation algorithm.

How much users prefer an attraction, known as attraction score, can be calculated from total check-in times and the attraction score of the relevant category, as shown in Equation 2:

$$\text{Attraction Score}(u, a_n) = ATC(a_n) * \text{User Preference}(u, AC(a_n)). \qquad (2)$$

4.3   Multiple Days Trip Recommendation Algorithm

The second step of the online phase is the multiple days trip recommendation algorithm (MDTRA), which aids users in developing a personalized trip. The algorithm first employs a heap to identify attractions that would interest the user, and then uses the greedy concept to incorporate these localities into the given itinerary. The three-phase process of MDTRA is illustrated in Figure 3.

In Phase 1, the root nodes of each R-tree are placed in the heap and arranged from largest to smallest according to their node scores. Node score can be calculated using Equation 3:

$$\text{Node Score}(u, n) = \text{max\_c}(n)*\text{User Preference}(u, NC(n)). \tag{3}$$

where $NC(n)$ represents the category to which node n belongs.

In phase 2, MDTRA extracts the most interesting attraction to the user from the heap. This is done by selecting the highest scoring element e(e may be a node of R-tree or an attraction) from the heap for scoring. There are two approaches to processing depending on whether e is a node or an attraction. If e is a node, then the algorithm extracts all the attractions, or child nodes, in this node. If the node contains attractions, then Equation 2 is used to calculate their attraction scores. If the node contains child nodes, then Equation 3 is employed to calculate their node scores. These child nodes or attractions are then re-inserted into the heap according to their node scores or attraction scores. If e is an attraction, it is the one most interesting to the user, because it is rated higher than the other node scores and attraction scores in the heap. At this point, the algorithm moves on to the next phase and incorporates e into the trip table.

Assuming that in the previous phase, the algorithm identified the most interesting attraction to the user as α, then in Phase 3 we see whether α can be worked into the trip table, considering the following four case scenarios:

Case 1:α can be worked into a suitable time interval in the trip table, in accordance with the availability of the traveller and the opening hours of the attraction. The surrounding time intervals are already full with other bookings. If our scenario is in accordance with Case 1, then the algorithm does not need to consider Cases 2-4. Case 1 is the best-case scenario, because finding a suitable time interval is not easy. If there are multiple slots to choose from, the algorithm will choose the one that minimizes any increase in travel distance.

Case 2:α can be worked into a suitable time interval in the trip table, in accordance with the availability of the traveller and the opening hours of the attraction. Only one of the intervals before and after this interval is allocated to other activities. Case 2 is the second best scenario. If there are multiple slots to choose from, the algorithm will choose the one that minimizes any increase in travel distance.

Case 3:α can be worked into a suitable time interval in the trip table, in accordance with the availability of the traveller and opening hours of the attraction. None of the surrounding time intervals have been booked for any other activity. In this case, the algorithm will schedule the attraction into the earliest slot available, because travellers usually want to visit places that interest them as soon as possible.

Case 4:There is no suitable time interval to schedule α into the trip table, in which case the algorithm eliminates this attraction. After the algorithm has determined whether α can be worked into the trip table, it checks to see whether the table is full. If so, the process is complete; if not, the algorithm returns to Phase 2.

5   Performance Evaluation

In this chapter, numerous tests were conducted to verify the validity of the algorithm. Gowalla's dataset [8] was used and Stockholm of Sweden was chosen as the city to be tested. Stockholm consists of 41950 attractions and the dataset comprises longitudinal coordinates, latitudinal coordinates, and the real data of the total number of check-ins. attraction category, opening hours, and durations were randomly generated. The parameters and rages are shown in Table 3, where the default parameters were noted in boldface. The questions were innovative questions and no similar algorithm existed in the past; therefore, intuitively, we applied the brute-force method (BF) to solve this problem. The BF finds all the attractions within the query range, calculates attraction scores for each attraction, and sorts the attraction score. Moreover, the BF, according to the scores, lists all possible combinations of attraction itineraries. In addition to the BF and MDTRA that we proposed, we conducted experiments on three more algorithms for future references. The first algorithm is known as the Random algorithm, which randomly selects and puts attractions that satisfy the demands of opening hours and durations into the itineraries. The second one is known as the Score algorithm, which does not consider the opening hours of attractions; instead, it chooses and recommends attractions receiving the highest scores to users. The algorithm is expected to obtain the maximum trip score that satisfies user demands. The third algorithm is known as the Distance algorithm, which chooses the attraction receiving the highest score on each day. Subsequently, centered around the attraction, the next closest attraction, a_{next}, is found, through which another closest attraction, a_{next}, is found. These procedures are repeated until the trip table is completed. This method is expected to obtain the minimum trip distance that satisfies user demands.

Table 3. Configuration parameters.

Parameter	Range of value
Day(k)	2、 4、 **6**、 8、 10
Time interval(i)	2、 3、 **4**、 5、 6
Category(c)	**6**

In the simulation, three experiments were conducted. In the first experiment, users could define the performance influence of the number of days, k, on individual algorithms. In the second experiment, the influence of time interval, i, on individual algorithms was concerned. To compare the find the superior algorithm, the performance was evaluated using Equation 4–8. The evaluation could consider four parameters, including average execution time(AET), average trip score (ATS), average trip distance (ATD), and average attraction access times (AAAT).

$$Max_{AET} = Max(AET_{BF}, AET_{MDTRA}, AET_{Random}, AET_{Score}, AET_{Distance}), \qquad (4)$$

$$Max_{ATS} = Max(ATS_{BF}, ATS_{MDTRA}, ATS_{Random}, ATS_{Score}, ATS_{Distance}), \qquad (5)$$

$$\text{Max}_{\text{ATD}}= \text{Max}(\text{ATD}_{\text{BF}}, \text{ATD}_{\text{MDTRA}}, \text{ATD}_{\text{Random}}, \text{ATD}_{\text{Score}}, \text{ATD}_{\text{Distance}}), \tag{6}$$

$$\text{Max}_{\text{AAAT}}= \text{Max}(\text{AAAT}_{\text{BF}}, \text{AAAT}_{\text{MDTRA}}, \text{AAAT}_{\text{Random}}, \text{AAAT}_{\text{Score}}, \text{AAAT}_{\text{Distance}}), \tag{7}$$

$$\text{Trip evaluation(Query)}=\frac{1}{4}\times\left(\frac{\frac{\text{Max}_{\text{AET}}-\text{AET}_{\text{Query}}}{\text{Max}_{\text{AET}}}+\frac{\text{ATS}_{\text{Query}}}{\text{Max}_{\text{ATS}}}}{+\frac{\text{Max}_{\text{ATD}}-\text{ATD}_{\text{Query}}}{\text{Max}_{\text{ATD}}}+\frac{\text{Max}_{\text{AAAT}}-\text{AAAT}_{\text{Query}}}{\text{Max}_{\text{AAAT}}}}\right)\times100\% \tag{8}$$

Query in Equation 8 is the algorithm to be evaluated. $\text{AET}_{\text{Query}}$ is the average execution time of the algorithm to be evaluated. $\text{ATS}_{\text{Query}}$ is the average trip score of the algorithm to be evaluated. $\text{ATD}_{\text{Query}}$ is the average trip distance of the algorithm to be evaluated. $\text{AAAT}_{\text{Query}}$ is the average attraction access times of the algorithm to be evaluated. AET, ATD, and AAAT are the smaller the better. This measure finds the percentage with which the algorithm to be evaluated outweighs the worst item. Moreover, ATS is the larger the better, and that was why it was written as the performance percentage of the optimal ATS.

Each experiment was provided with the results of the proposed trip evaluation equations. The construction of the R-tree and user preference calculations in each experiment were regarded as preprocess. All the experiments were completed in the following environment: Intel Core i7-3770 3.4GHz processor, 4GB main memory, and Windows 7 64-bit version. All the programs were compiled by C.

5.1   Impact of the Number of Days

In this section, we will discuss the influence of providing users with various numbers of days on the performance of the recommended trip algorithms. Figure 4(a) is a definition of the execution time required by the algorithms with an increased number of days. The figure shows that the BF needs much more time than the other four algorithms. When $k > 4$, the algorithms require more than 2500 s of execution time because the BF would need to permute and combine the extracted attraction, resulting in large consumption of time. In addition, because of the time-consuming nature of the BL, users might not utilize this method for trip selection. Thus, subsequent experiments would not discuss results yielded by the BF.

Figure 4(b) differentiates between the rests of the algorithms by removing the BF data. The Distance algorithm, due to its purpose of finding the closest attraction to the attraction with the highest score on each day, consumes considerable time on searching for the closest attraction when arranging for the everyday schedule. Therefore, the execution time increases with the number of days. The time of the other three algorithms does not increase with days because the execution time of these algorithms is merely related to the times, instead of the days, of attraction score calculations. The MDTRA outperforms the other three, for the calculation of attraction scores does not require the calculation of the score of each attraction, and the sorting results for all attractions are unnecessary.

Figure 4(c) shows the influence of trip days on the trip score of any attraction. Besides the Random algorithm, the trip scores generated by other algorithms reduce with the defined number of days. As an explanation, the number of recommended attractions increases with the defined number of days, and therefore attractions with lower scores are also included into the trip, thereby lowering the trip score. This figure indicates that attractions recommended by the Score algorithm have a higher trip score because the algorithm considers only the score but not the opening hours and distance. The trip score generated by the MDTRA, although lower, is closer to that recommended by the Score algorithm. The MDTRA could actually approach a higher score result. The Distance algorithm is lower than the Score and MDTRA algorithms because it finds only nearby attraction and does not consider scores. Thus, the overall trip score is lower than the MDTRA.

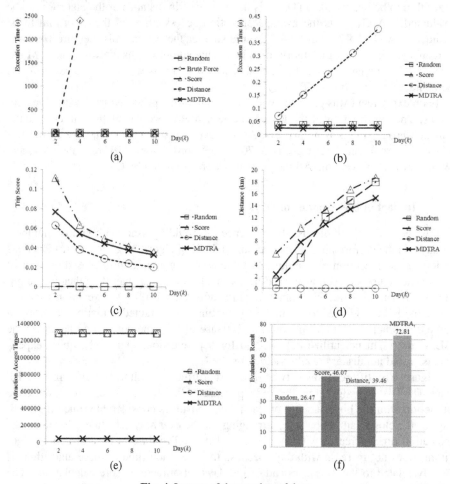

Fig. 4. Impact of the number of days.

Figure 4(d) displays the relationship between the trip distance and days recommended by the algorithms. According to the figure, except the Distance algorithm, the distance suggested by other methods significantly increases with the days because all attractions recommended by the Distance will be closest to the previous attraction. The Score algorithm does not consist of any mechanism that checks the distance. Therefore, the distance calculated by it is the longest amongst all four algorithms. When $k=2$ and $k=4$, the Random algorithm is superior to the MDTRA because the small number of randomly selected points does not greatly affect the distance. Nonetheless, as the number of days increases and available points become more, the Random can be worse than the MDTRA.

Figure 4(e) defines the number of days and the times of access to the attractions required by the algorithms. Except the MDTRA algorithm coupled with an R-tree, other algorithms do not need the assistance of index structures and, therefore, have to load all the attractions in Stockholm for judgment. The MDTRA only have to check attractions that interest the users.

Figure 4(f) shows the results of the proposed trip evaluation equation. The evaluation scores showed a descending order from the MDTRA, Score, Distance, to the Random. The MDTRA was much superior to other algorithms in terms of data loading times, and provided the optimal performance in execution time. Moreover, the trip score was only slightly worse than the Score algorithm. Although its trip distance was worse than the Distance, with its advantages in data loading times and execution time, its results of the trip evaluation equation were the highest. The Score algorithm significantly outweighed the Distance algorithm in execution time, and was superior to the Distance in trip score. Despite being inferior to the Distance in trip distance, the final evaluation favored the Score. The Distance algorithm, however, outperformed the Random in the trip score and trip distance and won the overall evaluation results.

5.2   Impact of the Number of Time Intervals

In this section, the influence of providing users with various time intervals on the performance of recommended trip algorithms. We did not employ the BF in this experiment, for the method has been known to be extremely time-consuming and was therefore excluded. Figure 5(a) shows the influence of increased time intervals on the execution time of the algorithms. Similar to Experiment 5.1, the Distance algorithm has execution time increasing with the needed time intervals in a single day because it not only has to find the attraction with the highest score each day, but also to find the closest attraction to the one previously mentioned. Time of the other three algorithms does not increase with time intervals because a direct proportion between the execution time and the times of attraction score calculations is identified. The MDTRA is the best algorithm amongst the four. It could generate itineraries without calculating the preference scores of all attractions.

Figure 5(b) indicates the influence of the number of time intervals on the trip score. The trip scores of all algorithms decreased with the increased time intervals. When the time intervals increase, recommended attractions also increase, and attractions with lower scores will be included into the itinerary, hence decreased the trip scores. In this figure, the Score algorithm demonstrated the highest attraction average scores, for it took only scores into account as a reference of recommendation; however, it abandoned opening time and durations. The trip scores were followed by the MDTRA, Distance, and Random. The Distance algorithm showed a lower score because it simply found adjacent attractions with the shortest distance, but it did not consider the score.

Figure 5(c) indicates the influence of the number of time intervals on the trip distances. Except the Distance algorithm, others showed significantly increased trip distances as time intervals increased. This is caused by the fact that the Distance schedules the itinerary only by finding the nearest attractions. Therefore, its itinerary distance is less influenced by the number of time intervals. According to the figure, as i increases, distance recommendations provided by the MDTRA can be less satisfactory than other algorithms because it could only arrange the attractions so that the trip distance is least changed. Because of the random selection, the Random algorithm showed a fluctuating trip distance. The Score algorithm considers only the attraction scores, does not have a mechanism that checks the distance, and in most scenarios has a longer trip distance.

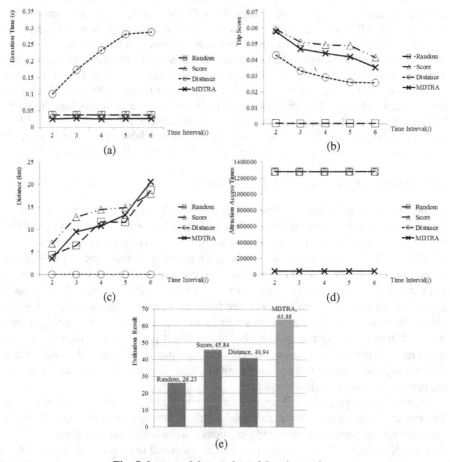

Fig. 5. Impact of the number of time intervals.

Figure 5(d) shows the number of time intervals and the times of access to attractions required by all the algorithms. As shown in Experiment 5.1, except the

MDTRA, other four algorithms must load all the attractions before calculating the scores. Thus, the access times equal the total number of attractions. The MDTRA was the optimal choice of the four.

Figure 5(e) indicates the results of the proposed trip evaluation equation. The highest scores could be attributed to the MDTRA, followed by Score, Distance, and Random. As shown in Experiment 5.1, the MDTRA demonstrated significantly better results in data loading times and execution time than other algorithms. Therefore, the trip scores were slightly lower than the Score algorithm. Although the trip distance is not as much as that generated by the Distance algorithm, its scores of the trip evaluation equation were the highest because of the advantages in data loading times and execution time. Similarly, the Score algorithm outperformed the Distance by a greater range, and the trip scores were also better. Therefore, even if the Score algorithm was inferior to the Distance in trip distance, it still obtained the optimal evaluation result. Moreover, the Distance algorithm outperformed the Random in regard to the trip scores and trip distance, and the overall evaluation results bettered the Random.

6   Conclusions

In this thesis, problems that might have occurred in the trip recommendation system were analyzed, including the inability to handle multiple days trip, user demands, user preferences, and trip distances. Existing trip recommendation systems calculate scores for all attractions and result in excessive expenses in data I/O. To solve these, we analyzed users' previous check-in records and suggested the innovative time interval concept and multiple days trip recommendation algorithm to overcome difficulties involving the abovementioned four dimensions. Moreover, the R-tree was used to help the algorithm reducing data access times to significantly increase the performance. Finally, an evaluation equation was proposed to serve as a reference for algorithm performance, and the experiments proved the validity of the proposed algorithm.

Acknowledgments. This work was supported in part by the Ministry of Science and Technology of Taiwan, R.O.C., under Contracts MOST 103-2218-E-035-018 and MOST 103-2221-E-006-198.

References

1. Bao, J., Zheng, Y., Mokbel, M.F.: Location-based and preference-aware recommendation using sparse geo-social networking data. In: SIGSPATIAL (2012)
2. Chiang, H.S., Huang, T.C.: User-adapted travel planning system for personalized schedule recommendation. Information Fusion **21**, 3–17 (2015)
3. Hsu, F.M., Lin, Y.T., Ho, T.K.: Design and implementation of an intelligent recommendation system for tourist attractions: The integration of EBM model, Bayesian network and Google Maps. Expert Systems with Applications **39**, 3257–3264 (2012)

4. Huang, Y., Bian, L.: A Bayesian network and analytic hierarchy process based personalized recommendations for tourist attractions over the Internet. Expert Systems with Applications **36**, 933–943 (2009)
5. Lee, C.S., Chang, Y.C., Wang, M.H.: Ontological Recommendation Multi-Agent for Tainan City Travel. Expert Systems with Applications **36**, 6740–6753 (2009)
6. Levandoski, J.J., Sarwat, M., Eldawy, A., Mokbel, M.F.: LARS: a location-aware recommender system. In: ICDE (2012)
7. Lu, E.H.C., Chen, C.Y., Tseng, V.S.: Personalized trip recommendation with multiple constraints by mining user check-in behaviors. In: SIGSPATIAL (2012)
8. Lu, E.H.C., Lee, W.C., Tseng, V.S.: A Framework for Personal Mobile Commerce Pattern Mining and Prediction. IEEE TKDE **24**, 769–782 (2012)
9. Lu, E.H.C., Lin, C.Y., Tseng, V.S.: Trip-mine: an efficient trip planning approach with travel time constraints. In: MDM (2011)
10. Ye, M., Yin, P., Lee, W.C., Lee, D.L.: Exploiting geographical influence for collaborative point-of-interest recommendation. In: SIGIR (2011)
11. Ying, J.J.J., Lu, E.H.C., Kuo, W.N., Tseng, V.S.: Urban point-of-interest recommendation by mining user check-in behaviors. In: UrbComp (2012)

Social Media on a Piece of Paper: A Study of Hybrid and Sustainable Media Using Active Infrared Vision

Thitirat Siriborvornratanakul[✉]

Graduate School of Applied Statistics, National Institute of Development Administration (NIDA), 118 SeriThai Rd., Bangkapi, Bangkok 10240, Thailand
thitirat@as.nida.ac.th

Abstract. In this world of digital and social media booms, a number of people spend their valuable times burying heads in smartphones, resulting in unintentional increased gaps in physical relationship with people nearby. A hybrid digital-physical medium is a possible solution for this problem by means of externalizing social media data and integrating them into a physical medium somehow. In this way, using social media will simultaneously connect us with both virtual and physical worlds.

This paper presents our first step towards implementation of such a hybrid system, using a mobile projector to project social media's digital data onto a physical color printed paper. With active infrared vision as an engine, we try to correlate physical printed colors with their infrared greyscale vision. A machine learning of multilayer perceptron is used to learn from samples whether there exists any reliable behavior that can be repeatedly used in the future. Nine color components from three well-known color models are combined, tested and evaluated before experimental results and future works are concluded and discussed.

Keywords: Color printed paper · Mobile projector · Active infrared · Machine learning

1 Introduction

During the past decade, our world has been forever changed by the emergence of social media, one of the most powerful tools that not only have infiltrated our daily life but also have triggered many world historical events. Apparently, these social media do connect people across the world by virtually shortening physical distances between one person and the others far away. But as an unintentional consequence, the social media physically add virtual distances between that one person and the others nearby. From a technology point of view, one possible solution for this problem is to externalize the world of social media from digital to physical, and from inside to outside display monitors. In this way, social networking will no longer result in complete disconnection between physical and digital worlds.

© Springer-Verlag Berlin Heidelberg 2015
L. Wang et al. (Eds): MISNC 2015, CCIS 540, pp. 331–340, 2015.
DOI: 10.1007/978-3-662-48319-0_26

Fig. 1. Our future vision of a hybrid medium where a color printed pamphlet is interactively augmented by facebook's data projected from a mobile projector.

To externalize digital social media onto a physical world, there are a number of ways that can be done in an Augmented Reality (AR) or Mixed Reality (MR) manner. To begin with, we focus on externalizing the digital social media onto a physical color printed paper (i.e., magazine, newspaper, advertising pamphlet) using a mobile projector and a camera. Our final vision is to create an AR system that is capable of projecting interactive digital data of social media onto a physical paper printed with color texts, images or illustrations. This system will not only decrease gaps between social media and physical world but also allow those waste papers to be reused in an eco-friendly and sustainable manner.

Accomplishing the goal mentioned in the above paragraph mainly involves three main concerns: a set of well-configured and well-calibrated devices (including at least projector, camera and infrared light), a proper visual detection or recognition algorithm, and a set of well-design interactions. A role of active infrared vision in this, is a hidden but very important one—to separate the light spectrums between projection's and visual analysis's. Without active infrared vision, projected imagery overlaid on a physical printed paper becomes visual noises that interfere any vision-based algorithm from correctly detecting or recognizing the physical printed contents. Major drawbacks of using active infrared vision include a limited workspace of indoor only, a need of an external infrared light source, and a miss of visual color features.

As a first step towards our goal, this paper presents a study aiming to predict how an active infrared camera sees colors printed on a physical paper. Contributions of doing this include:

1. To overcome the color limitation of infrared vision and allow both color and monotone printed papers to be used in our next steps of developing an

interactive and sustainable paper where the social media's digital data can be externalized onto.

2. To enable a controllable infrared vision where visibilities of printed contents as seen by an infrared camera can be controlled at will by careful selection of printed colors.

2   Related Works

The work presented in this paper involves researches in several fields. To narrow down the scope, only three domains are discussed here: mobile projector vs. social media, mobile projector vs. physical object, and active infrared in projection.

First of all, [6] and [8] propose two similar ideas of wearable devices with pico projector included. The pico projectors are responsible for externalizing wearer's social media data and showing them to other people in vicinity, introducing a new personal expression layer that encourages social interaction between the wearer and onlookers. The main difference between the two works is that they use different social media: Facebook for [6] and Twitter for [8]. In addition, [8] allows physical followers to become virtual followers by taking a picture of a projected tweet. An interesting point shared by these two works is that, they both choose to project data onto a ceiling which is considered an unpopular choice of projection surface according to [1,15]. This disagreement is probably because the works of [1,15] involve personal and group projections, whereas [6,8] focus on attracting onlookers' attention in order to trigger social interaction. To continue the work of mobile projector and social media, [8] points out that ceiling projection will cause confusion when tweets belonging to many wearers are projected onto the same proximity. Besides, the study in [6] suggests that selecting the online data for projection is crutial because most wearers in their study seem to become very privacy-sensitive once their social media's data (publicly shown online) are externalized. By the way, unlike [6,8], our expected system regarding social media is not supposed to use for attracting bystanders but to augment general color printed papers with additional information from social media. Therefore, ceiling projection is obviously not our choice. Nevertheless, in the future, we will follow the advice of [6] and be very careful when choosing social media's contents for projection.

Secondly, we discuss the domain of our long term goal of augmenting a printed paper with social media projection (a.k.a., physical object augmentation using mobile projector). Regarding this domain, there are a large number of related researches as the ability to physically augment or annotate a tangible object is sometimes said to be a considerbly more powerful feature of a projector than being a big display. Some interesting discoveries about this domain are written in [1,4]. In [1] where blackboard is the most popular choice of projection for kids, the researchers mention that most of the times there exists physical contents drawn on the blackboard during projection. For us, this can be implied that one natural reaction of using a mobile projector is to try mixing digital projection with something physical or tangible. The study of [4] strengthens our implication as their observation reveals that users prefer doing projection on physical objects on a table rather than on the big flat table itself; their reported reason is interesting—"because the physical object provides natural projection framing." According to these studies, it seems reasonable for us to create a system where a mobile projector is used to augment a piece of printed papers instead of a blank paper or a big surface like wall or floor.

Continuing from the above paragraph, a physical paper is one target of augmentation that has been used by many researchers trying to merge conventional paper-based documents with digital documents. It is suggested by [15] that, excluding a wall which is the most popular surface, paper and computer monitor are two potentially interesting projection surfaces in their user study. More detailed study in nomadic settings of LightBeam [4] also reports users' comment saying that projecting on deformable objects (e.g., papers) is a perfect choice of taking a peek into the projection beam. In our opinion, "paper as a projection surface for social media" is reasonable because both papers and social media share a similar sense of being something we can find almost everywhere. Hence, using both of them together should enhance and fulfill each other, enabling a good hybrid medium in a long run.

The last domain to discuss involves uses of active infrared in projection. Recently, there are many projection-based works utilizing active infrared. Two frequently found usages include invisibly encoding an unknown environment and lighting up an environment with infrared light. The first usage as in [2,4,5,7,14] is mostly for the purpose of real-time depth sensing using Kinect 360 sensor or some device with similar function. The second usage as presented in [7,10,12,13] uses the invisible illumination concept to create systems that are invariant to ambient light changes and involve no visual interference problem (explained earlier in section 1) at the same time. Despite of these projection systems using active infrared, to the best of our knowledge, there are very limited number of researches studying relationship between physical color and their active infrared vision. Interesting related works include ideas of color-infrared double information layer and infrared art in [11], as well as the world first industrial color infrared camera in [3]. The idea presented in our work here is more-or-less similar to those of [11] where a single object is designed to have two layers of different visual information (i.e., color and infrared layers). However, the approaches are different as ours tries to study a one-way relationship between physical colors (printed on a paper) and their infrared greyscale values using machine learning methodology.

3   Proposed Method

In order to predict greyscale infrared values of physical printed colors, one possibility is to derive an equation that involves all related factors like distance from a camera, infrared reflectance and absorption. Unfortunately, two important factors—infrared reflectance and absorption—are features unique to each type of color and paper, and most of the times, these two information are not publicly provided by their manufacturers. Hence, instead of a top-down approach, we apply a bottom-up approach using machine learning to create a model from well categorized examples.

Our inputs to machine learning are nine color components from three popular color models: RGB, HSV and YCrCb. By choosing one, two or three color components at a time ($n\_component = \{1, 2, 3\}$), there are 129 input combinations tested in our experiments, including 9 one-component, 36 two-component, and 84 three-component combinations. As for outputs, first, we categorize all 256 greyscale infrared values ranging from 0 to 255 into classes. In order to allow some flexibility, the maximum number of greyscale values per one class ($nval$) is varied from 1 to 10. This means that the number of possible classes (n_class) can be varied from 26 (i.e., $nval = 10$) to 256 (i.e., $nval = 1$). After categorization, we then assign an integer number to each class starting from 0; the smaller the class's number, the smaller the greyscale values in that class. For example, when $nval = 5$, n_class will equal to $\lceil 256/5 \rceil = 52$ including class0:{0,1,2,3,4}, class1:{5,6,7,8,9}, ... , class50:{250,251,252,253,254}, class51:{255}. Output of our machine learning is a classification result represented by the class's number.

In this paper, the chosen machine learning is Multilayer Perceptron (MLP)—a feedforward artificial neural network model. Our MLP includes three layers: one input layer, one hidden layer and one output layer. The number of nodes in the input layer equals to the number of color component inputs ($n\_component$); each node represents each color component input. The number of nodes in the output layer is equal to the number of classes (n_class); each node represents each class number. For the hidden layer, as there is no optimized choice known so far, we use the number of nodes equal to $((n\_component + n\_class) * (2/3))$. Base on OpenCV [9] implementation, our MLP is trained by a back-propagation algorithm with an activation function of symmetrical sigmoid. The output node who yields the maximum possibility will then be chosen as the output of our machine learning.

Figure 2 illustrates an example of how 256 infrared grayscale values are categorized into output classes and how a greyscale infrared value of arbitrary RGB color is predicted with our proposed MLP.

4   Experimental Setup

To collect a set of sample data showing correspondences between physical printed colors and their greyscale infrared values, we use devices shown in figure 3; a CMOS camera with long pass infrared filter is firmly fixed with an infrared ring light in a coaxial manner. As for the physical color printed papers, office 80gsm white A4 papers of the same brand are printed with 266 unique RGB colors (one paper per one color); the whole printing is done by the same color laser printer, computer, software and color profile. The 266 colors printed include 216 colors (equally distributed in RGB cube) for training and 50 random RGB colors for testing. When any color components other than R, G and B are required as inputs of machine learning, the original RGB color is converted to HSV and/or YCrCb using OpenCV's `cv::cvtColor` function.

Regarding one predefined RGB color, we sample the color printed paper placed at five different distances from the camera: 40, 57, 75, 92 and 110 cm.

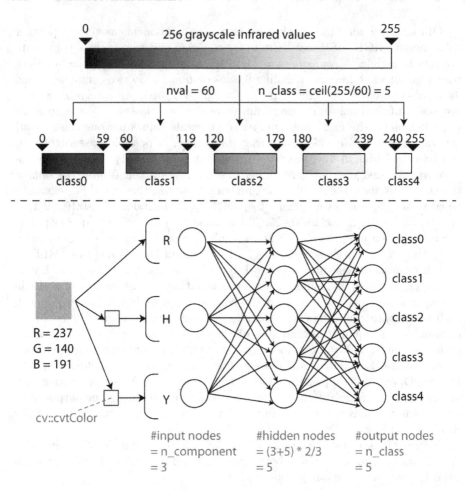

Fig. 2. Example of output categorization (top) and MLP prediction (bottom) when *nval* equals 60 and R, H and Y color components are used as inputs for training and testing MLP.

The middle distance of 75 cm is used to conduct a pilot study presented in this paper, whereas the other four are used to experimentally confirm whether the results from pilot study are consistent and truly repeatable.

All experiments are conducted on a HP Pavilion laptop with Intel(R) Core(TM) i7-4510U CPU running at 2 GHz on 8GB RAM. The main developer tools include Qt 5.4.0 MSVC2013 OpenGL 32bit and OpenCV 2.4.10 libraries, operating in Windows 8.1 Enterprise N 64-bit operating system.

940nm Infrared Ring Light

Fig. 3. Devices used for collecting sample data.

5   Experimental Results

To compare true values (manually measured) and values predicted by our MLP, we use four indicators to evaluate precision, imprecision and time. The four indicators include *correctHit* for percentage of correct classifications, *missAvg* and *missSD* for absolute class distance of incorrect classifications, and *time* for time used per one prediction. Table 1, 2 and 3 show experimental results from the pilot distance. Because of limited pages, these three tables only include the best values selected from 9 one-component combination inputs, 36 two-component combination inputs, and 84 three-component combination inputs; each combination input is repeatedly tested for *nval* ranging from 1 to 10. The best values in table 1 and 3 are chosen by maximum percentages of *correctHit* and minimum values of *time* respectively. As for table 2, the lower the *missAvg* value, the better. But in order to be fair for those with small *nval* values, the comparison is not done based on an absolute class distance represented by *missAvg* but *missAvg* * *nval*; this is to make the comparison more reflecting actual infrared greyscale values.

According to table 1 and 2, the highest precision achieved is belonging to the GBV input of 60% correct prediction at *nval* = 10, and the least imprecision is the SV input of 1.25 absolute class distance at *nval* = 8 which equals to *missAvg* * *nval* = 10. These two results, although not completely unacceptable, are not very good. Our only discovery is that they have V component in common and both use approximately the same amount of time according to table 3. But before drawing more conclusion, we use these two combination inputs selected from the pilot distance (i.e., GBV at *nval* = 10 and SV at *nval* = 8) and repeat the same experiments at the other four distances; the comparison results are shown in table 4. According to table 4, although the experimental distances are equally distributed, there seems to be no sign of consistent behavior in the results.

At present, our attempt of using MLP to learn the pattern of greyscale infrared values from printed colors provides no repeatable characteristics across

Table 1. Experimental results regarding the highest *correctHit* percentages measured at the pilot distance of 75 cm. * marks the best result.

nval =	1	2	3	4	5	6	7	8	9	10
1-component	0% all	2% (S)	14% (S)	32% (V)	22% (R) (Y)	16% (R) (G) (B) (S) (Y) (Cr)	50% (G)	36% (Cb)	54% (G)	54% (Cr)
2-component	0% all	2% (RG)	16% (RS)	26% (BS)	36% (BV)	42% (RB)	52% (VCr)	54% (VCb)	58% (GV)	56% (BV)
3-component	0% all	2% (RHS) (GBH) (GYCb) (GCrCb)	10% (SVCr)	28% (RGS) (GSCb)	34% (VYCr)	40% (RHY)	54% (YCrCb)	54% (RSV)	58% (GBCb)	* 60% * (GBV)

Table 2. Experimental results regarding the lowest *missAvg* values measured at the pilot distance of 75 cm. * marks the best result of the lowest $missAvg * nval$.

	nval =	1	2	3	4	5	6	7	8	9	10
1-component		(B)	(Cb)	(S)	(S)	(Cr)	(G) (S) (Y)	(V)	(V)	(S)	(R) (G) (H) (S) (Y) (Cb)
	Avg	109.24	19.34	12.86	3.02	2.56	2.24	1.51	1.53	1.93	1.48
	SD	16.71	13.90	10.60	3.71	2.86	2.40	1.00	0.79	1.22	1.28
	Avg * nval	109.24	38.68	38.58	12.08	12.80	13.44	10.57	12.24	17.37	14.80
	SD * nval	16.71	27.80	31.80	14.84	14.30	14.40	7.00	6.32	10.98	12.80
2-component		(HCb)	(VCr)	(RB)	(VCr)	(VCr)	(BV)	(VY)	(SV)	(VY)	(VY)
	Avg	106.24	19.98	3.93	2.53	2.05	1.69	1.68	1.25	1.57	1.20
	SD	18.00	14.05	4.92	3.09	1.62	1.03	1.32	0.66	0.87	0.62
	Avg * nval	106.24	39.96	11.79	10.12	10.25	10.14	11.76	* 10.00 *	14.13	12.00
	SD * nval	18.00	28.10	14.76	12.36	8.10	6.18	9.24	5.28	7.83	6.20
3-component		(BHCb)	(BSCr)	(HYCr)	(GBS)	(SVY)	(BSV)	(HVCr)	(SVY)	(HVCr)	(GSV)
	Avg	105.12	13.26	5.41	2.61	2.22	1.76	1.54	1.33	1.50	1.12
	SD	17.12	13.82	7.12	2.66	1.65	1.23	0.92	0.79	0.71	0.40
	Avg * nval	105.12	26.52	16.23	10.44	11.10	10.56	10.78	10.64	13.50	11.20
	SD * nval	17.12	27.64	21.36	10.64	8.25	7.38	6.44	6.32	6.39	4.00

Table 3. Experimental results regarding average times used per one MLP prediction. All data are collected at the pilot distance of 75 cm.

nval =	1	2	3	4	5	6	7	8	9	10
1-component	31.52 ms	9.32 ms	4.77 ms	3.10 ms	2.61 ms	2.18 ms	2.00 ms	1.87 ms	1.84 ms	1.68 ms
2-component	31.82 ms	9.62 ms	4.70 ms	3.28 ms	2.64 ms	2.32 ms	2.07 ms	1.95 ms	1.84 ms	1.95 ms
3-component	31.31 ms	9.45 ms	4.82 ms	3.26 ms	2.65 ms	2.29 ms	2.03 ms	1.90 ms	1.82 ms	1.76 ms

different distances. Future revision, like using other machine learning or approximation methods, is definitely required before we can proceed to the next step of implementing an interactive mobile projection system for social media.

Table 4. Experimental results regarding two best results selected from table 1 and 2 in five distances. * marks the best results chosen from the pilot distance of 75 cm.

	Distance (cm)	nval	correctHit (%)	missAvg (class distance)	missSD (class distance)	missAvg * nval (grey distance)	missSD * nval (grey distance)	time (ms)
GBV	40	10	**86**	3.86	3.98	38.60	39.80	1.75
	57	10	**10**	1.73	1.16	17.30	11.60	1.77
	75	10	* **60** *	1.65	1.35	16.50	13.50	1.73
	92	10	**72**	1.57	0.73	15.70	7.30	1.73
	110	10	**44**	1.29	0.59	12.90	5.90	1.71
SV	40	8	86	6.86	5.72	**54.88**	45.76	1.81
	57	8	30	4.34	5.11	**34.72**	40.88	1.82
	75	8	20	1.25	0.66	* **10.00** *	5.28	1.82
	92	8	58	1.81	1.05	**14.48**	8.40	1.81
	110	8	40	1.27	0.77	**10.16**	6.16	1.78

6   Conclusion

In order to develop a mobile system that interactively projects social media's digital data onto a physical color printed paper, this work presents our first step of investigating a one-way relationship between physical printed colors and their greyscale values as seen by an active infrared camera. Machine learning of multilayer perceptron is used to learn from a set of data collected at five different distances from the camera. The collected data include inputs of printed RGB color components and outputs of corresponding greyscale infrared values. The RGB color component inputs are converted to HSV or YCrCb when needed in order to increase choices of color component inputs to the machine learning. At present, experimental results show some good signs when the training and testing data are collected from the same distance. However, those good signs do not show repeatable behaviors when the distance is changed. Future works include revising this work with other computational or learning methods, and then proceeding to the next step of software and hardware development.

References

1. Akerman, P., Puikkonen, A.: Prochinima - using pico projector to tell situated stories. In: Proceedings of the 13th ACM International Conference on Human Computer Interaction with Mobile Devices and Services (MobileHCI 2011), pp. 337–346 (2011)
2. Harrison, C., Benko, H., Wilson, A.: Omnitouch: wearable multitouch interaction everywhere. In: Proceedings of the 24th Annual ACM Symposium on User Interface Software and Technology (UIST 2011), pp. 441–450 (2011)
3. Hornyak, T.: Sharp's security camera captures color video when it's pitch black, November 4, 2014. http://www.pcworld.com/article/2843032/sharps-color-security-camera-shoots-in-the-dark.html (accessed on February 10, 2015)
4. Huber, J., Steimle, J., Liao, C., Liu, Q., Muhlhauser, M.: Lightbeam: interacting with augmented real-world objects in pico projections. In: Proceedings of the 11th ACM International Conference on Mobile and Ubiquitous Multimedia (MUM 2012) (2012)

5. Jones, B., Benko, H., Ofek, E., Wilson, A.: Illumiroom: peripheral projected illusions for interactive experiences. In: Proceedings of the SIGCHI Conference on Human Factors in Computing Systems (CHI 2013), pp. 869–878

6. Leung, M., Tomitsch, M., Moere, A.: Designing a personal visualization projection of online social identity. In: Proceedings of the ACM Conference on Human Factors in Computing Systems (CHI 2011), pp. 1843–1848 (2011)

7. Molyneaux, D., Izadi, S., Kim, D., Hilliges, O., Hodges, S., Cao, X., Butler, A., Gellersen, H.: Interactive environment-aware handheld projectors for pervasive computing spaces. In: Kay, J., Lukowicz, P., Tokuda, H., Olivier, P., Krüger, A. (eds.) Pervasive 2012. LNCS, vol. 7319, pp. 197–215. Springer, Heidelberg (2012)

8. Ng, W., Sharlin, E.: Tweeting halo: clothing that tweets. In: Proceedings of the ACM Symposium on User Interface Software and Technology (UIST 2010), pp. 447–448 (2010)

9. OpenCV: Open source computer vision. http://opencv.org/

10. Riemann, J., Zhalilbeigi, M., Dezfuli, N., Muhlhauser, M.: Stacktop: hybrid physical-digital stacking on interactive tabletops. In: Proceedings of the ACM Conference on Human Factors in Computing Systems (CHI 2015), pp. 1127–1132 (2015)

11. Vujic, J., Stanimirovic, I., Medugorac, O.: Hidden information in visual and infrared spectrum. Informatologia 45(2), 96–102 (2012)

12. Willis, K., Shiratori, T., Mahler, M.: Hideout: mobile projector interaction with tangible objects and surfaces. In: Proceedings of the 7th International Conference on Tangible, Embedded and Embodied Interaction (TEI 2013), pp. 331–338 (2013)

13. Wilson, A.: Playanywhere: a compact interactive tabletop projection-vision system. In: Proceedings of the 18th Annual ACM Symposium on User Interface Software and Technology (UIST 2005), pp. 83–92 (2005)

14. Wilson, A., Benko, H., Izadi, S., Hilliges, O.: Steerable augmented reality with the beamatron. In: Proceedings of the 25th Annual ACM Symposium on User Interface Software and Technology (UIST 2012), pp. 413–422 (2012)

15. Wilson, M., Craggs, D., Robinson, S., Jones, M., Brimble, K.: Pico-ing into the future of mobile projection and contexts. Journal of Personal and Ubiquitous Computing 16(1), 39–52 (2012)

Characteristics and Related Factors for the Proportion of Third-Party Evaluations of Japanese Nursing Homes (Kaigo-Roujin-Fukushi-Shisetsu and Kaigo-Roujin-Hoken-Shisetsu)

Suguru Okubo[1(✉)] and Megumi Kojima[2]

[1] BMS Yokohama Inc., Yokohama, Japan
sokubo@bms-yokohama.co.jp
[2] Ritsumeikan University, Osaka, Japan

Abstract. We explored the characteristics and related factors for the proportion of third-party evaluations for long-term care facilities in Japan (Tokuyo and Rouken). We used data previously published by the Japanese government or prefectures. The median of the proportion of the third-party evaluations in Tokuyo was 14.6% and in Rouken it was 8.9%. Prefectures that have referred to third-party evaluations in the 5th Prefectural Insured Long-term Care Service Plan had a higher proportion of third-party evaluations in Tokuyo and Rouken. The difference of proportion of third-party evaluation among prefectures would be wider and wider gradually.

Keywords: Long-term care insurance · Nursing homes · Third-party evaluation

1   Objective

Third-party evaluation is one of the most important methods for quality improvement of long-term care. We explored the characteristics and related factors for the proportion of third-party evaluations for long-term care facilities in Japan (Tokuyo and Rouken).

2   Methods

2.1   Data Collection

We used data previously published by the Japanese government or prefectures (Table 1), and calculated the proportion of the third-party evaluations for Tokuyo and Rouken. Tokuyo and Rouken are long-term care facilities for the elderly that are reimbursed by long-term care insurance. The activities of daily living for Tokuyo residents are higher compared to those of Rouken residents. Rouken residents aim to return home from Rouken as Rouken provides more intensive medical care and rehabilitation services to residents compared to Tokuyo.

© Springer-Verlag Berlin Heidelberg 2015
L. Wang et al. (Eds): MISNC 2015, CCIS 540, pp. 341–353, 2015.
DOI: 10.1007/978-3-662-48319-0_27

Table 1. Data Used in the Research

Data item	Date	Source
Proportion of third-party evaluations for Tokuyo and Rou-ken in 47 prefectures	Nov 2014	Publication System of Long-term Care Service Information [1]
Population	Oct 2010	Population Census of Japan
Proportion of people over 65 years old	Oct 2010	Population Census of Japan
Number of review organizations (per 100,000 people over 65 years old)	Mar 2014	Third-party evaluation of welfare services [2]
Public upskilling programs for reviewers	Apr 2012 to Mar 2013	Third-party evaluation of welfare services [2]
References to third-party evaluations	Apr 2011 to Mar 2012	Prefectural Insured Long-term Care Service Plan

Public Upskilling Programs for Reviewers. Prefectures have implemented initial training courses and upskilling programs for reviewers. All reviewers must finish the initial training courses, but upskilling programs are for experienced reviewers [3] and not all prefectures have implemented these. The authors assumed that prefectures that have positive attitudes toward third-party evaluations tend to implement upskilling programs.

References to Third-Party Evaluations in Prefectural Insured Long-Term Care Service Plans. All prefectures must develop insured long-term care service plans once every three years according to Section 118 of the Long-Term Care Insurance Act. The plans need to include a support plan for the smooth payment of insurance. In this research, we searched for references to third-party evaluations in the 5th Prefectural Insured Long-term Care Service Plan (FY 2012 to FY 2014). All plans were collected from the prefectural websites.

2.2   Analysis

First, we calculated the minimum values, quartile points, and maximum values of the proportion of third-party evaluations of Tokuyo and Rouken in 47 prefectures. Second, we conducted bivariate analysis between the related factors and proportion of third-party evaluations. In the bivariate analysis between the continuous variables, we calculated Pearson's correlation coefficients and Spearman's correlation coefficients. In the bivariate analysis between the categorical variables and continuous variables, we conducted a Mann-Whitney U test. The level of significance was two-sided and set at 0.05. IBM SPSS Statistics Version 22 was used for all analyses.

3   Results

3.1   Proportion of Third-Party Evaluations in Each Prefecture

Table 2 and Table 3 shows the distribution of the proportion of third-party evaluations in all prefectures. The median of the proportion of the third-party evaluations in Tokuyo was 14.6% and in Rouken it was 8.9%.

Table 2. Proportion of third-party evaluations in 47 prefectures

	Tokuyo numbers	Proportion of third-party evaluations	Rouken numbers	Proportion of third-party evaluations
All	7,052	23.2%	4051	12.1%
Hokkaido	325	10.2%	188	4.8%
Aomori	91	26.4%	62	11.3%
Iwate	103	18.4%	63	0.0%
Miyagi	140	31.4%	85	11.8%
Akita	109	0.9%	55	1.8%
Yamagata	96	8.3%	46	2.2%
Fukushima	134	9.0%	90	8.9%
Ibaraki	206	5.8%	117	9.4%
Tochigi	118	5.9%	63	6.3%
Gunma	153	13.7%	89	15.7%
Saitama	313	4.8%	161	3.7%
Chiba	295	10.5%	136	11.0%
Tokyo	454	90.5%	184	21.7%
Kanagawa	375	19.7%	186	17.2%
Niigata	185	7.6%	104	1.9%
Toyama	78	12.8%	51	13.7%
Ishikawa	69	33.3%	45	26.7%
Fukui	61	19.7%	37	2.7%
Yamanashi	56	8.9%	32	3.1%
Nagano	160	12.5%	97	5.2%
Gifu	121	14.9%	72	5.6%
Shizuoka	208	26.9%	116	6.9%
Aichi	235	16.6%	185	21.1%
Mie	138	29.0%	68	29.4%
Shiga	76	7.9%	32	0.0%

Table 2. (*continued*)

Kyoto	146	78.8%	69	52.2%
Osaka	369	26.6%	200	11.5%
Hyogo	292	35.6%	164	21.3%
Nara	86	11.6%	45	6.7%
Wakayama	82	14.6%	41	0.0%
Tottori	42	42.9%	56	23.2%
Shimane	85	15.3%	38	7.9%
Okayama	145	4.1%	83	7.2%
Hiroshima	172	32.0%	111	15.3%
Yamaguchi	96	41.7%	66	1.5%
Tokushima	60	1.7%	53	3.8%
Kagawa	81	6.2%	51	0.0%
Ehime	94	12.8%	64	4.7%
Kochi	54	14.8%	33	9.1%
Fukuoka	281	22.1%	177	14.7%
Saga	57	21.1%	40	27.5%
Nagasaki	107	10.3%	61	8.2%
Kumamoto	134	33.6%	95	10.5%
Oita	83	31.3%	72	15.3%
Miyazaki	83	7.2%	43	34.9%
Kagoshima	147	8.8%	81	6.2%
Okinawa	57	14.0%	44	9.1%

Table 3. Distribution of the Proportion of Third-Party Evaluations (n = 47)

	Tokuyo	Rouken
Minimum value	1.0%	0.0%
25 percentile	8.8%	3.8%
Median	14.6%	8.9%
75 percentile	26.9%	15.3%
Maximum value	91.0%	52.0%

The proportion of third-party evaluations for Tokuyo had significant positive correlations with those for Rouken (Figure 1).

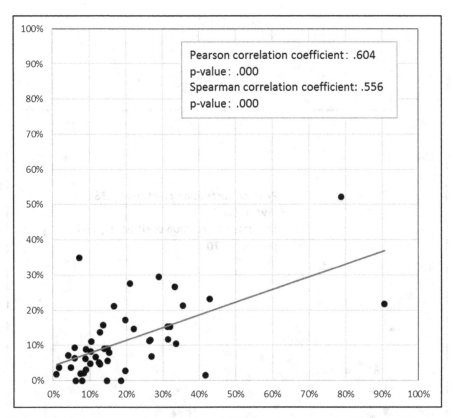

Fig. 1. Correlation between the proportion of third-party evaluations for Tokuyo and Rouken

3.2   Proportion of Third-Party Evaluations and Related Factors

The Pearson's correlation coefficient for the proportion of third-party evaluations in Tokuyo and prefectural populations was .388 (p-value = .007), while Spearman's correlation coefficient was .164 (p-value = .270).

The Pearson's correlation coefficient for the proportion of third-party evaluations in Rouken and prefectural populations was .185 (p-value = .214), while Spearman's correlation coefficient was .253 (p-value = .087).

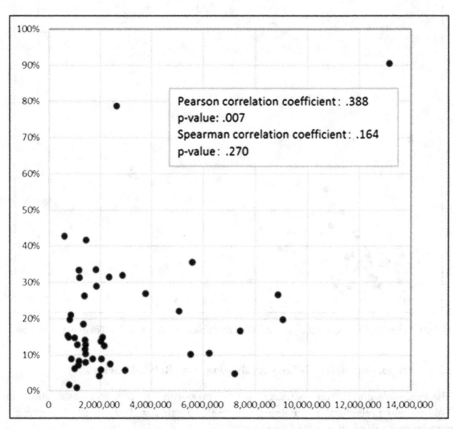

Fig. 2. Correlation between Proportion of Third-Party Evaluations for Tokuyo and Prefectural Populations

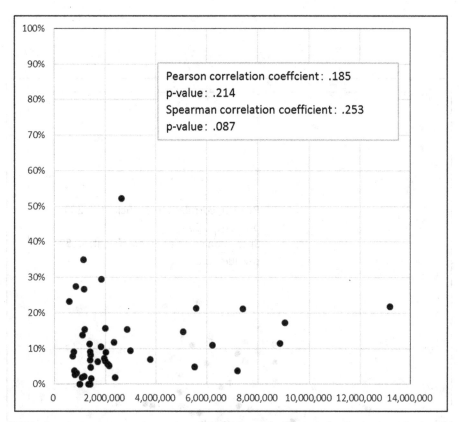

Fig. 3. Correlation between Proportion of Third-Party Evaluations for Rouken and Prefectural Populations

3.3 Proportion of People Over 65 Years Old

The Pearson's correlation coefficient for the proportion of third-party evaluations in Tokuyo and people over 65 years old was -.204 (p-value = .168), and Spearman's correlation coefficient was -.159 (p-value = .286).

The Pearson's correlation coefficient for the proportion of third-party evaluations in Rouken and people over 65 years old was -.263 (p-value = .074), while Spearman's correlation coefficient was -.420 (p-value = .003).

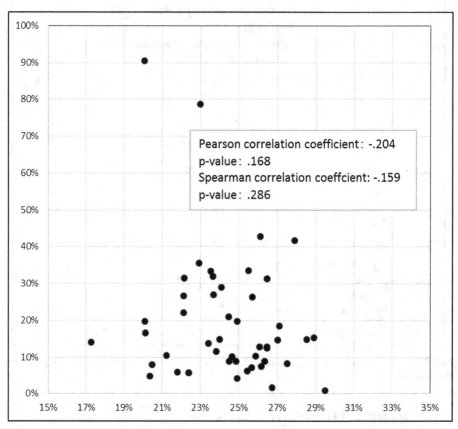

Fig. 4. Correlation between the Proportion of Third-Party Evaluations in Tokuyo and Prefectural Population over 65 Years Old

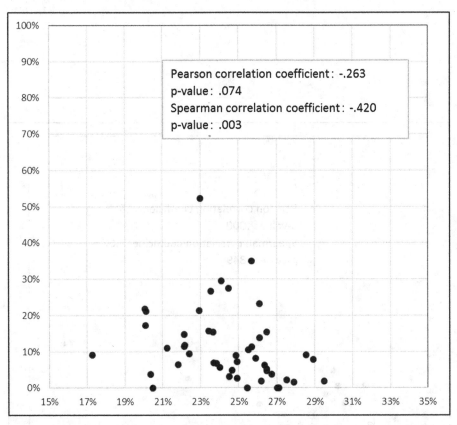

Fig. 5. Correlation between the Proportion of Third-Party Evaluations in Rouken and Prefectural Population over 65 Years Old

3.4   Number of Review Organizations (per 100,000 People over 65 Years Old)

The Pearson's correlation coefficient for the proportion of third-party evaluations in Tokuyo and the number of review organizations (per 100,000 people over 65 years old) was ·.555 (p-value = .000), and Spearman's correlation coefficient was .029 (p-value = .849).

The Pearson's correlation coefficient for the proportion of third-party evaluations in Rouken and the number of review organizations (per 100,000 people over 65 years old) was .348 (p-value = .017), while Spearman's correlation coefficient was .125 (p-value = .404).

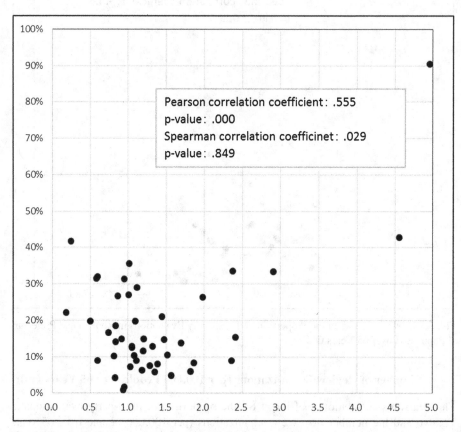

Fig. 6. Correlation between the proportion of third-party evaluations in Tokuyo and number of review organizations (per 100,000 people over 65 years old)

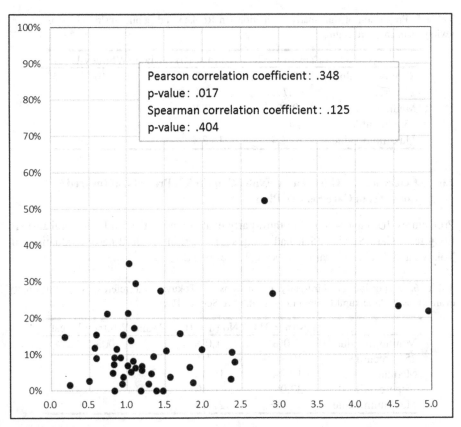

Fig. 7. Correlation between the proportion of third-party evaluations in Rouken and number of review organizations (per 100,000 people over 65 years old)

3.5 Public Upskilling Programs for Reviewers

Prefectures that have conducted public upskilling programs for reviewers had a higher proportion of third-party evaluations in Tokuyo (p-value = .014) (Tables 4 and 5).

Table 4. Proportion of third-party evaluations in Tokuyo and public upskilling programs for reviewers in each prefecture

	Yes (n = 37)	No (n = 10)	Mann-Whitney U test
Minimum value	5.9%	0.9%	.014
25 percentile	10.2%	3.5%	
Median	15.3%	7.1%	
75 percentile	28.0%	18.9%	
Maximum value	90.5%	32.0%	

Table 5. Proportion of third-party evaluations in Rouken and public upskilling programs for reviewers in each prefecture

	Yes (n = 37)	No (n = 10)	Mann-Whitney U test
Minimum value	0.0%	1.8%	.310
25 percentile	4.7%	3.3%	
Median	9.1%	6.9%	
75 percentile	19.1%	10.9%	
Maximum value	52.2%	15.3%	

3.6   References to Third-Party Evaluations in 5th Prefectural Insured Long-Term Care Service Plan

Prefectures that have referred to third-party evaluations in the 5th Prefectural Insured Long-term Care Service Plan had a higher proportion of third-party evaluations in Tokuyo and Rouken (p-values = .003) (Tables 6 and 7).

Table 6. Proportion of third-party evaluations in Tokuyo and references to third-party evaluations in Prefectural Insured Long-term Care Service Plan

	Yes (n = 31)	No (n = 16)	Mann-Whitney U test
Minimum value	6.0%	1.0%	.003
25 percentile	10.3%	5.1%	
Median	19.7%	10.3%	
75 percentile	32.0%	14.8%	
Maximum value	91.0%	27.0%	

Table 7. Proportion of third-party evaluations in Rouken and references to third-party evaluations in Prefectural Insured Long-term Care Service Plan

	Yes (n = 31)	No (n = 16)	Mann-Whitney U test
Minimum value	0.0%	0.0%	.003
25 percentile	5.6%	2.8%	
Median	11.5%	5.7%	
75 percentile	21.3%	7.2%	
Maximum value	52.0%	16.0%	

4   Discussion and Conclusion

The study limitations include the cross-sectional design and data period, which was inconsistent, meaning that we cannot refer to causal relationships. For example, although prefectures that had referred to third-party evaluations in the Prefectural Insured Long-term Care Service Plan had a higher proportion of third-party evaluations, we cannot state that the positive attitude of the prefectures accelerates the number of third-party evaluations. It is reasonable that prefectures where more facilities had already received third-party evaluations can easily include the objectives of third-party evaluations in their service plans.

However, the research indicated that more facilities had received third-party evaluation in prefectures where had more positive attitude toward it. The difference of proportion of third-party evaluation among prefectures would be wider and wider gradually. It is necessary to discuss how to accelerate third-party evaluation in prefectures where the proportion of third-party evaluation is low.

References

1. Publication System of Long-Term Care Service Information. http://www.kaigokensaku.jp/
2. Japan National Council of Social Welfare. Third-party evaluation in welfare services. Current status of third-party evaluation in each prefecture. http://shakyo-hyouka.net/evaluation5/
3. Japan National Council of Social Welfare. Third-party evaluation in welfare services. Third-party evaluation About review organization and reviewer. http://shakyo-hyouka.net/evaluation3/

Exploring Users' Information Behavior on Facebook Through Online and Mobile Devices

I-Ping Chiang[✉] and Sie-Yun Yang

Graduate Institute of Information Management,
National Taipei University, New Taipei City, Taiwan
ipchiang@mail.ntpu.edu.tw, syyang26@gmail.com

Abstract. With the increased use of mobile devices and social network services (SNSs), people can browse SNSs anytime and anywhere. This study used Facebook as the core sites for analyzing web users' information behavior online and on mobile devices by collecting users' clickstream data of three months.

Keywords: Information behavior · Social network services (SNSs) · Mobile usage · Internet marketing

1 Introduction

1.1 General Background Information

Since the development of the Internet, surfing the web has become a major part of people's daily life behavior. According to the Taiwan Network Information Center (TWNIC) report of 2014, the number of Taiwan's Internet users was more than 17 million [27]. Regarding regular Internet population, Foreseeing Innovative New Digiservices indicated that the number of Taiwan's Internet users has reached 11 million [10]. Fig. 1. shows the growth of Taiwan's regular Internet population over the years.

With the surge of social network services (SNSs), the usage of SNS has been one of the main behavior of web users. According to the TWNIC survey in 2014, the use of SNSs was higher than that of web browsing, and more than 64.32% of the respondents reported having used SNSs. In 2011, comScore's white paper showed that 82% of global Internet use SNSs. In addition, it showed that Internet users browse the SNSs every one-fifth of a minute, thus, showing an extremely high frequency. Comparing the Nielsen data of 2012 with that of 2011 showed that the number of people in the United States who use SNSs had grown by 37% (121 billion minutes). Thus, as mentioned, SNSs have become a major part of people's daily lives.

The recent growth in smartphones has changed people's lifestyles. A 2013 survey, conducted by Google and Ipsos MediaCT, in which 1000 people (18-64 years old) were interviewed, demonstrated that smartphone penetration exhibited a substantial growth from 19% to 51%. In addition, the survey indicated that these smartphone users were heavy users of social networks: 93% of users used SNS daily, from which 61% used SNSs at least once a day and 60% viewed the sites by using smartphones [12]. Therefore, smartphones have become a central part of daily lives.

© Springer-Verlag Berlin Heidelberg 2015
L. Wang et al. (Eds): MISNC 2015, CCIS 540, pp. 354–362, 2015.
DOI: 10.1007/978-3-662-48319-0_28

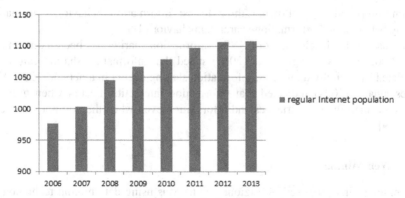

Fig. 1. Taiwan's regular Internet population

Since the invention of smartphones, people's web surfing behavior has changed. Moreover, SNSs are major part of people daily lives. Therefore, our study used the data mining technique to compare people's information behavior in using SNSs online and on mobile devices.

2   Literature Review

2.1   Information Behavior

Since the 1960s, researchers have focused on information-seeking activities from users' perspectives [11]. "Information Needs and Uses", written by Dervin and Nilan(1986), acknowledges to an crucial literature that clarifies information behavior to be user-oriented rather than system-oriented [8]. Since 2000, several studies on information behavior patterns and methodologies have been conducted, such as [16], [22], [28].

Since Pettigrew (2001) first proposed the online information behavior terms, they have been the focus of attention of scholars [22]. Border (2002) focused on search engines and discovered that the Internet searching entails three types: navigation-oriented, information-oriented and transaction-oriented [2].

Bates (2010) considered that information behavior was a relatively common term for describing the interactions between people and information, in particular, people seek and use information processes or methods [1].

Earlier studies have mostly explored information-searching and information-seeking behavior. Krikelas (1983) assumed that information-seeking behavior would change because each person or each circumstance is different [18]. Dervin and Nilan (1986) and Dervin (1999) considered that the information-seeking behavior would make sense if it was constructed in the reality of the background [8][7]. In 1999, Wilson stated that a user may feel "demand," and would further exhibit different patterns of information-seeking behavior, which may involve several uncertainties that could result in a wider range of information search, to "meet the needs [30]." In 2000,

Wilson proposed the following three terms: information behavior, information-seeking behavior and information-searching behavior [31].

Because of the development of SNSs, information sharing has been the main concern of social networking. Rioux (2005) stated that information sharing can not be completed by an individual, thus, information sharing occurs in a network [23]. Wang & Fesenmaier (2004) indicated that information interaction occurs when users are willing to share their experiences and information even when they do not know each other [29].

2.2   Web Mining

Web mining, first proposed by Etzioni (1996), was using data mining technology to find and extract implicit information in web documents or services [9]. On this basis, Cooley et. al. (1997) divided web mining into web content mining and web usage mining [5]. In 2000, Kosala and Blockeel proposed web structure mining, which is based on the network structure, as the third type of web mining [17]. Therefore, web mining comprises three components as shown in Fig. 2.

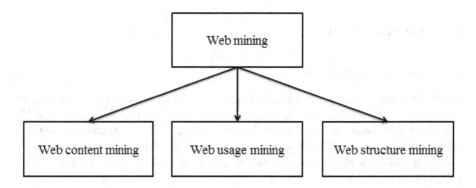

Fig. 2. Web mining taxonomy

Web content mining is also known as text mining, and is used for mining the page text, images, web pages, and content of various compositions for evaluating the correlation between the web page and a search query, and it is mainly used for information retrieval.

Web structure mining entails using hyperlinks to identify relationships among web pages and then analyzing this relationship to cluster or classify web pages.

Web usage mining typically involves using clickstream data to explore and analyze patterns automatically. Srivastava et al. (2000) considered web data as the source of web usage mining, otherwise, Cooley et al. (1999) considered users' session files to be the input of web usage mining [6], [24]. Commonly used algorithms are association rule, sequential pattern, and clustering algorithms. In web usage mining, association rule was used to access a number of web pages to find the implicit rules of data, sequential pattern involved using time-stamp to find data patterns [20], [25]; and clustering was used to cluster similar users or data [15], [21]. Therefore, when conducting web usage mining, the main sources have been web log files; thus, Cooley et al. (1999) proposed a comprehensive processing approach to processing web log files [6]. This processing approach comprises five steps: data cleaning, user identification, session identification, path completion, and formatting. This study applied these steps to pre-process our data.

2.3   Web Usage Mining Application

Because users using keyboards and mise to perform web searches, a period typically called timeout must be selected to appropriately cut the users' browsing path into sessions. He et al. (2013) used session analysis on LinkedIn to analyze users' search behavior and determined that once session identification is accurately implemented, users' search behavior can be analyzed [13].

Web log files record users' clickstream data; hence, they are used for analysis in web usage mining because the log files contain users' click records among web pages. Huberman (1998) performed the earliest clickstream data analysis and observed that different visitors accessed web page distribution [14]. Chou et al. (2010) considered collecting the clickstream data, which facilitated discovering could help us find precise information from user browsing behavior [4]. Bucklin and Sismeiro (2009) integrated previous academic studies of clickstream data and focused on two type of clickstream data: user and site-centric [3]. User-centric data focuses on understanding users' cross-site activity records, but the samples in a single site might be inadequate. Site-centric data pertains to users' browsing behavior at a particular Web site but not necessarily to users' background information. Story (2007) summarized clickstream data according to different measurements, including a web visitor's age and gender, browsing time analysis, and the relationship between the site and the number of pages viewed [26][26]. By referring to Stats Center[1] measurements, this study used web visitors' browsing time to analyze user-centric data.

3   Method

3.1   Research Design

According to the literature review, online information behavior studies typically use clickstream data, We collected 3 months' clickstream data and then used the user-centric data to determine Facebook users' gender, age, and period of week's browsing time for determining the general browsing behavior. Next, a t-test was used to explore the general browsing behavior of people using SNSs online and on mobile devices.

---

[1] http://www.web-stat.com/stats/checkstats.pl?loginID=demo

3.2   Data Collection

For this study, we cooperated and collected data from InsightXplorer Limited, Taiwan, which maintains an online panel for Taiwan's cybermarket and tracks panelists' online clickstream data. According to InsightXplorer, the term "online" implies people using a personal computer or laptop to access web pages. In an online dataset, we obtained data of 294 Facebook users, and in mobile devices we obtained data of 39 Facebook users.

4   Data Analysis

The sample data of Facebook users' profiles is shown in Table 1. We infer that the main gender of our population·online and on mobile devices is female, and the main age is 31–40 years.

Table 1. Facebook sample users' profiles

		Online	Mobile Devices
Unique users' sample size		294	39
Gender	Male	114	12
	Female	180	27
Age	20 and below	10	1
	21 to 30	98	10
	31 to 40	127	17
	40 and above	59	11

4.1   General Information Behavior

We used age and gender to determine users' average browsing time. Table 2 shows that the number of female users, both online and on mobile devices, surpasses that of male users, but the results of the average browsing time are contradictory. Next, Table 3 shows that the average period of week's browsing time is high for both online and mobile devices. Finally, Table 4 shows that the average online browsing time is higher for the group aged 31–40 years, whereas that on mobile devices is higher for the group aged 41 years and older. According to the data in these tables, we conclude that the average browsing time for mobile devices is higher than that for online devices.

Table 2. Average browsing time for males and females on online and mobile devices (unit: seconds)

	Online		Mobile Devices	
	Unique users' size	Average browsing time	Unique users' size	Average browsing time
Male	114	6219	12	17017
Female	180	1073	27	4214

Table 3. Average period of week's browsing time for using online and mobile devices (unit: seconds)

	Online		Mobile Devices	
	Unique users' size	Average browsing time	Unique users' size	Average browsing time
0 to 6	72	0.125458	34	0.367647
7 to 12	187	0.365749	34	0.956588
13 to 18	209	1.018416	31	1.557097
19 to 24	186	0.751699	20	0.95685

Table 4. Age-related average browsing time for using online and mobile devices (unit: seconds)

	Online		Mobile Devices	
	Unique users' size	Average browsing time	Unique users' size	Average browsing time
20 and below	10	439	1	6225
21 to 30	98	1237	10	1825
31 to 40	127	5086	17	5373
41 and above	59	2213	11	18446

4.2   Usage Analysis

We used a t-test to determine gender-related and weekly usage of Facebook. As shown in Tables 5 and 6, for both online and mobile devices, the t value is not significant for gender; thus, we conclude that gender does not affect Facebook usage. Next, we used a single factor ANOVA to determine the usage of week's period time and all ages. A shown in Tables 7 and 8, we found that the p value of week's period time usage is approached significant in mobile devices; thus, week's period time may do affect the usage. As shown in Tables 9 and 10, the result showed that the p value is not significant; thus, age does not affect usage either.

Table 5. Independent sample tests on gender for online devices

		Levene's Test for Equality of Variances		t-test for Equality of Means		
		F	Sig.	t	df	Sig.(2-tailed)
usage	Equal variances assumed	6.427	.012	1.351	292	.178
	Equal variances not assumed			1.076	113.468	.284

Table 6. Independent sample tests on gender for mobile devices

		Levene's Test for Equality of Variances		t-test for Equality of Means		
		F	Sig.	t	df	Sig.(2-tailed)
usage	Equal variances assumed	8.643	.006	1.597	37	.119
	Equal variances not assumed			1.140	12.007	.276

Table 7. A week's period time single factor ANOVA

Source of Variation	SS	df	MS	F	P-value	Fcrit
Between Groups	65.17305	3	21.72435	1.653322	0.175874	2.618608
Within Groups	8540.883	650	13.13982			
Total	8606.056	653				

Table 8. A week's period time single factor ANOVA

Source of Variation	SS	df	MS	F	P-value	Fcrit
Between Groups	22.95534	3	7.651781	2.550902	0.0591	2.683499
Within Groups	344.9583	115	2.999637			
Total	367.9137	118				

Table 9. Online users' age single factor ANOVA

Source of Variation	SS	df	MS	F	P-value	Fcrit
Between Groups	8.86E+14	2	4.43E+14	0.42023	0.657308	3.027898
Within Groups	2.96E+17	281	1.05E+15			
Total	2.97E+17	283				

Table 10. Mobile device users' age single factor ANOVA

Source of Variation	SS	df	MS	F	P-value	Fcrit
Between Groups	1.18E+14	2	5.88E+13	0.507578	0.606315	3.267424
Within Groups	4.05E+15	35	1.16E+14			
Total	4.17E+15	37				

5   Results and Disscussion

In this study, we sought to determine the information behavior of people using social network sites online and on mobile devices, and we chose Facebook as our main site. We used age, gender, and the average weekly browsing time as our general information behavior criteria. First, we determined that although female users are higher in number, their average browsing time is not long, whether online or on mobile devices; thus, we infer that female users use Facebook for specific purposes such as posting. Second, we used period of week's browsing time and age to explain that the period of week's browsing time centralizes between one o'clock p.m. to three o'clock p.m., and the age is between 21 and 40 years. This may be because such people are not only college students or office workers. In particular, people in the group aged 41 years and older show an greater average browsing time on mobile devices, implying that such people may be unfamiliar with personal computers or laptops; thus, they often use mobile devices to perform various tasks and may use Facebook to connect with others.For further analysis, we used a *t*-test and ANOVA to determine usage, and discovered that age, gender, do not affect Facebook usage in online and mobile devices, however, period of week's browsing time does affect Facebook usage in mobile devices; thus, we infer that in addition to the convenient of mobile devices, people can use Facebook anytime and anywhere.

Through this study, we found that that mobile devices are gradually replacing personal computers and laptops as media for browsing social network sites. Therefore, if other Web sites wish to increase their exposure, they can set up a fanpage on Facebook.

We analyzed only Facebook in this study; hence, we suggest that other researchers use other Web sites such as news or shopping sites. In addition, we used only user-centric data as our framework, but the addition of site-centric data can facilitate determining which two sites have a strong relationship, thus exemplifying a promising direction of research.

References

1. Bates, M.: Encyclopedia of Library and Information Sciences. CPC Press, New York (2010)
2. Broder, A.: A taxonomy of web search. ACM Sigir Forum **36**(2), 3–10 (2002)
3. Bucklin, R.E., Sismeiro, C.: Click Here for Internet Insight: Advances in Clickstream Data Analysis in Marketing. Interactive Marketing **23**(1), 35–48 (2009)
4. Chou, P.H., Li, P.H., Chen, K.K., Wu, M.J.: Integrating web mining and neural network for personalized e-commerce automatic service. Expert Systems with Applications **37**(4), 2898–2910 (2010)
5. Cooley, R., Mobasher, B., Srivastava, J.: Web mining: information and pattern discovery on the world wide web. In: Proceedings of the 9th IEEE International Conference on Tools with Artificial Intelligence, pp 558–567 (1997)
6. Cooley, R., Mobasher, B., Srivastava, J.: Data preparation for mining world wide web browsing patterns. Knowledge and Information Systems **1**(1), 5–32 (1999)
7. Dervin, B.: On Studying Information Seeking Methodologically: The Implications of Connecting Metatheory to Method. Information Processing and Management **35**(6), 727–750 (1999)
8. Dervin, B., Nilan, M.: Information needs and uses. Annual Review of Information Science and Technology **21**, 3–33 (1986)
9. Etzioni, O.: The World-Wide Web: quagmire or gold mine? Communications of the ACM **39**(11), 65–68 (1996)
10. Foreseeing Innovative New Digiservices: Taiwan's Internet users Summary Report of October 2013 Survey (2014). http://www.find.org.tw/market_info.aspx?n_ID=7208 (retrieve May 22, 2015)
11. Gonzalez-Teruel, A., Abad-Garcia, M.F.: Information needs and uses: An analysis of the literature. Library and Information Science Research **29**(1), 30–46 (2007)
12. Google and Ipsos MediaCT: Our Mobile Planet: Taiwan, Understanding the Mobile Consumer (2013). http://services.google.com/fh/files/misc/omp-2013-tw-en.pdf (retrieve May 22, 2015)
13. He, R., Wang, J., Tian, J., Chu, C.T., Mauney, B., Perisic, I.: Session analysis of people search within a professional social network. Journal of the American Society for Information Science and Technology **64**(5), 929–950 (2013)

14. Huberman, B.A., et al.: Strong regularities in world wide web surfing. Science **280**(5360), 95–97 (1998)
15. Kaufman, L., Rousseeuw, P.J.: Finding groups in data: an introduction to cluster analysis, vol. 344. John Wiley & Sons (2009)
16. King, D.W., Tenopir, C.: Using and reading scholarly literature. Information Science and Technology **34**, 423–477 (1999)
17. Kosala, R., Blockeel, H.: Web mining research: A survey. ACM Sigkdd Explorations Newsletter **2**(1), 1–15 (2000)
18. Krikelas, J.: Information-Seeking Behavior: Patterns and Concepts. Drexel Library Quarterly **19**(2), 5–20 (1983)
19. Kuhlthau, C.C.: Seeking meaning: a process approach to library and information services. Ablex, Norwood (1994)
20. Mannila, H., Toivonen, H., Verkamo, A.I.: Discovery of frequent episodes in event sequences. Data Mining and Knowledge Discovery **1**(3), 259–289 (1997)
21. Ng, R.T., Han, J.: Efficient and effective clustering methods for spatial data mining. In: Proceeding of the 20th International Conference on Very Large Data Bases, pp 144–155 (1994)
22. Pettigrew, K.E., Fidel, R., Bruce, H.: Conceptual frameworks in information behavior. Annual Review of Information Science and Technology **35**, 43–78 (2001)
23. Rioux, K.S.: Theories of information behavior. Information Today, Inc. (2005)
24. Srivastava, J., Cooley, R., Deshpande, M., Tan, P.N.: Web usage mining: Discovery and applications of usage patterns from web data. ACM SIGKDD Explorations Newsletter **1**(2), 12–23 (2000)
25. Srikant, R., Agrawal, R.: Mining sequential patterns: Generalizations and performance improvements. Springer, Heidelberg (1996)
26. Story, L.: How many site hits? Depends who's counting. New York Times October (2007). http://www.nytimes.com/2007/10/22/technology/22click.html?_r=0 (retrieve May 22, 2015)
27. Taiwan Network Information Center: Wireless Internet Usage in Taiwan Summary Report of October 2014 Survey (2014). http://www.twnic.net.tw/download/200307/20150202d.pdf (retrieve May 22, 2015)
28. Wang, P.: Methodologies and methods for user behavioral research. Annual Review of Information Science and Technology **34**, 53–100 (2001)
29. Wang, Y., Fesenmaier, D.R.: Modeling participation in an online travel community. Journal of Travel Research **42**(3), 261–270 (2004)
30. Wilson, T.D.: Models in information behaviour research. Documentation **55**(3), 249–270 (1999)
31. Wilson, T.D.: Human information behavior. Informing Science **3**(2), 49–56 (2000)

Issues Involving Social Networking and Senior Citizens

Harry Carley[✉]

Matsuyama University, Matsuyama, Ehime, Japan
pm333@air.ocn.ne.jp

Abstract. This paper discusses Internet Communication Technology (ICT) and in particular social networking in relationship to usage by senior citizens. Social networking as a percentage has been steadily increasing among all age groups including seniors. While the elderly may develop varying physical discrepancies such as auditory and optic complications these should not hinder their abilities to utilize social media devices. Not only does social networking allow for spontaneous intercommunication but for the elderly it can benefit overall wellbeing of the participants. Additionally, physical dexterity may be improved through continued use. It has been readily accepted that lifelong learning can lead to a decrease in mental decrepitude. The overall health benefits go beyond just communicating with grandchildren. Educational institutions and marketing establishments have yet to fully capitalize on the potential of senior social networking.

Keywords: Elderly · Internet · Seniors · Social networking

1   Introduction

The popularity of social networking has continued to escalate since its introduction into the general public. For many individuals it has become a daily necessity to be afforded the opportunity to be in constant contact with ones companions. Lack of Internet access could be comparable to absence of a television signal when TV first came into existence. In the United States, users of smartphones have continued to expand beyond the original purpose of browsing the Internet or calling and texting acquaintances. These days' smartphones are being utilized to assist in a diverse number of life issues. Many have scholars noted [1, 2 & 3] that the use of social networks may actually be isolating people from real world relationships. Contrary to this the nonpartisan fact tank the Pew Research Center [4] which conducts public opinion polls encompassing America and the world, states that as of Sept 2014, American adult usage consisted as:

- Multi-platform use is on the rise: 52% of online adults now use two or more social media sites, a significant increase from 2013, when it stood at 42% of internet users.
- For the first time, more than half of all online adults 65 and older (56%) use Facebook. This represents 31% of all seniors.

© Springer-Verlag Berlin Heidelberg 2015
L. Wang et al. (Eds): MISNC 2015, CCIS 540, pp. 363–371, 2015.
DOI: 10.1007/978-3-662-48319-0_29

- For the first time, roughly half of internet-using young adults ages 18-29 (53%) use Instagram. And half of all Instagram users (49%) use the site daily.
- For the first time, the share of internet users with college educations using LinkedIn reached 50%.
- Women dominate Pinterest: 42% of online women now use the platform, compared with 13% of online men.

In fact the dependence for many people and their smartphone has become absolutely essential to seemingly subsist in the Internet Age.

> Social media is now a parallel world. Here, users all around the globe have built personal lives, have developed relationships, have shared experiences, and have also grew businesses. Domains like marketing and advertising have a completely new meaning since social media existed. Today, in 2015, the saying is that if a person doesn't have a social media account, he or she simply doesn't exist [5].

There has been a continuous inclusive increase in usage by all age's most notably young people. Those 55 and over until recently, have not embraced smartphones as strongly as other demographic sections of society. Only in recent years have there been noticeable differences in the statistics of elder members of society and their relationship to social networking. In the United States;

> Six in 10 seniors (~59 percent) now go online (up from 53 percent in 2013), and just under half have broadband. Among those who go online, 71 percent are daily or near daily users, while 11 percent do so three to five times per week [6].

For pensioners social networking can add a feeling of belonging and acceptance as they retire from the work force. Concurrently, as elder citizens may be downsizing their lives and moving from independent living to a more structured retirement community, social networking can offer freedom to travel; albeit the Internet highway. Online social networking has the potential to enrich the lives of the elderly by providing them with an easy way to stay in touch with friends and family [7]. Unfortunately though;

> Older Americans have often been left out of the discussion when it comes to internet or social media usage; however, they are active participants in both these technologies. Senior citizens go online with increasing frequency, but are a largely ignored demographic [8].

There are many motivating factors why seniors should consider integrating social networking into their golden years. Ease of rapid communication allows not only family and friends to be reached but should a medical emergency arise assistance can be called upon with the swipe of a finger or one or two key strokes.

2   Trends

While there are currently many social networking platforms with multifariously more being added every day. Similar to seniors across the globe America's seniors have historically been late adopters to the world of technology compared to their younger compatriots, but their movement into digital life continues to deepen [9]. It has been shown that seniors in general have particular preferences when it comes to networking to keep abreast of their families and latest events. Not only is Facebook the overall leader among networking media sites globally but the 65+ group has also embraced the Internet hangout for many of the similar reasons that those of younger age groups have. Facebook allows for easy access to anyone the user is 'friends' with. Almost any form of communication is possible, be it pictures, videos, or simple text. Facebook use at nursing homes is also on the rise, as it lets patients talk to other patients via private messaging in the same building, without having to use the phone [10].

While Facebook can seem a little complicated at times and it does take a while to get used to Twitter on the hand is more simplistic and straight to the point. Basically nothing more than text messaging, seniors can quickly send a 'tweet' and receive an almost instantaneous reply from the recipient. Elderly people, especially those considered vulnerable and living in care homes can improve their mental health and boost brain power by learning to browse the web [11].

Other social networks can also fill specific desires for users. Assistance with understanding medicine for example, can be contacted through blogs or other resources as needed. All too often as senior's age their memories start to fade. This can become hazardous for those required to take daily medication. Through apps on their social networking devices pensioners can be reminded of when to take specific pills along with which ones they may have already consumed and at what time.

3   Life Adjustments

Many life changing adjustments arrive in the golden years for those advancing in age. For most the hustle and bustle of a work day schedule are concluded with the advent of retirement. For some people this can be a major reconcilement. One day they may be in a high position of authority when the next they are seemingly shut out of the workplace.

> As adults move into older age, the spatial and social barriers they encounter start taking their toll. Isolation, loneliness, and depression are commonly experienced as family and friends move away and are less accessible, and as individual mobility and independence start to decline [12].

Additionally as seniors age even further, there most likely will involve the loss of freedom of the road. The cancelation of a driver's license due to vision, hearing or even medications that negate the ability to drive can cause a feeling of helplessness and dependency on others. This life event for those who have been driving for forty or fifty years can be extremely disheartening. Social networking can help fill in the potholes on the road of life by offering seniors a road less travelled but just as enjoyable and entertaining. This new path can be filled with new and exciting elements to add meaning to their lives. The Internet offers a world of information, sights and sounds to help lift the spirits of elderly who have just lost the freedom of the open road. Well-being is also associated with better health, higher levels of social and civic engagement, and greater resilience in the face of external crises [13].

Moreover, simply the thought of being confined in a retirement community for many seniors can bring a thought of dread. They may have been home owners with a large circle of neighbours and friends. For countless members of the elderly populace life within a closed community could resemble a prison sentence. Social networking can be a key to unlock the isolation and despair. According to Kiel [14] most elders are receptive to learning how to use computers and are looking for methods to stay connected and be informed [15]

4   Perceptions and Attitudes

While many social networking sites have essentially been designed with bringing various people from all walks of life together to interact online, not every age group has the same desires or ideals of what social networking should be. There is a need to understand how the elderly perceive social networks, what benefits they could realize from the use of such networks and what might hinder their intention to use social networks to enrich their lives [16]. Whereas downloading the latest trendy videos from YouTube may seem appealing to the under 30 crowd for those 65+ this would most likely be a non-issue. On the other hand any modifications in retirement payment plans or increases in taxes for the elderly would be of consequential importance. Senior centres and other elder care institutions have much to gain from encouraging elders to adopt social networking websites [17].

5   Instructional Methods

Due to the fact that many individuals over 55 have not grown up as digital natives the use of computers and social networking in particular could be daunting and somewhat foreboding. The phrase "social networking" is usually a turnoff for seniors because the word "networking" itself sounds like you have an agenda [18]. The whole endeavour can quickly devolve into an epic struggle against the machine itself. Seniors navigating the strange new world of social media and mobile devices have to contend with new gizmos, ever-changing memes and shifting notions of privacy. [19]. It is clear that the design requirements for persons older than 85 years differ from those for persons of 56–65 years of age [20]. Instructional lessons and methods should therefore take into consideration the age and overall computer experience of the students. Many suggest that for older individuals who may be initially introduced to computer applications a one-to-one instructional approach may be suitable. Unfortunately, for numerous instructors this may not be feasible due to the sheer numbers of learners and limited time for teachers.

6   Benefits

Benefits for seniors utilizing social networking can come in many different forms, tangible as well as intangible. The ability to contact and stay in touch with loved ones would most likely be at the top of the list. Gone are the days of children writing their grandparents thank you letters for presents. Social networking allows for real time communication for any occasion.

As more and more educational institutions go online and offer distance learning the chance to explore topics and courses is every expanding. Most learning establishments offer senior discounts for elder citizens. The overall benefit to keeping the mind active is not only physical well being but that of mental sharpness too. Being active on social media not only enables and enhances social connections for seniors: it provides them with regular positive interactions with friends and family. All of this adds up to improved health and potentially reduced risk of dementia [21]

Associated with mental limitations those of advancing age also begin to encounter numerous physical limitations. Particularly for older people with limited mobility, seniors restricted to wheel chair or bed ridden may encounter a feeling of isolation. Social networking can enable people connections to still thrive. These connections may lead to a feeling of usefulness and association with others. Research has also found that those senior who engage in online social activity are more are likely to mingle in more personal relationships. Independent studies suggest that aging adults who were active on the Internet experienced a 30% decrease in depression, plus learning new things keeps older adult young at heart and engaged [22]

As is common at this age many married couples may lose their lifelong partner. This can be an extremely emotional and depressing time. It is not possible to say that social networking can take the place of a loved one. Beyond the despair though there is the possibility to go on with life. Visiting old places or interacting with fellow elders who may be in the same situation through the use of a chat room or group page is possible to bring some relief.

7   E-Commerce

Regarding marketing on the Internet the opportunity to expand retailing to seniors via social networking has yet to be fully accomplished. Marketers and advertisers are anxious to capture the attention and buying power of this demographic through this new channel. Quite simply there are a lot of them and they have money. [23]. Marketers regard the 65-and-older population as a sleeping giant when it comes to digital usage. And for now, at least the giant still mostly dozes [24].

The realization that there is a market is one thing, getting the sellers and buyers coherent of each other are another. E commerce offers seniors shopping opportunities that are more attractive than only buying books or renting videos. Most advertising has been typical of the elderly age group. But older people do buy things other than pharmaceuticals, adult diapers, and scooters [25].

Marketing in general is geared toward young people but in most instances, those between the ages of 18-30 most commonly have the least amount of money. Those nearing retirement or having already reached that milestone on the other hand may have accumulated wealth and pension income. This is occurring at the same time that most elderly likely having limited expenses due to the fact that their children are leading independent lives. Additionally, depending on family relations there may also be children or grandchild to expend funds for. These facts have not gone unnoticed by marketers.

Advice for those considering targeting senior purchasers in their marketing approaches should simply consider the age of their consumers. Ideally those pages with small font should be enlarged for elder citizens with vision impairments. Also, most seniors most likely do not want to see young models advertising clothes. More age appropriate sales models would offer more realistic ideas on how the clothes might look on the purchaser. Additionally, if 'pop up' windows are annoying to younger individuals they will more than likely be irritating and confusing to older audiences. One thing that almost all Baby boomer and senior consumers look for when making purchases online is ease of use and accessibility [26].

8   Conclusion

Although this paper has pursued the use of social networking for seniors on a global scale there are cultural attitudes and norms of the physical world that can still transfer over into the Internet world. These cultural upbringings may help or hinder the acceptance and application of certain social networks regardless of the age of the user. Each country may have specific cultural styles of approaching people and interacting. For instance, these methods may involve the interaction regarding levels of politeness among strangers. On commuter modes of transportation some countries citizens may prefer a text style of social networking over voice activated so as not to disturb others. Japan in particular adheres to this approach in most public places. While 'manner mode' may be a basic feature on most smart phones, some societies may strictly adhere to its basic operation more than others. For marketing and social networks looking to expand into other countries cultural practices are a key to interpreting what type of audience may be interested in participating.

Regardless of social norms the use of social networking offers a viable opportunity to connect, explore, and expand the world for senior citizens. A majority of retired citizens may be physically restricted to senior centres and retirement communities but with social networking they are not completely isolated from their family, friends, or acquaintances. The possibility of adding to their current network through Internet contacts the world over is boundless. Opportunities for instant communication and the latest news from the local to global arena can be instantly accessed. Material on a more personal scale such as pharmaceutical data or other health related info can be called up with an inconsequential number of key strokes. Shopping and educational opportunities abound to keep one active and thinking.

Unfortunately research and study into seniors and social networking is still lacking in depth and content. There is much opportunity for scholars to explore this matter. Future findings can only benefit the elderly and more importantly those who assist with their care and supervision as senior's age. Software development is still mainly geared to the younger population in most instances.

As more and more citizens who have been accustomed to utilizing computers enter the aging section of the population there will be more familiarity of social networking devices. As Baby Boomers become part of the senior demographic, they will keep their internet connections with them [27]. The potential for growth in social networking development, education, and management can only be seen as an area of future expansion.

References

1. Wilkinson, R., Marmot, M.: Social Determinants of Health: The Solid Facts, 2nd edn. World Health Organization, Europe (2011)
2. Hawton, A., Green, C., Dickens, A.P., Richards, S.H., Taylor, R.S., Edwards, R., Greaves, C.J., Campbell, J.L.: The impact of social isolation on the health status and health-related quality of life of older people. Springer Science+Business Media B.V. 2010 (2010). http://link.springer.com/article/10.1007/s11136-010-9717-2#page-1
3. Seeman, T.E.: Social ties and health. The benefits of social integration. Annals of Epidemiology 6(5), 442–451 (1996)
4. Smith, A.: Older Adults and Technology Use. Pew Research Center. Internet, Science & Technology (2014). http://www.pewinternet.org/2014/04/03/older-adults-and-technology-use/
5. Duggan, M., Ellison, N.B., Lampe, C., Lenhart, A., Madden, M.: Social Media Update 2014. Pew Research Center (2015). http://www.pewinternet.org/2015/01/09/social-media-update-2014/
6. Ahmad, I.: Fascinating #SocialMedia Stats 2015: Facebook, Twitter, Instagram, Google+. Digital Information World, Covering the world of marketing, social media, technology and infographics, January 2, 2015. http://www.digitalinformationworld.com/2015/02/fascinating-social-networking-stats-2015.html
7. Charles, M.: Senior Citizens' use of Facebook, Other Social Networks on the Rise. Social News, April 3, 2014. http://socialnewsdaily.com/34049/senior-citizens-use-of-facebook-other-social-networks-on-the-rise/
8. Lewis, S., Ariyachandra, T.: Seniors and Online Social Network Use. In: 2010 CONISAR Proceedings Conference on Information Systems Applied Research, vol. 3(1552), pp. 1–15 (2010). http://proc.conisar.org/2010/pdf/1522.pdf
9. How Senior Citizens Use Technology: trends in Internet and Social Media Use. Einsight. The Scott Public Relations Healthcare, Insurance and Technology Public Relations Blog. http://scottpublicrelations.com/how-senior-citizens-use-technology-trends-in-internet-and-social-media-use/#comment-80745

10. Social Media for Senior Citizens: A Rising Trend: Senior TV, Solutions Today, Technology for Tomorrow, April 24, 2014. https://seniortv.com/social-media-for-senior-citizens-a-rising-trend/

11. Woollaston, V.: Could Twitter help the elderly feel less lonely? Social media boosts mental health and confidence in older people, December 15, 2014.
http://www.dailymail.co.uk/sciencetech/article-2874531/Could-Twitter-help-elderly-feel-lonely-Social-media-boosts-mental-health-confidence-older-people.html

12. Zafar. A.: Facebook for Centenarians: Senior Citizens Learn Social Media. The Atlantic, August 31, 2011. http://www.theatlantic.com/technology/archive/2011/08/facebook-for-centenarians-senior-citizens-learn-social-media/244357/

13. Alpass, F.M., Neville, S.: Loneliness, health and depression in older males. Aging and Mental Health 7, 212–216 (2003)

14. Kiel, J.M.: The digital divide: Internet and e-mail use by the elderly. Informatics for Health and Social Care 30(1), 19–23 (2005)

15. Ariyachandra, T., Crable, E.A., Brodzinski, J.D.: Seniors' Perceptions of the Web and Social Networking. Issues in Information Systems X(2), 324 (2009).
http://iacis.org/iis/2009/P2009_1261.pdf

16. Cooper, C., Field, J., Goswami, U., Jenkins, R., Sahakian, B.: Mental Capital and Wellbeing. Wiley-Blackwell, Oxford (2010)

17. Ariyachandra, T., Crable, E.A., Brodzinski, J.D.: Seniors' Perceptions of the Web and Social Networking. Issues in Information Systems X(2), 324 (2009).
http://iacis.org/iis/2009/P2009_1261.pdf

18. Farley, J.: Q&A with an Expert who Teaches Technology to Local Seniors. Metrofocus (2012). http://www.thirteen.org/metrofocus/2012/01/qa-silver-surfers-the-brave-new-to-them-world-of-facebook/

19. Farley, J.: Q&A with an Expert who Teaches Technology to Local Seniors. Metrofocus (2012). http://www.thirteen.org/metrofocus/2012/01/qa-silver-surfers-the-brave-new-to-them-world-of-facebook/

20. Nef, T., Ganea, R.L., Muri, R.M, Mosimann, U.P.: Social networking sites and older users – a systematic review. International Psychogeriatics, 1–13 (2013).
http://www.cclm.unibe.ch/unibe/philhuman/cclm/content/e200922/e200932/e273258/linkliste273259/Nefetal.-2013-Socialnetworkingsitesandolderusersasystem_ger.pdf

21. Can Social Media Improve Mental Health in seniors? Crozer Keystone Health System, March 25, 2015. http://www.crozerkeystone.org/news/press-releases/2015/march/can-social-media-improve-mental-health-in-seniors/

22. Safety Tips: Senior Citizens. Seniors stay connected with friends and family through social media technology. Sievers Security. https://www.sieverssecurity.com/specials-safety-tips/safety-tips.asp?a=Article&id=128

23. Kaplan, M.: Ecommerce Merchants Should not Ignore Older Shoppers. PracticalEcommerce, December 16, 2013. http://www.practicalecommerce.com/articles/61984-Ecommerce-Merchants-Should-Not-Ignore-Older-Shoppers

24. Seniors Still Lukewarm on Web Activity. eMarketer. March 26, 2013.
http://www.emarketer.com/Article/Seniors-Still-Lukewarm-on-Web-Activity/1009757

25. Kaplan, M.: Ecommerce Merchants Should not Ignore Older Shoppers. PracticalEcommerce, December 16, 2013.
http://www.practicalecommerce.com/articles/61984-Ecommerce-Merchants-Should-Not-Ignore-Older-Shoppers

26. eCommerce Site Development & Management. Coming of Age, the Baby Boomer & Senior Marketing Agency. http://www.comingofage.com/services/ecommerce-site-development-management/
27. Ariyachandra, T., Crable, E.A., Brodzinski, J.D.: Seniors' Perceptions of the Web and Social Networking. Issues in Information Systems **X**(2), 324 (2009). http://iacis.org/iis/2009/P2009_1261.pdf

Toward Automatic Assessment
of the Categorization Structure
# of Open Data Portals

Hsin-Chang Yang$^{(\boxtimes)}$, Cathy S. Lin, and Po-Han Yu

National University of Kaohsiung, Kaohsiung, Taiwan
{yanghc,cathy}@nuk.edu.tw, skycloud7777@gmail.com

Abstract. Governments worldwide have been releasing their owned data
recently for public usage and arising lots of novel applications and services. Issues on open data were also intensively discussed from researchers
and practitioners. One of the key issues in adopting open data is the accessibility of the data, which are generally collectively provided in open data
portals. Open data portals categorize open datasets according to their
domains, providers, formats, and other properties for better accessibility of the data. However, these portals did not follow a conforming standard in establishing their categorization structures. In this work, we try
to assess the goodness of categorization structures of open data portals
automatically by investigating the coherence of the datasets in the same
category. The detailed methodology is described but preliminary experiments on Taiwan's open data portals are still undergoing.

Keywords: Open data · Open data portal · Quality assessment ·
Categorization structure

1 Introduction

Open data has been a popular topic in many areas such as politics, computer
science, sociology, etc. in last decade. A commonly accepted definition of open
data was given in the Open Data Handbook[1] by Open Knowledge Foundation as
'Open data is data that can be freely used, re-used and redistributed by anyone -
subject only, at most, to the requirement to attribute and sharealike.' Basically,
anyone can publish his data as open if he complies with the open definition
above. However, governments own enormous amount of data collected during
the governance. For example, the government will acquire the income data of all
citizens for collecting taxes. Such government data are generally prohibited from
access by general public for national security or other reasons. However, more
and more people are demanding the disclosure of government data for political
transparency, citizen engagement, or even personal interest. Government open
data is thus arisen from such demands. In fact, government open data dominates

---

[1] http://opendatahandbook.org/

© Springer-Verlag Berlin Heidelberg 2015
L. Wang et al. (Eds): MISNC 2015, CCIS 540, pp. 372–380, 2015.
DOI: 10.1007/978-3-662-48319-0_30

the amount and need of open data and makes it synonymous with the general 'open data' term. Following such convention, we will use 'open data' as short for 'government open data' and use them interchangeably in the following text.

To provide the maximal accessibility of open data, many governments and organizations created lots of portals to allow the public accessing the data. For example, US government created the DATA.GOV portal to allow people accessing the US government's open data. There are over 130,000 datasets available there until 2015. Taiwan government also created data.gov.tw portal where the amount of datasets triples to more than 7,000 datasets in 2015. Almost all open data portals provide some sorts of categorization structure besides search capability to allow users easy access of interested datasets. Category structures are widely used in Web portals such as Yahoo!. Users can browse through such structures to find sites related to interested topics. The structure is generally hierarchical to allow users unveiling their goal sites in a coarse to fine manner. For a portal containing gigantic amount of data, the design of its categorization structure will prominently affect the accessibility of its data. Without proper structure, the users will spend lots of time in searching the sites they want. Therefore, categorization structure construction plays a major role in portal design.

The quality of a categorization structure can be measured in two aspects. First, is the structure comprehensive enough to reach to goal data easily? Second, is the categorization correct enough to have related data in the same category? In this work, we try to develop automatic schemes for evaluating these issues. The fitness of the structure will be evaluated by comparing the portal's structure to a gold standard structure. On the other hand, the correctness of the categorization can be measured by investigating the similarity of data within the same category. We plan to evaluate Taiwan's open data portals using the proposed schemes to reveal the fitness of their categorization structures.

The remain of this article is organized as follow. Sec. 2 describes some works related to our research. In Sec. 3 we will introduce the proposed categorization structure assessment schemes. Sec. 4 gives the experimental setup and preliminary result. Finally, we give conclusions and discussions in the last section.

2   Related Work

The concept of open data has emerged in last decade and still lacks universally acceptable definition. Generally, open data refers to the data that are not constrained by any copyrights, patents, or administrative regulations and can be freely accessed and distributed [1]. A widely accepted definition by Open Knowledge is that 'Open data and content can be freely used, modified, and shared by anyone for any purpose.' [2] Some major characteristics of open data include:

- Availability and accessibility: The data must be complete and only needs a reasonable amount of reproduction cost. It is preferred to be available by downloading from Internet in easily modifiable format.
- Reusability and redistribution: The data should be provided in a way to allow being reused, redistributed, and migrated with other datasets.

- Universal participation: The data can be used, reused, and redistributed by anyone. There is no obligation to any application fields, persons, or communities. For example, non-commercial use only or educational-purpose only is not allowed.

The issue of open data quality assessment have not been addressed until recently. Basically, such assessment on open data is indistinguishable to general type of data. Measures on data quality [3,4] can then be applied in open data domain. Open data domain-specific data quality criteria and measurements emerged recently to evaluate the quality of open datasets as well as portals. A famous criterion proposed by Tim Berners-Lee is the 5-star scale scheme for linked open data [5]. In this scheme, the quality of an open dataset is measured and noted by different numbers of stars. The more the stars given, the open the dataset is. However, this scheme is rough and no operational definition is given. Zaveri et al. [6] gives a comprehensive review on quality assessment schemes for linked open data. Behkamal [7] proposed a metric-driven framework to predict the quality of open datasets. He selected six quality characteristics which each is assessed by a set of metrics. Knuth et al. [8] identify the key challenges of linked data quality and summarize the efforts in tackling these challenges in aspects such as data validation, data cleansing, data creation and reuse, and data versioning.

The quality assessment of open data portals is seldom discussed until recently. Since the nature of open data coheres to general Web data, quality measures of Web portals were commonly adopted. An example is the PoDQA project [9] that evaluates the quality of a Web Portal by 42 quality characteristics. To cope with the openness of data, OPQUAST project compiles a checklist containing 72 items for evaluating open data portals [2]. Umbrich et al. [10] developed the Open Data Portal Watch project which monitors the data quality of 90 open data portals. They applied six quality metrics on these portals and found that combined metrics will produce better understanding about the data. Colpaert et al. [11] devised the 5 stars of open data portals which resembles to the 5-stars scheme for linked open datasets by Tim Bernes-Lee [5]. The stars are used to represent the main function or affordance that the data portal is built or used for, ranging from one star for dataset registry to five stars for data hub. Unlike other schemes, no specific metric is devised in assessing the stars.

3   Categorization Structure Assessment of Open Data Portals

Two types of assessment are applied on a open data portal to evaluate the quality of its categorization structure. The first type, namely structural quality assessment, will identify the fitness of the categorization structure according to the similarity to a standard structure. The second type, namely categorization quality assessment, will measure the coherence of the data within the same

---

[2] https://data.oqs-cdn.com/checklists/589/Opquast-Opendata_20120831.pdf

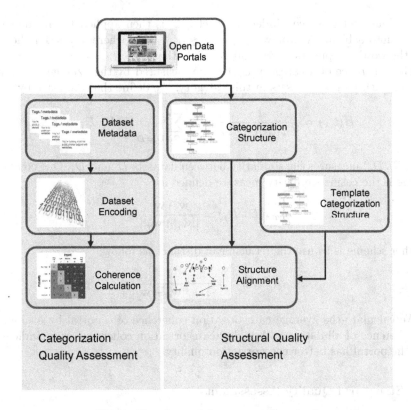

Fig. 1. Flowchart of the proposed method.

category. Fig 1 depicts the flowchart of the proposed method. We will describe each step in detail in the following subsections.

3.1 Categorization Quality Assessment

To evaluate the quality of the categorization by an open data portal, we shall find the coherence, i.e. similarity, of datasets in a category. In this work, the similarity between two datasets is measured by the similarity of their accompanied metadatas. The metadata of a dataset contains descriptive and characteristic information about the dataset. Although some regulations have been advised for metadata creation, different portals still adopt various fields or properties to describe the datasets in their metadatas. In this work, we will not differentiate the properties in a metadata but use a bag-of-word approach. We collect all words occurred in the metadata to represent the dataset. Let D_i be a dataset and M_i be its metadata. The set of unique words appeared in M_i will be collected in W_i. We then assemble all words across all W_i into the vocabulary V, i.e. $V = \bigcup_i W_i = \{v_j | 1 \le j \le |V|\}$. Note that we will ignore the metadata attribute/field names to reduce the redundancy since all metadata contain such words. The Vector Space Model approach [12] is then applied to encode each dataset into a binary vector $\mathbf{w}_i = \{w_{ij} | 1 \leq j \leq |V|\}$, where $w_{ij} = 1$ indicates that the word v_j appears in M_i and $w_{ij} = 0$ otherwise.

The coherence of a category C_k can be measured by the average similarity between each pair of datasets in this category. We define the coherence by

$$H(C_k) = \frac{1}{|C_k|(|C_k| - 1)} \sum_{D_i, D_j \in C_k, D_i \neq D_j} S(D_i, D_j). \tag{1}$$

$S_{cate}(D_i, D_j)$ measures the similarity between datasets D_i and D_j. One common scheme is the cosine similarity measure defined as

$$S_{cate}(D_i, D_j) = \frac{\mathbf{w}_i \cdot \mathbf{w}_j}{||\mathbf{w}_i|| ||\mathbf{w}_j||}. \tag{2}$$

Another scheme is to use their Euclidean distance as follow:

$$S_{cate}(D_i, D_j) = \frac{1}{1 + ||\mathbf{w}_i - \mathbf{w}_j||}. \tag{3}$$

We calculate the average categorization coherence of a portal by averaging the coherence of all datasets. A higher categorization coherence value indicates that the portal has better categorization quality.

3.2   Structural Quality Assessment

Although the quality of a categorization structure can be measured by some measurements involving self-properties such as depth and category size, we will adopt another approach by comparing this structure to a template structure. Selection of the template structure is an arduous task since we shall first define the fitness of a structure. Ideally, a good categorization structure should allow the users finding their intended goals as quickly and easily as possible. However, the determination of a template structure is beyond the scope of this work. Here, a template structure is presumed to exist. Basically, a categorization structure can be described by a tree $T(V, \mathcal{E})$ where V and $\mathcal{E}$ denote the vertices and edges, respectively. In this regard, finding the structural similarity between a candidate structure T_c and a template structure T_t can be mapped to the tree alignment or ontology alignment problems.

To compare two categorization structures, says T_c and T_t, we shall first define the similarity between two categories across these structures. Let $\mathcal{V}_i^c$ and $\mathcal{V}_j^t$ be categories (vertices) of T_c and T_t, respectively. Note that both vertices should be leaf vertices which haves no children since internal vertices general contain no dataset. This constraint is also held in the remaining text. That is, when we mention vertices, we are referring to leaf vertices unless otherwise specified. The similarity between these categories is defined as:

$$S_c(\mathcal{V}_i^c, \mathcal{V}_j^t) = S_{cate}(\overline{D}_i^c, \overline{D}_j^t), \tag{4}$$

where $\overline{D}_i^c$ is the mean dataset of category $\mathcal{V}_i^c$, i.e. its vector form is the average of all dataset vectors in $\mathcal{V}_i^c$:

$$\overline{\mathbf{w}_i^c} = \frac{1}{|\mathcal{V}_i^c|} \sum_{D_k \in \mathcal{V}_i^c} \mathbf{w}_k. \tag{5}$$

The mean dataset of $\mathcal{V}_i^t$, i.e. $\overline{D}_i^t$, is similarly defined.

We should find the most similar category in T_t for each category in T_c. Let $\mathcal{V}_{\hat{i}}^t$ be the most similar vertex in T_t of the vertex $\mathcal{V}_i^c$ in T_c. We then have

$$\hat{i} = \underset{k}{\mathrm{argmin}} \, ||\overline{\mathbf{w}}_k^t - \overline{\mathbf{w}}_i^c||. \tag{6}$$

The structural similarity between T_c and T_t can be measured by

$$S_{struc}(T_c, T_t) = \frac{1}{N_c(N_c - 1)} \sum_{\forall \mathcal{V}_i^c, \mathcal{V}_i^c \in T_c} |PL(\mathcal{V}_i^c, \mathcal{V}_j^c) - PL(\mathcal{V}_{\hat{i}}^t, \mathcal{V}_{\hat{j}}^t)|, \tag{7}$$

where $PL(V_1, V_2)$ returns the shortest path length between two vertices V_1 and V_2. N_c is the number of leaf vertices in T_c. Note that we allow a category in T_t being matched to multiple categories in T_c.

The structural similarity defined in Eq. 7 works on general tree structures. However, most of the open data portals have single layer categories without proper hierarchical structures. In such case, we will create a pseudo root category that has all categories as its children. Note that the path length between any two categories will be the same as 2. In fact, such pseudo root vertices will be created for categorization structures of every portals to create a single tree structure that allows measuring path lengths between two vertices.

4   Experiment Setup and Result

In this section, we will describe the setup of the experiments. We selected 12 open data portals in Taiwan as the candidate portals. Table 1 lists all candidate datasets used in the experiments.

For categorization quality assessment, we will calculate the average category coherence of all datasets in each portal. We have obtained a preliminary result on Taitung County Open Data Portal which contains only 16 datasets. Table 2 shows the result of the assessment.

For structural quality assessment, we will compare all candidate portals to the template portal (data.gov.tw). However, the experiments are still undergoing on the time of submission. We expect to measure the similarity between portals using Eq. 7. We will also investigate the fitness of the structure and give suggestions on categorization structure design for open data portals.

Table 1. The open data portals used in the experiments. The 'T' and 'C' in the 'Role' field denote template and candidate portals, respectively. The statistics are accessed in May 23, 2015.

Portals	Role	URL	Number of leaf categories	Number of datasets
National Open Data Portal of Taiwan	T	http://data.gov.tw/	16	7994
Taipei City Open Data Portal	C	http://data.taipei/	18	1377
New Taipei City Open Data Portal	C	http://data.ntpc.gov.tw/	21	237
Hsinchu County Open Data Portal	C	https://data.hsinchu.gov.tw/	10	53
Hsinchu City Open Data Portal	C	http://opendata.hccg.gov.tw/	18	153
Taichung City Open Data Portal	C	http://data.taichung.gov.tw/	11	154
Nantou County Open Data Portal	C	http://data.nantou.gov.tw/	15	72
Tainan City Open Data Portal	C	http://data.tainan.gov.tw/	10	334
Ilan County Open Data Portal	C	http://opendata.e-land.gov.tw/	10	108
Taitung County Open Data Portal	C	http://www.taitung.gov.tw/opendata/	10	16
Kaohsiung City Open Data Portal	C	http://data.kaohsiung.gov.tw/Opendata/	10	236
Kinmen County Open Data Portal	C	http://data.kinmen.gov.tw/	18	46

Table 2. The result of categorization quality assessment on Taitung County Open Data Portal.

Size of vocabulary V	126
Average category coherence using Eq. 2	0.0378
Average category coherence using Eq. 3	0.0241

5 Conclusions and Discussions

In this work, we proposed schemes to assess the quality of open data portals' categorization structures. Both quality of categorization and structure are evaluated. The categorization quality is measured by finding the coherence among all datasets within a category. The structural quality is assessed by comparing a portal's categorization structure to that of a template portal. Measures for such assessments were devised. Experiments were conducted on a set of 12 portals and over 10,000 datasets to verify the fitness of these measurements. Some preliminary results were obtained but the experiments are still undergoing and expected to be finished recently.

We selected a template portal in evaluating structural quality. One key question is on the selection of such template portals. Such selection is rather difficult if we had not surveyed the portals for the fitness of their categorization structures. We will always face the argument of defining an objective measure for the fitness of a categorization structure for open data portals. One possible solution without human intervention is to use an automatically generated template structure to evaluate the quality of a candidate structure. Since such generated structures are always generated according to some conformity measurements, it is expected that such automatic schemes will produce structures with better quality. This may minimize the effort of selecting the template structures and produce much credible results.

References

1. Auer, S., Bizer, C., Kobilarov, G., Lehmann, J., Cyganiak, R., Ives, Z.: DBpedia: a nucleus for a web of open data. In: Aberer, K., et al. (eds.) ISWC/ASWC 2007. LNCS, vol. 4825, pp. 722–735. Springer, Heidelberg (2007)
2. Open Knowledge Foundation: Open data handbook (2012). http://opendatahandbook.org/ (accessed May 15, 2015)
3. Pipino, L.L., Lee, Y.W., Wang, R.Y.: Data quality assessment. Communications of the ACM **45**(4), 211–218 (2002)
4. Batini, C., Cappiello, C., Francalanci, C., Maurino, A.: Methodologies for data quality assessment and improvement. ACM Computing Surveys **41**(3), 16:1–16:52 (2009)
5. Berners-Lee, T.: Linked data (2010). http://www.w3.org/DesignIssues/LinkedData.html (accessed July, 10 2014)
6. Zaveri, A., Rula, A., Maurino, A., Pietrobon, R., Lehmann, J., Auer, S.: Quality assessment for linked data: A survey. Semantic Web Journal (2015)
7. Behkamal, B.: Metrics-driven framework for lod quality assessment. In: Presutti, V., d'Amato, C., Gandon, F., d'Aquin, M., Staab, S., Tordai, A. (eds.) ESWC 2014. LNCS, vol. 8465, pp. 806–816. Springer, Heidelberg (2014)
8. Knuth, M., Kontokostas, D., Sack, H.: Linked data quality: identifying and tackling the key challenges. In: Knuth, M., Kontokostas, D., Sack, H. (eds.) 1st Workshop on Linked Data Quality (LDQ). Number 1215 in CEUR Workshop Proceedings, Aachen (2014)

9. Calero, C., Caro, A., Piattini, M.: An applicable data quality model for Web portal data consumers. World Wide Web **11**(4), 465–484 (2008)

10. Umbrich, J., Neumaier, S., Polleres, A.: Towards assessing the quality evolution of open data portals. In: Proceedings of ODQ2015: Open Data Quality: from Theory to Practice Workshop, Munich, Germany (2015)

11. Colpaert, P., Joye, S., Mechant, P., Mannens, E., Van de Walle, R.: The 5 stars of open data portals. In: Álvarez Sabucedo, L., Anido Rifón, L. (eds.) Proceedings of the 7th International Conference on Methodologies, Technologies and Tools Enabling e-Government, pp. 61–67. Universida de Vigo (2013)

12. Salton, G., McGill, M.J.: Introduction to Modern Information Retrieval. McGraw-Hill, New York (1983)

The Dimensions of Service Quality at College Computer Center: Scale Development and Validation

Yu-Lung Wu[1], Yu-Hui Tao[2], Deniel-Y. Chang[3], and Ting-Mu Huang[4(✉)]

[1] Department of Information Management, I-Shou University,
No.1, Sec. 1, Syuecheng Rd., Dashu District, Kaohsiung City 84001, Taiwan, R.O.C.
wuyulung@isu.edu.tw
[2] Department of Information Management, National University of Kaohsiung,
700, Kaohsiung University Rd., Nanzih District, Kaohsiung 811, Taiwan, R.O.C.
ytao@nuk.edu.tw
[3] Department of Information Applied English, I-Shou University,
No.1, Sec. 1, Syuecheng Rd., Dashu District, Kaohsiung City 84001, Taiwan, R.O.C.
dchang@isu.edu.tw
[4] Department of Information Engineering, I-Shou University,
No.1, Sec. 1, Syuecheng Rd., Dashu District, Kaohsiung City 84001, Taiwan, R.O.C.
hdm2314@gmail.com

Abstract. In this study, a multi-stage scale development procedure was conducted in order to identify and disclose the factor structure for assessing the service quality of college computer center that is lacking in the literature. The result shows that the service quality of college computer center contains five factors: Tangibles, Responsiveness, Contact, Access, and Security. These five dimensions not only can be used to control the attention paid by users for continuously improving service quality, but also will fulfill the research gap in the service quality literature of college computer center. We believe that this research outcome provides an effective and robust evaluation tool to both college computer centers for conducting appropriate service quality assessment to benefit the general users, and researchers for further investigating service quality issues in college computer center, as we see the LibQual scale to college library in the literature.

Keywords: Computer center · Service quality · Scale

1 Introduction

Computer center is a division related to information technology in a college, and has played an important role in larger colleges for over five decades. Specifically, a computer center provides professional services on computer hardware, software, Internet, and communication facility to the faculty, staff, and students [1]. In recent years, Internet and mobile networks have been accelerating the development of information technology for further influencing the user operations. For example, the Center for Learning and Teaching (CLT) in large U.S.A. universities is a faculty

© Springer-Verlag Berlin Heidelberg 2015
L. Wang et al. (Eds): MISNC 2015, CCIS 540, pp. 381–392, 2015.
DOI: 10.1007/978-3-662-48319-0_31

support service organization with a pool of experts for providing a variety of services, including instructional design, the teaching of the use of learning technologies, multimedia production, and faculty development [2], which functions like a specialized or a subdivision of computer center in Taiwan. Therefore, administrators, faculties, and students all receive huge benefits and help from the computer center, such as administrative routine process and information delivery, grading management, and network learning system. In general, through the campus network, information can be easily transferred to achieve an efficient decision-making, management, and learning purpose.

However, when the users express their opinions regarding computer center service process, normally they can't get appropriate responses. Even when the staffs at the computer center would like to consider improving these problems, they don't know where to start due to no available measurement tool of their quality of service. Furthermore, the good or bad of computer center service quality (CC-SQ) will continually influence school's efficiency on administration, education, and service. Therefore, how to assess and motivate the service quality of computer center is one critical issue.

As of May this year, among 156 universities in Taiwan, 96 universities set up the computer centers while the other 60 universities combined computer centers with. To our knowledge, only a few computer centers have assessment tools of service quality, and measurement scales are not consistent and as a result of a robust development procedure. Therefore, this study addresses the CC-SQ assessment issue by designing a robust and systematic procedure to develop service-quality scale for covering adequate computer center service items. However, the scale needs to be repeatedly tested in order to establish its stability. Particularly, the service contents and items will change corresponding to the evolving information technology and users' needs. If the contents of the service quality scale were published, it might help continuous study to extend and update the service quality scale.

2   Related Concepts

2.1   Service Quality Model

In the literature, measuring customers' perception on service quality has identified characteristics consisting of intangibility, heterogeneity, perishable and inseparability [3]. Due to the inaccurate estimation of the qualitative-service quality, no consensus has been reached among researchers until Parasuraman, Zeithaml and Berry (1985) proposed Service Quality Model (SERVQUAL) and gradually revised its contents and models [4, 5, 6, 7]. The SERVQUAL consisted of customers' initial expectation and perception of service quality, which was based on customers' responses of perceived services for exploring the gap of business services and perceived dimensions of service quality. Therefore, SERVQUAL becomes the most well-known among the measurement scales and is widely used.

Although the SERVQUAL seems to play a leading role, there are still other measurement scales proposed to challenge Parasuraman, Zeithaml and Berry's Service Quality Model. Cronin & Taylor [8] considered the model of SERVPERF should directly focus on consumers' subjective performance. The rationale is that it is a model of perceived service quality, which does not need to compare with consumers' expectation of services. In addition, the items of the original SERVQUAL have improved by scholars Cronin & Taylor from 44 items to 22 items. They made empirical results explaining the SERVPERF was better than SERVQUAL. Finally, Parasuraman, Zeithaml & Berry [9] also proposed amendments for its scale, integrated dimensions of tangibles, reliability, responsiveness, assurance and empathy consisting of 22 items.

Parasuraman, Zeithaml and Berry's service quality conceptual model and SERVQUAL scale are better applied to traditional stores [10, 11, 12, 13, 14]. Thus, they also advised that the SERVQUAL scale should be modified by considering the situation and characteristics of different industries. For example, Yeo and Li [15] explored the influences of service quality in higher education, and how they contribute to the overall performance of a higher learning institution in Singapore. They found three key aspects of service standards, including customer orientation, course design/delivery, and support services. In addition, it also considers the virtual characteristics of the site to develop different services-quality scale.

2.2   E-Service Quality

With the rise of the Internet, many studies explored key dimension of web service quality, such as Loiacono's WebQual [16] is composed of twelve dimensions as a measurement scale on website, which is based on redefining the service quality for online store sites and thus become a conceptual model of for website service quality. Zeithaml, Parasuraman & Malhotra [17] proposed seven dimensions for influencing website service quality, including efficiency, reliability, fulfillment, privacy, responsiveness, compensation, and contact, which divides the service quality into two before use and after use. Parasuraman et al. [14] had even combined many concepts of online services quality in the literature, such as Loiacono [16], Yoo & Donthu [18], and Wolfinbarger & Gilly [13], for constructing the ES-QUAL scale for measuring the service quality of online store. The scale was classified into two categories, core services (ES-QUAL) and remedial services (E-RecS-QUAL) which were composed of 33 question items.

Ladhari [19] reviewed the website service quality in the literature and proposed six common dimensions of service quality, including reliability/fulfillment, responsiveness, web design, ease of use/ usability, privacy/security, and information quality/benefit. He also suggested that the establishment of special dimensions on website classification was required.

Furthermore, Huang, Lin, and Fan [20] developed mobile service quality, concluded with five factors, i.e., contact, responsiveness, fulfillment, privacy, and efficiency), for the supporting services in the process of virtual product shopping and four factors, i.e., contact, responsiveness, fulfillment, and efficiency, for the supporting services in the process of physical product shopping.

With the above-mentioned literature and Yee et al. [21], we found out that most of the targets are transaction-based retailers in either physical or website service-quality research, which is different from the computer center as concerned in this research.

The tasks of computer center can be divided in three categories, including general affairs, software and hardware, and security, in order to mainly support the teaching, administrative, and service [22]. College computer center does not contain any commercial transaction mechanism, such as the ordering of goods and after-sale service. In addition, college computer center services include physical entity, such as hardware setted, support, and maintenance, and virtual aspects, such as remote services, online counseling, download, and install. Thus, despite the above literature has established good foundation of the service-quality measurement scales for business-oriented physical entities or websites, they cannot fully apply to the measurement of college computer centers.

2.3   Libraries Service Quality

In terms of non-profit organizations, the service quality of library is the most relevant research, and highly related to the business of the computer center. Association of Research Libraries (ARL) cooperated with Texas A & M University Library in 1999 develop library service quality scale (LibQUAL $^{+TM}$) based on SERVQUAL. The 22 core survey items fall into three dimensions including affect of service, information control, and library as place. The survey requires users to indicate their minimum service level, desired service level, and perceived service performance for all 22 items. Additional items addressing information literacy outcomes, library use, and general satisfaction are also included with open-ended comments box. In 2008, the ARL/Texas A&M research and development team tested a shorter form called Lib-QUAL$^{+®}$ Lite, which used item sampling methods to gather data on all 22 Lib-QUAL$^{+®}$ core items, but required individual users to respond to only a subset of the 22 core questions [23].

Roy et al. [24] used data collected via LibQUAL^{+TM} scale from academic library users, and established that Affect of Service (AS), a dimension of library service closely related to the empathy dimension in SERVQUAL, has the greatest effect on overall satisfaction with the library. Fagan [25] used LibQUAL^{+TM} responses to test three-different models of user perceptions of library service quality and the global fit indices showed that both two- and three-factor models were empirically supported, while the three-factor model had better theoretical support. Furthermore, Kim [26] identifies distinct university library website resources (ULWR) usage patterns and preferred sources of information across user groups, and responses to library usage factors varied across user groups. These variances may be derived from users' distinct academic tasks. One of the unanticipated findings was the users' general perceptions on the complexity of the website design.

Even though the service quality of a computer center seems to be a non-profit oriented research, the unique characters of computer center need to be considered. Modifying SERVQUAL scale or ES-QUAL scale only changed the meaning and connotation, and will still be constrained by the framework of SERVQUAL scale. Furthermore, certain computer-center service items may not be adequately covered. To our knowledge, measurement scale of service quality for college computer centers has not been researched in related literature. Due to the importance of computer center to the operational efficiency of any modern college nowadays, its service quality needs to be independently developed like the library service quality measurement did.

3   Research Design

3.1   Item Generation

This study summarized the literature, and initially accessed the factors that affect the computer-center service quality. After sorting out related literature on the dimensions of SERVQUAL, E-SQ and LibQUAL^{+TM}, the current study generated a total of 62 items in an item list. To make these items more accurate and meet actual situations, feedback from experts and users was sought in two stages.

In the first stage, one-on-one interviews with five administrators and ten staffs of computer center were recorded. The administrators and staff were asked to give advice and amend the questionnaire items. Based on the results of the interviews, similar items were combined, several relevant items were added, and several irrelevant items were deleted. The final total number of items was 46.

In the second stage, a pre-test of the questionnaire on 10 university teachers, 10 staffs, and 30 university students was conducted. The fitting results indicated that these items were not a good fit for students, but more suitable to teacher and staff. The respondents were then asked to compare the importance of the questionnaire items and then made amendments and suggested items to be deleted. In the end, 33 items remained. The initially retained scale items and their references are shown in Table 1.

Table 1. Initial questionnaire items on computer center service quality.

Questionnaire Items	Sources Scales	Dimensions
Q01. Space arrangement for computer center	SERVQUAL	Tangibles, Li-
Q02. Space arrangement for computer laboratory	LibQUAL^{+TM}	brary as Place
Q03. Quietness in computer laboratory		
Q04. Comfortable seating in computer laboratory		
Q05. Easy operation of broadcasting system, in computer laboratory		
Q06. Clearness of overhead projector in computer laboratory		
Q07. Clearness of computer screen display in computer laboratory		
Q08. Temperature controlling system in computer laboratory		
Q09. Lightness in computer laboratory		

Table 1. (*continued*)

Q10. The staff has professional knowledge on computer equipment when problems occur	SERVQUAL E-SQ LibQUAL^{+TM}	Reliability Affect of Service
Q11. Fast Internet connection provided by computer center		
Q12. The staff will inform users estimated available time when computer equipment problems occur	SERVQUAL E-SQ LibQUAL^{+TM}	Responsiveness Affect of Service
Q13. The staff will be glad to help users when computer equipment problems occur		
Q14. The staff will quickly respond to user needs when computer equipment problems occur		
Q15. The staff is with courtesy	SERVQUAL LibQUAL^{+TM}	Assurance Affect of Service
Q16. The staff can be trusted.		
Q17. The service attitude of the staff when computer equipment problems occur	SERVQUAL E-SQ LibQUAL+ ^{TM}	Empathy Compensation Contact Affect of Service
Q18. Computer classroom borrowing process		
Q19. Content and design of school home page		
Q20. Older equipment donation and recycling.		
Q21 Staff efficiency when computer-related equipment problem occur	E-SQ LibQUAL^{+TM}	Efficiency Affect of Service
Q22 Efficiency of staff consulting service		
Q23. Speed and stability of Internet connection when used on campus	E-SQ LibQUAL^{+TM}	Fulfillment Information Control
Q24. Speed and stability of the Internet connection when used off campus		
Q25. Speed and stability of the campus wireless network		
Q26. Easy use of campus information systems		
Q27. Stability of campus information systems		
Q28. Functional integrity of campus information systems		
Q29. Authorized teaching software installation in computer laboratory	E-SQ LibQUAL^{+TM}	Privacy Information Control
Q30. Authorized anti-virus software installation in campus		
Q31. E-mail use and management		
Q32. Door access control of computer center		
Q33. Door access control of computer laboratory		

3.2   Data Collection

As for the collection of empirical data, this study took online survey as a research tool. The questionnaire is divided into two parts. The first part was on the respondents' background, including age, sex, identity (teacher, staff or student) and university location. The second part was on the 33 service-quality items of computer center, which followed the argument of Cronin and Taylor [27] that the service effects subjectively perceived by users represent the computer center's service quality. The responses were measured using a five-point Likert scale, with 5 representing strong agreement and 1 representing strong disagreement.

The questionnaire was setup on a Web server with a fixed IP address. To make the sampling as random as possible, we randomly drew the sample list from universities Websites in Taiwan, invited volunteers to fill up the questionnaire by using telephone or email. Those who filled out the questionnaire were given a random drawing to receive one of several gifts to increase the questionnaire response rate. Questionnaire data was saved in the database server setup for future analysis.

There were totally 312 questionnaires recorded in the database, after deducting 55 invalid or nonconforming ones, 257 valid records were retained for the following analyses, as summarized in Table 2. In the data set, the male to female ratio is 54.6% to 45.4%, and the 18-25 age group accounted for 51.6%, followed by the 26-35 age group, which accounted for 20.9%. Students were the largest group, which accounted for 55.4%, followed by staffs and teachers with 25.5% and 19.1%, Universities located in the north accounted for 41.7% of the respondents while in the Central and south are 22.3% and 31.7%.

Table 2. Profiles of the respondents.

Item	Frequency (%)	Item	Frequency (%)
Gender		Occupation	
Male	140(54.6)%	Students	142(55.4%)
Female	117(45.4)%	Staffs	66(25.5%)
Age		Teachers	49(19.1%)
18–25 years old	133(51.6%)	University location	
26–35 years old	54(20.9%)	North Taiwan	107(41.7%)
36–45 years old	42(16.4%)	Central Taiwan	57(22.3%)
Over 46 years old	28(11.1%)	South Taiwan	82(31.7%)
		East Taiwan	11(4.3%)

Before implementing the exploratory factor analysis (EFA) in accordance, the values of the Bartlett test of sphericity (5732.57) and the corresponding P-value (0.000) from the sample data were calculated. Since both values reached significance, the data set was suitable for performing EFA. Additionally, the Kaiser-Meyer-Olkin value was calculated as 0.867, very close to 1, indicating that the data sampling was appropriate. To decrease the number of items effectively, a reliability analysis was performed on the sorted 33 items according to the basic concepts previously discussed. Items Q17 and Q20, with Cronbach's α at the construct level lower than 0.7, were then deleted, leading to a total of 31 highly reliable questionnaire items.

4   Data Analysis and Results

The factor of eigenvalue greater than 1 was extracted by principal component analysis with varimax rotation, each involving the elimination of items with low loadings (below 0.5) on all factors or high cross-loadings on two or more factors [14, 28]. The five dimensions resulting from the EFA and comprising 28 items explained 75.12% of the variability. Cronbach's α for each dimension was between 0.832 and 0.935 are shown in Table 3, which shows that the items within each dimension have high consistency. The final model confirmed all model paths, thus indicating a good degree of fit between the model and the data, the questionnaire items in Table 3 are our final validated measurement scale. The 28-item, five-factor confirmatory factor model for CC-SQ resulted in a significant chi-square value (p < 0.001), with RMSEA, GFI, AGFI, NFI, and CFI values of 0.034, 0.935, 0.901, 0.960, and 0.99, indicating CFA results revealed excellent overall fit.

In Table 4, we can find that no pair of measures had correlations exceeding the criterion of 0.9 as suggested by Hair et al. [28], which implies that no multicollinearity existed among these constructs. We also employed the opinion of Kline [29], discriminant validity can be established when an inter-factor correlation is below .85. Fornell and Larcker [30] suggested a more robust way of measuring discriminant validity in which a correlation between two constructs should be lower than the squared root of AVE value for any one of the two constructs. According to these suggestions, all the constructs have discriminant validity.

The results of the structural model are shown in the Fig 1. The Five dimensions are all significant (p<0.001) with a higher order factor of CC-SQ, which positively influence users' satisfaction.

Table 3. EFA and CFA results.

| Item | CFA | | EFA | | | | |
	Factor loading	T value	Tangibles	Responsiveness	Contact	Access	Security
Tangibles (α=0. 933)							
Q02/TA01	0.830	16.268	0.673				
Q03/TA02	0.819	16.089	0.724				
Q04/TA03	0.792	15.109	0.681				
Q05/TA04	0.835	16.495	0.731				
Q06/TA05	0.764	14.673	0.742				
Q08/TA06	0.865	17.251	0.697				
Q09/TA07	0.837	16.002	0.673				
Responsiveness (α=0.832)							
Q10/RE01	0.868	16.069		0.752			
Q12/RE02	0.764	14.713		0.711			
Q13/RE03	0.729	14.776		0.733			
Q14/RE04	0.802	17.324		0.817			
Q19/RE05	0.812	15.426		0.823			
Q21/RE06	0.852	16.016		0.743			

Table 3. (*continued*)

Contact (α=0.911)				
Q15/CO01	0.817	16.341	0.765	
Q16/CO02	0.814	17.647	0.821	
Q18/CO03	0.831	17.637	0.772	
Q22/CO04	0.782	14.300	0.717	
Access (α=0.935)				
Q11/AC01	0.835	16.279	0.681	
Q23/AC02	0.824	16.121	0.727	
Q24/AC03	0.800	15.214	0.692	
Q25/AC04	0.841	16.505	0.739	
Q26/AC05	0.799	14.672	0.738	
Q27/AC06	0.871	17.314	0.692	
Q28/AC07	0.845	16.125	0.663	
Security (α=0.902)				
Q30/SC01	0.847	16.546		0.676
Q31/SC02	0.879	17.531		0.735
Q32/SC03	0.743	13.425		0.829
Q33/SC04	0.773	14.629		0.799

Goodness of fit			
x^2 =194.301	GFI=0.935	NFI=0.960	CFI=0.99
DF=151	AGFI=0.901	TLI(NNFI)=0.982	RMSEA=0.034

Table 4. Discriminant validity of all dimensions.

Latent Variable Correlations	Tangibles	Responsiveness	Contact	Access	Security
Tangibles	*0.514*				
Responsiveness	0.251	*0.460*			
Contact	0.510	0.555	*0.565*		
Access	0.269	0.147	0.075	*0.374*	
Security	0.721	0.389	0.614	0.254	*0.558*

* The squared root of average variance extracted (AVE) is depicted in bold italic type on the diagonal.

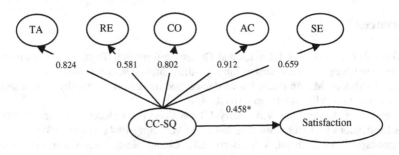

Fig. 1. Result of Structural Model

5   Conclusions and Limitations

In the current study, CC-SQ measurement scale is identified as an important research issue in the Introduction and Related Concepts sections, where a gap between the practical needs and support from academic research is likewise presented. To address the issues on the comprehensiveness and lack of empirical validation, this study constructed a multi-stage scale development procedure by integrating both qualitative and quantitative research methods to examine and identify the factors for evaluating the service quality of college computer center.

The major research findings can be derived from two perspectives. First, among the five dimensions, access and tangibles are the most critical in CC-SQ. The meaning of access dimension is close to the fulfillment in E-SQ and information control in LibQUAL^{+TM}, which implies the strong needs by users on Internet access and information Systems. The meaning of tangibles is close to the tangibles in SERVQUAL and the library as place in LibQUAL^{+TM}, which reflect majority of student users have expectation on adequate equipment and environment. Responsiveness is important but is of the least concern among the five dimensions. Second, these five dimensions can be deemed as a higher order factor CC-SQ that positively affects the user's satisfaction.

Despite all the efforts spent on this research, it still has some limitations, which future researchers should be aware of before applying the above research results. First, this study did not make further efforts to identify the college users. Thus, there is no way to distinguish the perceived differences between users with different characteristics. Therefore, future researchers should make insightful cross-comparison analyses on users with different characteristics to identify the typology of users and to probe into the implications hidden behind the typology of users. Second, as mentioned in the context of our first data collection, without accessing the internal databases of colleges, perfect random sampling was not possible. Future research should try to collaborate with college s to achieve a more appropriate random sample from colleges' databases.

Acknowledgments. The authors are grateful to National Science Council of the Republic of China for financially supporting this research under NSC 102-2410-H-214-017-MY2.

References

1. Weber, D.C.: University Libraries and Campus Information Technology Organizations: Who Is in Charge Here? Journal of Library Administration **9**(4), 6 (1988)
2. He, W., Abdous, M.: An online knowledge-centred framework for faculty support and service innovation. VINE **43**(1), 96–110 (2013)
3. Parasuraman, A., Zeithaml, V.A., Berry, L.L.: A Conceptual Model of Service Quality and Its Implications for Future Research. Journal of Marketing **49**(4), 41–50 (1985)
4. Parasuraman, A., Zeithaml, V.A., Berry, L.L.: Communication and Control Processes in the Delivery of Service Quality. Journal of Marketing **52**(2), 35–48 (1988)

5. Parasuraman, A., Zeithaml, V.A., Berry, L.L.: SERVQUAL A Multiple-Item Scale for Measuring Consumer Perceptions of Service Quality. Journal of Retailing **64**(1), 35–48 (1988)
6. Parasuraman, A., Zeithaml, V.A., Berry, L.L.: Refinement and Reassessment of the SERVQUAL Scale. Journal of Retailing **67**(4), 420–450 (1991)
7. Parasuraman, A., Zeithaml, V.A., Berry, L.L.: Reassessment of expectations as a comparison standard in measuring service quality: implications for further research. Journal of Marketing **58**(January), 111–124 (1994)
8. Cronin Jr., J.J., Taylor, S.A.: Measuring Service Quality: A Reexamination and Extension. Journal of Marketing **56**, 55–66 (1992)
9. Parasuraman, A., Zeithaml, V.A., Berry, L.L.: Alternative scales for measuring service quality: A comparative assessment based on psychometric and diagnostic criteria. Journal of Retailing **70**(3), 201–230 (1994)
10. Agarwal, R., Venkatesh, V.: Assessing a firm's web presence: a heuristic evaluation procedure for the measurement of usability. Information Systems Research **13**(2), 168–186 (2002)
11. Jayasuriya, R.: Measuring Service Quality in IT Services: Using Service encounters to Elicit Quality Dimensions. Journal of Professional Services Marketing **10**(1), 11–23 (1998)
12. Palmer, J.W.: Web site usability, design, and performance metrics. Information Systems Research **13**(2), 151–167 (2002)
13. Wolfinbarger, M., Gilly, M.C.: eTailQ: Dimensionalizing, measuring and predicting etail quality. Journal of Retailing **79**(3), 183–198 (2003)
14. Parasuraman, A., Zeithaml, V.A., Malhotra, A.: A Multiple-Item Scale for Assessing Electronic Service Quality. Journal of Service Research **7**(3), 213–233 (2005)
15. Yeo, R.K., Li, J.: Beyond SERVQUAL: The competitive forces of higher education in Singapore. Total Quality Management & Business Excellence **25**(2), 95–123 (2014)
16. Loiacono, E.T.: Webqual: A Web Site Quality Instrument, Ph.D. thesis, University of Georgia (2000)
17. Zeithaml, V.A., Parasuraman, A., Malhotra, A.: Service Quality Delivery Through Web Sites: A Critical Review of Extant Knowledge. Academy of Marketing Science **30**(4), 362–375 (2002)
18. Yoo, B., Donthu, N.: Developing a scale to measure the perceived quality of an Internet shopping site (SITEQUAL). Quarterly Journal of Electronic Commerce **2**(1), 31–46 (2001)
19. Ladhari, R.: Developing e-service quality scales: A literature review. Journal of Retailing and Consumer Services **17**, 464–477 (2010)
20. Huang, E.Y., Lin, S.W., Fan, Y.C.: M-S-QUAL: Mobile service quality measurement. Electronic Commerce Research and Applications **14**, 126–142 (2015)
21. Yee, R.W.Y., Yeung, A.C.L., Cheng, T.C.E.: An empirical study of employee loyalty, service quality and firm performance in the service industry. Int. J. Production Economics **124**, 109–120 (2010)
22. Woodsworth, A., Williams, J.F.: Computer Centers and Libraries: Working Toward Parterships. Library Administration and Management **2**(2), 88 (1988)
23. Association of Research Libraries (2015). http://libqual.org/about/about_lq/general_faq
24. Roy, A., Khare, A., Liu, B.S.C., Hawkes, L.M., Swiatek-Kelley, J.: An Investigation of Affect of Service Using a LibQUAL+™ Survey and an Experimental Study. The Journal of Academic Librarianship **38**(3), 153–160 (2012)
25. Fagan, J.C.: The dimensions of library service quality: A confirmatory factor analysis of the LibQUAL+ instrument. Library & Information Science Research **36**, 36–48 (2014)

26. Kim, Y.M.: Users' perceptions of university library websites: A unifying view. Library & Information Science Research **33**, 63–72 (2011)
27. Cronin, J., Taylor, S.A.: Measuring service quality: A reexamination and extension. Journal of Marketing **56**(3), 55–68 (1992)
28. Hair, J., Anderson, J.R., Tatham, R., Black, W.: Multivariate data analysis. Prentice-Hall, New Jersey (2006)
29. Kline, R.B.: Principles and Practice of Structural Equation Modeling, 2nd edn. Guilford, NY (2005)
30. Fornell, C., Larcker, D.: Evaluating structural equation models with unobservable variables and measurement error. Journal of Marketing Research **18**(1), 39–50 (1981)

SNS Opinion-Based Recommendation for eTourism: A Taipei Restaurant Example

August F.Y. Chao[1] and Cheng-Yu Lai[2(✉)]

[1] Department of Management Information System,
National Chengchi University, Taipei, Taiwan
aug.chao@gmail.com
[2] Department of Business Administration,
Chung Yuan Christian University, Taoyuan, Taiwan
cylai@cycu.edu.tw

Abstract. By the use of Internet technology in the travel and tourism industry, tourists are considered to play more significant role in the process of planning and designing tourism-related products and services. The amount of information that can acquire from Internet may far exceed one can handle, and makes the decision considerations in the travel planning process fairly complicated. Yu [3] proposed an integrated functional framework and design process for providing web-based personalized and community decision support services, and argue to extract user experiences by using case-based reasoning. However, to construct patterns from case-based reasoning among gigantic amount of user-generated content is a heavy-loading task. In this study, we adopted latent semantic analysis (LSA) [8, 9], which is constructed language pattern and discover semantic relationship between topics in big data scenario, to recommend restaurant according to desiring for similar experience. Both academic and practical implications of proposed approach are also discussed.

Keywords: Big data · Etourism · Latent semantic analysis · Recommendation · Social networking site · User-generated content

1 Introduction

The use of Internet technology in the travel and tourism industry has led eTourism to become a leading market in the B2C arena [1]. As eTourism applications evolved and enhanced, in addition to acquiring tourism information, travelers would demand more personalized services on planning leisure or business trips, building travel packages, and ordering other tourism services. In other words, users are expected to play a more active role in the process of designing their own tourism-related products and services [1, 2].

In order to leverage existing opinions and experiences on the Internet, Yu (2005) [3] proposed a consumer-oriented intelligent decision support system (CIDSS), which aimed to facilitate the event-based tourism related personalized and community decision services. However, with the existing easy-access Web 2.0 technologies, online user-generated content, contains personal experiences and opinions, become crucial

© Springer-Verlag Berlin Heidelberg 2015
L. Wang et al. (Eds): MISNC 2015, CCIS 540, pp. 393–403, 2015.
DOI: 10.1007/978-3-662-48319-0_32

references for travelers to choose accommodations [4] and services, as well as assist vacation planning before journey take place [5]. Online user-generated content, different from notes enlisted in public opinion sites, are distributed over Web 2.0 sites, like blogs or social network sites (SNSs), searched according to designated keywords, and presented in variety formats [7]. It is very important in tourism industry that user-generated content, which contains valuable customer feedback, can generate digital interpersonal word-of-mouth communication, but to read these searched results from Internet in a limited time are difficult. And most importantly, in CIDSS proposed by Yu's [3], the case-based reasoning methods were used to capture user's experiences to facilitate decision making. However, it would be insufficient by gigantic mount of user opinions from Internet search due to lake of existing reasoning model for user tourism behaviors in practices.

The purpose of this study is to combined latent semantic analysis approach (LSA) [8, 9] with Yu's [3] CIDSS framework. We adopt LSA method and take restaurant recommendation as example to show how this framework works in big data era. By building language model of user-generated contents based on what dining experiences people desire to have or think of, this study proposed a method that can calculate the similarity of collected opinions to each restaurant, instead of using case-based reasoning, and recommend proper restaurant according to user location and similarity results. The collected opinions were gathered from a SNS, and were shared by certain user-to-user online relationship, like tweeter following or physical acquaintance. There are two differences between case-based reasoning and proposed LSA similarity: (1) We use language structure model of opinions instead of building case bank (or called mediator architecture [10]), because what worth to be told and shared are all presenting in user-generated contents; (2) Language structure model, in our method, was built from opinions according to words that author used and organized, as well as required require less human intervention. Therefore, our method can be use in different scenarios, as long as sufficient quantity of domain opinions.

The rest of this paper is structured as follows. In section 2, we discuss some theoretical foundations and review the literatures related to this research. Then, the functional framework and a prototype system are presented in section 3. Finally, conclusions and suggestions for further research are made in section 4.

2   Literature Review

In this section, we will first discuss Yu's [3] CIDSS framework and case-based reasoning recommendation that adopted in current framework, following by reviewing the latent semantic analysis in big data scenario and related studies.

2.1   Yu's CIDSS Framework and Case-Based Reasoning

Yu (2005) presents CIDSS that aim on leveraging community collaborative knowledge to facilitate personal need in tourist planning, including several services [3]: Personalized Data and Model Management, Information Search and Navigation,

Product/Vendor Evaluation and Recommendation, Do-It-Yourself Travel Planning and Design, Community and Collaboration Management, Auction and Negotiation, Trip Tracking and Quality Control. CIDSS is a comprehensive services architecture framework for eTourism, including managing personal preferences, criteria of recommendation, buying process, and feedback. Despite of after-buying procedures, recommendation method is the key to this framework success.

In order to adopting community tourism knowledge and compile to recommendation criteria, Ricci *et al.* [10] present a case-based reasoning approach for a web-based intelligent travel recommender system to support users in travel-related information filtering and product bundling. After construct structural user experience case models, semi-structure queries can be performed in system. For example:

Select all TA where Activity="lodging" and Service.type = "hotel" and hotel.cost < 60 and Location="Rome".

Or,

Select all TA where Activity="canoeing" and Location.type="high-mountain or deep-canyon".

Fig. 1. Case-based reasoning query samples

In Fig. 1, queries for case-based reasoning were like SQL structure and can use range descriptions for querying suitable candidates, as well as every criterion, like *Activity*, *(Service|Location).type*, *hotel.cost* and *Location*, was considered as an important attribute and had to be identified before compile into user experience model. In the end, the collection of user experience models was user knowledge of selecting tourist products or services.

The major problems of using case-based reasoning in tourism recommendation process were hard to retrieve appropriate case from tourist decision process and how to adapt compiled model to fit new situations [11]. Accordingly, the first issue addressed in this study was regarding to the difficulty of gathering user decision process behavior while choosing a specific tourism product or service, (the knowledge from tourism experts). Besides, it was also hard to incorporate forthcoming tourism stores or services based on the limited compiled *case bank*. Finally, the key mechanism in CIDSS, recommendation services, fails apart due to lacks of efficient candidate outcomes, and leads it hard to implement in eTourism business.

2.2   Big Data for Tourism and Latent Semantic Analysis

Tourism is an industry rooted in the promotional power of communication [5]. Emerging Internet technologies not only deliver an interpretive viewpoint of travelers' feelings toward their journey, but also create positive and negative word-of-mouth that can influence the loyalty, product/service evaluations, and purchase decisions of future consumers [12]. Because of the low costs of maintaining SNSs, the sharing mechanisms available on SNSs and accurate search engines, this type of word-of-mouth spreads faster than ever on a global scale [13, 14]. Currently, tourism user-generated contents was scattered in different Internet channel and possessing big data characters [15], and large quantity of travelogues which contained traveler's experienced allow us to extract more information from it. Consequently, a content analysis approach was encouraged by Stepchenkova *et al.* [16]. In Stepchenkova's *et al.* [16] study, two software tools were constructed to analysis textual data, identify variables of interest, counting occurrences of interest variables in texts, retrieve and store statistical results for tourism usage. In addition, Pang *et al.* [17] proposed a summarizing mechanism for summarizing tourist destinations with both textual and virtual descriptions. Although, previous studies suggest that should look into what textual user-generated content contains to facilitate user needs in practice, but lack of neither providing proper tourism recommendation model nor taking personal preference into account. As results, it was hard to apply those studies to facilitate Yu's CIDSS framework.

Considering textual opinions were a representing format of user experience, and both choosing words and organizing sentences in texts explain authors' thoughts and feelings toward tourism products and services. It is reasonable to align what user need with tourism opinions and semantically recommend similar experience according community reviews, so we adopts LSA [8, 9] to compare what user need and existing opinions in SNSs. LSA represents the words used in opinion and any set of these words as points in very high-dimensional "semantic space", and that is applicable to text corpora approaching the volume of relevant language experienced by people [8]. Several studies had conducted research on building recommendation system by using LSA [18, 19]. Choi *et. al.* [18] extracted relevant knowledge according semantic linkage in opinions and establish tourism ontology as recommendation rules. On the other hand, Cantador and Castells [19] proposed a automatically identify Communities of Interest from the tastes and preferences expressed by users in personal ontology-based profiles, and semantic content-based collaborative recommendations can be proceed according to the model group profiles. Although, current studies show LSA can be used in recommendation system; but lack of proper overview for tourism industry. However, the ability for extracting crucial information from massive textual reviews make it suitable to adopted into Yu's CIDSS framework.

3   LSA Recommendation Prototype

This study put attention on proposing tourism recommendation system as a substitution of case-based reasoning recommendation in Yu's [3] CIDSS framework. In order to aligning tourist buying behaviors and what user want, we adopt LSA over collected textual opinions and leveraging the knowledge from SNSs to suggesting a restaurant that can make user have similar experience. A prototype web service had been built in this study, and construction detail explains in the next part including collecting opinions, data preparation, building similarity model, and prototype description.

3.1   Collecting Opinions and Data Preparation

In order to build our prototype and prove our concept, we take the liberty of native language, Chinese, and satisfied domestic needs of choosing Taipei restaurant as data collection subjects. And in order to using social network knowledge, we careful choose media channel and selected a specific SNS as opinion collecting source which is a heading media channel of restaurant reviews [20] and members can read what friends shared and share opinions to friends or public in its services.

In order to conduct LSA on collected opinions, a sufficient amount of opinions is required in this study, as well as sufficient number of reviews to a restaurant to possess enough language features. We considered higher rank (rank value larger than 40 over 60) opinions are positive, and collected all opinions that is assigned under same restaurants having more than 3 positive opinions. At the end, we selected 5,420 Chinese opinions among 318 restaurants in Taipei city. As reminder, those restaurants do have negative opinions, but considering recommending what people need instead of what people avoid, we use only positive opinions.

After we collected all textual opinions, all texts were segmented using SINICA CKIP for part-of-speech tagging[1], and tagged data provided information regarding parts-of-speech at the proper level of information granularity for continuing analysis. After tagging, it is possible to process chunked texts for negation words. Negation words are the most important class of semantic meaning shifters and influencing the following words before seeing sentence break [21]. For the purposes of this research, the following were considered negation words: '不', '沒有', '不要', '不能', '沒', '無', '不會', '但是', '但'. We take general tourism opinions about pricing, "這件東西一點都不貴" (this is not expansive at all), as a part-of-speech and negation example. We can retrieved processed sentence from CKIP as "這(Nep) 件(Nf) 東西(Na) 不(D) 貴(VH) 。" The parenthesis marked notations are the part-of-speech role of original sentence: initial-N are nouns, initial-V for verb, and D for adverse. After applying negation to this chunked sentence, it influences following words "貴" and will be marked as "貴-" for continuing LSA analysis.

In general, Chinese words tagged as nouns, verbs and adjectives possess more semantic meaning than words tagged as adverbs and others. Therefore, we isolated words that contain more semantic meaning in opinions and filter out those are irrelevant for interpretation tourism experiences. In order to gaining more matching words in LSA procedure, we have to extend word forms by Chinese synonym list. The Chinese Synonym Forest, Tongyici Chilin (同義詞詞林) [22]. It is a collection of Chinese Synonyms for 70 thousand morphemes, terms, phrase, idioms, and archaisms [23], as well as this word list is very good semantic extension refereeing list. After negation, semantic filtering and extension, we manage to process all 5,420 opinions into 65,389 tokens (unique words) for building LSA model.

---

[1] A Part-Of-Speech Tagger is software that reads text and designates each word as a part of speech (and other token), such as noun, verb, adjective. The Part-of-speech tools from SINICA CKIP are available at http://ckipsvr.iis.sinica.edu.tw/.

3.2   Building LSA Model and Prototype Implement

Building LSA model for recommendation requires sufficient amount of natural language processed opinions and well extended language resources as we explained previously. Next step is converting those processed opinions into a bag of word (BOW) matrix that is a sparse 5,420 by 65,389 matrix. Then, we use gensim [24] to calculate the LSA model and connecting to other web application usage. The benefit of using gensim library for building LSA model is it can work under parallel mode for big data usage and update existing LSA model for forthcoming opinions. In general, we compiled and conducted natural language technique on all collected opinions, convert results into a LSA model for querying. The prototype LSA model is configured using 200 topic numbers, and topic samples of LSA model shown as following:

Table 1. Topic Samples of Trained LSA model and Given Semantic Meaning

Topic	Occurring words and weightings	Given Meaning
Topic 1:	0.035*"蛋糕"+ 0.022*"麵包" + 0.022*"鬆餅"+ 0.021*"烤 +...	bakery
Topic 2:	-0.080*"辣味" + -0.080*"麻辣"+ -0.080*"辛" + -0.062*"毒辣"...	spicy
Topic 3:	0.054*"毆鬥" + 0.054*"毆打" + 0.054*"拳打腳踢" + 0.054*"毆"+ ...	fighting
Topic 4:	0.061*"直拉" + 0.061*"拉拉" + 0.061*"抻" + 0.061*"拉絴"+ ...	stretch
...	...	...

From Table 1, we can see LSA model, under topic number equal to 200, is consistent with different semantic topics that can be described by occurring words and different weightings, and we can label each topic according to occurring words in general. However, it is hard to explain topic in detail, because the co-occurring behaviors of each words were hard to understand. The practical usage of this model is aligning query opinions with this LSA model, so we can retrieve the similarity results of query opinions each collected opinions. To calculate the similarity between query opinion and each collected opinions, we preformed the same process of natural language process over query opinions, (including CKIP part-of-speech tagging, negation, tag filtering), convert to BOW format as a vector, project this vector into LSA model, and finally conduct a cosine similarity among projected query opinion vector and projected collected opinions.

Using this similarity results to performing recommendation, we calculate average opinion similarity results of each restaurant as general desired experience for each query, and also we design an algorithm for incorporating similarity results and user current location, pseudo procedure illustrated as following:

For *restaurants* in Taipei then:
  desired_similarity_list = new Array();
  For *opinions* mention restaurant:
    opinion_vector = *LSA_vector_projection(opinion)*
    desired_ similarity = *cosine(query_opinion_vector, opinion_vector)*
    desired_ similarity_list.push(*desired_ similarity*)
  desired_expirence = average(desired_similarity_list) * distance_weight

Fig. 2. Algorithm for Proposed Recommendation Model

In Fig. 2, we use average similarity as general desired experience according what user input. Instead of using specific criteria in predefined model, like case-based reasoning, we consider what people input is what they desired experience. For example while user query simple string "咖啡"(coffee), it means user want to find some restaurant serving coffee; as well as querying "跟男朋友約會的餐廳" means user want to find a restaurant that can date with his/her boyfriends, and the gender criteria have to be taken into account. In our proposed model, every word in sentence, expecting filtered part-of-speech tagged, would be considered as a semantic meaning to explain what experience user want while choosing restaurant, and those words are expended by synonym list to gain maximum matching. Therefore, the "開心"(happy) in querying string"可以聊得很開心的餐廳" can also be matching "歡樂"、"愉快" and other words enlisted in the same synonym group meaning happy.

To summarize all proposed recommendation approach, we illustrated all processes including data preparation, cleaning, extension, latent semantic modeling and interface for recommendation requesting as following:

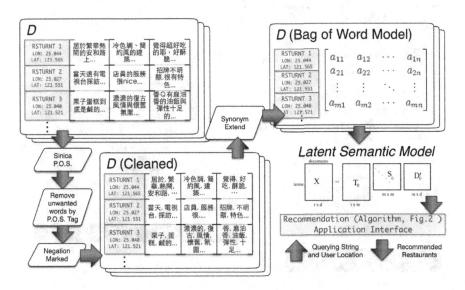

Fig. 3. Overview of Proposed Recommendation

In Fig. 3, we described all dependent processes. To prove our concept, we built a website at http://boo.fychao.info, which is a simple web page embedded Google Maps and can show recommendation results in order after user querying. To facilitate this service, we also constructed a web application interface (API) for serving recommendation results, and source code can be retrieved at https://github.com/fychao/LBS-SIM-TPE-RSTRNT-RCMND. The API service was run by simple HTTP service and return recommendation list with restaurant attributes after calling with user querying string and user location. By default, user will be set at Taipei City Hall (latitude: 25.041171, longitude: 121.565227) if user is not in Taipei region. Considering a mobile user access this service, geo-location can be retrieved by front-end browser built-in function. After user typing querying string, both querying string and geo-location information would be sent to recommendation API and processed. The recommendation API directly load existing LSA model, perform natural language process over querying string, convert into bag-of-words vector, calculate similarity among all restaurant opinions by using Fig. 2 algorithm, and finally return 9 candidates in JSON format. The front-end website was designed to retrieved JSON format from recommendation API, and placed information marker in Google Map frame including recommendation ranking order and restaurant. The front-end interface was in flexible width web design so user can have the same experience while using this service. The overview of prove of concept was showing as following:

In Fig. 4, web page shows results recommendation of querying "便宜聚餐", user can click ordered bottom above to trigger information marker within Google Map. In each information marker, it contains recommendation order, nearby street view from Google Map, address and further search click, which link to search engine.

Fig. 4. Prove-of-Concept Website in Mobile Device (left) and Notebook (right).

3.3   Proposed Recommendation Results

We have explained the recommendation mechanism in previous section, and how it work with front-end user interface. In this section, we use two different querying string as examples: (A) "帶男朋友約會"(date with boyfriend) and (B) "帶女朋友約會" (date with girlfriend). Both querying string contains almost the same words excepting "男"(boy) and "女"(girl), but recommending restaurant for user are supposed to be different. While users query (A), they want to accommodate a male need, therefore user would read restaurant opinions about male preferences; and while querying (B) vice versa. However, it is very difficult to formulate this kind of condition into case-based reasoning event if thinking closely. However, in our proposed service, the recommendation results compared all compiled restaurant opinions and suggesting ones according community's opinions, so every word, excepting lack of semantic meaning, would take into account. We enlisted recommendation results of querying both (A) and (B) at the same default location, showing as following:

In Table 2, recommendation results of querying (A) and (B) have 6 the same results and 3 different ones (asterisk marked). Considering male buying behavior, 平田壽司Togo外帶 (take-out restaurant), 龍緣號(low-cost local food restaurant), and 勵進餐廳-台電員工餐廳(hotpot restaurant) are easily understand quite suitable for males. On the other hand, female prefer to dine at 好-丘(bagel bistro), BIGTOM美國冰淇淋文化館(ice-cream shop), and A380空中廚房(theme restaurant). However this kind of results is very hard to get in conventional search engine, due to lack of considering all social network and community opinions.

Table 2. Querying Samples (A) and (B) from Prototype Website

Querying string	"帶男朋友約會"	"帶女朋友約會"
Recommendation Order (from 1 to 9)	于記杏仁豆腐-通化店	月島文字燒-台北忠孝SOGO店
	Imbiss歐式餐坊	郭家蔥油餅
	ARROW-TREE-亞羅珠麗-阪急店	Imbiss歐式餐坊
	平田壽司Togo外帶*	ARROW-TREE-亞羅珠麗-阪急店
	龍緣號*	好-丘good-cho-s*
	smith-hsu-忠孝店	于記杏仁豆腐-通化店
	月島文字燒-台北忠孝SOGO店	BIGTOM美國冰淇淋文化館-*
	勵進餐廳-台電員工餐廳*	smith-hsu-忠孝店
	郭家蔥油餅	A380空中廚房-信義店*

4   Conclusion and Discussion

Nowadays, it is impossible for travelers to read all relevant tourism information, so a consumer-oriented intelligent decision support system is absolutely required to recommend and management tourism products and services user. However, existing framework is not able to leverage existed knowledge on the Internet, especially in big data scenario due to lack of automatic case-based modeling mechanism. In this study, we proposed a SNS opinion-based recommendation approach as a substitute for

CIDSS framework [3]. The similarity algorithm is rooted in compiled opinions on SNSs and the detail processes in current framework were illustrated in the previous section. In order to testing our concept, we built a web service prototype and monitor the recommendation outcome in different querying situation. We also discussed the recommendation results from proposed approach were semantically similar to opinions from social network and community. Furthermore, the capability of opinion-based recommendation is possible to provide more accurate candidates if we compile more opinions and build more robust language for LSA.

The limitations of this study are narrow scope and limited opinions, and also the fundamental problems are caused by Chinese natural language processing. In addition, the interactions between proposed recommendation approach and Yu's CIDSS were not discussed yet. The implications of this study can be directly introduced as a web service, like prove-of-concept website, as well as be incorporated by other tourism domains, like sightseeing spot suggestion. In future, we will put our focus on studying the interface of this proposed approach and Yu's CIDSS, as well as introducing real big data into model.

Acknowledgments. The authors would thank the Ministry of Science and Technology, Taiwan, for financially supporting this research under contract MOST 103-2410-H-033-045 -.

References

1. Werthner, H., Ricci, F.: E-Commerce and Tourism. Communications of ACM **47**(12), 101–105 (2004)
2. Fodor, O., Werthner, H.: Harmonise: A Step toward an Interoperable E-Tourism Marketplace. International Journal of Electronic Commerce **9**(2), 11–39 (2005)
3. Yu, C.-C.: Personalized and community decision support in eTourism intermediaries. In: Andersen, K.V., Debenham, J., Wagner, R. (eds.) DEXA 2005. LNCS, vol. 3588, pp. 900–909. Springer, Heidelberg (2005)
4. Ye, Q., Law, R., Gu, B.: The Impact of Online User Reviews on Hotel Room Sales. International Journal of Hospitality Management **28**(1), 180–182 (2009–1)
5. Pudliner, B.: Alternative Literature and Tourist Experience: Travel and Tourist Weblogs. Journal of Tourism and Cultural Change **5**(1), 46–59 (2007)
6. Gretzel, U., Yoo, K.: Use and Impact of Online Travel Reviews. In: O'Connor, P., Höpken, W., Gretzel, U. (eds.) Information and Communication Technologies in Tourism 2008, pp. 35–46. Springer, Heidelberg (2008)
7. Xiang, Z., Gretzel, U.: Role of Social Media in Online Travel Information Search. Tourism Management **31**(2), 179–188 (2010)
8. Landauer, T.K., Foltz, P.W., Laham, D.: An Introduction to Latent Semantic Analysis. Discourse Processes **25**(2–3), 259–3284 (1998)
9. Thomas, H.: Collaborative filtering via gaussian probabilistic latent semantic analysis. In: the 26th Annual International ACM SIGIR Conference on Research and Development in Information Retrieval. ACM Press, New York (2003)
10. Ricci, F., Werthner, H.: Case Base Querying for Travel Planning Recommendation. Information Technology & Tourism **4**(3–4), 3–4 (2001)

11. Leake, D.B.: Experience, Introspection and Expertise: Learning to Refine the Case-based Reasoning Process. Journal of Experimental & Theoretical Artificial Intelligence 8(3–4), 319–339 (1996)
12. Pan, B., MacLaurin, T., Crotts, J.C.: Travel Blogs and the Implications for Destination Marketing. Journal of Travel Research 46(1), 35–45 (2007)
13. Dellarocas, C.: The Digitization of Word of Mouth: Promise and Challenges of Online Feedback Mechanisms. Management Science 49(10), 1407–1424 (2003)
14. Litvin, S.W., Goldsmith, R.E., Pan, B.: Electronic Word-of-Mouth in Hospitality and Tourism Management. Tourism Management 29(3), 458–468 (2008)
15. Chen, H., Chiang, R.H., Storey, V.C.: Business Intelligence and Analytics: From Big Data to Big Impact. MIS Quarterly 36(4), 1165–1188 (2012)
16. Stepchenkova, S., Kirilenko, A.P., Morrison, A.M.: Facilitating Content Analysis in Tourism Research. Journal of Travel Research 47(4), 454–469 (2009)
17. Pang, Y., Hao, Q., Yuan, Y., Hu, T., Cai, R., Zhang, L.: Summarizing Tourist Destinations by Mining User-Generated Travelogues and Photos. Computer Vision and Image Understanding 115(3), 352–363 (2011)
18. Choi, C., Cho, M., Choi, J., Hwang, M., Park, J., Kim, P.: Travel ontology for intelligent recommendation system. In: Third Asia International Conference on Modelling & Simulation, AMS 2009, pp. 637–642. IEEE Press, New York (2009)
19. Cantador, I., Castells, P.: Extracting Multilayered Communities of Interest from Semantic User Profiles: Application to Group Modeling and Hybrid Recommendations. Computers in Human Behavior 27(4), 1321–1336 (2011)
20. Cheng, Y.H., Ho, H.Y.: Social Influence's Impact on Reader Perceptions of Online Reviews. Journal of Business Research 68(4), 883–887 (2015)
21. Liu, B.: Sentiment Analysis and Opinion Mining. Synthesis Lectures on Human Language Technologies 5(1), 1–167 (2012)
22. Mei, J.J., Zhu, Y.M., Gao, Y.Q., Yin, H.X.: Tongyici Cilin (Dictionary of Synonymous Words). Shanghai Cishu Publisher, Shanghai (1983)
23. Sun, Y., Chen, C., Liu, C., Liu, C., Soo, V.: Sentiment classification of short chinese sentences. In: The 26th Conference on Computational Linguistics and Speech Processing, Jhongli (2010)
24. Řehůřek, R., Sojka, P.: Software Framework for Topic Modelling with Large Corpora. In: The LREC 2010 Workshop on New Challenges for NLP Frameworks, Valletta, Malta, pp. 45–50 (2010)

Improving Supply Chain Resilience with Employment of IoT

Yu Cui[✉]

Otemon Gakuin University, 2-1-14 Nishiai, Ibaraki-shi, Osaka, Japan
yucui@otemon.ac.jp

Abstract. With the establishment of highly advanced and sophisticated iSC, enterprises are able to improve their competitiveness, and meanwhile, effortlessly overcome the issues which cannot be resolved under the traditional supply chain operation. When we intensely show our concern to the construction of iSC, the fragility of supply chain caused by continuous pursue of high efficiency cannot be neglected. Resilience resembles to the immune system of supply chain, the more we pay attention to it, the more robust and stable the supply chain is. Henceforth, through the synergy effect of the acquired experience during the enhancement of resilience and the accumulation of information and network technology which progresses during the establishment of supply chain, operation of supply chain will turn out to be more simple and secure. Meanwhile, improvement of customer satisfaction degree will also benefit from it. In this paper, we propose a relationship model that is utilized to define the mechanism of improving Supply Chain Resilience with employment of IoT.

Keywords: Supply chain resilience · IoT · Intelligent supply chain

1 Introduction

The origin of supply chain informatization dates back to the 70s of last century. As the first generation of supply chain information platform, EDI emerged and settled the problem of cross-system information exchange and process control. When it comes to the 80s, with the establishment of supply chain concept, information exchange had been attached more and more importance for the integration management from upstream to downstream. In the 90s, after discovering the bullwhip effect developed in supply chain, researchers and practitioners started to realize that enterprises not only have to possess channels for information exchange, but also need the platform for sharing information. After 21st century, through the emergence and improvement of systems such as ERP, WMS and TMS, demand orientated supply chain operation was gradually consented by business operators. In the mean time, supply chain of types such as pull type and push + pull type (Simchi-Levi, et al., 2007) had also been confirmed in succession.

Nowadays, with the expanding of information and development of network technology, informatization and transformation of supply chain are confronted with new challenges. As the advent of new technologies such as mobile communication, big

© Springer-Verlag Berlin Heidelberg 2015
L. Wang et al. (Eds): MISNC 2015, CCIS 540, pp. 404–414, 2015.
DOI: 10.1007/978-3-662-48319-0_33

data, social networking sites, cloud computing brings strikes and impacts to enterprises successively, the internet of things came into being, furnishing enterprises and supply chain executors with new inspirations and solutions. Furthermore, supply chain informatization is propelled to a new step, rendering intelligent supply chain(iSC), which used to be regarded as an unreachable target by supply chain executors be actualized virtually.

The concept of iSC was initially put forward in a report written by IBM in 2009 which enumerates numerous problems confronted by supply chain enterprises from industrial point. The faster business environment transfers, the more diversified and complicated supply chain demands are. When traditional supply chain operation modes are incompetent to settle the challenges faced by enterprises, the establishment of iSC will assist enterprises to apply uncertainty flexibly so as to enhance their competitive and realize the stable and sustainable development.

As a matter of fact, before the advent of iSC, concepts such as Smart Electric Power System, Smart Environmental Protection and Smart Logistics have already come out in succession. In manufacturing industry, especially, being represented by iMake, the concept of Smart Production was proposed back in the 70s of last century. Nevertheless, what was stressed at that time was the utilization of robots and the automation during manufacturing process. Afterwards, along with the emergence of intellectual products including intelligent electric appliances, intelligent medical equipments, and etc, intelligent tools which can make a judgment and choice for human became prevalent, intelligent of that time not only indicated the automation of process, but also the combination of automation and informatization.

When it comes to the Internet era of today, merely adopting automatic collection technology to make a judgment and choice has become far from enough. Apart from that, connecting with network and applying it to transfer collected information to data process and service center (abbreviated as data center) and making real-time adjustment, it is such a combination of dynamic management, control and selecting can be called intelligent of this era. In other words, today's intelligent manufacturing has to embody the characteristic of automation, informatization and networking.

What represents the intelligent management of supply chain (iSCM) includes: 1, whole process of management; 2, centralization of information; 3, dynamics of system.

Whole-process of management is to fulfill the principle of integration and outsource upheld by supply chain. On one hand, information system of downstream vendor needs to be connected; purchase and supply system of upstream needs to be connected. On the other hand, supply chain system inside enterprise has to be unblocked for the whole-process management.

Centralization of information is to realize the new demand of enterprise internal integration. Based on the consideration of efficiency and vitality, data is orientated to adjust the contradiction between resource management and decision making authority of the whole enterprise so as to elevate the level of resource management and optimization.

Dynamic of system is to take identifying information and spatial-temporal information as the information source of supply chain dynamic management in order to raise real-time adjusting ability, which means it is possible to make judgment and adjustment at any time in accordance to external changes and satisfy the requirement from complicated and rapid developed external environment for enterprises. The basis of dynamic of system is locating information, so when enterprises contemplate how to achieve the above characteristics ought to be possessed by iSC, the best choice for them is the integration of ever maturer Internet of Things and cloud computing.

2   Technical Architecture of Internet of Things

Internet of Things (IoT) is the internet connecting things. In detail, it is based on conventional agreements, through information sensing equipment concluding RFID, smartsens, GPS, laser scanning technology to connect things with internet for information exchange and communication, realizing intelligent recognition, location, supervision, tracking and management (Zelbst et al., 2010).

Network Architecture

Network architecture of IoT is constituted by Sensing Layer, Network layer and Application Layer.

Being similar to sense organs of human being, sensing layer is composed by sensors, sensor gateways, digital tags and RFID readers, which enable it to acquire information of objects at any time as the origin of object identification and information collection in Internet of Things.

Network layer is responsible for the delivery of information acquired by Sensing Layer and it is formed by private networks, internet, wired and wireless communication networks and network management system.

Table 1. Key technologies of Internet of Things

Name of technology	Brief Introduction
Coding Technology	Codes for differentiating enterprises, products and series.
RFID Technology	Wireless radio frequency identification technology. During high-speed mobilization, it is possible to read information saved by tags through transmission of materials. Mainly be applied in product follow-up and mobile payment.
Sensor Network	Wireless network formed by tiny nodes according to rules, being able to carry out information collection, judgment and upload.
M2M Technology	Assembly of technologies for enhancing communication and network capacity of machines and equipments.

Application Layer is the interface between IoT and users. After being analyzed and processed through cloud computing, information captured by sensing layer provides users with identification, location, supervision, tracking and management of objects in accordance with demands of different industries, realizing smart utilization of Internet of Things.

In practical application, RFID has been applied in product logistics tracking, product counterfeit proof, and etc; GPS technology has been applied in location, tracking, detection and networking of objects; WSN (Wireless Sensor Network) has been utilized for monitoring logistic equipment and warehouse surroundings (Sánchez et al., 2012). Key technologies of IoT is indicated in Table 1.

3   Technical Architecture of Cloud Computing

Cloud computing is a mode of making on demand visits to configurable and shared computing resource pools which are based on networks. Concluding network, server, storage, application and service, these configurable and shared resource pools can be provided or released through the most economically efficient management or interaction with service providers (Ferguson, et al., 2011).

After the advent of computing concepts such as distributed computing, grid computing, utility computing, P2P computing and market-orientation computing, cloud computing was presented by Google CEO Eric Schmidt based on the fully development and popularization of internet (Schrodl, et al., 2012).

Key technologies of cloud computing are presented in Table 2.

Table 2. Key technologies of Cloud Computing

Name of Technology	Brief Introduction
Virtualization Technology	Principal technical basis for realizing cloud computing. Being capable of blocking transformation caused by diversifications of physical equipments.
Mass data distributed storage technology	Adopt redundance storage technology to guarantee the reliability of stored data.
Mass data management technology	Process and analyze large dataset and provide efficient services to users.

4   Through IoT and Cloud Computing Technology to Realize iSC

The technology of IoT was applied in supply chain management long time ago. What restraints big scale application of the new technology is how to process the huge amount of information collected by Sensing layer of IoT and to actualize intelligent control on objects through analysis on collected information. The advent of cloud computing solves the technical bottleneck of IoT and accelerates massive application of the technology of IoT.

All the links in supply chain are source of data. In detail, products with RFID tags or all the information regarding changes of position, amount and property of parts will be recorded in tags by terminal device, in the meantime, be uploaded to database located at cloud through network. System can monitor all the steps throughout production in accordance to predetermined conditions so as to guarantee the security and order of production.

Enterprises on supply chain can apply the mode of mixed cloud. To be specific, core data that enterprise must keep confidential be processed by private cloud; data and operations that are related to supply chain coordination be processed by public cloud. Based on public cloud, data base which can be shared by all the members on supply chain is built up, and an appropriate link between private cloud and public cloud is establish for the coordination of data flow in supply chain.

Technology of IoT applied in supply chain mainly plays the role of product real-time tracking and product process efficiency acceleration, while cloud computing technology primarily conducts the storage and process of mass information gathered through technology of IoT.

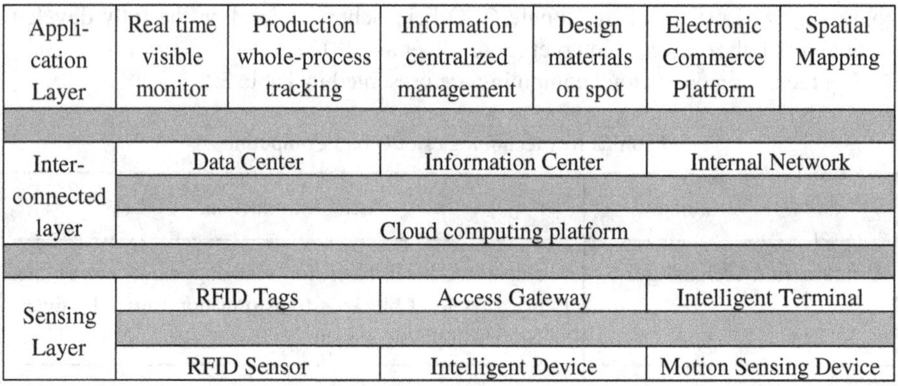

Appli-cation Layer	Real time visible monitor	Production whole-process tracking	Information centralized management	Design materials on spot	Electronic Commerce Platform	Spatial Mapping
Inter-connected layer	Data Center		Information Center		Internal Network	
	Cloud computing platform					
Sensing Layer	RFID Tags		Access Gateway		Intelligent Terminal	
	RFID Sensor		Intelligent Device		Motion Sensing Device	

Fig. 1. Layer architeture of iSC

If we divide iSC throughout practical operation into three layers to construct. These layers can be set up as Sensing layer, Interconnected layer and Application layer.

Sensing Layer
Enterprises can promptly sense all kinds of information that is related to them through sensing network. Sensing network aggregates information from various enterprises, users, procedures, devices and systems. At the same time, sensing network serve for diverse enterprises, enabling the range of information that enterprises sense extends and the depth of information deepens. The scope and accuracy rate of information search ultimately be enhanced to a great extent.

Interconnected Layer

Enterprises can utilize internet, wireless network and Internet of Things to actualize the connection of enterprise internal and external information. By doing so, precise and prompt satisfaction of user demands, optimum use of resources, maximum work efficiency, minimum waste, energy saving and emission reduction can all be achieved successfully.

Application Layer

With the establishment of relationship between knowledge and the relationship between knowledge and people, enterprise knowledge network becomes complete and ordered, and innovation network becomes more intelligent. The more innovation network is applied, the smarter it becomes. Innovation network will automatically propel the information and knowledge needed throughout innovation process, in the meantime, massive repetitive work during innovation process is automatically finish by innovation network.

The establishment of iSC enables enterprises and even the whole supply chain to react immediately to complicated and variable market, strengthening the adaptability of supply chain. Meanwhile, industry sharing of information is boosted by the acceleration of enterprise informatization and networking.

For IoT, for the movement and circulation of objects in internet, the incessant development of data transfer technology, trading market has to be constructed, and software system has to be developed. We need to attach more importance to the network services which presents Internet of Things with communication function. In the meantime, standardized services also need to be equipped, with standard setting organizations such as CEN, ISO and ETS play a crucial role.

Real-time dynamic indicators of supply chain, such as inventory analysis, customer analysis and data of this sort, have to be transferred to server through network after the detection of sensor. Data of huge amount are delivered to centralized cloud computing data center of Internet of Things. From here, data and information from various channels are collected through cloud computing. With the assistance of business analysis software, informational strategies that ignites a great profit are extracted.

The application of Internet of Things can improve the visibility of supply and information transparency of supply chain management. With the fully utilization internet and various technologies, RFID technology can effectively recognize tagged products in batch and acquire accurate product information. Products from raw materials, semi-finished products to finished products are conducted real-time supervision throughout the processes of delivery, storage, distribution, sales and recycling. By doing so, product information can be acquired at any time, automation degree can be raised, and rate of error can be diminished, ultimately, transparency of supply chain managements comes true.

Agility and integration of SCM and compact conformity of supply chain links are also realized through Internet of Things. Supply chain information system based on IoT can coordinate and integrate internal and external production activities of enterprise. Through the execution of automatic production line, production situation is acquired timely. Enterprise can make replenishment information on the basis of manufacturing schedule so as to realize the balance of assembly-line, achieving better flexibility of production. High level agility and integration of SCM incessantly lower the inventory of enterprise and supply chain channel, productivity is also elevated without cease.

5   Supply Chain Resilience

During the construction of iSC, not only should we consider the issues mentioned before, such as information sharing, standardized service, excellent cost-effectiveness, visibility, transparency, agility and integration which have been studied and carried out as primary topics of supply chain, but also should we strive to resolve the fragility problem of supply chain which currently agitates enterprise most.

Advanced management modes such as globalization purchasing pattern, JIT production, and concepts of precise and agile supply chain have been applied and promoted in numerous enterprises. Though efficiency of supply chain operation is enhanced, risk coping capability of supply chain is weakened in the meantime. Upon events that affect regular production occur at any link of supply chain, upstream and downstream enterprises will be influenced immediately, further than that, the impact will spread to all the member enterprise on supply chain.

Ultimately, partial invalidation event occurred in joint enterprises will disrupt ordinary operation of supply chain and even cause the termination of supply chain. Accordingly, operation cost of supply chain will rapidly raise, and customer service level will be lowered. Examples of this sort arise from time to time recently, for instance, the 3.11 Great East Japan Earthquake of 2011 and the severe flood occurred in Thailand in the same year seriously struck enterprises. In the meantime, due to the cascade effect of supply chain, countless big-scale manufacturing enterprises suffered a blow more or less.

Integrating with the managerial theory and technique of network technology and modern supply chain management such as IoT and cloud computing, iSC is a comprehensive and integrated system of technology and management which is constructed in or among enterprises for the intelligence, networking and automation of supply chain. iSC of future not only can effectuate the better efficiency of supply chain, but also the robustness of supply chain operation. The robustness of today's supply chain is not merely embodied in the attainment of crisis intervention of supply chain enterprises, but also is reflected in its self-healing ability, in other words, it is an ability reflecting the degree of completion of supply chain resilience (SCR).

The Research on SCR
The research on SCR started from the beginning of 21st century. The first widespread study on SCR began in the United Kingdom, following transportation disruptions from fuel protests in 2000 and the outbreak of the Foot and Mouth Disease in early 2001 (Pettit et al., 2010). Christopher and Peck (2004) developed an initial framework for a resilient supply chain and asserted that SCR can be established through four key principles: (1) resilience can be built into a system before the occurrence of a disruption (i.e., re-engineering), (2) high level of collaboration and cooperation are necessary to identify and manage risks, (3) agility is essential to react

quickly to unforeseen events, and (4) the culture of risk management is a necessity. Other characteristics such as agility, availability, efficiency, flexibility, redundancy, velocity, and visibility are treated as secondary factors.

During the occurrence of emergencies, core enterprises in supply chain utilize information technology to establish information sharing online trading system for distribution enterprises and downstream customers. The information sharing platform joins downstream customers and upstream suppliers into the supply chain management system and equips downstream customers and upstream suppliers with functions such as online order, new product inquiry, information feedback, process monitor, account settlement and sales promotion. In the mean time, core enterprises can target at downstream customers of different scales and set up appropriate services on the basis of information platform so as to promote the sales and services for customers. Through information platform, upstream suppliers can join resilient supply chain management system. When emergencies happen, the platform can perform arrangement and analysis on inventory and sales data and assist in the realization of automatic remind of order quantity, variety and distribution in resilient management system. In addition, enterprise can realize trans-regional and multi-layer distribution management with the utilization of information technology. The overall integration of operation data of enterprise and the predication of risks of supply chain can be also actualized.

A Concept Framework for Achieving SCR

In the way of taking measures concretely, the supply chain is supposed to equip capabilities which are able to exert the resilience. We would restate these essential capabilities briefly through reviewing some study outcomes so far.

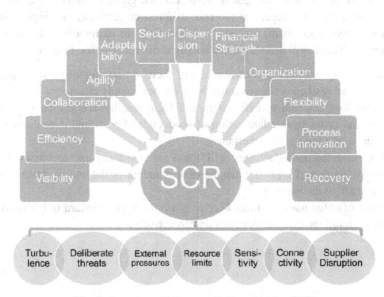

Fig. 2. A conceptual framework for achieving SCR

In the figure 2, there are necessary capabilities for achieving SCR upside, and key points which represent supply chain vulnerabilities are lined up downside. As we mentioned in previous section, supply chain vulnerabilities are disclosed in accordance with internal and external interruptions of supply chain, so that barriers may interrupt operations of the whole supply chain and the recovery from supply chain disruptions. In response, with equipping essential capabilities for achieving SCR, it is possible to remove the supply chain vulnerabilities and promptly recover from emergencies and damages of the whole supply chain. In addition, it can be expected to evolve much superior situation.

6   Improving Supply Chain Resilience with Employment of IoT

To sum up, through the conduction of continuous online self-assessment, resilient supply chain predicts eventualities and occurrent events during supply chain operation and adopts measures to restrain or rectify promptly. To diminish the discontinuity between supply and service of production to the maximum, applying data obtaining technique fully and implementing decision support algorithm are imperative. By doing so, discontinue frequency and duration of product supply will be decreased, and production and supply will be normalized after the occurrence of discontinuity. In essence, resilience is the immune system of smart supply chain and embody the most substantial characteristic of IoT.

Through the relation table of the degree of intelligence and resilience below, we illustrate the mutual effect between them and their influence on supply chain.

The two dimensions on the graph are Resilience grade (x-axis) and Smart grade (y-axis) that we believe represent the core differences between, say, high-tech electronics and consumer chemicals. These differences are extremely instructive when considering iSC and will drive different supply chain resiliencies. In detail:

- Engineering-oriented manufacturers will focus on intelligence and connectivity efforts and resiliency improvement both on their supply chain.
- Technology-oriented manufacturers will focus intelligence and connectivity efforts and promote standardization of products and parts and decentralization in order to provide against emergency.
- Asset-oriented manufacturers will focus on resiliency improvement on their supply chains and keep to grasp the basic standard of the intelligent and connective grade in the industry.
- Brand-oriented manufacturers keep to grasp the basic standard of both and gradually strengthen their weak points on respectively.

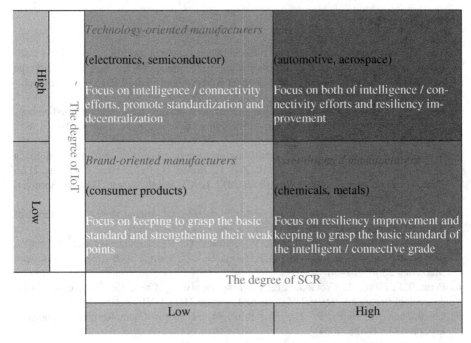

		The degree of SCR	
		Low	High
High	Technology-oriented manufacturers (electronics, semiconductor) Focus on intelligence / connectivity efforts, promote standardization and decentralization	(automotive, aerospace) Focus on both of intelligence / connectivity efforts and resiliency improvement
Low	Brand-oriented manufacturers (consumer products) Focus on keeping to grasp the basic standard and strengthening their weak points	(chemicals, metals) Focus on resiliency improvement and keeping to grasp the basic standard of the intelligent / connective grade

Fig. 3. The blending effects of Integlligence and Resilience within Supply Chain

The purpose of this examination is to suggest to the top management persons of enterprises to "find themselves" on the Figure 3. Do you indeed experience delay of iSC establishment or resiliency improvement (some fortunate businesses do not) and do those delays drive problems that are manifested through your supply chain?

7   Conclusions

With the establishment of highly advanced and sophisticated iSC, enterprises are able to improve their competitiveness, and meanwhile, effortlessly overcome the issues which cannot be resolved under the traditional supply chain operation. When we intensely show our concern to the construction of iSC, the fragility of supply chain caused by continuous pursue of high efficiency cannot be neglected. Resilience resembles to the immune system of supply chain, the more we pay attention to it, the more robust and stable the supply chain is. Henceforth, through the synergy effect of the acquired experience during the enhancement of resilience and the accumulation of information and network technology which progresses during the establishment of supply chain, operation of supply chain will turn out to be more simple and secure. Meanwhile, improvement of customer satisfaction degree will also benefit from it.

References

1. The smarter supply chain of the future: Global chief supply chain officer study, IBM Institute for Business Value (2009). http://www.ibm.com/common/ssi/fcgi-bin/ssialias?infotype=PM&subtype=XB&appname=GBSE_GB_TI_USEN&htmlfid=GBE03163USEN&attachment=GBE03163USEN.PDF
2. Ferguson, D.F., Hadar, E.: Optimizing the IT business supply chain utilizing cloud computing, International Conference & Expo on Emerging Technologies for a smarter World, NY, August 2011
3. Schrödl, H.: Adoption of cloud computing in supply chain management solutions: a SCOR-aligned assessment. In: Wang, H., Zou, L., Huang, G., He, J., Pang, C., Zhang, H.L., Zhao, D., Yi, Z. (eds.) APWeb Workshops 2012. LNCS, vol. 7234, pp. 233–244. Springer, Heidelberg (2012)
4. López, T.S., Ranasinghe, D.C., Harrison, M., Mcfarlane, D.: Adding sense to the Internet of Things. Personal and Ubiquitous Computing 16(3), 291–308 (2012)
5. Zelbst, P.J., Green Jr., K.W., Sower, V.E., Baker, G.: RFID utilization and information sharing: the impact on supply chain performance. The Journal of Business & Industrial Marketing 25(8), 582–589 (2010)
6. Pettit, T.J., Fiksel, J., Croxton, K.L.: Ensuring supply chain resilience: development of a conceptual framework. Journal of Business Logistics 31(1), I–VII (2010)
7. Christopher, M., Helen, P.: The Five Principles of Supply Chain Resilience. Logistics Europe 12(1), 16–21 (2004)

Factors Influencing the Adoption of Using Mobile Banking on a Smartphone: an Empirical Case Study in Bangkok, Thailand

Suthathip Suanmali[✉]

School of Management Technology, Sirindhorn International Institute of Technology, Thammasat University, 131 M. 5, Tiwanon Road, Bangkadi, Pathum-Thani, Thailand
ssuanma@gmail.com

Abstract. Mobile banking is similar to internet banking, where banking application provides all the capabilities of mobile Web banking. The use of such innovative technology has changed the way banking business is conducted; consumers with a smart phone can manage their account anytime and anywhere. Banks typically reduces their operating costs by keeping customers out of the branches. However, there are factors that cause consumers in Thailand to question the acceptance of internet banking on a smart phone. This paper is aimed to explore the factors that influence Thai consumer acceptance of mobile banking on a smart phone. The findings are significant in that confidence, convenience, and security risk. They are key attributes in the adoption of this new trend. This outcome provides useful information for bank management in formulating marketing strategy.

Keywords: Internet banking · Mobile banking · Regression analysis · Technology adoption

1 Introduction

Banks and Financial institutions are focused to gain competitive advantage by using internet technology to provide value added services to meet consumer needs at lower their operating costs. Mobile banking is similar to Internet banking, where consumers accessing their bank accounts through their cell phones or smartphones. The main benefit is that cell phone is more portable and banking apps provide all the capabilities of mobile Web banking. They are usually designed especially for specific bank and phone model or operating systems such as iPhone as well as smart phones with Blackberry and Android operating systems. The idea of anytime-anywhere account access on mobile device has become attractive for consumers worldwide, and widely seen as the most important and most popular delivery channel for banking services. The fast growing internet and smartphone users have conspired most banks and financial institutions to offer a mobile banking.

The number of mobile subscribers in Thailand had exceeded its population since 2010 (106.6 %), reported by National Statistics Office (NSC). The 3G licenses

© Springer-Verlag Berlin Heidelberg 2015
L. Wang et al. (Eds): MISNC 2015, CCIS 540, pp. 415–424, 2015.
DOI: 10.1007/978-3-662-48319-0_34

granted for all the main telecommunication operators; Dtac, AIS, and True Move. The mobile penetration rate in Thailand is growing even faster. According to [1], the 3G subscribers in the first quarter of 2014 are over 55 million. In 2013, Thailand set the world record in migrating 2G subscribers to 3G in only five months. The market shares of 3G subscribers are divided among three major telecommunication firms in Thailand, where AIS occupies 44.1% of the market and the rest of the market is divided between 29.36 % of Dtac and 24.26% of True Move. The report from [2] indicated that time spent on the internet via mobile phones accounted for 49 % of all media use. In addition, the media consumption on average is 7.4 hours per day, and of those hours, 2 hours and 56 minutes or almost 40% of his or her time is spent on their mobile device [3]. According to [4], the number of mobile devices sold in Thailand is 23.24 million mobile handsets in 2013 and 9.67 million are classified as smartphones. In 2013 on the fourth quarter alone, Thailand experienced 63.6% increase in the number of smartphones shipped.

Thailand is described by the World Bank as a middle-income country, while more than 90 percent of the population has a bank account [5]. A competitive banking industry is driving access through the addition of branches as well as through electronic channels such as ATMs, POS devices, and the Internet. However, mobile banking on smartphones is not a mature technology, as it is not widely used, and there is still a strong need for face-to-face banking and personal contact. Security has been one of the biggest risks in any kind of internet banking [6]. The anxiety of customers is linked directly to security problem. As adoption of mobile banking increases, banks in various countries are also paying increasing attention to the specific risks brought by the use of the mobile channel. It is possible to offset the increase in risk caused by using less secure mobile technologies by introducing operational controls such as password or PIN number.

This report focuses on the assessment of the specific operational risks of mobile channel that can influence on the adoption of mobile banking in Thailand. Operational risk is referred to as the risk of loss arising from the failure of operational procedures. It is related to the choice of technology include: internal fraud, including theft and unauthorized activity, external fraud including theft and systems security, business disruption and system failures, failures in the execution and maintenance of transactions, and failures on the part of vendors and suppliers [6]. The purpose of this paper is to investigate the factors that influence the customers' acceptance and use of mobile banking, in a specific operational context.

2   Overview of Mobile Banking in Thailand

The Bank of Thailand (BOT) regulates and supervises financial institution businesses engaged in deposit taking and lending as well as those operating such businesses as commercial banks, finance companies and credit fanciers in Thailand. Currently, many banks have launched the mobile apps offered online services to consumers in addition to other banking channels such as ATMs, telephone banking and internet banking. The first form of internet banking was established in 1999 by one of the leading commercial banks, Siam Commercial Bank (SCB). There are 28 commercial banks in Thailand and 15 of them have offered their services through mobile apps with no additional fees. Even though smart phone users are soaring, 56% of those users do not yet use their mobiles for data or internet. Only 4% of mobile users in Thailand have performed mobile banking activities on their phones. The most interest group in the potential of mobile banking is those who are between 16 to 30 years old. In addition, the top three features on a smartphone that owners use are taking photos/videos, listening to music, and playing games [7].

According to Bank of Thailand (2012), there are 4.85 million accounts online that set up for internet banking, but there are less than one million accounts for mobile banking. The number of transactions performed using mobile banking is 7.46 million transactions, which is four times less that the number performed via internet banking [8]. Intuitively, there is opportunity for mobile banking to grow in Thailand. Commercial banks realize the cost reduction when they can discourage face-to-face transactions and keep their customer out of the branches as tradition services at a bank branch do require labor costs. Meanwhile many smartphones users are cautious about moving from traditional channels of banking services to mobile channel of banking services. This raises the question what factors influence such decisions.

3   Technology Adoption Theories and Research Framework

Researchers in the area of technology adoption have determined to explain the intension to use the technology for many years. It is found, base on well known theories - Technology Acceptance Model (TAM) and Theory of Planned Behavior (TPB), that personal beliefs and attitude of users generate the intension to adopt the technology [9]. According to [10], two external factors which are perceived risk and perceived benefits are important factors influencing customer's adoption of *i*-banking. One of the reasons, mentioned by [11], that some banks are hesitate of using e-banking channel is that their customers are afraid of risk and security issues. Perceived risks of *i*-banking consist of performance risk, social risk, financial risk, privacy risk, time risk, and physical risk. These risks were carefully reviewed to see if they are suitable for this mobile banking study. Since this research is emphasized on risks in the context of operational, which are risks of loss arising from the failure of operational procedures. Hence, time risk, physical risk and social risk will not be considered in this research.

In addition, cultural and social elements are another factor that can drive smartphone users to adopt the mobile banking. The authors of [12] pointed that social can pressure or influence on a potential user's belief about adopting the internet banking. Typical theories, like TAM and TPB, reflect that users' attitudes, perceived usefulness, perceived ease of use, perceived behavior control are related to the adoption of technology. TAM is well employed by many researchers in the past to determine how IT innovation is accepted by an individual. It is focusing on the perception of usefulness and ease of use, which gives the general perspective of i-banking acceptance. Hence, other external factors are added to this research. The model mentioned in this paper is focusing on other influencing factors as well. The summary of attributes that is included in this study is shown below.

- Perceived Usefulness
  - o Accomplishment of financial transactions using mobile banking more quickly
  - o Easier to carry out financial transactions via mobile banking
  - o Realize the advantage of using mobile banking on a smart phone
- Perceived Ease of Use
  - o The ease of learning and using mobile banking on a smartphone
  - o The requirement of no mental effort when operating mobile banking on a smartphone
- Social Influence
  - o Influence by various people
- Perceived Behavior Control
  - o Ability to operate transaction using mobile banking on a smartphone well
  - o Ability to control the entire banking procedures via a smartphone
  - o Complete resources and knowledge to use mobile banking on a smartphone
- Operational Risks
  - o Financial risk
  - o Privacy risk

3.1 Method of Approach and Data Collection

Two research methods are employed in this study. One is documentary research from literature surveys, journals, articles, and previous research works. These data are collected from research published in credible national international journals. Second is survey research. Surveys are done by means of questionnaire surveys for smartphone users who have a bank account and interested in or currently using mobile banking.

The survey is conducted in such a way that questionnaires are randomly distributed to respondents who are willing to answer the questions. Simultaneously, interviews are administered to middle level management of commercial banks and volunteers from respondents. Topics addressed are the effect of technology on banking transactions, concerns or any obstacle to accomplish the use of mobile banking.

The population is a group of Thai people who has a bank account and currently using a smartphone with internet access. The Yamane sampling technique is utilized to identify the require sample size of no less than 200 people.

3.2   Materials and Procedure

Instruments used to collect and record data for this study are open-ended interview questions and survey questionnaires (closed questionnaires, 5-point Likert scale, and open-ended questions for suggestions and recommendations). Survey questionnaires attempt to address the overall issues of adopting the mobile banking, for example, usefulness – accomplishment of financial transactions in less time using mobile banking, ease of use - the ease of learning and using mobile banking, behavior control – ability to control the entire banking procedures on via mobile banking, Social influence – people/relatives/friends, operational risks – financial, privacy and social risks. In addition, data related to the intension to adopt the mobile banking on a smartphone is obtained from the last part of the developed self-administered questionnaires. Each item/question is measured by using a 5-point Likert scale ranging from one to five where

1 stands for strongly disagreed,
2 stands for disagreed,
3 stands for neutral,
4 stands for agreed
5 stands for strongly agreed.

The collected data are analyzed using Statistical Package for Social Science (SPSS) version 20.0. Factor analysis is also utilized to determine the underlying structure of the original 17 items toward the adoption of mobile banking. Lastly, multiple regression analysis is employed to predict significant factors, which affect the adoption of mobile banking on a smart phone.

3.3   Factor Analysis and Multiple Regression

Exploratory Factor Analysis is a statistical procedure that used to identify a small number of factors that can be used to represent relationships among sets of interrelated variables. The Kaiser-Meyer-Olkin — KMO statistic is used to determine if the variables (17 items) share common factor(s). If they share common factor(s), then the partial correlations should be small and the KMO should be close to 1.0 [13].

Regression analysis is a predictive analysis for the relationships between dependent variable and many independent variables. In general the Estimated Multiple Regression Model with k number of independent variables is shown in equation (1).

$$\hat{y} = b_0 + b_1 x_1 + b_2 x_2 + \dots + b_k x_k \tag{1}$$

In addition, the variance inflation factor (VIF) is used to indicate the multicollinearity problem as it can lead to incorrect interpretation as the variables unstable and difficult to interpret. The VIF provides an index that measures how much the variance of an estimated regression is inflated as compared to when the predictor variables are not linearly related. the VIF that is less than 5 for any independent variables shows the uncorrelated relationship which can be imply that the multicollinearity is not considered as a problem for that variable. On the other hand, if the VIF is greater than 5, there is a correlation between two independent variables; therefore, those independent variables should be considered dropping out from the model [14].

4   Result and Data Analysis

4.1   Descriptive Results

The characteristics of 400 smartphone users indicated that 298 of them are interested in using mobile banking on their smartphones. Currently, 317 of them have experienced using mobile banking on a smartphone. Most of the respondents are between 20 up to 40 years old; they are accounted for 66.5% of all respondents. More than half of them earn on average between, the minimum salary for entry level position with a bachelor degree, 15,000 - 30,000 Baht. Our respondents are young; they are either university students or considered as a working class. The summary of demographic information is summarized in table 1.

4.2   Reliability Test and Factor Analysis

Prior to multiple regression analysis, the Principal Component Analysis was conducted from 27 determinants with orthogonal rotation (varimax). The value of Kaiser-Meyer-Olkin (KMO) value is 0.863 out of 1, this indicate that samples are adequacy to proceed on factor analysis. On the Bartlett's test of Sphericity, the approximation of Chi-square and the p-value is 4497.30 and 0.000, respectively. These indicated that correlations between items are sufficiently large for Principal Component Analysis (PCA). In addition, the overall Cronbach's Alpha coefficient of all factor dimensions is 0.875, which is higher than 0.875. In accordance to [15], the generally accepted from 0.80 is appropriate for cognitive tests. As seen in the table 2, all factor loading scores are higher than 0.50. The first factor consists of items under the perceived usefulness and perceived ease of use. There are referred to as convenience of using mobile banking on a smartphone. The second factor is social influence, measuring the level of adoption to mobile banking when a respondent is pressured by certain people in his or her life. The third factor is linked with perceive behavior control. It is referred to as confidence - respondent's confidence to operate and manage the mobile banking transactions. The fourth and fifth factors involve operating risks associated with using mobile banking.

4.3   Multiple Regression Analysis

The result of significant factors affecting the adoption of mobile banking on a smartphone is summarized in table 3, were significant factors are confidence, convenience and security risk. Observe that all VIF values are below 5, indicating no collinearity within our data.

Table 1. Demographic information

Smartphone Users	Item	f	%
Gender	Male	194	48.50
	Female	206	51.50
Age	Under 20	41	10.25
	20-29	177	44.25
	30-39	89	22.25
	40-49	49	12.25
	≥50	44	11.00
Education Level	College/university	116	29.00
	Bachelor	194	48.50
	Master	51	12.75
	PhD	4	1.00
	High school	30	7.50
	Others	5	1.25
Occupation	Own Business	86	21.50
	Government official	28	7.00
	Private company	85	21.25
	Freelance	36	9.00
	Retired	7	1.75
	University students	156	39.00
	Others	2	0.50
Income (Thai Baht)	Less than 15,000	78	19.50
	15,001 - 30,000	201	50.25
	30,001 - 45,000	64	16.00
	More than 45,000	57	14.25
Are you interested in using mobile banking on your smart phone?	Yes	298	74.50
	No	102	25.50
Have you used mobile banking on a smartphone before?	Yes	317	79.25
	No	83	20.75

Table 2. Results of Factor Analysis

Factors[a]		Factor Loading	Variance Explained (%)
FACTOR 1: Convenience			21.04
	Ability to complete transaction in less time	0.891	
	Ease to carry out transactions	0.899	
	Convenience to use at anytime and anywhere	0.847	
	Ease to learn and use the features	0.510	
	No mental effort when operating mobile banking	0.709	
FACTOR 2: Social Influence			16.23
	Pressure to use mobile banking from family	0.840	
	Pressure to use mobile banking from friends	0.887	
	Pressure to use mobile banking from other who are matter	0.882	
FACTOR 3: Confidence			14.52
	Confidence in the ability to perform transactions on mobile banking	0.688	
	Confidence in the ability to control and manage the entire procedures	0.868	
	Confidence in resources and knowledge to use mobile banking	0.810	
FACTOR 4: Financial Risk			13.61
	Concern of losing money due to careless mistakes	0.861	
	Concern of no compensation due to transaction errors	0.889	
	Concern of financial information is recorded illegally	0.741	
FACTOR 5: Security Risk			13.05
	Trust of giving personal information over mobile banking	0.843	
	Trust of being protected by mobile apps/system	0.862	

[a] Principal component factors with iterations: Varimax rotation.

Table 3. Collinearity Statistics and Regression Result

FACTOR	Adoption of Mobile Banking ($n = 400$)			
	Unstandardized coefficient	Standardized coefficient	t	VIF
FACTOR 1: Convenience	0.237	0.218	3.375***	1.758
FACTOR 2: Social Influence	0.087	0.091	1.542	1.459
FACTOR 3: Confidence	0.253	0.223	3.554***	1.658
FACTOR 4: Financial Risk	-0.071	-0.066	-1.170	1.326
FACTOR 5: Security Risk	0.234	0.182	3.246***	1.333

Adjusted R Square   =   0.25
Durbin-Watson   =   1.534
F =   21.045***

* $p<0.10$, **$p<0.05$, and ***$p<0.01$

5   Discussion and Conclusion

Traditionally, commercial banks in Thailand have concentrated on ATM penetration to service all segments nationwide, and they offer a range of services, including money transfers, bill payments, insurance payments, and cash deposit and withdraw. To engage in different dimension of banking transaction, banks have been offering internet banking and eventually moving toward mobile banking. Mobile banking industry is growing and opportunities have been highlighted in this paper earlier. The data collected in this research are based on smartphone users. The study obtained here indicates that three factors that are key drivers toward the adoption of using mobile banking is confidence, convenience, and security risk. All of them have positive coefficients, so they move in the same direction as the adoption variable – higher level of each dependent variable will result in a higher chance to adopt the mobile banking on his or her smartphone.

Observe that the standardized coefficients suggest the influential weights of each significant factor. Both self-confidence of each user and the acknowledgement of the convenience perceived from using mobile banking contribute to the adoption of this new banking feature about the same weights, 22.3% and 21.8% respectively. Security risk is an issue and has caused for concern among users. The coefficient is positive which suggest that if smartphone users can trust the entire mobile banking module, he or she is likely to adopt it on their smartphones.

Mobile banking is an innovated channel that one can carry out the transactions anywhere-anytime effortlessly. With the growing markets of smartphones and 3G internet subscribers, both banks and clients can mutually gain positive impact from proper implementation of mobile banking module. Mobile app is offered to clients with free of charge, and it certainly reduces the Banks' operating cost. Banks have to ensure that the apps are equipped with excellent security standard to mitigate the security risks and generate trust among clients.

Further study is suggested to be done on different determinants such as physical risk and social risk. Government policies may influence the adoption of mobile banking as well. Furthermore, this study do omit the factor about cost of internet subscription. This factor should be carefully reviewed in the future because accessing mobile banking on a smartphone depends of the strength of internet connectivity, and clients solely bear the cost of high-speed internet.

References

1. Rasmussen, A.T.: Thailand's Mobile Market information. Yozzo Pub. (2014)
2. The Nielsen Company: Mobile Consumer Report (2013). http://www.nielsen.com/us/en/insights/reports/2013/mobile-consumer-report-february-2013.html
3. Southgate, D.: Global Mobile Behavior. Millward Brown (2014)
4. International Data Corporation (IDC) Thailand: Thai PC Market Falls to Lowest Point in 7 Years (2015). http://www.idc.com/getdoc.jsp?containerId=prTH25515615
5. International Finance Corporation: Mobile Money, World Bank Group (2011)
6. Bankable Frontier Associates: Managing the Risk of Mobile Banking Technologies. Technical Report, Organization for Economic Co-operation and Development (2008)
7. Sakawee, S.: Thai Mobile Market. TECH in Asia (2013)
8. Numnonda, T.: ICT Overview & Opportunity in Thailand, IMC Institute (2013)
9. Riffai, M.M.M.A., Grantb, K., Edgarc, D.: Big TAM in Oman: Exploring the Promise of On-line Banking - Its Adoption by Customers and the Challenges of Banking in Oman. International Journal of Information Management (2011)
10. Knight, S., Moschou, S., Sorell, M.: Analysis of sensor photo response non-uniformity in RAW images. In: Sorell, M. (ed.) e-Forensics 2009. LNICST, vol. 8, pp. 130–141. Springer, Heidelberg (2009)
11. Howcroft, B., Hamilton, R., Hewer, P.: Consumer Attitude and the Usage and Adoption of Home-based Banking in United Kingdom. The International Journal of Banking Marketing **20**(3), 111–121 (2002)
12. Chan, S.C., Lu, M.: Understanding Internet Banking Adoption and Use Behavior: A Hong Kong Perspective. Journal of Global Information Management **12**(3), 21–43 (2009)
13. Ghamdi, A.S.A.M.A.: Sustainable Saudi Business Tourism (SBT) Innovation: Improving, The Position of SBT Coping with Information System. International Journal of Computer Science and Network Security **10**(7), 300–310 (2006)
14. Rogerson, P.A.: Statistical Methods for Geography. Sage, London (2001)
15. Field, A.P.: Discovering Statistics Using IBM SPSS Statistics. Sage, London (2013)

Negativity Bias Effect in Helpfulness Perception of Word-of-Mouths: The Influence of Concreteness and Emotion

Chih-Chien Wang[1(✉)], Feng-Sha Chou[1], Chiao-Chieh Chen[1], and Yann-Jy Yang[2]

[1] Department of Business Administration, National Taipei University, 151,
University Rd., San Shia District, New Taipei City 23741, Taiwan
wangson@mail.ntpu.edu.tw, {choufengsha,chiaoc620}@gmail.com
[2] College of Tsing Hua, National Tsing Hua University, 101, Sec. 2, Kuang-Fu Rd.,
Hsinchu City 30013, Taiwan
yj-yang@mx.nthu.edu.tw

Abstract. Word-of-mouth (WOM) is a powerful information resource for consumers to judge a product or service. WOM has persuasive power when audiences regard it as helpful. There is a psychological phenomenon, called the negativity bias, in which negative WOMs are usually regarded as more helpful than positive ones. However, not all previous studies supported the existence of negativity bias. This paper examines the negativity bias effect of WOM and explores the moderating effect of length and emotional content on the helpfulness perception of WOM. This study adopted experimental designs of 2 (positive/negative WOMs) X 2 (long/short WOMs) X 2 (WOMs with strong emotion/weak emotion) of 139 subjects. The empirical survey results reveal that length and emotional content of WOMs are influential factors for helpfulness perception of WOMs. Moreover, both length content and emotional content of WOMs also moderate the negativity bias effect of helpfulness perception. The negativity bias effect appears when WOMs are concrete and emotional.

Keywords: Word-of-Mouths · Negativity bias effect · Product review

1 Introduction

During the purchase decision-making process, the customers search for information about products/services. The information can be provided by the firms (advertising) as well as by other consumers (Word-of-mouths, WOMs). Literatures advocate that WOM is more persuasive than traditional advertising [1-3]. People tend to accept the recommendations when the message is from WOMs of unbiased authority, and reject the advertising claim when they believe the message is a sales tool rather than information and guidance [4]. WOM is an informal way of person-to-person communication regarding a brand, a product, service, or their providers [5]. WOM has significant influence on a consumer's purchase decision [6, 7] and is a powerful information resource that helps shape consumers' attitude toward a product or service (Katz & Paul, 1996).

© Springer-Verlag Berlin Heidelberg 2015
L. Wang et al. (Eds): MISNC 2015, CCIS 540, pp. 425–436, 2015.
DOI: 10.1007/978-3-662-48319-0_35

WOMs were traditionally disseminated by face-to-face oral communication, but with the rise of the Internet they now can be spread and preserved by a variety of online applications, such as e-mail, online communities, product review websites, and social networking websites [8]. The WOMs spread online are called Electronic Word-Of-Mouths (EWOMs), which is defined as encompassing the opinions and consumption experiences online about specific companies, products, or services [9, 10]. Many online shopping websites (such as amazon.com, eBay.com, hotels.com, tripadvisor.com) provide online platforms for consumers to express their opinions toward products and services. Due to the high penetration rate of the Internet as well as the advance of information technology, consumers now can use Internet to obtain lots of reviews regarding a product or service for their purchase decisions.

A WOM has persuasive power when audiences regard it as helpful. Literature reported the influence on WOMs' persuasiveness effect of volume, valence, and distribution (dispersion) of WOMs. The valence (positive or negative) of WOMs reveals the quality of product/services [1, 11]. Positive WOM (negative WOM) will help consumers generate more positive (negative) attitudes toward the product or service, which will increase sales of products by affecting consumers' buying intentions and shopping decisions [12]. However, consumers usually pay greater attention to negative WOMs than to positive ones [13]. Thus, negative WOMs are more influential than positive ones, and will reduce consumers' purchasing intentions [14]. When WOM includes comments from different viewpoints, it provides people another approach to evaluating the product. Since most of the WOMs are positive in most situations, negative WOMs may provide unique comments that are different from the positive ones. Thus, negative WOMs may be considered as more helpful than the positive ones.

The term negativity bias is used to describe the phenomenon that negative WOMs are regarded as more helpful and with more persuasive power. Psychological literature advocated that there is asymmetry in the influence of positive and negative information [15]. The psychological effect of negative bias reveals that negative information usually outweigh other positive information [16, 17]. Based on the argument of negative bias, negative WOMs are more influential than positive ones and will reduce consumers' purchasing intentions [14]. Negative WOMs are reported as more helpful than positive ones based on results of some empirical studies [i. e. 3, 13, 14, 18, 19-21]. However, not all research supported the existence of negativity bias in WOM communication. Wu [22] reported the inexistence of negativity bias in a virtual experiment and in real WOM data when controlling for the quality of WOMs. There is no consistent conclusion in the literature regarding the existence of negativity bias effect in helpfulness perception of WOMs. Thus, the current research aims to reveal the existence or non-existence of negativity bias and consider the valance of WOM as an antecedent for helpfulness perception.

In addition, consumers may like to know details of others' consumption experiences before making their purchase decisions. However, not all WOMs are detailed. Some WOMs are short and with only brief or few comments. Detailed information is needed to convince other consumers not to consume the product or service. Thus, people may rate the detailed (long) WOMs more helpful than abstracted (short) ones.

Schellekens, Verlegh and Smidts [23] found that the concreteness of negative WOMs is a factor influencing individuals' perception of their helpfulness. Hence, this study considers the length of WOMs as a potential moderator between valence and usefulness perception.

Negative WOMs usually includes some unpleasant consumption experiences. However, overly emotional content may cause WOM readers to doubt the reviewer's motivation, resulting in perceiving the WOM as less helpful [24]. Helpfulness perception of information may decrease when there is too much personal feeling contained in the comment [25]. A WOM will be considered as too irrational when it contains too much emotional content, which will lead the WOMs to be perceived as less helpful. Thus, this study considers the emotional level of WOMs as a potential moderator of the negativity bias effect.

The paper aims to explore the negativity bias effect of WOM and reveal the moderating effect of length and emotion strength of WOM on the negativity bias effect. The paper attempts to answer the following research questions:

1. Will people perceive negative WOMs as more helpful than positive ones?
2. Will length and emotion of WOMs moderate people's perception of the helpfulness of positive and negative WOMs?

2   Theoretical Background and Hypotheses

2.1   Negativity Bias

The asymmetry of the persuasive power of positive and negative information is a general psychological phenomenon [15]. Generally speaking, when both good and bad things are presented, the bad ones outweigh the good ones [22, 26]. In psychology, this phenomenon is called the "negativity bias", which argues that negative information exerts more impact on peoples' judgment and perceptions than positive information [2, 27, 28]. The negativity bias does not mean that bad things will always outweigh good ones. It argues that, generally, bad things have a higher impact than good ones when they are of equal or similar importance [26].

Tversky and Kahneman [29] used the concept of loss aversion in the riskless choice to explain the negativity bias effect, which revealed that people would focus more attention on negative information than positive information. People evaluate products or services more negatively because of the disproportionate amount of attention, elaboration, and emphasis on the negative information [30]. Negatively valenced messages have a greater influence on individuals than positive valence ones [31].

Negativity bias is a common terminology which is widely used in many research fields. In WOM research, negativity bias effect can be used to explain the WOM generation as well as persuasion effect of WOM. Some previous literature focused on the impact of negativity bias in WOM generation, which regarded the negativity bias as the phenomenon that low levels of satisfaction resulted in greater volume of WOMs than high levels [32, 33]. Shin, Song and Biswas [19] used the team negativity bias to describe the phenomenon that consumers with negative consumption experience had the higher intention to generate WOMs than ones with positive experience. The negativity bias in WOM generation occurs due to the asymmetrical response to positive and negative events [13, 33].

The current study discusses the negativity bias effect in helpfulness perception, which reveals the phenomenon that consumers usually perceive negative WOMs as more helpful than positive WOMs [13]. Positive WOMs are usually more common than negative ones [13]. All products or services have at least some positive attributes [34-36]. Therefore, the presence of positive WOMs is not necessarily be linked to high quality perception, since all products or services have at least some positive attributes [34-36]. On the contrary, negative information is usually more distinctive, and, therefore, more diagnostic [29]. Thus, literature found that negative information is more influential than positive information [16]. The negative information receives more attention from consumers and distracts them from positive information, subsequently giving negative information a disproportionately strong persuasive power and resulting in the negativity bias [29].

Although negativity bias is a common psychological tenet, not all research supports the existence of negativity bias in WOM communication. Some previous empirical studies confirmed that negative WOMs were more helpful than positive ones [3, 13, 14, 18-21]. However, other studies argued that negativity bias existed only under some conditions [32, 36-39]. They examine negativity bias and found that the widely-held belief that "bad is stronger than good" is overturned. Thus, the existence or non-existence of negativity bias is an unanswered question. The first hypothesis of the current study is to confirm the negative relationship between valence and helpfulness perception, as follows:

H1: WOM's valence is negatively influenced by helpfulness perception and behavioral intention. People perceive negative WOMs as more helpful than positive ones.

2.2   Length

Based on the linguistic category model, the language used in interpersonal communication can be divided into abstract and concrete language [23, 40]. When discussing the negativity bias effect of WOM, there could be some consideration of how the WOM length affects the helpfulness perception. The concreteness of WOMs refers to how detailed are the messages the WOMs contain. The length of WOMs varies from single words or phrases to thousands of words. Some long WOMs come with detailed information about the product or service. However, other short WOMs are abstract, with only overall comments but no details. Consumers need detailed comments about the product or service for their purchase decision. Compare with shorter WOMs, longer WOMs with concrete information may provide more diagnostic information to influence the consumer's decision [41]. Thus, when people are requested to rate the helpfulness of WOMs, they may rate the detailed (long) WOMs rather than abstract (short) WOMs as helpful. An empirical study of Mudambi and Schuff [42] confirmed the argument that longer WOMs were considered as more helpful that the short ones. Wu [22] found that the length of WOMs on amazon.com was positively related with helpfulness perception. Schellekens, Verlegh and Smidts [23] found that the persuasion power of WOMs is associated with language abstractness in WOMs. Based on the above discussion, we propose the following hypothesis:

H2: Length of WOMs will influence the helpfulness perception of WOMs and moderate the influence of valence on helpfulness perception.

2.3 Emotion

Customers' positive and negative emotions come with their satisfactory or unsatisfactory consumption experience. Unpleasant consumer experience often result in negative emotions, allowing consumers to conduct a series of follow-up correspondence, including posting negative WOMs [43-45]; on the other hand, positive emotion would lead to positive WOMs.

The emotional content of WOMs refers to subjective consumption experience of the product or service while less emotional content refers to the objective assessment of pros and cons of a product or service.[46]. WOMs usually contain the subjective consumption experience of WOM senders. The content of WOMs can be divided into two categories: objective and subjective [8]. Objective reviews describe clear and impersonal facts, such as product specification and pricing [47]. In contrast, subjective reviews are usually based on personal experience with the product or service, which may contain emotions of WOMs senders. Consumers may infer a sender's motivation from the emotional expression in the WOMs. Banerjee and Chua (2014) revealed that overuse of emotional words may cause readers to doubt the reviewer's motivation, resulting in perceiving the WOM as less helpful.

Petty and Cacioppo [48] revealed that objective and easily understood messages have a stronger influence than subjective or affective messages. Cheung and Lee [8] argued that consumers have an obvious preference for objective information. When a WOM comes with excessively personal feelings, people will regard it as less objective and irrational, which will reduce the value perception of it. Based on the above discussion, we propose the following hypothesis:

H3: Emotional of WOMs will influence the helpfulness perception of WOMs and moderate the influence of valence on helpfulness perception.

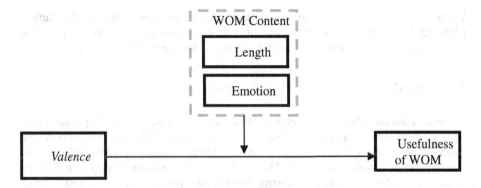

3   Methodology

3.1   Research Design

This study used an experimental design to investigate the existence or non-existence of the negativity bias effect, and whether it has an impact on subjects' helpfulness perception regarding WOMs. The movie *The Hobbit: The Battle of the Five Armies*, which ranked at the top of the box office list during the experimental period of November 2014, was used in the experimental design. We adopted an experimental design of 2 (positive/negative WOMs) X 2 (long/short WOMs) X 2 (WOMs with strong emotion/weak emotion). WOMs used in the study were collected and revised from the real WOMs posted on the movie review website Yahoo!Movie Taiwan (https://tw.movies.yahoo.com/). All the WOMs are similar to original WOMs, so we expect messages that will disclose posters' personal identity data.

3.2   Procedure

Participants first reported their preference levels regarding the subject movie prior to reading all reviews. Then, we asked participants to view and rate the perceived helpfulness of all the WOMs. After they had read the WOMs about the subject movie, participants were requested to rate their preferences for this movie again. We provided a small souvenir worth approximately 3 US dollars as an incentive for all qualified responses.

The previous preferences and subsequent preferences regarding the movie were measured by asking participants: "What is the possibility that you will watch this movie?" A five-point scale was used (1=not at all likely, 5=very much likely). The helpfulness perception of WOM was measured by asking participants: "Is this review helpful to you?" A five-point scale was used (1=not at all helpful, 5=very helpful). The measurement item was developed by Wu (2013).

3.3   Samples

We recruited 139 Taiwanese college students to join the experiment. Of the participants, 55 were male (39.6%) while the other 84 participants are female (60.4%). The ages of the participants ranged from 16 to 30 years (Mean=19.30; S.D =2.43).

4   Result

To understand the interaction effects of WOM characteristics on the helpfulness perception of WOM, we adopted the multivariate analysis of variances (MANOVA) for examination. Table 1 listed a multivariate analysis of variances (MANOVA) results. The dependent variable in MANOVA analysis is helpfulness perception while independent variables are the length (abstract/detail), emotion (strong/weak), and valence (positive/negative) of WOM.

According to MANOVA analysis results, the main effects on helpfulness perception of valence (F=2.474, p=0.118), length (F=40.825, p<0.001), and emotional content (F=38.188, p<0.001) of WOMs are significant. H1 was supported. We found an interaction effect between length and emotion (F = 15.297, p<0.001), length and valence (F= 6.411, p=0.012). Moreover, we also find an interaction effect between emotion and valence (F = 5.468, p=0.021). It reveals that the helpfulness perception of WOM was affected by WOM characteristics.

Table 1. ANOVA analysis of helpfulness perception by length, emotion, and valence of WOMs

Variables	F	P-value
Length	11.302	0.001
Emotion	43.324	0.000
Valence	0.978	0.324
Length x Emotion	15.431	0.000
Length x Valence	29.444	0.000
Emotion x Valence	5.468	0.021
Length x Emotion x Valence	1.127	0.290

*** p<0.001; ** p<0.01; * p<0.05   *Dependent variable: usefulness*

To clarify the presence or not of the negativity bias, we compare the perceptual helpfulness of three WOM characteristics (valence, length, and emotion). Table 2 shows the helpfulness perception according to valence, length, and emotion. As Table 2 revealed, subjects' helpfulness perception to positive WOMs did not differ from negative ones, which not supporting the argument for a negativity bias. In addition, long WOMs were perceived as more helpful than short ones ($\bar{x}_{long}$ = 3.43 > $\bar{x}_{short}$ =3.27, p < 0.001). WOMs with strong emotion were considered as less helpful than those with weak emotion ($\bar{x}_{strong}$ = 3.16<$\bar{x}_{weak}$ =3.46, p < 0.001).

Table 2. t-test results of helpfulness perception by WOM valence, length, and emotion.

		Helpfulness perception Means	SD	t-value	p-value	Comparison
Valence	Positive	3.31	0.98	0.785	0.433	
	Negative	3.27	0.93			
Length	Detailed	3.37	0.95	-2.867	0.004	Long > Short
	Abstract	3.21	0.95			
Emotion	Emotional	3.08	0.95	-7.361	0.000	Less Emotional
	Less emotional	3.50	0.98			> Emotional

To confirm the negativity bias existence, length of WOM was subdivided into detail (long) and abstract (short) WOMs to consider the influence of valence of WOMs on helpfulness perception. Regarding the long WOMs, helpfulness perception of positive WOMs did not significantly differ from that of negative ones (p =0.068). But regarding short WOMs, helpfulness perception of positive WOMs was significantly higher than that of negative ones ($\bar{x}_{positive}$= 3.33>$\bar{x}_{negative}$ =3.09, p = 0.003). The negativity bias effect did not exist in both long and short WOMs. In contrast, the positivity bias effect existed in the short WOMs; subjects perceived positive short WOMs were more helpful than the negative short ones.

Table 3. t-test results for moderate effect of WOM length on negativity bias

| WOM Length | Valence | Helpfulness perception | | t-value | p-value |
		Means	SD		
Detailed	Positive	3.30	0.97	-1.829	0.068
	Negative	3.45	0.93		
Abstract	Positive	3.33	0.99	2.964	0.003
	Negative	3.09	0.89		

In Table 4, we considered the influence of emotion content on helpfulness perception of WOMs. In the emotional WOMs, the perceived helpfulness was not significantly different between positive and negative WOMs (p =0.561). Moreover, in less (weak) emotional WOM content, there was no difference between positive and negative content of emotional WOMs (p =0.079).

Table 4. t-test results for moderate effect of WOM emotion on negativity bias

| Content of WOM | Valence | Helpfulness perception | | t-value | p-value |
		Means	SD		
Emotional	Positive	3.06	0.98	-0.581	0.561
	Negative	3.11	0.92		
Less emotional	Positive	3.56	0.92	1.760	0.079
	Negative	3.43	0.91		

Table 5 reveals the interaction effects of emotion, length and valence of WOM content on helpfulness perception. Results show the negativity bias effect is only present in the detailed (long) WOMs with emotional content ($\bar{x}_{negative}$ = 3.20 > $\bar{x}_{positive}$ =2.91, p = 0.009). But in the less emotional cases, the helpfulness perception is not significantly different between positive and negative WOM (p=0.946). Moreover, the negativity bias effects did not exist in abstract (short) WOMs, no matter whether the WOMs are emotional or less emotional.

Table 5. t-test results for moderate effect of WOM emotion and length on negativity bias

| Content of WOM | | Valence | Helpfulness perception | | t-value | p-value |
			Means	SD		
Detailed	Emotional	Positive	2.91	0.93	-2.620	0.009
		Negative	3.20	0.90		
	Less emotional	Positive	3.68	0.86	-0.068	0.946
		Negative	3.69	0.89		
Abstract	Emotional	Positive	3.21	1.01	1.672	0.096
		Negative	3.01	0.93		
	Less emotional	Positive	3.45	0.97	2.565	0.011
		Negative	3.17	0.85		

5   Discussion

This study aims to explore the negativity bias effect of WOMs and reveal the moderating effect of length and emotion of WOMs on the perceived helpfulness of WOMs. The empirical study result reveals that the negativity bias is only present in long WOMs with emotional content. Some researchers advocated that negative WOMs were more helpful than positive ones [3, 13, 14, 18-21]. However, some other literatures argued that negativity bias existed only under some conditions [32, 36-39]. The current study found no direct influence of valence on helpfulness perception. But the negativity bias effect exists when WOMs were long and with emotional content. Thus, this study supports the argument that negativity bias effect only exists under some conditions. It implies that the negativity bias effects require specific conditions to be triggered and that purely negative WOM content did not drive these conditions.

Length of WOMs is a moderator for the relationship between helpfulness perception and WOM valence. We find that compared with abstract WOMs, the detailed (long) WOMs were perceived as having higher helpfulness. Detailed WOMs make individuals feel more informed about the product or service. Moreover, the negativity bias did not exist in interaction effects of valence and length of WOM, and the interaction effects were only present in abstract (short) WOMs. But individuals perceived negative WOMs with short content as less helpful than long ones. There was no difference in helpfulness perception between negative and positive WOMS when the WOMs and not consider the moderate effect of emotional content.

There was a negative relationship between emotional content and helpfulness perception. People regarded WOMs with weak emotional content as more helpful than ones with strong emotional content. WOMs with strong emotional content may be treated as irrational and subjective while WOMs with weak emotional content may be treated as objective. Thus, WOMs with too much emotional content will be regarded as less helpful.

The moderating effects of emotional WOM on the negativity bias effect do not exist. However, the current study found the negativity bias effects present in the detailed (long) WOMs with emotional content.

This study extended the negativity bias and WOM literature. We found that the negativity bias effects of WOM should be considered in the context of contingency factors. The length and emotional content both are contingency factors. This study also found that helpfulness perception of a WOM will be affected by emotion and length of the WOM.

However, the influence on helpfulness perception of interaction effects between valence, concreteness, and emotional content is not clear. The existence or non-existence of negativity bias effect is also unclear. More efforts are needed to reveal the reasons behind the inconsistent results.

The current research aimed to explore the influencing factors of WOM helpfulness perception. However, this study used only one product (movie) in the empirical study. Product is a potential moderator of the negativity bias effect. Future studies can focus on the moderating effect of product type on negativity bias effect.

References

1. Godes, D., Mayzlin, D.: Using Online Conversations to Study Word-of-Mouth Commun. Market. Sci. **23**, 545–560 (2004)
2. Herr, P.M., Kardes, F.R., Kim, J.: Effects of Word-of-Mouth and Product-Attribute Information on Persuasion: An Accessibility-Diagnosticity Perspective. J. Consum. Behav., 454–462 (1991)
3. Sen, S., Lerman, D.: Why Are You Telling Me This? An Examination into Negative Consumer Reviews on the Web. J. Interact. Market. **21**, 76–94 (2007)
4. Dichter, E.: How Word-of-Mouth Advertising Works. Harvard Bus. Rev. **44**, 147–160 (1966)
5. Harrison-Walker, L.J.: The Measurement of Word-of-Mouth Communication and an Investigation of Service Quality and Customer Commitment as Potential Antecedents. J. Serv. Res. **4**, 60–75 (2001)
6. Arndt, J.: Role of Product-Related Conversations in the Diffusion of a New Product. J. Market. Res., 291–295 (1967)
7. Nowak, K.L., McGloin, R.: The Influence of Peer Reviews on Source Credibility and Purchase Intention. Soc. **4**, 689–705 (2014)
8. Cheung, C.M., Lee, M.K.: What Drives Consumers to Spread Electronic Word of Mouth in Online Consumer-Opinion Platforms. Decis. Support. Syst. **53**, 218–225 (2012)
9. Hennig-Thurau, T., Gwinner, K.P., Walsh, G., Gremler, D.D.: Electronic Word-of-Mouth Via Consumer-Opinion Platforms: What Motivates Consumers to Articulate Themselves on the Internet? J. Interact. Market. **18**, 38–52 (2004)
10. Hennig-Thurau, T., Walsh, G., Walsh, G.: Electronic Word-of-Mouth: Motives for and Consequences of Reading Customer Articulations on the Internet. Int. J. Electron. Com. **8**, 51–74 (2003)
11. Cui, G., Lui, H.-K., Guo, X.: The Effect of Online Consumer Reviews on New Product Sales. Int. J. Electron. Com. **17**, 39–58 (2012)
12. Zhu, F., Zhang, X.: Impact of Online Consumer Reviews on Sales: The Moderating Role of Product and Consumer Characteristics. J. Market. **74**, 133–148 (2010)
13. Melián-González, S., Bulchand-Gidumal, J., López-Valcárcel, B.G.: Online Customer Reviews of Hotels as Participation Increases, Better Evaluation Is Obtained. Cornell. Hosp. Q. **54**, 274–283 (2013)
14. Liu, T.C., Wang, C.Y., Wu, L.W.: Moderators of the Negativity Effect: Commitment, Identification, and Consumer Sensitivity to Corporate Social Performance. Psychol. Market. **27**, 54–70 (2010)
15. Peeters, G.: The Positive-Negative Asymmetry: On Cognitive Consistency and Positivity Bias. Eur. J. Soc. Psychol. **1**, 455–474 (1971)
16. Rozin, P., Royzman, E.B.: Negativity Bias, Negativity Dominance, and Contagion. Pers. Soc. Psychol. Rev. **5**, 296–320 (2001)
17. Ito, T.A., Larsen, J.T., Smith, N.K., Cacioppo, J.T.: Negative Information Weighs More Heavily on the Brain: The Negativity Bias in Evaluative Categorizations. J. Pers. Soc. Psychol. **75**, 887 (1998)
18. Sridhar, S., Srinivasan, R.: Social Influence Effects in Online Product Ratings. J. Market. **76**, 70–88 (2012)
19. Shin, D., Song, J.H., Biswas, A.: Electronic Word-of-Mouth (Ewom) Generation in New Media Platforms: The Role of Regulatory Focus and Collective Dissonance. Market. Lett. **25**, 153–165 (2014)

20. Qiu, L., Pang, J., Lim, K.H.: Effects of Conflicting Aggregated Rating on Ewom Review Credibility and Diagnosticity: The Moderating Role of Review Valence. Decis. Support. Syst. **54**, 631–643 (2012)
21. Bae, S., Lee, T.: Gender Differences in Consumers' Perception of Online Consumer Reviews. Electron. Commer. Res. **11**, 201–214 (2011)
22. Wu, P.F.: In Search of Negativity Bias: An Empirical Study of Perceived Helpfulness of Online Reviews. Psychol. Market. **30**, 971–984 (2013)
23. Schellekens, G.A., Verlegh, P.W., Smidts, A.: Language Abstraction in Word of Mouth. J. Consum. Behav. **37**, 207–223 (2010)
24. Banerjee, S., Chua, A.Y.: Understanding the process of writing fake online reviews. In: 2014 Ninth International Conference on Digital Information Management (ICDIM), pp. 68–73. IEEE Press, Thailand (2014)
25. Parrott, W.G.: But Emotions Are Sometimes Irrational. Psychol. Inq. **6**, 230–232 (1995)
26. Baumeister, R.F., Bratslavsky, E., Finkenauer, C., Vohs, K.D.: Bad Is Stronger Than Good. Rev. Gen. Psychol. **5**, 323 (2001)
27. Fiske, S.T.: Attention and Weight in Person Perception: The Impact of Negative and Extreme Behavior. J. Pers. Soc. Psychol. **38**, 889 (1980)
28. Mittal, V., Ross Jr., W.T., Baldasare, P.M.: The Asymmetric Impact of Negative and Positive Attribute-Level Performance on Overall Satisfaction and Repurchase Intentions. J. Marketing, 33–47 (1998)
29. Tversky, A., Kahneman, D.: Loss Aversion in Riskless Choice: A Reference-Dependent Model. Q. J. Econ., 1039–1061 (1991)
30. Lane, V.R., Keaveney, S.M.: The Negative Effects of Expecting to Evaluate: Reexamination and Extension in the Context of Service Failure. Psychol. Market. **22**, 857–885 (2005)
31. Cameron, K.S., Caza, A.: Introduction Contributions to the Discipline of Positive Organizational Scholarship. Am. Behav. Sci. **47**, 731–739 (2004)
32. Lang, B.: How Word of Mouth Communication Varies across Service Encounters. Manag. Serv. Qual. **21**, 583–598 (2011)
33. Davidow, M., Leigh, J.H.: The Effects of Organizational Complaint Responses on Consumer Satisfaction, Word of Mouth Activity and Repurchase Intentions. J. Consum. Satisf. Dissatif. Compaining Behav. **11**, 91–102 (1998)
34. Lee, M., Youn, S.: Electronic Word of Mouth (Ewom) How Ewom Platforms Influence Consumer Product Judgement. Int. J. Advert. **28**, 473–499 (2009)
35. Skowronski, J.J., Carlston, D.E.: Social Judgment and Social Memory: The Role of Cue Diagnosticity in Negativity, Positivity, and Extremity Biases. J. Pers. Soc. Psychol. **52**, 689 (1987)
36. Purnawirawan, N., Dens, N., De Pelsmacker, P.: Balance and Sequence in Online Reviews: The Wrap Effect. Int. J. Electron. Com. **17**, 71–98 (2012)
37. Chen, Z., Lurie, N.H.: Temporal Contiguity and Negativity Bias in the Impact of Online Word of Mouth. J. Marketing Res. **50**, 463–476 (2013)
38. Zhang, J.Q., Craciun, G., Shin, D.: When Does Electronic Word-of-Mouth Matter? A Study of Consumer Product Reviews. J. Bus. Res. **63**, 1336–1341 (2010)
39. Daugherty, T., Hoffman, E.: Ewom and the Importance of Capturing Consumer Attention within Social Media. J. Marketing Commun. **20**, 82–102 (2014)
40. Semin, G.R., Fiedler, K.: The Cognitive Functions of Linguistic Categories in Describing Persons: Social Cognition and Language. J. Pers. Soc. Psychol. **54**, 558 (1988)
41. Schindler, R.M., Bickart, B.: Perceived Helpfulness of Online Consumer Reviews: The Role of Message Content and Style. J. Consum. Behav. **11**, 234–243 (2012)

42. Mudambi, S.M., Schuff, D.: What Makes a Helpful Review? A Study of Customer Reviews on Amazon. Com. MIS Quart. **34**, 185–200 (2010)
43. Mattila, A.S., Ro, H.: Discrete Negative Emotions and Customer Dissatisfaction Responses in a Casual Restaurant Setting. J. Hosp. Tour. Res. **32**, 89–107 (2008)
44. Zeelenberg, M., Pieters, R.: Beyond Valence in Customer Dissatisfaction: A Review and New Findings on Behavioral Responses to Regret and Disappointment in Failed Services. J. Bus. Res. **57**, 445–455 (2004)
45. Wetzer, I.M., Zeelenberg, M., Pieters, R.: "Never Eat in That Restaurant, I Did!": Exploring Why People Engage in Negative Word-of-Mouth Communication. Psychol. Market. **24**, 661–680 (2007)
46. Yin, D., Bond, S., Zhang, H.: Anxious or Angry? Effects of Discrete Emotions on the Perceived Helpfulness of Online Reviews. MIS Quart. **38**, 539–560 (2014)
47. Xia, L., Bechwati, N.N.: Word of Mouse: The Role of Cognitive Personalization in Online Consumer Reviews. J. Interac. Advert. **9**, 3–13 (2008)
48. Petty, R.E., Cacioppo, J.T.: The Effects of Involvement on Responses to Argument Quantity and Quality: Central and Peripheral Routes to Persuasion. J. Pers. Soc. Psychol. **46**, 69 (1984)

The Role of Facebook in the 2014 Greek Municipal Elections

Georgios Lappas[1(✉)], Amalia Triantafillidou[1], Prodromos Yannas[2],
Anastasia Kavada[3], Alexandros Kleftodimos[1], and Olga Vasileiadou[1]

[1] Department of Digital Media and Communication,
Technological Education Institute of Western Macedonia, Kastoria Campus, Fourka Area,
GR52100, Kastoria, Greece
{lappas,a.triantafylidou,kleftodimos,
o.vasileiadou}@kastoria.teikoz.gr
[2] Department of Business Administration, Piraeus University of Applied Sciences,
Petrou Ralli & Thivon 250, GR12244 Aigaleo, Greece
prodyannas@teipir.gr
[3] Department. of Journalism and Mass Communication, University of Westminster,
Watford Road, Northwick Park Harroe, MIDDX HA1 3TP, UK
a.kavada@westminster.ac.uk

Abstract. The purpose of this study is to examine the use of Facebook by candidates running for the 2014 Greek Municipal Elections by addressing the following questions: (1) which factors affect Facebook adoption by municipal candidates?, and (2) whether Facebook usage along with the popularity of candidates' Facebook pages influence candidates' vote share. Results indicate that Facebook is not a very popular campaigning tool among municipal candidates in Greece. This implies that Greek candidates still rely on traditional ways to lure their voters. Furthermore, findings reveal that candidates running in large municipalities who hadn't been elected before are more likely to utilize Facebook as a means of political marketing. Despite the low adoption rate, results suggested that candidates who made use of Facebook won more votes compared to non-Facebook candidates. Moreover, it was found that a candidate's Facebook page popularity is a good indicator of the candidate's vote share.

Keywords: Facebook · Political marketing strategy · Vote share · Greek municipal elections · Quantitative analysis

1   Introduction

Throughout the years the Internet has become an important vehicle for political campaign activities. In 1996 candidates incorporated websites in their election campaigns in order to provide top-down communication to voters; in 1998 they used emails for contacting with voters and in 2003 blogs became an important part of their online campaign activities [1]. Web 2.0 entered the political marketing arena, with Facebook paving the way, during the 2008 presidential elections in the US [2]. Until now, polit-

© Springer-Verlag Berlin Heidelberg 2015
L. Wang et al. (Eds): MISNC 2015, CCIS 540, pp. 437–447, 2015.
DOI: 10.1007/978-3-662-48319-0_36

ical marketers have acknowledged the value of Web 2.0 tools as a cost-effective method of political promotion [3, 4]. According to Pena-Lopez [5], Web 2.0 gives candidates the opportunity to produce and promote customized messages for their targeted voters.

Despite research enthusiasm on e-campaigning, most of the studies have been conducted in candidate-centric electoral systems. These electoral systems tend to favor the quick adoption of new media [6] allowing for personalized campaign styles. In contrast, little is known about how candidates originating from party-centric electoral systems and adopting a Western European campaigning style [7] have realized the potential of new media in reaching voters [8].

Not too long ago, the Greek political campaigning style was characterized by its party focus. Greek candidates during the pre-election period of national elections relied on the party's communication strategy. In the 1996 national elections, political parties were in charge of managing and controlling the communication activities of their candidates using primarily television [9]. Although the use of the internet in election campaigns was introduced in Greece by the Panhellenic Socialist Movement (PASOK) in the parliamentary elections of 1996 its adoption did not change the party orientation of the Greek electoral campaigns. In the 2000 national election all nine political parties had an online presence through their websites whereas only 1 out of 6 members of the Greek Parliament had a campaign web site [10]. Until 2000, Greek candidates followed the party's communication strategy and made no use of Internet for personal promotion [11].

In the following years, Greek politics faced a number of critical changes. In the process of appealing to voters of various ideological persuasions, parties adopted a catch all orientation without a clear basis of differentiation and voters became more volatile in their preferences. [9]. These changes combined with the new opportunities for more individualistic forms of political campaigning offered by the Internet, impelled Greek candidates to increase their personal online visibility. In the 2004 national elections approximately 1 out of 3 candidates representing the two major parties was a web candidate (33%). From the 2000 to the 2004 Parliamentary elections, the number of e-campaigning politicians doubled, underlying the importance of the web in politicians' campaign strategy. Web 2.0 tools were first adopted by Greek political parties in the October 2009 national elections [12]. Although, websites were still the main online stream for parties and candidates in the 2009 national elections, among Web 2.0 tools (YouTube, Twitter etc) Facebook was the most popular medium.

This study examines candidates' use of Facebook in the Greek Municipal elections of 2014. Specifically, the purpose of the present study is twofold. First, to decipher Whether characteristics of candidates such as gender, incumbency, status, municipality size have an effect on candidate's Facebook use, and second to identify whether Facebook use by candidates along with citizens' awareness of the Facebook pages of candidates are important predictors of a candidates vote share.

2   Literature Review

Researchers point out that the use of Internet has brought changes in the way politicians promote themselves and voters participate in politics. Norris [13] has postulated three political campaign models using as a yardstick the most prevalent medium used in the campaign. The first model is known as the pre-modern campaign model in which parties/candidates use mainly the press and interpersonal communication to persuade voters. The next model is referred to as the modern campaign model where parties/candidates promote themselves via television news and commercials. The usage of new technologies by parties and candidates has given rise to the third postmodern model. Vergeer et al. [6] adds a fourth model to highlight the increasing use of social media as a medium for political campaigning. The fourth model is referred as the personal campaign model considering the fact that social media campaigns allow for personalized promotion and candidate-centric races. Tops et al. [14] introduces the term cyber-democratic model of democracy to draw attention to the significant role of internet and electronic networks in shaping politics. Social networking sites such as Facebook and Twitter let candidates differentiate themselves on a personal basis, increasing thus their awareness to their online supporters [8]. What is more, social media enable candidates to provide "not just top-down communication, but also network-based horizontal communication" [7, p. 6].

In a number of recent studies Facebook has been found to play a significant role in political campaigns across the globe. For example, almost 40% of candidates had a Facebook presence in the 2009 Norwegian parliamentary elections [7] while 29 out of the 31 investigated candidates in the 2011 local elections in Norway used Facebook as a promotional vehicle [8]. In the context of the 2007 Australian Parliamentary elections, Chen [15] indicated that Facebook was the most prominent social media campaign tool. Gulati and Williams [16] investigating the 2012 US congressional elections found that 97% of the candidates running for the Senate seats and 90% of the candidates for the House of Representatives had a Facebook presence.

3   Conceptual Framework

3.1   Factors Affecting Facebook Adoption

Adoption of Facebook by candidates is affected by several factors related to the personal characteristics of the candidate as well as other electoral system-related factors. For example, Strandberg [17] found that gender was a significant predictor of Facebook adoption by candidates contesting in the 2011 Finnish Parliamentary elections. Specifically, males were more likely to be Facebook adopters. Williams and Gulati [18] investigating candidates running for the House of Representatives in 2006 found that a candidates' incumbency status affects the use of Facebook. Specifically, they report that challengers who were not holders of any political position were more likely to implement Facebook campaigns.

The effect of the size of the candidate's electoral district on Facebook adoption has not been clearly identified by prior research. For example, in the context of the 2008 Congressional elections Gulati and Williams [16] found no effect of the district urbanization on the Facebook usage of candidates. On the contrary, Strandberg [17] found a positive effect of the urbanization level on the Facebook adoption of Finnish candidates in the 2011 pre-election period. Significant differences were found in the level of Facebook usage between Greek candidates running in different peripheries in the 2010 Greek local elections. Specifically, candidates with a Facebook profile came from large peripheries and highly urbanized areas (i.e., Attiki, Central Macedonia) [19, 20]. These inconsistent results call for a further investigation of the relationship between a candidate's Facebook presence and the size of the district in which the candidate is contesting. Toward this end it is suggested that candidates running in large electoral districts will be active Facebook users compared to candidates belonging to small constituencies. Thus, the following hypotheses are developed:

H1: Gender will significantly influence Facebook use among candidates.

H2: Incumbency status will significantly influence Facebook use among candidates.

H3: Candidates contesting in electoral districts with high population densities will differ significantly in the level of Facebook usage compared to those in small electoral districts.

3.2   Factors Affecting Facebook Effectiveness

A politician's involvement with social media can have a positive impact on his vote share [21]. Recently, a number of studies support the fact that Facebook can be regarded as an important campaign activity capable of contributing to election success. The significant role of Facebook on the effectiveness of political marketing was best described by the term "Facebook election" used by Johnson and Perlmutter [22] to refer to the 2008 presidential elections in the US when the candidate Barak Obama incorporated Facebook efficiently in his campaign. Empirical evidence derived from the 2010 local elections in Greece suggested that candidates with a Facebook profile doubled their winning odds compared to non-Facebook adopters [12]. Another study, in the US showed that 55% of the congressional candidates who incorporated Facebook in their campaigns won a seat compared to 11% of elected candidates who were non-users. Based on the aforementioned results, the following hypothesis is developed:

H4: Facebook usage by candidates will significantly influence in a positive manner the number of votes received.

Much more empirical research is needed in order to shed more light on those aspects of Facebook activity that can become significant predictors of a candidate's vote share. Williams and Gulati [23] investigating the 2006 U,S, Congressional elections found that the number of an incumbent's Facebook friends is a significant indicator of his relative vote share. The contribution of Facebook on the electoral success of a candidate could be attributed to the fact that young voters prefer to support a political candidate with whom they have created a Facebook friendship [24].Other studies point out that the "like" feature of Facebook posts can reflect the "real" vote share of candidates. For example, in the 2009 Greek National elections the winning party PASOK outpaced the other parties in the number of the "likes" received by its Facebook friends[12]. Similar results were found in the context of the Finnish national election where the winning party (Finns party) came first in the number of Facebook "likes" received [25]. The Prime Minister Julia Gillard who won the Australian election in the 2010 was also the winner in the Facebook liking arena [26]. Based on the aforementioned, the following hypothesis is developed:

H5: The number of Facebook "likes" will significantly influence in a positive manner the number of votes received by a candidate.

4   Methodology and Results

The analysis presented in this paper is based on a sample of 1,318 candidates who ran for the May 2014 Greek Municipal elections. The data for this study came from sources such as official state records and various online platforms. For each candidate data were collected regarding the size of the municipality in which he/she was contested, the number of votes he/she received, the gender, the age, and the previous experience of the candidate.

Then researchers examined whether each candidate had an official Facebook page. For every candidate that owned a Facebook page, the total number of "page likes", talking about and followers that appeared on his/her Facebook page were collected. Following data collection, the statistical package for social sciences SPSS 17.0 was utilized in order to test the research hypotheses.

Based on the analysis, almost 30.2 percent of candidates (398 candidates) had a Facebook page. A closer look at the Facebook pages of the candidates indicates that an average Facebook page of candidates running for the municipality seats received 1,159.52 "page likes", 117.04 "talking about" and had 651.51 followers.

Most of the candidates were male 89.7% while only 10.3% of them were females. Of the 398 candidates that had a Facebook page 367 of them were males (92.2%) while only 31 of them were females (7.8%). However, these numbers do not indicate that male candidates are more likely to use Facebook compared to females, since males are overrepresented in the sample. Regarding, the incumbency status of candidates, 68% were regarded as challengers and 32% were incumbents. 60.1% of the candidates with a Facebook page were challengers and 39.9% were incumbents.

In order to test whether a candidate's gender and incumbency status predicts the use of Facebook a binary logistic regression was performed. Binary logistic regression was used to answer the first two hypotheses (H1 and H2). This type of regression was used since the dependent variable of interest – Facebook use - is a dichotomous categorical variable. The results of the regression analysis are shown in Table 1. The -2Log-likelihood value of the model is 1,588.81. Moreover, the significance level of the chi-square statistic is small (χ^2= 19.30, p = 0.000), thus, it can be concluded that the model is significantly better than the intercept only model. Hence, the model explains well the variations in the Facebook usage. The regression model was also evaluated by using the goodness-of-fit test proposed by Hosmer and Lemeshow. The chi-square value of the Hosmer and Lemeshow test was insignificant (χ^2= 0.006, p= 0.997) indicating a good fit for the data.

Table 1. Binary logistic regression results for facebook use variable

Heading level	Exp (B)
Constant	0.265*
Gender	1.444
Incumbency	1.634*
-2Log Likelihood	1,588.81
Chi-Square	19.30*

*Significant at the p=0.05 level.

As Table 1 shows only the incumbency status coefficient is statistically significant. Exp (B) for incumbency status is 1.634 which means that a candidate is 1.634 times more likely to have a Facebook page if she/he is a challenger. Hence, H1 is rejected while H2 is accepted.

H3 assumes that candidates running in electoral districts with high population densities are more likely to use Facebook as a campaigning tool compared to candidates who run in small municipalities. In order to test H3, independent t-test was used. Results of the tests are shown in Table 2.

Table 2. Results for Independent t-test for Municipality Population.

	Mean number of a municipality's inhabitants	T-Statistic/ Significance
Candidates with a Facebook Page	69,037.92	
Candidates without a Facebook Page	46,135.58	-3.747/0.000

Findings indicate that significant differences (p<0.05) exist between Facebook candidates and non-Facebook candidates in regards to the population density of the municipality in which they contest (t=-3.747, sig=0.000). Specifically, candidates with a Facebook page contest in larger municipalities in terms of inhabitants (M=69,037.92) compared to candidates who do not make use of Facebook as a campaigning tool (M=46,135.58). Thus, it can be argued that candidates running in large

electoral districts are more likely to use Facebook in order to get noticed and communicate with their citizenry compared to candidates who contest in small districts.
Hence, H3 is accepted.

H4 examines whether Facebook usage is a factor that has a significant effect on the
vote share of a candidate. Independent samples t-test was used again to test H4 (Table 3).

Table 3. Results for Independent t-test for Vote Share.

	Mean number of Votes	T-Statistic/ Significance
Candidates with a Facebook Page	5,985.16	-8.267/0.000
Candidates without a Facebook Page	3,569.58	

Based on Table 3, significant differences ($p<0.05$) were found between Facebook
candidates and non-Facebook candidates in terms of vote share ($t=-8,267$, $sig=0.000$).
In particular, the mean score of votes for candidates with a Facebook page
($M=5,895.16$) is higher compared to candidates without a Facebook page
($M=3,569.58$). Hence, it can be argued that the usage of a Facebook page could be a
factor that might exert an influence on the vote share of candidate. H4 was supported.

In order to test H5, which implies that citizens' awareness of a candidate's Facebook page as reflected in the number of Facebook "page likes" is an important factor
that influences the candidate's vote share, a correlation analysis was conducted using
Pearson's coefficient. Pearson's correlation coefficient was utilized since the two
variables - number of "page likes" and number of votes received - were continuous
and Pearson's coefficient measures the strength and direction of relationship between
two continuous variables. Moreover, the values of Pearson coefficient range between
-1 to +1. Results indicate that there is a significant ($p<0.05$) positive correlation between the number of Facebook "page likes" and the number of vote share of candidates ($r=0.583$). The significant correlation found could be characterized as moderate
since the value of Pearson's coefficient was below 0.70. Hence, one can conclude that
as the number of a candidate's Facebook "page likes" increases then his/her vote
share will increase as well, in a moderate level. In other words, as citizens become
aware of a candidate's Facebook page and support it by pressing the "page like" but-
ton then the chances that citizens might vote for that candidate might increase as well.
H5 was supported.

5   Discussion

The present study examined the use of Facebook by Greek candidates running for the
2014 local elections. Moreover, the factors that affect Facebook implementation by
candidates were investigated. Finally, the impact of Facebook on candidates' vote
share was assessed.

Exploitation of Facebook during the municipal election was pretty low since the majority of candidates were not present on Facebook. Only 30.2% candidates owned a Facebook page. Evidence suggests that candidates of the local elections are moving towards the implementation of Facebook campaigns, albeit slowly. A small increase in the use of Facebook was found in the 2014 elections compared to the municipal elections of 2012 in which 26.4 of candidates utilized Facebook [20]. Possible reasons for this low exploitation of Facebook tools by Greek local municipal governments could be attributed to the fact that municipal candidates continue to rely on traditional forms of campaigning such as face-to-face communication. However, one should bear in mind that Facebook penetration in Greece is 41% while 26% of social networking sites' users access social media via their smart-phones [27]. Hence, candidates might have not utilized extensively Facebook as a tool for election campaigning due to the fact that the majority of Greek citizens are not using Facebook.

Findings of the present study confirmed the significant impact of urbanization on Facebook adoption. These results are similar with those reported in the study of Strandberg [17] in the context of Finnish elections. Thus, Facebook is becoming an important platform for targeting voters in densely populated urban areas. Facebook use also depended on a candidate's prior political experience. According to Williams and Gulati [18] challengers utilize Facebook in order to overcome the advantages of incumbents who have established supporters and contacts.

Facebook is a political marketing tool used by candidates to send messages to potential voters [28], to interact with them, to influence their thoughts and attitudes and finally to win their votes [29]. The present research revealed a significant relationship between Facebook use and vote share. Hence, Facebook was identified by the present study as a powerful political marketing tool that can boost a candidate's vote share. Politicians can win votes if they are active users of Facebook. Moreover, a candidate's visibility on Facebook as reflected in the amount of "page likes" received can also be regarded as an indication of his/her ability to reach voters offline. This study found that Greek voters will connect on Facebook with the candidate that they will vote. Therefore, the more "page likes" a candidate receives the wider its offline impact would be in terms of vote share. The above findings are consistent with the earlier study of Williams and Gulati [23] which found a significant relationship between a candidate's Facebook supporters and his/her vote share. The positive link found in the present study between the number of "page likes" and the number of votes could be attributed to the content posted on the Facebook pages of candidates. Perhaps, candidates with higher scores than their opponents on the number of "page likes" , succeeded in engaging their followers by publishing interesting content. As a consequence Facebook engagement was then transformed into offline support.

For example, in the municipality of Piraeus, the candidacy of Yannis Moralis ("page likes": 14,524), who won the elections differentiated his Facebook campaign from his main competitor Michaloliakos Vasileios ("page likes": 7,048). Specifically, Yannis Moralis posted content regarding his priorities focusing on young people, athletics and sports. His posts also included photos with children as well as young volunteers. In addition, a number of the posts published were photos and information about one of his municipal candidate councilors Evangelos Marinakis who is a famous shipping magnate and president of Piraeus football club Olympiacos. Hence, Yannis Moralis used his candidate councilor as an endorsement for his campaign. However, one of the most important features of his Facebook campaign was the communication style he used. Most of the posts were "we messages" with positive tone that aimed to build relationships with users. Furthermore, a number of posts prompted users to share the messages with other users while Yannis Moralis engaged in dialogue with his users by replying to their comments. On the contrary, Vasileios Michaloliakos provided mainly one way information to his Facebook users. Specifically, his posts informed users about his appearances and interviews in television, his speeches, and his meetings with professional groups and candidates.

Another example of Facebook campaign strategy differentiation was that of Patoulis Georgios who contested in the municipality of Marousi against his competitor Vlachos Konstantinos. Patoulis Georgios was an incumbent ("page likes": 6,719) and won again the 2014 elections. The majority of his posts on Facebook included information about his proposed social policies towards the poor and elderly. Moreover, his Facebook campaign focused on profiling his candidate councilors as well as his accomplishments as a mayor. On the other hand, his competitor Vlachos Konstantinos ("page likes": 386) concentrated on publishing posts about his speeches, meetings with citizens and associations. Another important feature of his Facebook page was negative campaigning since a large number of posts referred to the negative aspects of the main opponent Mr. Patoulis Georgios. Hence, it can be concluded that the way a candidate promotes his/her self on Facebook could affect the awareness users have of his/her Facebook page which in turn would influence the number of votes received.

Several practical implications arise from the study's findings. Political marketers should take serious consideration on Facebook current impact on election campaigns and underscore to their clients the crucial role of maintaining a Facebook presence, especially regarding the creation of Facebook friendships. Facebook friendships can increase a candidate's visibility and awareness.

What was the aftermath of Facebook campaigns following the 2014 Greek Municipal elections? Answering this question is not a simple task since Facebook is one out of the many campaign tools Greek candidates used to lure their voters (i.e. TV ads, speeches) and voting decision is affected by other micro and macro factors as well (i.e. political ideology, economic climate). In the present study, the impact of Facebook usage on voting share was assessed without taking under consideration other indicators that might influence voters. Jungherr et al. [30] highlights the danger of reporting biased results by excluding certain variables from the models that predict voting share. Hence, future research should continue to test the impact of Facebook alongside with other traditional and online media.

Acknowledgments. This research has been co-financed by the European Union (European Social Fund – ESF) and Greek national funds through the Operational Program "Education and Lifelong Learning" of the National Strategic Reference Framework (NSRF) - Research Funding Program: ARCHIMEDES III. Investing in knowledge society through the European Social Fund.

References

1. Cornfield, M., Rainie, L.: The Impact Of The Internet On Politics. Pew Internet and American Life Project (2006). http://www.pewinternet.org/ppt/pip_internetandpolitics.pdf
2. Vesnic-Alujevic, L.: Political Participation and Web 2.0 in Europe: A Case Study Of Facebook. Public Relations Review 38(3), 466–470 (2012)
3. Gueorguieva, V.: Voters, Myspace, and Youtube: The Impact Of Alternative Communication Channels On The 2006 Election Cycle And Beyond. Social Science Computer Review. 26(3), 288–300 (2008)
4. Williamson, D.: Social Network Ad Spending: 2010 Outlook. e-Marketer (December 2009)
5. Pena-Lopez, I.: Striving behind the shadow – the dawn of Spanish politics 2.0. In: Van der Hof, S., Groothuis, M. (eds.) Innovating Government. Normative, Policy and Technological Dimensions of Modern Government, pp. 111–128. IGI Global, Hershey (2011)
6. Vergeer, M., Hermans, L., Sams, S.: Online Social Networks and Microblogging in Political Campaigning: The Exploration of a New Campaign Tool and New Campaign Style. Party Politics 19(3), 477–501 (2011)
7. Karlsen, R.: A Platform for Individualized Campaigning? Social Media and Parliamentary Candidates in the 2009 Norwegian Election Campaign. Policy & Internet 3(4), 1–25 (2011)
8. Enli, G., Skogerbø, E.: Personalized Campaigns in Party-Centred Politics. Information, Communication & Society 16(5), 757–774 (2013)
9. Papathanassopoulos, S.: Election Campaigning in the Television Age: The Case of Contemporary Greece. Political Communication 17(1), 47–60 (2000)
10. Kotsikopoulou, V.: Ekloges Kai Diadiktyo: I Periptosi Tou Inomenou Vasileiou Kai Tis Elladas [Elections and the Internet: The cases of the United Kingdom and Greece]. In: Demertzis, N. (ed.) I Politiki Epikoinonia Stin Ellada [Political Communication in Greece], pp. 173–210. Papazissis Publishers, Athens (2002)
11. Demertzis, N., Diamantaki, K., Gazi, A., Sartzetakis, N.: Greek Political Marketing Online: An Analysis of Parliament Members' Web Sites. Journal of Political Marketing 4(1), 51–74 (2005)
12. Lappas, G., Kleftodimos, A., Yannas, P.: Greek Parties and Web 2.0. Paper Presented at the Workshop: Elections, Campaigning and Citizens Online. Oxford Internet Institute (2010)
13. Norris, P.: A Virtuous Circle: Political Communications in Postindustrial Societies. Cambridge University Press, Cambridge (2000)
14. Tops, P., Horrocks, I., Hoff, J.: New technology and democratic renewal: the evidence assessed. In: Tops, P. (ed.) Democratic Governance and New Technology. Routledge, London (2000)
15. Chen, P.J.: Candidates' New Media Use in the 2007 Australian National Election. Record of the Communications Policy & Research Forum 2008, pp. 62–78 (2008). http://www.networkinsight.org/verve/_resources/Record_CPRF08.pdf
16. Gulati, G.J., Williams, C.B.: Social Media and Campaign 2012: Developments and Trends for Facebook Adoption. Social Science Computer Review (2013)
17. Strandberg, K.: A Social Media Revolution or Just a Case of History Repeating Itself? The Use of Social Media in the 2011 Finnish Parliamentary Elections. New Media & Society (2013)
18. Williams, C.B., Gulati, G.J.: Social Networks in Political Campaigns: Facebook and The Congressional Elections of 2006 and 2008. New Media & Society 15(1), 52–71 (2013)

19. Yannas, P., Kleftodimos, A., Lappas, G.: Online Political Marketing in the 2010 Greek Local Elections: The Shift From Web to Web 2.0 Campaigns. Paper presented at the 16th International Conference on Corporate and Marketing Communication. Athens, Greece, April 2011

20. Lappas, G., Triantafillidou, A., Yannas, P.: Social media campaigning by candidates. In: The 2010 Greek Municipal Elections. Paper presented at the International Conference on Contemporary Marketing Issues. Thessaloniki, Greece, June 2012

21. Effing, R., van Hillegersberg, J., Huibers, T.: Social media and political participation: are facebook, twitter and youtube democratizing our political systems? In: Tambouris, E., Macintosh, A., de Bruijn, H. (eds.) ePart 2011. LNCS, vol. 6847, pp. 25–35. Springer, Heidelberg (2011)

22. Johnson, T., Perlmutter, D.: Introduction: The Facebook Election. Mass Communication and Society **13**, 554–559 (2010)

23. Williams, C., Gulati, G.: Social networks in political campaigns: Facebook and the 2006 midterm elections. Paper Presented at the 2007 Annual Meeting of the American Political Science Association. Chicago, Illinois, August 2007

24. Towner, T., Dulio, D.: The Web 2.0 Election: Does The Online Medium Matter? Journal of Political Marketing. **10**(1), 165–188 (2011)

25. Leskinen, H.: The Greens of Finland in Social Media: Facebook as a Communication Forum. Unpublished Bachelor's Thesis. Retrieved from Theseus.fi database (2012)

26. Macnamara, J., Kenning, G.: E-electioneering 2010: Trends in Social Media Use in Australian Political Communication. Media International Australia, Incorporating Culture & Policy **139**, 7–22 (2011)

27. European Digital Landscape 2014, http://147.102.16.219/demo1/attachments/124_european%20digital%20landscape%202014.pdf

28. Andersen, K.N., Medaglia, R.: The use of facebook in national election campaigns: politics as usual? In: Macintosh, A., Tambouris, E. (eds.) ePart 2009. LNCS, vol. 5694, pp. 101–111. Springer, Heidelberg (2009)

29. Utz, S.: The (Potential) Benefits of Campaigning via Social Networking Sites. Journal of Computer Mediated Communication. **14**(2), 221–243 (2009)

30. Jungherr, A., Jürgens, P., Schoen, H.: Why the Pirate Party Won the German Election of 2009 or the Trouble with Predictions: A Response to Tumasjan, A., Sprenger, To, Sander, PG, & Welpe, IM Predicting Elections with Twitter: What 140 Characters Reveal About Political Sentiment. Social Science Computer Review **30**(2), 229–234 (2012)

Perceived Usefulness of Word-of-Mouth: An Analysis of Sentimentality in Product Reviews

Chih-Chien Wang[1(✉)], Ming-Zhe Li[1], and Yolande Y.H. Yang[2]

[1] Graduate Institute of Information Management, National Taipei University,
151 University Rd, Sansia District, New Taipei City 237, Taiwan
wangson@mail.ntpu.edu.tw, rock_battle320@hotmail.com
[2] Department of Business Administration, National Taipei University,
151 University Rd, Sansia District, New Taipei City 237, Taiwan
Yolande@mail.ntpu.edu.tw

Abstract. People may regard some word-of-mouths (WOMs) as more useful than others. Review valance, emotion, and concreteness are the antecedents which may influence people's perceived usefulness of the WOMs. This study collected online customer reviews from Amazon.com and conduct regression analysis and Partial Least Squares (PLS) analysis to analyse the data. The findings revealed that negativity bias does exist, with negative reviews deemed as more useful than positive reviews. Besides, the research also revealed that online customer reviews with more positive emotion expression are perceived as less useful. However, online customer reviews with more negative emotion expression are perceived more useful. In addition, online customer reviews with longer review length are perceived as more useful than shorter online customer reviews.

Keywords: Word-of-mouth (WOM) · Review usefulness · Negativity bias · Emotion

1 Introduction

The consumers need adequate information to solve the problems they face [1]. With the prevalence of networks and mobile devices, the consumers have more channels to search information to make their purchase decisions. Compared to traditional word-of-mouth marketing, the consumers can receive both positive and negative messages from the network, and be influenced by those messages. Park and Kim [2] argued that online reviews are normally expressed in words, so the viewers can easily evaluate the contents and the network provides more and richer information than the real world to the consumers.

There is asymmetric effect between positive and negative information [3]. People regard criticism more than complements, and pay more attention to bad events than to good news [4]. Park and Lee [5] also pointed out that in online WOM, the negative messages are more effective than the positive ones. But Wu [6] overturned the result

© Springer-Verlag Berlin Heidelberg 2015
L. Wang et al. (Eds): MISNC 2015, CCIS 540, pp. 448–459, 2015.
DOI: 10.1007/978-3-662-48319-0_37

about the bias towards negative messages, arguing that the consumers may not regard the negative messages as more useful. The existence of the bias towards negative message has now become a topic worth exploring.

The information contained in online WOM includes the over-all rating of the product (certain number of stars or whether the customer is satisfied), detailed ratings for individual items, the experience of using the product, the feelings after reading a novel or watching a movie, or the service of a restaurant. Though credibility is increased by the written words, the sales of the reviewed products and the number of responses are affected by the online review being abstract or explicit, and the readability [7]. Negative WOM often carry negative emotional words[8]. As argued by Weun, Beatty and Jones [9], negative WOM is strong negative emotion felt by the customer during the service failure. Excessive negative emotional words make the reader doubt the motivation of the reviewer, and lower the usefulness rating of the review[8].

The current study focuses on explore the antecedents of the perceived usefulness of WOM. The current study aims to answer the following questions :

1. Will positive customer reviews convince customer more than negative ones?
2. Will emotion contents in customer reviews influence usefuleness perception to WOMs?
3. Will review concreteness influence usefulness perception?

2   Literature Review and Hypotheses

2.1   Online Word-of-Mouths and Reviews

Dichter [10] defined the word-of-mouth (WOM) as the informal verbal communication between individuals, based on non-commercial intentions, to discuss and exchange information about some products, services or brands. Electronic word-of-mouth (e-WOM) is defined as the informal communications through websites such as forums, chat rooms, bulletins, blogs, messaging services or virtual communities. Consumers can use these channels to gather the experiences, opinions and product information from other consumers, and express their own experiences, opinions, knowledge and information. eWOM can be regarded as a new type of WOM[11]

With the advancement of the Internet and the prevalence of mobile devices, consumers are relying on online consumer reviews as a reference of purchase deceisions [12]. Web 2.0 has added social interaction to eWOM for readers to affirm the reviews or add their own comments[13].

Online reviews have attracted much attention from researchers (e.g.,[14-16]. Some of those focused on the way reviews are transmitted, and the influences of online reviews[17, 18]. Others focused on the richness of reviews and anonymity [14]. But few addressed how the consumers reduce the uncertainty and extract the information they want.

2.2   Perceived Usefulness

Perceived usefulness of reviews is defined as how the consumer finds a review is useful in helping the purchase decision. As an example, the Amazon.com review interface contains three main elements: (a) the number of stars given by the reviewer (commonly known as the review rating), (b) the review content, which is displayed in the body of the text along with review meta characteristics such as the author's pseudonym (user name or nickname) and the date of submission, and (c) the usefulness of the review displayed in a textual way.

A main axis of past researches is the relationship between usefulness and the characteristics of the reviews [16, 19-22]. The length of the review and readability were identified as important variables of the review characteristics [7, 16]. The review length affects the credibility of perceived usefulness of the review [16, 17, 23, 24] and the readability is positively related to usefulness[7].

2.3   Review Length

Chevalier and Mayzlin [14] found in a research of Amazon.com that the amount of information in reviews (in the number of words) is significantly related to the sale of the product. A possible explanation is that the length of reviews corresponds to the degree of concern of the reviewer on the particular product. The length of the reviews is reflected in the size occupied on the screen. Compared to shorter reviews, the longer reviews are less likely to be ignored because they stand out visually.

2.4   Readability

Readability is how easy a reader can read an article [22], which affects how one comprehends the product review [25]. Past researches also indicated that the formatting of a review might also affect readability [26]. Separating the article into paragraphs or using headers indicating positive and negative opinions may also aide the consumer to classify and comprehend the reviews.

Some other researches indicated the factors that influence the perceived usability being the source credibility, product categories, valence and ratings [27] [28] [16] [17, 24]. In general, readability is related to one's effort and level of education, representing the ability to understand the particular piece of text [25, 29].

2.5   Review Valence and Negativity Bias

Negativity bias was observed in past researches, implying negative information is more valuable and helpful in decision making than positive information[4]. Negative information is deemed to be more useful because it's more diagnostic [30] or more consequential [31]. Bronner and de Hoog [32] indicated that positive or neutral reviews are more catching because they are more frequent than negative ones [33, 34].

Back in 1967, Arndt pointed out that 57% of comsumers will buy a product when hearing positive WOM, but 82% of consumers won't buy it when hearing negative WOM. Similar conclusions were obtained by Hennig-Thurau, Gwinner, Walsh and Gremler [11]. But there are some scholars against the existence of the negativity bias [6]. Some other researches revealed that the negativity bias can vary according to temporal contiguity[35], conformity [36], and product categories [37]. So the effect of negativity bias is not conclusive yet. Therefore, we proposed following hypothesis:

H1: Review valence is negatively related to usefulness.

2.6   Emotion

Emotions are mental states of readiness that arise from appraisals of events or one's own thoughts [38] Emotions may result in related actions or behaviors [39]. Recently the emotion has been conceptualized in two dimensions: valance (positive or negative) and arousal (high or low) [40-42]. For example, if a consumer's review expresses anger, it may be classified as negative valence and high arousal.

Past researches pointed out that emotion contains implicit or explicit information that helps us make decisions quickly [43] and generate intense and effective reactions in an individual [44]. Emotional words may greatly affect the judgment of the reader [45] and frequently shared [46]. Recently, Yin, Bond and Zhang [47] confirmed the effect of discrete emotion of anxiety and anger on perceived helpfulness above and beyond ratings or information content alone. Cao, Duan and Gan [48] used text mining to uncover the role of emotions play in helpfulness. therefore, we proposed following hypothesis:

H2: Positive emotion is positively related to usefulness

H3: Negative emotion is negatively related to usefulness

2.7   Linguistic Inquiry and Word Count

Languages play an important role between the reviewers and the readers. In traditional word of mouth, there are many non-verbal clues such as facial expressions and body languages, and interactive exchanges that help convey the contents. In the online world, though, the words, grammar, and inflexions are used to convey the meanings.

Past researches focused more on the valence of WOM than on the contents of the words. Some recent researches are starting to focus on the abstractness of the language [49], figurative uses [50], temporal contiguity [35], and explaining language[51], which would affect the helpfulness ratings by the consumer[48] [35].

Semin and Fiedler [52] proposed the Linguistic Category Model (LCM) to divided the language into categories and used the categories to count the words in text. LIWC has been used to efficiently classify texts along psychological dimensions and to predict behavioral outcomes. We used the LIWC software[53] to compute the emotional words in the review.

H4. Review concreteness is related to review usefulness.

H5: Readability is positively related to review usefulness.

3   Empirical Study

3.1   Data and Procedure

The Research Process We used the software tool Easy Web Extract 3.0 to collect data from Amazon.com. The software can analyse the web pages, and extract the desired data (such as ratings, reviews, usefulness, links to related products). The collected data were further processed with Linguistic Inquiry and Word Count (LIWC) to parse the emotional contents. LIWC is a word parsing software that can distinguish between more than 70 categories to determine the perception, emotion and attitude. The output of the LIWC was fed to SPSS for analysis. Readability was obtained through the Readability function in the Visual Basic Application for Microsoft Word 2013. We collected 87,500 pieces of valid data from Amazon.com, of which 54,299 pieces were for search goods, collected from sub-categories of wearable devices, and 33,201 pieces were for experience goods, collected from sub-categories of computer games. The data were collected between January and Macrh, 2015.

In this research, we divided the reviews into two groups according to the number of words, using the average number as the boundary. For search goods, the average number is 85, and for experience good, the average number is 60. Tables 1 lists the variables in our study.

3.2   Data Analysis

Table 2 provide a descriptive statistics of online product review sample. The average length of reviews is 60 word (M=60.49, SD=122.285), each review varying from 0 to 4818 words. The average word number of positive emotion 17.09 %, and the average word number of negative emotion 1.312 % separately suggest that a review text contain 17.09 % positive emotion expression and 1.312 % negative emotion expression in total length of a review. The average readability score is 54.83 and, which indicate that most reviews are appropriate for general reading.

Figure 1 reveals the result of PLS model of all reviews. Both positive (path coefficient=-.020, P<0.5) and negative emotion (path coefficient=0.011,P<0.001)are significant. The relationship between emotion and usefulness is supported. Review valence is also negatively related to review usefulness (path coefficient=-0.053,P<0.5).

Figure 2 reveals the result of PLS model for the group with review length smaller than average length. Both positive (path coefficient=-0.042, P<0.001) and negative emotion (path coefficient=0.018, P<0.5) are significant. The relationship between emotion and usefulness is supported. Review valence is also negatively related to review usefulness (path coefficient=-0.083,P<0.5).

Table 1. Variable Description

Variable	Information collected from the website
Product Rating	The overall appraisal of the product, in number of stars
Total Useful Votes	Number of votes to the question "Is this review useful?"
Number Of "Useful" Votes	Number of "Useful" votes
Ratio Of "Useful" Votes	Total useful votes divided by the number of useful vote
Word Counts	Number of words of each review, as counted by Office
Flesch Reading Ease	The score computed by Office 2013
Flesch-Kincaid Grade Level	The score computed by Office 2013
Positive Emotion expression	The percentage of words with positive emotion as counted by LIWC
Negative Emotion expression	The percentage of words with negative emotion as counted by LIWC

Table 2. Descriptive Statistics

Variable	Minimum	Maximum	Mean	SD
Product Rating	1	5	4.38	1.195
Total Useful Votes	0	4818	3.64	32.834
Number Of "Useful" Votes	0	4818	8.86	50.771
Ratio Of "Useful" Votes	0	1	0.15	0.354
Word Counts	0	1253	60.49	122.285
Flesh Reading Ease	0	100	54.83	37.615
Flesch-Kincaid Grade Level	0	94	3.95	4.037
Positive Emotion expression	0	100	17.0900	22.2486
Negative Emotion expression	0	100	1.31260	3.760346
				1

Note: $N=33201$.

Figure 3 reveals the result of PLS model for the group with review length larger than average length. Review valence is also negatively related to review usefulness (path coefficient=-0.040,P<0.01). Review concreteness is positively related to review usefulness (path coefficient=0.019, P<0.01).

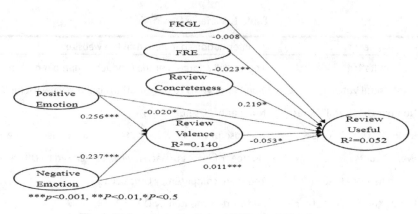

Fig. 1. PLS analysis result for All Word-of-Mouths

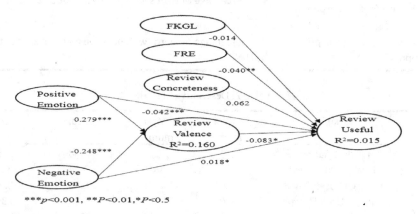

Fig. 2. PLS analysis result for short Word-of-Mouths

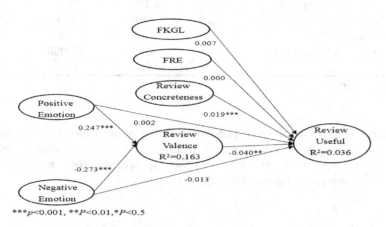

Fig. 3. PLS analysis result for long Word-of-Mouths

4   Discussion

This research examines the existence of negativity bias of eWOM and the impact on emotion and review valence, and finds that negative reviews are more useful than positive reviews through regression analysis result of different models. Customer reviews with lower rating acquire more votes of usefulness than customer reviews with higher rating, that is to say, negativity bias of eWOM does exist. The research result is consistent with literature that proved the existence of negativity bias of eWOM is true, instead of the inexistence of negativity bias of eWOM [6].

Second, the research investigates that whether affection influences review usefulness, finding that the fewer positive expression of emotion in online customer reviews make consumers perceive more usefulness in the review. The reason is that consumers probably regard those positive reviews with many positive emotional expression as forged by the online product vendor or other review writer hired from the product vendor.

Third, the research also finds that consumers perceive reviews with most negative expression as more useful. Those negative customer reviews coming from the experience of every customer themselves seems more convincing than eWOM with many positive expression of emotion. The possibility of disseminating negative eWOM by online product vendor themselves is rare, as a large amount of negative WOM will hurt the brand reputation of the product.

Fourth, the research finds that concrete customer review is more useful than abstract reviews. In addition, readability of the product review did not influence perceived usefulness.

Table 3. Hypotheses testing

Hypo-thesis	PLS Hypothesis Relationship	All eWOM	Coefficient/t-value Short eWOM	Long eWOM	Result
H1	Valence→ Useful	-.053(2.201)*	-.083(2.029)**	-.040(2.913)**	Support
H2	Pos.Emo→ useful	-.020(2.952)**	-.042(3.935)***	0.002(0.183)	Support
H3	Neg.Emo→ useful	.011(2.162)***	.018(2.152)**	-0.013(1.074)	Support
H4	W.C → Useful	0.219(1.966*	0.062(1.917)	0.092(11.940)***	Support
H5	FRE→ Useful	-0.0203(2.72)**	-0.004(2.698)**	0.000(0.040)	Not Support
	FKGL→ Ueful	-0.008(1.056)	-0.014(1.067)	-0.007(0.479)	Not Support

This study contributes to both theory and practices related to the usefulness of online reviews. Firstly, we showed that review helpfulness is affected by the negativity bias because negative reviews receive more useful votes. We also found that how this negativity bias effect is influenced by product categories: negative reviews are more important in experience goods than in search goods. We think this can be attributed to the evaluation of experience goods relying more on personal experience, feelings, and subjective judgments while the evaluation of search goods relying more on the features and performances of the product, in which personal experiences are less useful.

Secondly, we confirmed the influence of emotional factors on the usefulness of reviews, especially in negative reviews, where the use of emotional words increases the usefulness rating. But in positive reviews, the use of emotional words is negatively related to the usefulness rating. Negative emotional words add authenticity of the reviews. Because the companies may use various incentives to encourage consumers to leave online messages, excessive positive words may lead the consumer to doubt if the review is paid by the company.

Lastly, with data collected from a real online retailer, we provided empirical evidences to support our conceptual predictions. We collected a total of 87,500 online consumer reviews to validate the correlation between readability, length, and the usefulness.

This study indicates that using emotion expression in a product review could influence review valence and usefulness. Reviewers on website are supposed to realize that negative expression of emotions may cause readers to perceive reviews as more useful. From the prospects of marketing, product vendor needs to be aware of the volume of product review developing a negative trend which may rapidly attract consumers' eyes. In that way, audiences probably have a negative impression on the product.

References

1. Solomon, P.: Discovering Information Behavior in Sense Making. I. Time and Timing. J. Am. Soc. Inform. Sci. **48**, 1097–1108 (1997)
2. Park, D.H., Kim, S.: The Effects of Consumer Knowledge on Message Processing of Electronic Word-of-Mouth Via Online Consumer Reviews. Electronic Commerce Research and Applications **7**, 399–410 (2009)
3. Peeters, G.: The Positive-Negative Asymmetry: On Cognitive Consistency and Positivity Bias. Eur. J. Soc. Psychol. **1**, 455–474 (1971)
4. Baumeister, R.F., Bratslavsky, E., Finkenauer, C., Vohs, K.D.: Bad is Stronger Than Good. Rev. Gen. Psychol. **5**, 323 (2001)
5. Park, C., Lee, T.M.: Information Direction, Website Reputation and Ewom Effect: A Moderating Role of Product Type. J. Bus. Res. **62**, 61–67 (2009)
6. Wu, P.F.: In Search of Negativity Bias: An Empirical Study of Perceived Helpfulness of Online Reviews. Psychol. Market. **30**, 971–984 (2013)
7. Ghose, A., Ipeirotis, P.G.: Estimating the Helpfulness and Economic Impact of Product Reviews: Mining Text and Reviewer Characteristics. IEEE Transactions on Knowledge and Data Engineering **23**, 1498–1512 (2011)
8. Banerjee, S., Chua, A.Y.: A study of manipulative and authentic negative reviews. In: Proceedings of the 8th International Conference on Ubiquitous Information Management and Communication, pp. 76. ACM

9. Weun, S., Beatty, S.E., Jones, M.A.: The Impact of Service Failure Severity on Service Recovery Evaluations Andpost-Recovery Relationships. J. Serv. Mark. **18**, 133–146 (2004)
10. Dichter, E.: How Word-of-Mouth Advertising Works. Harvard Bus. Rev. **44**, 147–160 (1966)
11. Hennig-Thurau, T., Gwinner, K.P., Walsh, G., Gremler, D.D.: Electronic Word-of-Mouth Via Consumer-Opinion Platforms: What Motivates Consumers to Articulate Themselves on the Internet? J. Interact. Mark. **18**, 38–52 (2004)
12. Senecal, S., Nantel, J.: The Influence of Online Product Recommendations on Consumers' Online Choices. J. Retailing. **80**, 159–169 (2004)
13. Blazevic, V., Hammedi, W., Garnefeld, I., Rust, R.T., Keiningham, T., Andreassen, T.W., Donthu, N., Carl, W.: Beyond Traditional Word-of-Mouth. J. Serv. Manage. **24**, 294–313 (2013)
14. Chevalier, J.A., Mayzlin, D.: The Effect of Word of Mouth on Sales: Online Book Reviews. J. Marketing Res. **43**, 345–354 (2006)
15. Clemons, E.K., Gao, G.G.: Consumer Informedness and Diverse Consumer Purchasing Behaviors: Traditional Mass-Market, Trading Down, and Trading out into the Long Tail. Electronic Commerce Research and Applications **7**, 3–17 (2008)
16. Mudambi, S.M., Schuff, D.: What Makes a Helpful Review? A Study of Customer Reviews on Amazon. Com. MIS Quart. **34**, 185–200 (2010)
17. Sen, S., Lerman, D.: Why Are You Telling Me This? An Examination into Negative Consumer Reviews on the Web. J. Interact. Mark. **21**, 76–94 (2007)
18. Zhang, J.Q., Craciun, G., Shin, D.: When Does Electronic Word-of-Mouth Matter? A Study of Consumer Product Reviews. J. Bus. Res. **63**, 1336–1341 (2010)
19. Pang, B., Lee, L.: A sentimental education: sentiment analysis using subjectivity summarization based on minimum cuts. In: Proceedings of the 42nd annual meeting on Association for Computational Linguistics, p. 271. Association for Computational Linguistics
20. Ghose, A., Ipeirotis, P.G.: Designing ranking systems for consumer reviews: the impact of review subjectivity on product sales and review quality. In: Proceedings of the 16th Annual Workshop on Information Technology and Systems, pp. 303–310
21. Pavlou, P.A., Dimoka, A.: The Nature and Role of Feedback Text Comments in Online Marketplaces: Implications for Trust Building, Price Premiums, and Seller Differentiation. Inform. Syst. Res. **17**, 392–414 (2006)
22. Korfiatis, N., Rodríguez, D., Sicilia, M.-A.: The impact of readability on the usefulness of online product reviews: a case study on an online bookstore. In: Lytras, M.D., Damiani, E., Tennyson, R.D. (eds.) WSKS 2008. LNCS (LNAI), vol. 5288, pp. 423–432. Springer, Heidelberg (2008)
23. Doh, S.-J., Hwang, J.-S.: How Consumers Evaluate Ewom (Electronic Word-of-Mouth) Messages. Cyberpsychol. Behav. **12**, 193–197 (2009)
24. Willemsen, L.M., Neijens, P.C., Bronner, F., de Ridder, J.A.: "Highly Recommended!" the Content Characteristics and Perceived Usefulness of Online Consumer Reviews. J. Compu.-Mediat. Comm. **17**, 19–38 (2011)
25. Zakaluk, B.L., Samuels, S.J.: Readability: Its Past, Present, and Future. ERIC (1988)
26. Liu, J., Cao, Y., Lin, C.-Y., Huang, Y., Zhou, M.: Low-quality product review detection in opinion summarization. In: EMNLP-CoNLL, pp. 334–342
27. Cheung, C.M., Lee, M.K., Rabjohn, N.: The Impact of Electronic Word-of-Mouth: The Adoption of Online Opinions in Online Customer Communities. Internet Res. **18**, 229–247 (2008)

28. Forman, C., Ghose, A., Wiesenfeld, B.: Examining the Relationship between Reviews and Sales: The Role of Reviewer Identity Disclosure in Electronic Markets. Inform. Syst. Res. **19**, 291–313 (2008)
29. DuBay, W.H.: The Principles of Readability. Online Submission (2004)
30. Herr, P.M., Kardes, F.R., Kim, J.: Effects of Word-of-Mouth and Product-Attribute Information on Persuasion: An Accessibility-Diagnosticity Perspective. J. Consum. Behav. 454–462 (1991)
31. Rozin, P., Royzman, E.B.: Negativity Bias, Negativity Dominance, and Contagion. Pers. Soc. Psychol. Rev. **5**, 296–320 (2001)
32. Bronner, F., de Hoog, R.: Consumer-Generated Versus Marketer-Generated Websites in Consumer Decision Making. Int. J. Market. Res. **52**, 231–248 (2010)
33. Fiske, S.T.: Attention and Weight in Person Perception: The Impact of Negative and Extreme Behavior. J. Pers. Soc. Psychol. **38**, 889 (1980)
34. Peeters, G., Czapinski, J.: Positive-Negative Asymmetry in Evaluations: The Distinction between Affective and Informational Negativity Effects. Eur. J. Soc. Psychol. **1**, 33–60 (1990)
35. Chen, Z., Lurie, N.H.: Temporal Contiguity and Negativity Bias in the Impact of Online Word of Mouth. J. Marketing Res. **50**, 463–476 (2013)
36. Purnawirawan, N., De Pelsmacker, P., Dens, N.: Balance and Sequence in Online Reviews: How Perceived Usefulness Affects Attitudes and Intentions. J. Interact. Mark. **26**, 244–255 (2012)
37. Zhu, F., Zhang, X.: Impact of Online Consumer Reviews on Sales: The Moderating Role of Product and Consumer Characteristics. J. Marketing **74**, 133–148 (2010)
38. Bagozzi, R.P., Gopinath, M., Nyer, P.U.: The Role of Emotions in Marketing. J. Acad. Market. Sci. **27**, 184–206 (1999)
39. Lerner, J.S., Keltner, D.: Beyond Valence: Toward a Model of Emotion-Specific Influences on Judgement and Choice. Cognition Emotion **14**, 473–493 (2000)
40. Bradley, M.M., Lang, P.J.: Measuring Emotion: The Self-Assessment Manikin and the Semantic Differential. J. Behav. Ther. Exp. Psy. **25**, 49–59 (1994)
41. Cacioppo, J.T., Petty, R.E., Losch, M.E., Kim, H.S.: Electromyographic Activity over Facial Muscle Regions Can Differentiate the Valence and Intensity of Affective Reactions. J. Pers. Soc. Psychol. **50**, 260 (1986)
42. Russell, J.A.: A Circumplex Model of Affect. J. Pers. Soc. Psychol. **39**, 1161 (1980)
43. Bechara, A., Damasio, A.R.: The Somatic Marker Hypothesis: A Neural Theory of Economic Decision. Game Econ. Behav. **52**, 336–372 (2005)
44. Frijda, N.H., Mesquita, B.: The Social Roles and Functions of Emotions (1994)
45. Schindler, R.M., Bickart, B.: Perceived Helpfulness of Online Consumer Reviews: The Role of Message Content and Style. J. Consum. Behav. **11**, 234–243 (2012)
46. Stieglitz, S., Dang-Xuan, L.: Emotions and Information Diffusion in Social Media-Sentiment of Microblogs and Sharing Behavior. J. Manage. Inform. Syst. **29**, 217–248 (2013)
47. Yin, D., Bond, S., Zhang, H.: Anxious or Angry? Effects of Discrete Emotions on the Perceived Helpfulness of Online Reviews. MIS Quart. **38**, 539–560 (2014)
48. Cao, Q., Duan, W., Gan, Q.: Exploring Determinants of Voting for the "Helpfulness" of Online User Reviews: A Text Mining Approach. Decis. Support. Syst. **50**, 511–521 (2011)
49. Schellekens, G.A., Verlegh, P.W., Smidts, A.: Language Abstraction in Word of Mouth. J. Consum. Behav. **37**, 207–223 (2010)

50. Kronrod, A., Danziger, S.: "Wii Will Rock You!" the Use and Effect of Figurative Language in Consumer Reviews of Hedonic and Utilitarian Consumption. J. Consum. Behav. **40**, 726–739 (2013)
51. Moore, S.G.: Some Things Are Better Left Unsaid: How Word of Mouth Influences the Storyteller. J. Consum. Behav. **38**, 1140–1154 (2012)
52. Semin, G.R., Fiedler, K.: The Linguistic Category Model, Its Bases, Applications and Range. Eur. J. Soc. Psychol. **2**, 1–30 (1991)
53. Pennebaker, J.W., Booth, R.J., Francis, M.E.: Linguistic Inquiry and Word Count: Liwc [Computer Software]. liwc. net, Austin (2007)

The Smell Network

Ritesh Kumar[1,2(✉)], Rishemjit Kaur[3], and Amol P. Bhondekar[1,2]

[1] CSIR-Central Scientific Instruments Organisation, Chandigarh, India
riteshkr@csio.res.in
[2] Academy of Scientific and Innovative Research, New Delhi, India
[3] Nagoya University, Nagoya, Japan

Abstract. The smell of a molecule is subjective, because there is a variablity in its representative language. The reporting is done according to the vocabulary repertoire of human subjects and researchers concerned. The olfactory databases thus consist of molecules and their smell characteristics defined by words. In this paper, we have demonstrated a network based approach based on the words to understand the perceptual universe. We defined perceptual communities based on the normalized co-occurrence network and hence propose the perceptual classes. We find the characteristics of this perceptual social network. We have also proposed a generative LDA-based topic modeling approach for topic detection in olfactory databases. This is for the first time that an objective approach to defining perceptual classes has been carried out which confirms with many subjective analyses that has been done till now. This work may open new avenues towards understanding the relationship between language and olfaction besides objectively defining perceptual classes.

Keywords: Social network analysis · Clustering coefficient · Community detection · Latent diritchlet allocation · Olfaction · Data mining

1 Introduction

Recently, network analysis has grabbed much attention due to its clear representation in terms of entities and relationship and, they have provided some really interesting insights into the data they represent [1]. The present work aims at discovering inherent statistical structure in large chemical and perceptual databases available online in order to derive principles for predicting odor perception from the chemical structure of odorants. An adjacency list of odorants on the basis of perceived smell was created. Further, a *smell network* was created in which each odorant forms the node and weight of edge between them shows the normalized number of smells they share with each other. The results provide useful insights into the odor space such as the special characteristics of the smell network.

The last decade has seen a tremendous surge in natural language processing based software and their application to different areas. The mathematical techniques behind these works have given an impetus to even the social sciences and literature[2–4]. The most famous works have been to automatically detect topics and classify large corpus

© Springer-Verlag Berlin Heidelberg 2015
L. Wang et al. (Eds): MISNC 2015, CCIS 540, pp. 460–469, 2015.
DOI: 10.1007/978-3-662-48319-0_38

of documents into well organized corpus. Probabilistic topic modeling techniques provide comprehensive algorithms to define topics from which we can make a definite sense [5]. The techniques have been used on many kinds of data and have found application in finding patterns in genetic data, images and social networks etc. [6,7,8]. The last decade has seen a tremendous surge in natural language processing based software and their application to different areas. The mathematical techniques behind these works have given an impetus to even the social sciences and literature [2–4]. The most famous works have been to automatically detect topics and classify large corpus of documents into well organized corpus. Probabilistic topic modeling techniques provide comprehensive algorithms to define topics from which we can make a definite sense[5]. The techniques have been used on many kinds of data and have found application in finding patterns in genetic data, images and social networks etc. [6,7,8].

Olfaction has been a very subjective and contentious of our senses owing to large variations in its reporting which is partly due to the fact that people experience it very differently [9,10,11]. The only non-invasive method of its reporting in humans is via language. Here, our language abilities have either failed us or we have not dug enough [12]. There is no standard way of representation of the smell of an odour molecule. The words spoken by the subjects vary too much. Hence, it is imperative to understand the theme behind the words spoken by the individuals. There are a lot of potential benefits of such kind of analysis. First, it will let us understand the relationship between language and olfaction. Second, it will let us rationalize the enigma behind the smell representation. Third, it will give an objectivity to the olfactory perceptual representation, which we can be further utilised in designing odour molecules. It can also help in designing searchable databases having meaningful smell topics and their relationship with each other.

1.1   Database Description

The present study looks into 4 different databases available publically i.e. SuperScent [13], Leon and Johnson [14] (LJ), Flavornet [15] and GoodScents [16]. The prime feature of all these databases is odor molecules described by some words or percepts and most of the times their molecular references. It should be noted that the variance of perceptual descriptors in these databases cannot be fully established due to the lack of information related to the odor experiments for them. Some discrepancies in the molecular reference entries were validated and removed by comparing the given molecular references entries and the calculated molecular weights by software E-DRAGON("Molecular Descriptors for Chemoinformatics. 2 Edn. Wiley-VCH. 2009). Then, some curation was done on these databases for creation of the feature sets. Firstly, the words describing the perceptual qualities of the molecules were tokenized to result in a set of perceptual descriptors (eg: for acetal perceptual qualities were specified as "ether green nut earthy sweet vegetable" these were tokenized as "ether" "green" "nut" "earthy" "sweet" "vegetable"). Overall, from the tokens for all the molecules, conjunctions (e.g. "and"), adverbs (e.g. "less", "somewhat"), suffixes (e.g. "like", "note"), auxiliary verbs (e.g. "has") and some other words which dont convey qualitative olfactory information (e.g. "other", "over", "powerful" and "preserves") were further removed to get the perceptual descriptor vectors (eg: in this case it is "ether" "green" "nut" "earthy" "sweet" "vegetable" as it is). Further, the equivalent percepts from semantics point of view such as "fruit/fruity" , "alcohol/alcoholic"etc. were merged. We also removed the percept "odorless" and the corresponding molecules from the database. Thus, we created the perceptual vector of a molecule by an indicator function such that 1 is assigned to the descriptors that were used to describe the presence of an odour and 0 otherwise. These matrices here will be represented as perceptual matrix (P), where rows represent molecules (m) and columns percepts (p). Further, in the databases were combined into a common representation of matrix of 3001 molecules and 520 perceptual descriptors..

Olfaction has been a very subjective and contentious of our senses owing to large variations in its reporting which is partly due to the fact that people experience it very differently [9,10,11]. The only non-invasive method of its reporting in humans is via language. Here, our language abilities have either failed us or we have not dug enough [12]. There is no standard way of representation of the smell of an odour molecule. The words spoken by the subjects vary too much. Hence, it is imperative to understand the theme behind the words spoken by the individuals. There are a lot of potential benefits of such kind of analysis. First, it will let us understand the relationship between language and olfaction. Second, it will let us rationalize the enigma behind the smell representation. Third, it will give an objectivity to the olfactory perceptual representation, which we can be further utilised in designing odour molecules. It can also help in designing searchable databases having meaningful smell topics and their relationship with each other.

2   Methods

The perceptual representation can now be further exploited to understand the underlying distribution of perceptual descriptors. One of the biggest problems involving olfaction has been the subjectivity surrounding its understanding primarily because the language which is spoken to convey the olfactory information is too broad and sparse i.e. the topics which people talk about is not defined. We tried to understand this problem in twin strategies at first trying to understand the perceptual space using network analysis techniques. Further, we sought to find the topical distribution using the very well established latent dirichlet allocation (LDA) technique.

2.1   Network Analysis and Community Detection

The perceptual descriptor matrix A is converted into a co-occurrence matrix by a simple transformation $C = P^T * P$ such that c_{ij} = co-occurrence of perceptual descriptors i and j, $i \neq j$ and $c_{ij} = 0$ for i = j. Thus we have a 520 X 520 symmetric perceptual matrix. We can consider co-occurrence values as a measure of similarity between percepts. This kind of similarity though has certain caveats, as they get affected by the frequency of individual elements (size-effect) [18] i.e. more frequent perceptual descriptors start dominating the co-occurrence matrix. Therefore, the co-occurrence matrix was normalized by defining the inclusion index as $S_i = c_{ij}/\min(c_{ii}, c_{jj})$. This index takes into account the skewness caused by the more frequent words. This index gives more emphasis on the lower dimension mainly due to the reason that a molecule is represented by a very few number of percepts and the percepts are very sparsely distributed, so the rarely used percepts if used together with more frequent percept pull the weights between them to a higher value and hence can be part of the same group. The normalized co-occurrence matrix was then used to create a network such that the nodes are the perceptual descriptors and there is an edge between them if they have co-occurred in any molecular description, the edges are then weighted by the inclusion index as defined above.

The study of community detection in the networks involves partitioning the graph into communities based on some objective function where intra-community connections are dense and inter-community connections are very sparse. We conducted modularity maximization for community detection proposed by Blondel *et al.* [19]

2.2   Latent Dirichlet Allocation

Latent Dirichlet Allocation is a generative algorithm, which assumes each document as a mixture of some topics, and each word in the document is sampled from these topics. Mathematically, each document in the corpus can assumed to be a mixture of different topics, with each topic having a diritchlet prior and each topic is considered to be a multinomial distribution over |V| words in the vocabulary. A document is generated by sampling over the topics and then sampling words from the topic mixture. A document Latent Dirichlet Allocation is a generative algorithm, which assumes each document as a mixture of some topics, and each word in the document is sampled from these topics. Mathematically, each document in the corpus can assumed to be a mixture of different topics, with each topic having a diritchlet prior and each topic is considered to be a multinomial distribution over |V| words in the vocabulary. A document is generated by sampling over the topics and then sampling words from the topic mixture. A document of N words w = <w_1, w_2,....w_N> is generated by first sampling θ from the Dirichlet prior <$\alpha_1, \alpha_2,...\alpha_k$> such that $\theta_i \geq 0$ and $\sum_i \theta_i = 1$. Then for each word N , a topic $t_n \in$ <1,2,...k> is sampled from a multinomial distribution over θ, Mult(θ) such that $p(t_n = i | \theta) = \theta_i$. At last, each word w_n is sampled, conditioning on the topic t_n from the multinomial distribution, $p(w|t_n)$. It can be represented completely in the following way. Probability of a document is therefore as in eq. 1.

$$p(w) = \int (\prod_{n=1}^{N} \sum_{t_n=1}^{k} P(w_n|t_n \; ; \; \beta) P(t_n|\theta)) \, p(\theta; \alpha) d\theta \qquad (1)$$

Here, $p(\theta; \alpha)$ is Dirichlet, $P(t_n|\theta)$ is a multinomial parameterized by θ, and $P(w_n|t_n \; ; \; \beta)$ is a multinomial over the words. The model is parameterized by the Dirichlet parameters α = <$\alpha_1, \alpha_2,... \alpha_k$> and a k X |V| matrix β.

3   Results and Discussion

Table 1 describes the different network properties of normalized perceptual co-occurrence matrix. There are some important points to note in the properties. First, the average weighted clustering coefficient of the network is 14 times more than the similar random network created using Erdos-Reyni G(n,m) model [20]. Second, the degree distribution seems to be following a power law, which may mean the presence of hubs and scale free property of the network. We are investigating the implications of this property in the perceptual space sense.

Table 1. Network Properties

Network Properties	Value
#Nodes	520
#Edges	7495
Avg. Degree	28.44
Avg. Clustering Coeff.	0.772
Avg. Clustering Coeff. G(n,m) model	0.054
Power law coefficient	Xmin = 60, α = 2.8

Community detection was applied on the normalized co-occurrence matrix to obtain the perceptual classes. At first, we changed the resolution values to observe how the network broke into the different categories. Fig 1-3 shows the 2,3 and 4 categories break-up respectively. The perceptual classes first partitioned into "fruity, floral, sweet, herb, etc." percepts and "earth, meat, vegetable, nut etc.", broadly a bakery nature of smells and other fruit-sweet-floral smells. We went on till 15 communities mainly based on the previous literature and the exact algorithm to identify optimum number of communities is still going on. As observed, in the communities, we find that the similar kind of percepts have come together, e.g. floral, lily, lilac, geranium, rose, violet, blossom, gardenia, iris, petal, neroli, rue etc. have come together signifying a broad floral group. Likewise butter, alcohol, coconut, mushroom, cheese, coumarin oil etc. have come together, as suggested by Abe et al, Muller et al. and later Zarzo [21–23], these percepts are similar due to the common source and the short chained aliphatic aldehydes which produce them. The categories and network diagram can be viewed at http://odornetwork.com/network/index.html.

The perceptual matrix A, thus created following the procedure outlined above is subjected to LDA. We used Gensim Library[24] in python, developed for topic modeling, to find the broader topics in perceptual matrix. The molecules here act as the alias for documents and the percepts for the terms or words. We ran the LDA module for different number of topics and runs. We ran the algorithm for 100 passes and 20 topics.

Fig. 1. Two Categories breakup

Fig. 2. Three Categories breakup

Fig. 3. Four Categories breakup

Getting ourselves one step further, we sought to find how many of the topics are optimum i.e. how many categories of perceptual descriptors cover all the perceptual space. We applied the method proposed by Arun *et al.* [25], in which he applies the concept of Kullback–Leibler (KL) divergence and proposes that the minimum point of divergence after the first local maximum should be considered to be as the optimum number of topics. KL divergence is non-symmetric measure of difference between two probability distributions P & Q defined as in eq. 2 for continuous distributions

$$D_{KL}(P\|Q) = \int_{-\infty}^{\infty} p(x) \ln \frac{p(x)}{q(x)} dx \tag{2}$$

where, p(x) and q(x) are the probability density for P and Q. They have considered LDA as a matrix factorization mechanism, wherein a corpus C is split into two matrix factors M1, M2 as in, $C = M1_{d \times t} \times Q_{t \times w}$ having d number of documents and w number of words in the vocabulary. The quality of split depends upon the number of topics t. We ran the LDA algorithm for number of topics 1-100 and calculated KL divergence at every split. The optimum number of topics turned out to be 7.

The topic distribution in Table 2 gives us a glimpse of perceptual classes. The first topic may be associated with much broader green smell which indicates freshness. The perfumers use this smell to be worn in the morning or they are sometimes sold like that. The second topic relates broadly to floral smell or the smell which has the origin from flowers. Third topic majorly relates to smells originating from fruit source. The fourth topic is related to sweet smells. Fifth and sixth topics throw ambiguities, whereas seventh topic relates to pungent and strong smells. Although some of the positioning of percepts is very ambiguous requiring the manual intervention for deciding which topic some of the words belong to and perhaps more sophisticated methods, yet we see that there are some ambiguities in the result, but above results look promising. This also gives us a basis to cluster molecules according to their percepts and hence define a distance measure between them, which can be exploited to further design complex relationships between the physico-chemical properties of the molecules and their perceived smell. This also indicates towards a more intriguing aspect of smell. It is a well-known fact that vision has clear cut defined primary colors. Can we define and describe such systems in smell with methods and measures like this, of course we would we far stretching our logic, if we said we have done so, but with sufficient number of molecules and their proper annotations, who knows we might someday with methods like these at our disposal.

Table 2. Topic Distribution

Topic 1	Topic 2	Topic 3	Topic 4
green	floral	fruit	spice
nut	herb	fat	balsam
mint	wood	wax	caramel
vegetable	rose	sweet	sweet
earth	sweet	apple	butter
leaf	musk	pineapple	phenol
alcohol	aldehydic	berry	animal
fresh	jasmine	banana	almond
pepper	orange	melon	smoke
roasted	fresh	pear	cheese

Topic 5	Topic 6	Topic 7
citrus	earth	sulfur
fruit	wood	meat
sweet	camphor	onion
ethereal	oil	coffee
honey	vanilla	garlic
wine	cream	alliaceous
amber	coconut	roasted
grape	must	vegetable
lemon	sweet	fish
ether	mushroom	pungent

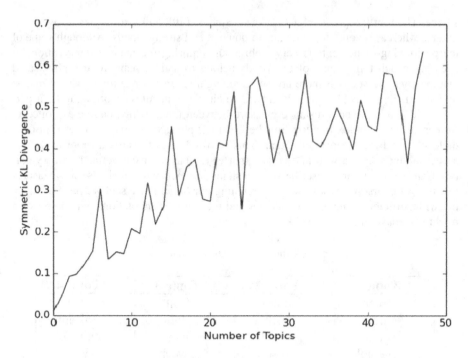

Fig. 4. Symmetric K-L divergence vs number of topics.

4   Conclusion

In this paper we demonstrated a smell network to understand the percepts and their relationships. We also proposed a model to find the optimum number of perceptual classes using LDA for odor molecules. The dataset and the visualization diagram is available at http://odornetwork.com/. The work is still under progress and we hope to come up with models to relate perceived smell of a molecule to its physico-chemical properties.

Acknowledgments. The authors would like to thank Dr. Benjamin Auffarth for his valuable suggestions and Dr. G P S Raghava for his insight into the work.

References

1. Ahn, Y.-Y., Ahnert, S.E., Bagrow, J.P., Barabási, A.-L.: Flavor network and the principles of food pairing. Sci. Rep. **1** (2011)
2. De Smet, W., Moens, M.-F.: Cross-language linking of news stories on the web using interlingual topic modelling. In: Proceedings of the 2nd ACM workshop on Social web search and mining, pp. 57–64. ACM (2009)

3. Hong, L., Ahmed, A., Gurumurthy, S., Smola, A.J., Tsioutsiouliklis, K.: Discovering geographical topics in the twitter stream. In: Proceedings of the 21st International Conference on World Wide Web, pp. 769–778. ACM (2012)

4. Purver, M., Griffiths, T.L., Körding, K.P., Tenenbaum, J.B.: Unsupervised topic modelling for multi-party spoken discourse. In: Proceedings of the 21st International Conference on Computational Linguistics and the 44th Annual Meeting of the Association for Computational Linguistics, pp. 17–24. Association for Computational Linguistics (2006)

5. Steyvers, M., Griffiths, T.: Probabilistic topic models. Handb. latent Semant. Anal. **427**, 424–440 (2007)

6. Chang, J., Gerrish, S., Wang, C., Boyd-graber, J.L., Blei, D.M.: Reading tea leaves: how humans interpret topic models. In: Advances in Neural Information Processing Systems, pp. 288–296 (2009)

7. Wang, W., Barnaghi, P., Bargiela, A.: Probabilistic topic models for learning terminological ontologies. Knowl. Data Eng. IEEE Trans. **22**, 1028–1040 (2010)

8. Shmulevich, I., Dougherty, E.R., Zhang, W.: From Boolean to probabilistic Boolean networks as models of genetic regulatory networks. Proc. IEEE **90**, 1778–1792 (2002)

9. Auffarth, B.: Understanding smell—The olfactory stimulus problem. Neurosci. Biobehav. Rev. **37**, 1667–1679 (2013)

10. Khan, R.M., et al.: Predicting odor pleasantness from odorant structure: pleasantness as a reflection of the physical world. J. Neurosci. **27**, 10015–10023 (2007)

11. Zarzo, M., Stanton, D.T.: Identification of latent variables in a semantic odor profile database using principal component analysis. Chem. Senses **31**, 713–724 (2006)

12. Gottfried, J.A., Dolan, R.J.: The nose smells what the eye sees: crossmodal visual facilitation of human olfactory perception. Neuron **39**, 375–386 (2003)

13. Dunkel, M., et al.: SuperScent—a database of flavors and scents. Nucleic Acids Res. **37**, D291–D294 (2009)

14. Leon, M., Johnson, B.: Glomerular response archive. (2008). http://gara.bio.uci.edu/index.jsp

15. Acree, T., Arn, H.: Flavornet (2004). http://www.flavornet.org/flavornet.html

16. Luebke, W.: The good scents company (1980). http://www.thegoodscentscompany.com/index.html

17. Molecular descriptors for chemoinformatics, 2nd edn. Wiley-VCH (2009) (3).pdf

18. Shaoul, C., Westbury, C.: Word frequency effects in high-dimensional co-occurrence models: A new approach. Behav. Res. Methods **38**, 190–195 (2006)

19. Blondel, V.D., Guillaume, J.-L., Lambiotte, R., Lefebvre, E.: Fast unfolding of communities in large networks. J. Stat. Mech. Theory Exp. **2008**, P10008 (2008)

20. Erdos, P., Rényi, A.: On the evolution of random graphs. Bull. Inst. Internat. Stat. **38**, 343–347 (1961)

21. Abe, H., Kanaya, S., Komukai, T., Takahashi, Y., Sasaki, S.: Systemization of semantic descriptions of odors. Anal. Chim. Acta **239**, 73–85 (1990)

22. Müller, J.: The H&R book of perfume: Understanding fragrance; origins, history, development; guide to fragrance ingredients. Glöss Verlag, Hambg (1992)

23. Zarzo, M.: Hedonic judgments of chemical compounds are correlated with molecular size. Sensors (Basel) **11**, 3667–3686 (2011)

24. Řehůřek, R., Sojka, P.: Software framework for topic modelling with large corpora (2010)

25. Arun, R., Suresh, V., Madhavan, C.E.V., Murthy, M.N.N.: In: Advances in Knowledge Discovery and Data Mining, pp. 391–402. Springer (2010)

A Social Network Analysis Based on Linear Programming-Shapley Value Approach for Information Mapping

António Oliveira Nzinga René[1]([⊠]), Eri Domoto[2], Yu Ichifuji[3], and Koji Okuhara[1]

[1] Department of Information and Physical Sciences, Graduate School of Information Science and Technology, Osaka University, 1-5 Yamadaoka, Suita, Osaka 565-0871, Japan
a-rene@ist.osaka-u.ac.jp
[2] Department of Media Business, Faculty of Economics, Hiroshima University of Economics, 5-37-1, Gion, Asaminami-ku, Hiroshima 731-0192, Japan
[3] National Institute of Informatics, 2-1-2 Hitotsubashi, Chiyouda-ku, Tokyo 101-8430, Japan

Abstract. The concept of network has different meanings in different disciplines. Consider, for instance, social sciences where, mostly, denotes a set of actors or other fields defining the concept as agents, or nodes, or points, or vertices that may have relationships with one another. One of the issues to solve in social network analysis is the problem of ranking the nodes. In this paper, we combine cooperative game theory framework and linear programming techniques and suggest an equivalent alternative model to Shapley value. We present a numerical illustration explaining how our model can be applied in problems related to social networks.

Keywords: Social network analysis · Cooperative games · Shapley value · Linear programming · DEMATEL

1 Introduction

Social network analysis (SNA) is the study of structural and communication patterns such as, degree distribution, density of edges, diameter of the network, etc. and usually partitioned into two principal classes [1], namely: *node and edge centric analysis* which includes *i*) centrality measures such as degree, betweeneness, stress and closeness; *ii*) anomaly detection; *iii*) link prediction, etc. The second class, related to *network centric analysis* includes the study of *i*) community detection; *ii*) graph visualition and summarization; *iii*) frequent subgraph discovery and *iv*) generative models, etc.

Due to its importance a lot of studies have been performed using conventional approaches such as graph theoretic techniques, spectral methods, optimization techniques, etc. to solve problems relating to nodes ranking, diversity among the nodes, link prediction, inferring social network from social events, influence

L. Wang et al. (Eds): MISNC 2015, CCIS 540, pp. 470–482, 2015.
DOI: 10.1007/978-3-662-48319-0_39

maximization, community detection, determination of implicit social hierarchy and as well design of incentives in networks, etc. Recent advances include data mining and machine learning techniques and game theory as well. For more details with respect to the main tasks of SNA, refer to [1–3].

In this work, we give particular attention to cooperative game theory (CGT) techniques. Precisely, we use an alternative equivalent Shapley value [9] model based on linear progarmming (LP) approach [4] to solve the problem of information sharing among the players. This combination makes our paper distinguish to others for the fact of using LP technique which, as stated by [5], has become presently, the mathematical technique most used in solving a variety of problems. It is a standard tool that has saved many thousands or millions of dolars for many companies or businesses of even moderate size in the various industrialized countries of the world, and its use in other sectors of society has been spreading rapidly. A major proportion of all scientific computation on computers is devoted to the use of linear programming, and published articles describing important applications now number in the hundreds. In fact any problem whose mathematical model fits the very general format for the linear programming model can be defined as an LP model. Interchangeably, within the paper, terms such as *actors*, *agents* and *players* have the same meaning. The motivation of this study is, specifically, the problem of ranking nodes which makes up an important issue in social network research. However, some inadequacies can be found using conventional methods since these approaches do not give particular attention to the dynamics of strategie interactions among the agents. As proved by [1], Game Theory offers a good framework to overcome these drawbacks. In terms of contribution, this paper suggest a model which combines CGT and LP to find an optimal solution set which is equal to the conventional Shapley value.

The paper is organized as follows: after this introduction, Section 2 describes basic concepts about social network analysis (SNA) lifting up the traditional tecniques used in this field. Section 3 introduces coalitional game, also called cooperative or transferable utility (TU) games, describes Shapley value as solution concept with some of its properties and presents an equivalent alternative model to this approach. The next section brings mathematics concepts on interactions between nodes, descibed as academic fields to estimate the amount of information shared. Through a numerical example, Section 4 uses the results from the previous section to compute the optimal solution for ranking the nodes. The work closes with some concluding remarks and direction to future studies.

2   An Overview on Social Network Analysis (SNA) Techniques

The word *network* has different meanings according to the context or discipline. In social sciences, for instance, mostly denotes a set of actors; in other fields the concept is applied to agents, or nodes, or points, or vertices that may have relationships (or link, or edges, or ties) with one another (Figure 1).

Few or many agents can be coupled through one or more kinds of relations between them in a network. A *simplex* network represents a single type of relation among the agents, whilst a *multiplex* networks denote those with more than one kind of relation, whose analysis can be done using different networks, one for each type. Each relation may be *directed*, that is, originates in a source actor and reaches a target agent, take for instance the network of social sciences presented in Fig. 1 regarding journals and citations; or it may be a relation representing an occurrence, co-presence, or a bounded-relation between the pair of agents, i.e. *indirected* [3].

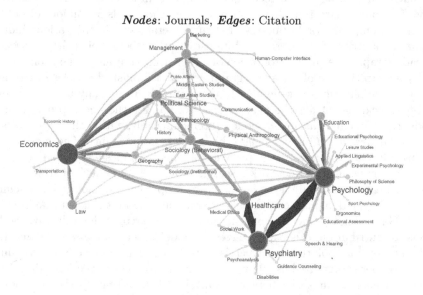

Fig. 1. Relationships among the Social Sciences. http://eigenfactor.org/

Formally, networks are studied in Graph Theory, a branch of mathematics. When referring to social sciences' aspects *network* is often named *graph* in Graph Theory, while the term is reserved for specific type of graphs.

Definition 1. *Precisely, in graph theory a directed graph* G *is defined as an ordered pair* $G := (V, A)$ *subject to the following conditions:*

- V *is a set whose elements are variously referred to as* **nodes**, **vertices**, *or* **points**.

- A *is a set of ordered pairs of vertices, called* **arcs**, **arrows**, *or* **directed edges**. *An edge* $e = (x, y)$ *is said to be directed from* x *to* y, *if* x *is the tail of* e *and* y *is the head of* e.

Comprehensively, SNA is the mapping and measurement of relationships and flows between people, groups, organizations, computers or other information processing entities, [2]. Describes how agents are located or embedded in the overall network and how the pattern of individual choices gives rise to more holistic patterns. SNA has been used in sociology, anthropology, information systems, organizational behavior, and many other disciplines.

Essentially, in SNA [2] individual people are inserted within networks of face-to-face relations with other persons. Families, neighborhoods, school districts, communities, etc. The interesting point for a social network analytic lies on how the individual is embedded within a structure and how this structure emerges from the relations between individuals. Referring to an organizational level this could be applied to describe how agents (employees, employers, departments, etc.) relate to each other by the interactions. A summary on the question why is SNA important can be listed as follows [1].

- SNA helps to understand complex connectivity and communication patterns among individuals in the network.
- Determine the structure of networks.
- Determine the influential individuals in social networks.
- Offers a deep understanding on how social network emerge.
- Supports designing effective viral marketing campaigns for targeted advertising

3   Coalitional Games

Through this section we intend to describe some basic concepts of coalitional game theory, but before that we recall that, generally speaking, game theory studies what happens when self-interested players (agents) interact. Saying players are self-interested does not necessarily mean that they want to cause harm to each other, or even that they care only about themselves. By contrary, it means that each agent has his own description of which states of the world he likes (which can include good things happening to other agents) and that he acts in trying to bring about these states of the world, or as also defined in [6] game theory is a mathematical branch of economic theory and analyses *decisions situations* that have the character of a game and go far beyond economics in their application. Its significance can also be seen in the award of the Noble prize in 1994 to the theoreticians John Forbes Nash, John Harsany and Reinhard Selten.

The basic assumptions that underlie the theory are that decision makers (DMs) pursue well-defined exogenous goals (they are rational) and take into account their knowledge or expectations of other DMs' behavior (they reason strategically)[7]. This approach is divided in two branches, namely cooperative and noncooperative games. Most of applications on game-theoretic are related to the latter one. In this paper, as said before, our focus is the former one. Therefore, the rest of the section is concentrated on this topic.

Given a set of players, a cooperative game with transferable utility (TU game), also known as coalitional game, defines how well each group (or *coalition*) of players can do for itself. The concern is not with how the players make choices within a coalition, how they coordinate, or any other such detail; the main idea is to take the payoff to a coalition as given. It is assumed that the payoffs to a coalition may be freely redistributed among its member, that is, the utility can be transferred among them. This assumption is satisfied when there is a universal currency that is used for exchange in the system, and means that each coalition can be assigned a single value as its payoff [8].

Definition 2. *(Cooperative game with transferable utility) A TU game is a pair (N, v), where:*

- *N is a finite set of players, indexed by i; and*

- *$v : 2^N \mapsto \mathbb{R}$ associates with each coalition $S \subseteq N$ a real-valued payoff $v(S)$ that the coalition's members can distribute among themselves. The function v is also called the characteristic function, and a coalition's payoff is also called its worth. We assume that $v(\emptyset) = 0$.*

Ordinarily, cooperative game answers two fundamental questions:

- Which coalition will form?

- How should that coalition divide its payoff among its member?

The answer to the first question is often "the grand coalition" - the coalition corresponding to all players in N however, this answer can depend on having made the right choice about the last question. Any subset S of the player set $N = \{1, 2, \cdots, n\}$ is called a coalition and S, aims at obtaining $v(S)$ which defines a *characteristic function* to describe the amount gained by each player. The $v(S)$, with $v(\emptyset) = 0$,

Definition 3. (Superadditive game). *A game $G = (N, v)$ is superadditive if for all $S, T \subset N$, $S \cap T = \emptyset$, holds $v(S \cup T) \geq V(S) + V(T)$.*

Superadditivity makes sense when coalitions can always work without interfering with one another; hence, the value of two coalitions will be less than the sum of their individual values. This property implies that the value of the entire set of players, i.e., the "grand coalition" is no less than the sum of the value of any nonoverlapping set of coalitions. In other words, the grand coalition has the highest payoff among all coalitional structures.

3.1   Shapley Value

Dividing the payoff to the grand coalition among the players is the main task in cooperative game. The grand coalition is the coalition achieving the highest payoff over all the coalitions, and so is expected it to form and on the other hand, there may be no choice for the players but to form the grand coalition.

Regarding the difficult to decide how this coalition should divide its payoffs there is a variety of solution concepts that propose different ways of performing this division such as the core, Shapley value, nucleolus, etc. In this paper, particular attention is given to Shapley value for the sake of being the solution concepts under which we want to build our approach.

The concept of Shapley value [9] lies on a set of axioms, i.e., is characterized as a unique mapping φ satisfying these axioms.

An n-vector $\phi(v)$ means the value of a game v, satisfying

Axiom 1 (Dummy player). *If S is any carrier of v, then*
$\sum_s \varphi_i(v) = v(S)$.

Axiom 2 (Symmetry). *For any permutation π, and $i \in N$,*
$\varphi_{\pi(i)}(\pi v) = \varphi_i(v)$.

Axiom 3 (Additivity). *If d and v are any games,*
$\varphi_i(d + v) = \varphi_i(d) + \varphi_i(v)$.

These axioms are sufficient to determine a value $var\phi$ uniquely, for all games, i.e, formula (1) gives the Shapley value, explicitly.

$$\varphi_i[v] = \sum_{\substack{T \subset N \\ i \in T}} \frac{(t-1)!(n-t)!}{n!}[v(T) - v(T - \{i\})] \tag{1}$$

where n denotes the number of players, $t \subset T$ denotes the size of the coalition and $\varphi_i[v]$ indicates the marginal contribution of player i in the game. By superadditivity the bracket is always at least equal to $v(\{i\})$ [9]. Hence

$$\varphi_i[v] \leq v(\{i\}) \tag{2}$$

3.2 An Alternative Equivalent Shapley Value Based on LP

Consider a game (N, v) whose imputation vector is given by $z(N, v) = (z_1(N, v), z_2(N, v), \cdots, z_n(N, v))$ and satisfying the following properties:

- Individual rationality: $z_d(N, v) \geq v(\{d\})$, $d = 1, 2, \cdots, N$
- Grand coalition rationality: $\sum_{d=1}^{N} z_d(N, v) = v(N)$.

Let $M_{r,s}$, $(r = 2, 3, 4, \cdots, R; s = 1, 2, \cdots, r-1)$ be a set of weights with nonnegative values but, has at least one positive value for each r. Attempt now to the following problem:

$$\text{Minimize} \sum_{S \subset R, \ S \neq R} M_{r,s} \left(v(S) - \sum_{d \in S} z_d(R, v) \right)^2 \tag{3}$$

$$\text{Subject to} \sum_{d \in R} z_d(R, v) = v(R)$$

A vector imputation $z_d^+(R, v)$ gives the solution for (4) through the following expression:

$$z_d^+(R, v) = \frac{1}{r} \left\{ v(R) + \sum_{i \in R} (c_{di} - c_{id'}) \right\} \tag{4}$$

and $\Gamma(d^+, d'^-) = \{S \subset R \mid d \in S, \ d' \notin S\}$. The elements of the set $M_{r,s}$, are defined as

$$M_{r,s} = \frac{1}{r-1} \left\{ _{r-2}C_{s-1} \right\}^{-1} \tag{5}$$

where r indicates the amount of players in the game and s the coalition size.

This process produces a vector imputation equal to Shapley value defined in (1). For more details on this topic, we refer readers to [4,10,11] and the references therein.

Moreover, imagine a sample d from a data set with x_i and y_i as values of the referred sample. An error among these elements can be obtained through

$$e_d = f_d(\boldsymbol{A}, \boldsymbol{M}, \boldsymbol{v}) = (\boldsymbol{A}^\mathrm{T} \boldsymbol{M} \boldsymbol{v} - \boldsymbol{A}^\mathrm{T} \boldsymbol{M} \boldsymbol{A} \boldsymbol{z})_d \tag{6}$$

where $(\ )_d$ denotes the selection of the $d - th$ row values. So $E = \sum_d |e_d|$ is the sum of all error functions using the multiple linear regression model that minimize the sum of the residuals' absolute values[10]. The problem described in (6) can be restated as an LP problem as follows

$$\begin{aligned}
\text{Minimize} \quad & \epsilon \\
\text{Subject to} \quad & \boldsymbol{A}^\mathrm{T} \boldsymbol{M} \boldsymbol{v} + \boldsymbol{s}^+ - \boldsymbol{s}^- = \boldsymbol{A}^\mathrm{T} \boldsymbol{M} \boldsymbol{A} \boldsymbol{z} \\
& \sum_{d \in K} z_d(K, v) = v(K) \\
& 0 \le \boldsymbol{s}^+ \le \epsilon, \ 0 \le \boldsymbol{s}^- \le \epsilon
\end{aligned} \tag{7}$$

where ϵ in the objective function, is the error to be minimized, $\boldsymbol{s}^+ = [s_1^+, s_2^+, s_3^+, \cdots, s_n^+]^\mathrm{T}$ and $\boldsymbol{s}^- = [s_1^-, s_2^-, s_3^-, \cdots, s_n^-]^\mathrm{T}$ are the vector sets of slack variables, $\boldsymbol{A}$ is a matrix constructed according to the coalition formation, $\boldsymbol{v}$ is a column matrix whose elements are the real values $v(S)$, i.e., the values of the characteristic function and $\boldsymbol{M}$ a diagonal matrix whose elements are the weights obtained through (5).

Example 1. Consider a 3-person game with $N = \{a, b, c\}$ the finite set of players or nodes. Then, we have

$$z = \begin{bmatrix} z_a(K, v) \\ z_b(K, v) \\ z_c(K, v) \end{bmatrix}, \qquad v = \begin{bmatrix} v(\{a\}) \\ v(\{b\}) \\ v(\{c\}) \\ v(\{a, b\}) \\ v(\{a, c\}) \\ v(\{b, c\}) \end{bmatrix}$$

and

$$M = \begin{bmatrix} M_{3,1} & 0 & 0 & 0 & 0 & 0 \\ 0 & M_{3,1} & 0 & 0 & 0 & 0 \\ 0 & 0 & M_{3,1} & 0 & 0 & 0 \\ 0 & 0 & 0 & M_{3,2} & 0 & 0 \\ 0 & 0 & 0 & 0 & M_{3,2} & 0 \\ 0 & 0 & 0 & 0 & 0 & M_{3,2} \end{bmatrix}$$

$$A = \begin{bmatrix} 1 & 0 & 0 \\ 0 & 1 & 0 \\ 0 & 0 & 1 \\ 1 & 1 & 0 \\ 1 & 0 & 1 \\ 0 & 1 & 1 \end{bmatrix}, \qquad A^{\mathrm{T}} = \begin{bmatrix} 1 & 0 & 0 & 1 & 1 & 0 \\ 0 & 1 & 0 & 1 & 0 & 1 \\ 0 & 0 & 1 & 0 & 1 & 1 \end{bmatrix} \tag{8}$$

4   Analyzing the Impact of the Relationship Between Different Nodes in a Network

Consider a network with five hypothetical nodes where each one represent an academic field. We seek to analyze the interaction between different nodes within a network. With this in mind we used decision-making trial and evaluation laboratory (DEMATEL) [12] techniques to analyze the influence of a node i over j while we work on the data in Table 1 where for instance we take some of the fields from the network in Fig. 1 to produce direct relation matrix $A : [a_{ij}]$ (Table 2).

Table 1. Nodes' data

Nodes	Management	Economics	Psychology	...	Healthcare
Management	1	1	2	1	1
Economics	111	112	111	113	111
Psychology	78	3	35	88	47
...	6	10	4	9	6
Healthcare	95	9	69	89	72

To recall that nodes represent players in the game. The ranking of these players will be described in 5 using Shapley value through the LP model proposed in 3.2. To express how strength is the direct influence between the nodes we initially computed the direct influence matrix $X : [x_{ij}]$ as

$$X = \frac{1}{\max\limits_{1 \le j \le n} \sum\limits_{i=1}^{n} |a_{ij}|} \cdot A \tag{9}$$

Table 2. Direct relation matrix **A**

Nodes	Management	Economics	Psychology	$\cdots$	Healthcare
Management	0	0	0	0	0
Economics	0	0	0	0	0
Psychology	0	0	0	1	0
$\cdots$	0	2	1	0	0
Healthcare	0	0	0	0	0

The reverse information obtained from X, i.e, the indirect influence is derived from (10) computing F : $[f_{ij}]$, the total relation matrix.

$$F = x + x^2 + \cdots + x^n = x(I - x)^{-1} \tag{10}$$

Table 3. Total relation matrix **F**

Nodes	Management	Economics	Psychology	$\cdots$	Healthcare	d	r	$d+r$
Management	0.009	0	0.002	0	0	0.011	0.069	0.08
Economics	0.017	0	0.004	0	0	0.021	0.201	0.222
Psychology	0.008	0.017	0.05	0.095	0	0.17	0.205	0.375
$\cdots$	0.01	0.184	0.116	0.011	0	0.321	0.109	0.43
Healthcare	0.025	0	0.033	0.003	0	0.061	0	0.061

In Table 3, column d represents the overall effect of a specific node over others, it is given by adding each field's corresponding row. To find the total effect a node receives from others, column r is computed by adding each element's corresponding column and, the last column, $d + r$, measures impact and non-impacts, i.e., it estimates if a field plays a central role in the network. Furthermore, the indices ij in A denotes the path to reach j from i and as well as the length of the path from i to j. Knowing the number of paths $A, A^2, \cdots, A^n$, one can get the reachability matrix R by (11).

$$R = \frac{I}{I - A} = I + A + A^2 + A^n \tag{11}$$

Now, to get information on the importance of a field over other one, assume that a table of scores consisting of n items $o_1, o_2, \cdots, o_n$, is given by (12) as regard to n industries $S_1, S_2, \cdots, S_n$.

$$E = e_j^i \tag{12}$$

with $e_j^i \in [0, 1]$ on $\mathbb{R}$. The amount of information (*entropy*) S can be obtained by

$$S = -p \log_2 p - q \log_2 q \tag{13}$$

where p is the ratio of successful fields and q simply the opposite. So, the amount of information about each item o_i is then given as

$$S_{(i)}^\alpha = -p_{(i)}^\alpha \log_2 p_{(i)}^\alpha - q \log_2 q_{(i)}^\alpha \qquad (14)$$

with $p_{(i)}^\alpha$ denoting the percentage of successful fields in α level of o_i, and $q_{(i)}^\alpha$ the percentage for the opposite case. The average of the amount of information is given by

$$S_{(i)}^- = \sum_{\alpha=1}^{\theta} S_{(i)}^\alpha p_{(i)}^\alpha \qquad (15)$$

where $S_{(i)}^-$ expresses the average of the entropy. The amount of information acquired by each node whether success or failure is given by the *total entropy* ΔS_i as described in (16), which defines the importance $I_1(i)$ of node i, i.e., it expresses numerically how it is important an specific node in the social network.

$$\Delta S_{(i)} = S - S_{(i)}^\alpha \qquad (16)$$

In Section 5 using Table 4 with (17) and (18) we computed Shapley values using an LP scheme to find the relationship between direct and indirect gains when cooperation occurs.

$$v(S) = \sum_{i \in S} v_i + \Delta v(S) \qquad (17)$$

$$\Delta v(S) = \sum_{i \neq j \in S} \tilde{r}_{ij}(v_i + v_j) \qquad (18)$$

Computational Issues. The process scheme described in this paper can be applied through the following stepwise procedure:

Step 1: **Apply the DEMATEL technique to analyze the influence of a node over other**: Analyze the influence of a node i over j through the data by computing the direct influence matrix $X : [x_{ij}]$ (9) and build its direct relation matrix $A : [a_{ij}]$

Step 2: **To determine the characteristic functions**: Consider the entropy for each node and define Eq.(18) as characteristic function to compute the contribution of each node or coalition of nodes in the game.

Step 3: **To determine the Shapley Value**: Compute the weights $M_{r,s}$ by employing Eq.(5). Using the characteristic function values of all coalitions and the weights, Shapley value and error function are calculated by solving (7).

5   Numerical Example

Through a numerical illustration we aim to estimate the ranking of the five nodes using data from previous tables. For such purpose we computed Shapley value using program (7).

Table 4. Entropies of the five nodes

Nodes	$S_{(i)}^1$	$S_{(i)}^2$	$S_{(i)}^3$	$S_{(i)}^4$	$S_{(i)}^5$	$S_{(i)}^-$	$\Delta S_{(i)}$
Management	0.845	1.0	0	0	0	0.887	0.113
Economy	0.918	0	0.918	1.0	1.0	0.818	0.182
Psychology	0.811	0.918	0.971	0	0	0.907	0.093
$\cdots$	1.0	0	1.0	0.918	0.811	0.800	0.200
Healthcare	0	0	0	0.918	0.971	0.507	0.493

Table 5. Characteristic Function $v(S)$ for single and two players' coalitions

Coalition	v(S)	Coalition	v(S)	Coalition	v(S)
$v\{A\}$	0.11341	$v\{AB\}$	0.29577	$v\{BD\}$	0.38236
$v\{B\}$	0.18236	$v\{AC\}$	0.2061	$v\{BE\}$	0.67505
$v\{C\}$	0.09269	$v\{AD\}$	0.31341	$v\{CD\}$	0.35445
$v\{D\}$	0.2	$v\{AE\}$	0.6061	$v\{CE\}$	0.6047
$v\{E\}$	0.49269	$v\{BC\}$	0.27505	$v\{DE\}$	0.69477

$$
\mathbf{A} = \begin{bmatrix}
1 & 0 & 0 & 0 & 0 \\
0 & 1 & 0 & 0 & 0 \\
0 & 0 & 1 & 0 & 0 \\
0 & 0 & 0 & 1 & 0 \\
0 & 0 & 0 & 0 & 1 \\
1 & 1 & 0 & 0 & 0 \\
1 & 0 & 1 & 0 & 0 \\
1 & 0 & 0 & 1 & 0 \\
1 & 0 & 0 & 0 & 1 \\
0 & 1 & 1 & 0 & 0 \\
0 & 1 & 0 & 1 & 0 \\
0 & 1 & 0 & 0 & 1 \\
0 & 0 & 1 & 1 & 0 \\
0 & 0 & 1 & 0 & 1 \\
0 & 0 & 0 & 1 & 1 \\
1 & 1 & 1 & 0 & 0 \\
1 & 1 & 0 & 1 & 0 \\
1 & 1 & 0 & 0 & 1 \\
1 & 0 & 1 & 1 & 0 \\
1 & 0 & 1 & 0 & 1 \\
1 & 0 & 0 & 1 & 1 \\
0 & 1 & 1 & 1 & 0 \\
0 & 1 & 1 & 0 & 1 \\
0 & 1 & 0 & 1 & 1 \\
0 & 0 & 1 & 1 & 1 \\
1 & 1 & 1 & 1 & 0 \\
1 & 1 & 1 & 0 & 1 \\
1 & 1 & 0 & 1 & 1 \\
1 & 0 & 1 & 1 & 1 \\
0 & 1 & 1 & 1 & 1
\end{bmatrix}
\tag{19}
$$

Table 6. Characteristic Function *v(S)* for three and four members' coalitions

Coalition	v(S)	Coalition	v(S)	Coalition	v(S)
v{ABC}	0.38846	v{ADE}	0,8061	v{ABCD}	0.58846
v{ABD}	0.49577	v{BCD}	0.47505	v{ABCE}	0.88115
v{ABE}	0.78846	v{BCE}	0.47505	v{ABDE}	0.98846
v{ACD}	0.4061	v{BDE}	0.76774	v{ACDE}	0.89879
v{ACE}	0.69879	v{CDE}	0.84714	v{BCDE}	0.96774
v{ABCDE}	1.08115	—		—		—	

A is a (30×5) matrix built according to the coalition formation in Tables 5 and 6.

$$\mathbf{A^T MAv} = \begin{bmatrix} 1.284484 \\ 1.353417 \\ 1.275668 \\ 1.38152 \\ 1.670617 \end{bmatrix} \tag{20}$$

$$\mathbf{A^T MA} = \begin{bmatrix} 2.083 & 1.0832 & 1.0832 & 1.0832 & 1.0832 \\ 1.0832 & 2.083 & 1.0832 & 1.0832 & 1.0832 \\ 1.0832 & 1.0832 & 2.083 & 1.0832 & 1.0832 \\ 1.0832 & 1.0832 & 1.0832 & 2.083 & 1.0832 \\ 1.3332 & 1.0832 & 1.0832 & 1.0832 & 2.083 \end{bmatrix} \tag{21}$$

Solving program (7) through a simplex tableau we got the following set of feasible solution:

- **Objective function to be minimized:** $\epsilon = 0.00586$
- **Shapley values:**
  - For the sake of simplicity we denote the nodes by **A**, **B**, **C**, **D** and **E** referring to management, economics, psychology, $\cdots$ and healthcare, respectively.
  - $\varphi_A = 0.10755$; $\varphi_B = 0.1765$; $\varphi_C = 0.09873$; $\varphi_D = 0.20461$ and $\varphi_E = 0.49376$
- **Slack variables:** $s_1^+ = s_2^+ = s_3^+ = s_4^+ = s_5^+ = s_6^+ = 0$ and $s_1^- = s_2^- = s_3^- = s_4^- = s_5^- = s_6^- = 0.00586$

Hence, from the obtained results we conclude that **healthcare**, corresponding to **node E** ranks first.

6   Concluding Remarks

In this paper, we presented an alternative equivalent Shapley value through an LP model in order to rank the nodes in social networks. A numerical example was presented to show the computation process behind our model. Further studies should be directed to aspects related to constraints's sensitivity in terms of uncertainty.

References

1. Narayanam, R.: Game Theory Models for Social Networks Analysis. IBM Research (2012).   http://www.imsc.res.in/~sitabhra/meetings/socialnetwork0212/talks/ Ramasuri_Narayanam_Game_Theoretic_Models_for_Social_Network_Analysis_I.pdf (retrieved March 13, 2015)
2. Liebowitz, J.: Linking social network analysis with the analytic hierarchy process for knowledge mapping in organizations. Journal of Knowledge Management **9**(1), 76–86 (2005)
3. Izquierdo, L.R., Hanneman, R.A.: Introduction to the Formal Analysis of Social Networks Using Mathematica. (Version 2) (2006). http://luis.izqui.org/papers/ Izquierdo_Hanneman_2006-version2.pd (retrieved May 20, 2015)
4. Rene, A.O.N., Okuhara, K., Domoto, E.: Allocation of Weights by Linear Solvable Process in a Decision Game. International Journal of Innovative Computing, Information and Control **8**(3), 907–914 (2014)
5. Hillier, F.S., Lieberman, G.J.: Introduction to Operations Research, International edn., 9th edn. McGraw-Hill Education, Singapore (2010)
6. Julmi, C.: Introduction to Game Theory (2012). http://www.bookboon.com (retrieve December 2013)
7. Osborne, M.J., Rubinstein, A.: A Course in Game Theory. MIT Press, Electronic version (1994)
8. Brown, K.L., Shoham, Y.: Essentials of Game Theory - A concise, Multidisciplinary Introduction. Morgan & Claypool, California (2008)
9. Owen, G.: Game Theory, 2nd edn. Academic Press Inc., London (1982)
10. Ruiz, L.M., Valenciano, F., Zarzuelo, J.M.: Some New Results on Least Square Values for TU Games. Sociedad de Estadistica e Investigacion Operativa **6**(1), 139–158 (1998)
11. Namekata, T.: Probabilistic Interpretation of Nyu-value (a Solution for TU game). Bulletin of the Economic Review at Otaru University of Commerce **56**(2–3), 33–40 (2005). (in Japanese)
12. Lee, H., Tzeng, G., Yeih, W., Yang, S.: Revised DEMATEL: Resolving the Infeasibility of DEMATEL. Applied Mathematical Modeling **37**(10–11), 6746–6757 (2013)

Automatic Geotagging for Personal Photos with Sharing Images on Social Media Networks

Been-Chian Chien[✉], Hua-Tung Shi, Chang-Hsien Fu, and Rong-Ming Chen

Department of Computer Science and Information Engineering,
National University of Tainan, Tainan, Taiwan, R.O.C.
bcchien@mail.nutn.edu.tw

Abstract. The information of location for digital photos is important for users to recall memory and retrieve photos with interested events. Geotagging is the process of identifying geographical metadata to digital photos. Since a large number of images may be generated in a journey or trip, traditional manual spatial annotation is time consuming and infeasible for personal collecting. In this paper, GPS signals on photographic devices and images of websites, or photo-sharing social media web are adopted to automatically identify geo-location of personal photos. We study and compare the effectiveness of two individual approaches and the combination approach. The results show that geographical coordinates are the most influential component for resolving geo-location. The geotagging process using sharing photos on social media service is also revealed and discussed.

Keywords: Image annotation · Geo-tagging · Social media network · Classifier

1 Introduction

With the popularity of mobile digital devices, producing and storing photographs are now far easier than before. The number of digital photos in personal collection has grown rapidly making photos' management and retrieval become an important challenge. To find a specific photo from a large amount of images, people usually try to recall their memory of activities or events and search semantic content in images visually one by one. It is definitely inefficient in this way. However, the semantic of events has to be annotated in advance if we want to search photos by keywords. It is also a hard work to build annotations manually [1][2].

Time and location are the important information for photographers and viewers to recall their memory and retrieve the interested photos. Date and time are basic data logged by digital camera. Some GPS and cellular technologies have been embedded in cell phone and camera to record location information in digital photos recently. It is possible and feasible to tag geographical location for digital photos automatically.

In this paper, the proposed system performs two major tasks, GPS geo-locator and geo-classifiers, to find the geo-location for digital photos. The first task uses GPS information equipped by photographic devices to map geo-locations. The second task

© Springer-Verlag Berlin Heidelberg 2015
L. Wang et al. (Eds): MISNC 2015, CCIS 540, pp. 483–494, 2015.
DOI: 10.1007/978-3-662-48319-0_40

collects tagged training images from social media networks to generate geographic classification model using machine learning approaches. The two geotagging methods are evaluated individually and compared with the combination scheme. The detailed experimental results are also shown and discussed in the following sections.

2   Related Work

The problem of geotagging in multimedia is an emerging issue which has potential of coupling many useful applications [3]. Tagging POI(point of interest) from images is a difficult task. A simple algorithm to estimate possible geographic locations using images was proposed in [4]. Toyama et. al. [5] brought an architecture of combining GPS information in digital images and local maps to acquire location tags. Naaman [6] first developed a three-pass clustering algorithm using time and images to generate event hierarchy and location hierarchy. Then, location names are annotated to the hierarchies by mapping geographic coordinates. A further study of image geotagging on social media webs is to investigate the geo-tag accuracy of Flickr [7]. They found that the accuracy of geo-tag is strongly dependent upon the number of images around the venue. More photos was taken at the location more accurate tags were given. The other related tagging issues based on social media network can be found in [8][9].

The advanced applications, such as automatic geotagging tools and systems [10] [11], are also released in prototype. The general image tags mining approaches [12] and automatic summarizing collections of photos [13] are interesting research topics in recent years.

3   The Design of System and Approaches

3.1   System Architecture

The system is composed of six major components including scene segmentation, feature extraction, geo-tag classifier, GPS geo-locator, geographical database, and geo-tag ranking, as shown in Fig. 1. The component functions are briefly described as follows:

Scene Segmentation. A collection of sequential photos generally contain different events or locations. The photos with the same event and location are usually shot in a near time. The segmentation process in this component tries to separate the set of photos into independent groups according to the photographing time.

Feature Extraction. Image visual features like color, shape, and texture are important content for image retrieval and classification. In our system, image features including CEDD [14], FCTH [15], AutoColorCorrelogram [16], JCH [17] are extracted and used for learning image classifiers.

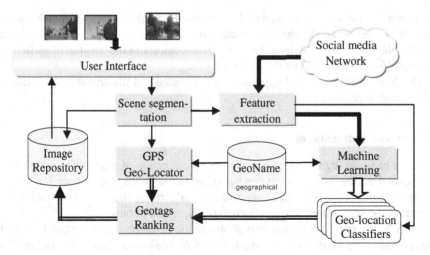

Fig. 1. The system architecture.

GPS Geo-Locator. If the digital photo was taken by cell phones or other devices with GPS, coordinate information such as latitude and longitude can help us to resolve location names by referring to the geographical database directly.

Geo-Tag Classifier. For each geo-tag, the training images can be gathered from social media networks like Flickr, Google images, Instagram. Geo-tagging classifiers will be built by using effective machine learning methods.

Geographical Database. The database supports global geographical names, location hierarchy, and corresponding coordinates. An abundant geo-tag database is needed for recognizing global place names. The web service like GeoNames [18] is a rich source of providing wide-range geographical ontology.

Geo-Tag Ranking. This component combines the tagging results from GPS geo-locator and geo-tag classifiers. The main function is to integrate the scores of geo-tags obtained in GPS geo-locator and geo-tag classifiers and decide the final rank of geo-tags.

At first, we must build the geographical database and gather the training images set from social media networks for the system. The geographical classification models (geo-tag classifiers) then were accomplished by machine learning technology on the training images. While a user needs to manage a collection of serial photos with some events, the system operation starts to segment the photo collections into different scene groups by adopting photographing time slot between two photos. The grouped scene photos represent their neighborhood of photographing locations. The photos in each scene group can share and utilize their GPS geographical information to others if they have. The GPS geo-locator will resolve the geographical name and tag them for the group of photos. In the meanwhile, the image visual features of photos are also extracted and classified by the geo-tag classifier whatever they have GPS coordinates. Finally, the photo tags obtained from the GPS-locator and the geo-tag classifiers are integrated and re-scored to be the final rank of tags.

The detailed functions and approaches are depicted and illustrated in the following subsections in detailed.

3.2   Scene Segmentation

The goal of scene segmentation is to separate a collection of sequential digital photos into several groups which the photos belong to the same geographic location in each. In this section, we give a scene segmentation algorithm based on time variance to determine the segmentation in a collection of sequential photos.

Let N be the number of sequential photos in the given photos collection C and t_i be taking time of the ith photo in C, where $1 \leq i \leq N$. We assume that $t_0 = t_1$ and the time slot between taking the ith and $(i-1)$th photos is $\Delta t_i = t_i - t_{i-1}$. We also define a window frame with size w which represents the number of photos we observed. The average time slot in the observed window frame is

$$\mu = \frac{1}{w-1} \sum_{k=i}^{w+i-1} \Delta t_k , \tag{1}$$

where i and $w+i-1$ stand for the index of the first and the last photos in the observed window of photos collection C. Then, the time slot Δt_i is less than the mean μ will be removed using the following characteristic function.

$$\Delta t'_i = \begin{cases} 0 & if \quad \Delta t_i < \mu , \\ \Delta t_i & if \quad \Delta t_i \geq \mu . \end{cases} \tag{2}$$

For the new time slots, $\Delta t'_i$, we re-compute the mean and variance of time slots in the window frame, as follows:

$$\mu' = \frac{1}{w-1} \sum_{k=i}^{w+i-1} \Delta t'_k , \tag{3}$$

$$\sigma' = \sqrt{\frac{1}{w-1} \sum_{k=i}^{w+i-1} (\Delta t'^2_k) - \mu'^2} . \tag{4}$$

The new mean time slot μ' and variance σ' are used to be the condition of segmenting a sequence of photos. The threshold θ can be set to cluster the photos in the window frame as group when the time slot satisfies $(\Delta t_k - \mu')/\sigma' < \theta$. Then, the starting photo of the sliding window will be moved to the next photo of the last segmented photos. If there is no segmentation inside the window, the window size will be enlarged and μ' and σ' will be re-computed. This scene segmentation is executed repeatedly until all of the photos being separated.

The detailed algorithm of scene segmentation is shown in Fig. 2.

3.3   GPS Geo-Locator

The digital photos generally can contribute almost exact coordinates where they were taken if photographic devices equip with GPS receiver or cellar technology. While resolving geographical name of locations from GPS signals, a geographical database containing location information, such as latitude, longitude, country, and district, is needed. GeoNames [18] is employed to be the location ontology of our system. Geo-Names is free of charge and consists of over 8 million geographical tags and related information. The photographing place can be reasoned from the geographical database once the photo's geographical coordinate is acquired.

A digital photo with GPS data can be used to find out the possible and the nearest location names by searching the coordinates in the geographical database. We define the tag confidence for the candidate location as the following formula:

$$G_i(x_p) = \begin{cases} 0 & \text{if } d(x_i, x_p) > R, \\ 1 - d(x_i, x_p)/R & \text{if } d(x_i, x_p) \leq R. \end{cases} \tag{5}$$

Algorithm: Scene Segmentation

Input: Window size w,   Threshold θ

Output: Segmentation points set S

$S = \varnothing$;

$i = split = 0$;

do

    for $k = i$ **to** $(w+i-1)$

        if $\Delta t_k < \mu$

            $\Delta t_k = 0$;

        end if

    end for

    $k = i$;

    compute μ', σ' by equ. (3) & (4);

    while $(k < (w+i))$ and $((\Delta t_k - \mu')/\sigma' < \theta)$

        $k = k + 1$;

    end while

    if $k < (w+i)$

        $split = k$;

        $S = S \cup \{split\}$;

        $i = split$;

    else

        $split = 0$;

        $w = w+1$;

    end if

while $(i < N)$

Fig. 2. The scene segmentation algorithm.

where x_p and x_i are the coordinates, latitude and longitude, recorded on the photo p and location tag i, respectively. R is the radius of searching range. The $d(x_i, x_p)$ is the direct distance between the two coordinates x_p and x_i.

For parts of photos in a sequence of scene photos without GPS information, they still can gain tag confidences from other photos with GPS information in the same group by interpolation of photographing time. However, no tag confidence will be produced if all photos in the same scene group had no GPS coordinates.

3.4   Geo-Location Classifiers

Flickr and Instagram are famous social media platforms of sharing photographs and associated photographic information. These Webs also provide images with locations for most of important landmarks and attractions. They can be easily collected using keyword queries. The classification model for each location tag in the geographical database can be constructed in advance from the collections of social media webs. The digital photos without latitude/longitude coordinates can try to discover candidate locations using geo-location classifiers.

The procedure of constructing geo-classifiers is as follows. First, the system uses the tags in geographical database as queries to acquire the relevant images from social media service webs, like Google image, Instagram and Flickr. The collection of images corresponding to the tag are considered as the positive samples of training set. Second, the image visual features including color, shape, and texture are extracted as a high-dimension feature sets. Then, we apply the machine learning approach, random forest [19], to learn the geo-location classifier for each geo-tag. Each classification model in the geo-tag classifier set will respond a value of real number $P_i(x_p)$ in the range of [0,1] to the corresponding tag representing the probability of an unknown photo p being at the location i.

3.5   Geo-Location Ranking and Tagging

After obtaining the confidence of geo-locator and the probability of geo-tag, we combine both two tag measurements to rank the most possible tags. The final rank of geo-location tags are determined by the following approach.

Let G be the set of location tags with $G_i(x_p) \neq 0$; that is, the tag $i \in G$ is a candidate inside the distance R of possible photographing location. The definitions of $G_i(x_p)$ and $P_i(x_p)$ are the same to the previous subsection 3.3. and 3.4. Let P is the set of location tags recognized by geo-location classifiers and $P_i(x_p) = 0$ if geo-tag classifier i can not identify the photo x_p (i.e. $i \notin P$). The final score of the photo x_p on location i is $L_i(x_p)$, where

$$L_i(x_p) = \begin{cases} [G_i(x_p) + P_i(x_p)]/2 & \text{if } i \in G \cap P, \\ G_i(x_p)/2 & \text{if } i \in G \setminus P, \\ \min_{i \in G} G_i(x_p) - [1 - P_i(x_p)] & \text{if } i \in P \setminus G. \end{cases} \quad (6)$$

The geo-tags are ranked by the values of $L_i(x_p)$ for $i \in G \cup P$.

4 Experiments and Evaluation

To evaluate the performance of the system, two experiments were performed. The first experiment analyzes the clustering results of scene segmentation algorithm. The second experiment were set up to evaluate and compare the effectiveness of the geotagging approaches.

We used 8 data sets totally for the two experiments. The number of photos and the number of scenes for each data set are shown in Table 1. These collections of photos are provided by tourists after their traveling. The locations including local landscapes in Taiwan, popular attractions and famous landmarks in Egypt and Europe. All photos have marked latitude/longitude coordinates. The average number of photos for each scene is about 9 to 28 photos.

Table 1. Summary of the 8 data sets in our experiments.

Datasets	DS1	DS2	DS3	DS4	DS5	DS6	DS7	DS8
# of photos	86	41	95	53	72	245	304	336
# of Scenes	3	4	4	4	8	16	29	33

In the first experiment, the window size w in the scene segmentation algorithm is set to be 50. The other parameter, threshold θ, is tested using 1.00, 1.50, 1.63, 1.96, and 2.00, respectively. The segmentation results are evaluated by two measures, *Micro accuracy* and *Macro accuracy*, defined as follows.

$$MicroAcc = \frac{\sum_{i=1}^{n} \frac{N_i}{n_i}}{n}, \qquad MacroAcc = \frac{\sum_{i=1}^{n} N_i}{N}, \qquad (7)$$

where n is the number of scenes of ground truth, n_i is the number of photos for the ith scene of ground truth; N_i is the number of photos segmented correctly for the ith scene, N is the total number of photos. *Micro accuracy* is the mean of average accuracy of segmentation for all scenes, and *Macro accuracy* stands for the average accuracy of segmentation for all photos.

Table 2. The results of *MicroAcc* for tested data sets.

θ	DS1	DS2	DS3	DS4	DS5	DS6	DS7	DS8
1.00	60.78%	90.00%	86.48%	98.21%	40.00%	79.46%	37.28%	48.93%
1.50	65.69%	100.0%	100.0%	98.21%	60.00%	68.53%	36.61%	45.77%
1.63	65.69%	100.0%	100.0%	98.21%	60.00%	68.53%	36.54%	45.77%
1.96	90.32%	100.0%	100.0%	100.0%	80.00%	68.53%	36.54%	46.76%
2.00	90.32%	100.0%	100.0%	100.0%	80.00%	68.53%	38.00%	46.76%

Table 3. The results of *MacroAcc* for tested data sets.

θ	DS1	DS2	DS3	DS4	DS5	DS6	DS7	DS8
1.00	81.54%	85.37%	86.17%	97.92%	52.78%	86.12%	46.71%	60.42%
1.50	89.23%	100.0%	100.0%	97.92%	56.94%	82.45%	52.30%	59.23%
1.63	89.23%	100.0%	100.0%	97.92%	56.94%	82.45%	51.98%	59.23%
1.96	90.77%	100.0%	100.0%	100.0%	66.67%	82.45%	51.98%	59.23%
2.00	90.77%	100.0%	100.0%	100.0%	66.67%	82.45%	51.98%	59.23%

Table 2 shows the results of *MicroAcc* which is the mean accuracy for each photo segmentation in different threshold θ. Table 3 shows the results of *MacroAcc* which is the accuracy of segmented photos in the data set. The two results show that a fix θ value can not fit all scenarios. The times slots between taking two photographs is strongly dependent on events and scenes. The discrepancy exists among the data sets. DS1, DS2, DS3, and DS4 have good segmentation results since their scene areas are local and narrow, but different scenes locate apart. Further, the numbers of scenes for these data sets are small. However, the results of DS5, DS7 and DS8 are not so ideal since most of the scenes are near in walking distance and the photographer did not have particular focus. In the case of DS6, each scene occupies a zone with wide area and distinct scenes are also far apart. The same scene is often segmented into small part of scene in such a case.

The second experiment tried to evaluate the performance of GPS geo-locator and geo-tag classifier. For GPS geo-locator, the searching range R is set as 5km to 10 km. The tags for the tested photo are ranked by the tag confidences in Eq.(5). A geo-tag classifier is learned using about 50 positive training images collected from social media web (Google image or Flickr), and 100 negative training examples are randomly selected from images belonging to other tags. The geo-tag classifiers will return the probabilities of an image belonging to the corresponding tags. Then, the tags were ranked by the probabilities. The final tag ranking of combining both of the two approaches uses the ranking score in Eq. (6).

Two measurements, average ranking(AR) and mean average precision(MAP), are used to evaluate the effectiveness of tagging. The two measures are defined as follows:

$$AR = (\sum_{i=1}^{N} r_i)/N, \qquad MAP_k = \frac{N_k}{N}, \qquad (8)$$

where N is the total number of photos in the data set, r_i is the rank of the correct tag of the photo i, and N_k is the number of photos which is ranked at top k correctly. Please note that we set $k = 10$ in our experiments since users usually consider no more than the top 10 recommendation. The rank r_i was given as 11 if the rank of correct tag for the photo i is larger than 10.

The experimental results of *AR* and *MAP* for individual GPS geo-locator and geo-tag classifiers are shown in Table 4, Table 5, respectively. For GPS geo-locator approach, we also observed the situations that the proportion of photos in data sets has no GPS information. We selected three incomplete data sets, 25%, 50%, and 75% randomly, and complete GPS information to test their effectiveness. The upper row of each dataset marked "single" means that the tags in the photo are resolved by the GPS information only itself. The lower rows of datasets in the tables are the resolving results sharing with the GPS information in the photo group of scene segmentation while some photos missing GPS information.

It is obvious that the more GPS information the better ranking for the GPS geo-locator. The GPS geo-locator is more effective than geo-tag classifiers even the data with only 25% GPS coordinates. Since the number geo-tag classification models is large, it is hard to identify the correct model from large number of classifiers. There are still many challenges on this issue.

Table 6 and Table 7 are the experimental results of *AR* and *MAP* combining GPS geo-locator and geo-tag classifiers together. The top 10 average precision *MAP* shows that the geo-tag classifier is helpful for case of single image without GPS-information. However, it seems not improve the effectiveness of grouping photos in the segmented scene. Furthermore, classifiers degrade the performance of precision in some cases. Although *MAP* was not improved obviously for grouping images, *AR* of single image was improved while the mass of GPS information is missing, such as the case of 25%.

In practical applications of real world, GPS information can not be acquired always. However, since users usually produced a sequence of photos for an event, the sharing of information is a possible way to improve the effectiveness of tagging photos. In our experiments, the correct geo-tag were found at the top 5 candidates in average after combining different tagging approaches. While recommending tags for users' labeling application on personal daily photos, the system can provide a scheme with high tagging precision.

Table 4. The respective *AR* results for geo-locator and geo-tag classifiers.

AR	single/group sharing	GPS Geo-locator (% of GPS)				Geo-tag Classifier
		25%	50%	75%	100%	
DS1	single	8.60	6.14	3.68	1.09	9.50
	group	1.11	1.11	1.09	1.09	
DS2	single	8.81	6.39	3.71	1.78	10.54
	group	1.73	1.76	1.76	1.78	
DS3	single	8.61	6.06	3.62	1.12	9.00
	group	1.12	1.06	1.06	1.12	
DS4	single	8.56	6.06	3.50	1.13	9.92
	group	1.90	1.06	1.00	1.13	
DS5	single	8.56	6.67	4.28	2.10	7.96
	group	2.06	2.06	2.10	2.10	
DS6	single	8.61	6.30	3.96	1.60	6.31
	group	1.74	1.80	1.67	1.60	
DS7	single	9.15	7.47	5.43	3.77	9.66
	group	4.73	4.17	3.89	3.77	
DS8	single	9.52	8.05	6.52	4.98	9.08
	group	6.20	5.05	4.93	4.98	

Table 5. The respective *MAP* results for geo-locator and geo-tag classifiers.

MAP_{10}	single/group sharing	GPS Geo-locator (% of GPS)				Geo-tag Classifier
		25%	50%	75%	100%	
DS1	single	0.246	0.492	0.738	1.000	0.200
	group	1.000	1.000	1.000	1.000	
DS2	single	0.220	0.463	0.732	0.927	0.049
	group	0.927	0.927	0.927	0.927	
DS3	single	0.245	0.500	0.745	1.000	0.277
	group	1.000	1.000	1.000	1.000	
DS4	single	0.250	0.500	0.750	1.000	0.208
	group	0.917	1.000	1.000	1.000	
DS5	single	0.250	0.444	0.694	0.917	0.431
	group	0.917	0.917	0.917	0.917	
DS6	single	0.249	0.498	0.747	1.000	0.629
	group	0.963	0.971	0.992	1.000	
DS7	single	0.214	0.418	0.651	0.849	0.207
	group	0.743	0.796	0.836	0.849	
DS8	single	0.167	0.345	0.515	0.690	0.298
	group	0.568	0.690	0.693	0.690	

Table 6. The *AR* results of final ranking after combining geo-locator and geo-tag classifiers.

AR	single/group sharing	GPS Geo-locator (% of GPS) + Geo-tag Classifier			
		25%	50%	75%	100%
DS1	single	7.40	5.60	3.51	1.37
	group	2.46	1.37	1.39	1.37
DS2	single	8.39	6.49	3.85	2.00
	group	2.00	2.00	1.98	2.00
DS3	single	7.01	5.23	3.43	1.46
	group	2.27	1.56	1.56	1.46
DS4	single	7.77	5.54	3.35	1.13
	group	1.27	1.27	1.44	1.13
DS5	single	6.33	5.21	3.63	2.29
	group	2.38	2.53	2.29	2.29
DS6	single	5.11	3.75	2.60	1.27
	group	1.22	1.30	1.27	1.27
DS7	single	8.20	6.81	5.12	3.84
	group	4.19	4.00	3.86	3.84
DS8	single	7.88	6.89	5.69	4.55
	group	5.18	4.70	4.55	4.55

Table 7. The *MAP* results of final ranking after combining geo-locator and geo-tag classifiers.

MAP_{10}	single/group sharing	GPS Geo-locator (% of GPS) + Geo-tag Classifier			
		25%	50%	75%	100%
DS1	single	0.415	0.585	0.785	1.000
	group	0.892	1.000	1.000	1.000
DS2	single	0.268	0.463	0.732	0.927
	group	0.927	0.927	0.927	0.927
DS3	single	0.468	0.628	0.809	1.000
	group	0.936	0.989	0.989	1.000
DS4	single	0.396	0.604	0.771	1.000
	group	0.979	0.979	0.938	1.000
DS5	single	0.569	0.681	0.819	0.917
	group	0.903	0.889	0.917	0.917
DS6	single	0.710	0.824	0.898	1.000
	group	1.000	0.996	1.000	1.000
DS7	single	0.362	0.523	0.717	0.852
	group	0.816	0.832	0.842	0.852
DS8	single	0.402	0.497	0.604	0.699
	group	0.664	0.699	0.699	0.699

5   Concluding Remarks

The problem of geo-tagging from images is difficult if there is no extra information and semantic knowledge. The geo-location in images still can not be found correctly even though additional information like GPS latitude/longitude coordinates are given. This paper proposes two geo-tagging approaches, GPS geo-location method and image classification models, to investigate the kernel issues of tagging geo-location problem using images. The effectiveness of the two individual approaches are tested and compared with the mixed method. The results show that sharing information within a collection of photos can effectively improve the accuracy of geo-tagging. The segmentation of photo scenes is also important for increasing the precision of tagging. Finally, although the images collected from social media networks create the geo-tag classifiers easily, they are not so effective. GPS information is still the most accurate data for tagging geo-location.

The challenge of tagging correct geo-location in an image through the help of social media networks without GPS information is the direction of future research.

Acknowledgments. This research was supported in part by the National Science Council of Taiwan, R. O. C. under contract MOST 103-2221-E-024-010.

References

1. Jeon, J., Lavrenko, V., Manmatha R.: Automatic image annotation and retrieval using cross-media relevance models. In: the 26th International ACM SIGIR Conference on Research and Development in Information Retrieval, pp. 119–126 (2003)
2. Carneior, G., Chan, A.B., Moreno, P.J., Vasconcelos, N.: Supervised Learning of Semantic Class for Image Annotation and Retrieval. IEEE Trans. on Pat. Anal. and Mach. Intell. **29**, 394–410 (2007)
3. Luo, J., Joshi, D., Yu, J., Gallagher, A.: Geotagging in Multimedia and Computer Vision—A Survey. Multimedia Tools and App. **51**(1), 187–211 (2011)
4. Hays, J., Efros, A.A.: IM2GPS: estimating geographic information from a single image. In: IEEE Conference on Computer Vision and Pattern Recognition, pp. 1–8 (2008)
5. Toyama, K., Logan, R., Roseway, A.: Geographic location tags on digital images. In: the Eleventh ACM International Conference on Multimedia, pp. 156–166 (2003)
6. Naaman, M., Song, Y.J., Paepcke, A., Garcia-Molina, H.: Automatic organization for digital photographs with geographic coordinates. In: Joint ACM/IEEE Conference on Digital Libraries, pp. 53–62. IEEE (2004)
7. Hauff, C.: A study on the accuracy of Flickr's geotag data. In: the 36th International ACM SIGIR Conference on Research and Development in Information Retrieval, pp. 1037–1040 (2013)
8. Moxley, E., Kleban, J., Manjunath, B.S.: Spirittagger: a geo-aware tag suggestion tool mined from Flickr. In: the 1st ACM International Conference on Multimedia Information Retrieval, pp. 24–30 (2008)
9. Ivanov, I., Vajda, P., Lee, J.S., Goldmann, L., Ebrahimi, T.: Geotag Propagation in Social Networks based on User Trust Model. Multimedia Tools and App. **56**(1), 155–177 (2012)
10. Abbasi, R., Grzegorzek, M., Staab, S.: Large scale tag recommendation using different image representations. In: Chua, T.-S., Kompatsiaris, Y., Mérialdo, B., Haas, W., Thallinger, G., Bailer, W. (eds.) SAMT 2009. LNCS, vol. 5887, pp. 65–76. Springer, Heidelberg (2009)
11. de Figueirêdo, H.F., Lacerda, Y.A., de Paiva, A.C., Casanova, M.A., de Souza Baptista, C.: PhotoGeo: A Photo Digital Library with Spatial-Temporal Support and Self-annotation. Multimedia Tools and App. **59**(1), 279–305 (2012)
12. Wang, X.J., Zhang, L., Li, X., Ma, W.Y.: Annotation Images by Mining Image Search Result. IEEE Trans. of Pat. Analy. and Mach. Intell. **30**(11), 1919–1932 (2008)
13. Jaffe, A., Naaman, M., Tassa, T., Davis, M.: Generating summaries and visualization for large collections of geo-referenced photographs. In: the 8th ACM International Workshop on Multimedia Information Retrieval, pp. 89–98 (2006)
14. Chatzichristofis, S.A., Boutalis, Y.S.: CEDD: color and edge directivity descriptor: a compact descriptor for image indexing and retrieval. In: Gasteratos, A., Vincze, M., Tsotsos, J.K. (eds.) ICVS 2008. LNCS, vol. 5008, pp. 312–322. Springer, Heidelberg (2008)
15. Chatzichristofis, S.A., Boutalis, Y.S.: Fcth: fuzzy color and texture histogram a low level feature for accurate image retrieval. In: 9th International Workshop on Image Analysis for Multimedia Interactive Services, Klagenfurt, Austria, pp. 191–196 (2008)
16. Huang, J., Kumar, S.R., Mitra, M., Zhu, W.J., Zabih, R.: Image indexing using color correlograms. In: Conference on Computer Vision and Pattern Recognition, San Juan, Puerto Rico, pp. 762–768 (1997)
17. Wallace, G.K.: The JPEG Still Picture Compression Standard. Comm. of the ACM **34**, 30–44 (1991)
18. Geonames. http://www.geonames.org
19. Ku, C.W.: A Hybrid Framework for Automatic Image Annotation. Master Thesis, National University of Tainan, Taiwan (2013)

Predicting the Popularity of Internet Memes with Hilbert-Huang Spectrum

Shing H. Doong[(✉)]

ShuTe University, 59 HenShan Rd., Yanchao District, Kaohsiung City, Taiwan
tungsh@stu.edu.tw

Abstract. This paper investigates the popularity of Internet memes exemplified by Twitter hashtags. A data set of more than 16 million tweets containing 690 thousand hashtags has been prepared by using the Twitter's Streaming API. Some early adoption properties of a hashtag are used to predict its later popularity level. One such property is the adoption time series of the tag. Differential series resulting from differences between two successive adoption timestamps indicates the diffusion speed of a tag. Mean and standard deviation of the differential series have been used to predict the popularity level of a hashtag in previous studies. However, the mean and standard deviation statistics cannot catch the oscillation property of a time series. This study employs the Hilbert-Huang Transform [1] to analyze the differential series. Experimental results show that the derived Hilbert-Huang spectrum can help predict the popularity level of a hashtag at the later stage.

Keywords: Internet memes · Twitter · Time series · Hilbert-Huang transform

1   Introduction

According to Merriam-Webster's collegiate dictionary, a meme is "an idea, behavior, style or usage that spreads from person to person within a culture." The diffusion of memes involves complicated mechanisms that sometimes cannot be fully understood. It is difficult to trace and analyze diffusion of memes in the physical world. However, due to technological advances in information and communication technology (ICT), Internet memes can be easily recorded and analyzed for their diffusion process.

Web 2.0 has created a distributed structure of Internet contents that are produced and consumed by average citizens of today. By fulfilling the love and belonging needs of the Maslow's hierarchy of needs and also due to the easy-to-use characteristics of modern ICT, online social networks have blossomed in the past decade. Twitter is one of the major microblogging services in the world. Registered users can send a tweet of no more than 140 characters. Since 2013, the service has reached the hallmark of 500 million tweets per day. As of May 2015, Twitter has more than 500 million users with a pool of 300 million active users [2]. Due to its speed of message communication and tweet characteristics, Twitter has been described as "the SMS of the Internet".

© Springer-Verlag Berlin Heidelberg 2015
L. Wang et al. (Eds): MISNC 2015, CCIS 540, pp. 495–508, 2015.
DOI: 10.1007/978-3-662-48319-0_41

Users of Twitter can create tweets by embedding topics in hashtags which are analyzed by Twitter to make a list of current trending topics. According to Twitter's own research, tweets with at least one hashtag receive 2 times more engagements than tweets with no hashtags. Here engagements include clicks, retweets, replies and favorites. Users can create their hashtags without any restrictions as long as there is no embedded space, and through the powerful follow network of Twitter, some hashtags may go viral in a short time.

In this study, we consider Twitter's hashtags as Internet memes and investigate the bursting behavior of hashtags. Viral hashtags are rare, but they may offer advantages to marketers. If a methodology can predict the later popularity level of hashtags by using their early adoption characteristics only, marketers can use this technique to try out competing hashtags and invest marketing resources in those predicted as viral hashtags.

Since hashtags are user-defined words, lexical content of the tags as well as context of tweets containing the tags have been used to predict their popularity [3]. In addition to these content-based properties, network characteristics of tag adopters have also been used [4], because Twitter's follow network provides an important conduit to propagate memes. The other type of tag characteristics involves the adoption time series of a hashtag [5], because differences between two successive adoption time-stamps indicate the diffusion speed of a tag.

In this study, we use Twitter's Streaming API to collect public tweets. According to Twitter, the Streaming API may sample up to 1% of the total available public tweets [6]. Our goal is to predict the popularity level of a hashtag by using its early adoption properties. By early, we mean the earliest few tweets and adopters of a hashtag. Our properties include the number of different early adopters, the number of mentions and the number of hyperlinks in early tweets, and the adoption time series of early tweets.

Regarding the adoption time series, previous studies often use simple statistics like the mean and standard deviation to describe the series [5]. We notice that these simple statistics cannot catch the oscillation properties of a time series, since they are unchanged no matter how we rearrange the time series. In order to get a good understanding of the oscillation, we use the Hilbert-Huang Transform (HHT) to decompose a time series into intrinsic mode functions (IMF) and compute Hilbert spectrum of IMF [1]. The Hilbert-Huang spectrum representing oscillation properties of the differential time series is used with other features to predict the popularity level of a hashtag.

This paper is organized as follows. Section 2 is devoted to a literature review of hashtag popularity prediction problems and HHT. We describe our methodology and data set in section 3. Section 4 presents the experimental results and discussions. We conclude the paper with remarks and future study directions in section 5.

2   Literature Review

We review literature related to hashtag popularity prediction and HHT in this section. Twitter, the world's largest microblogging service offers both weblog function and social network function. Though Kwak et al. [7] questioned the social network characteristics of Twitter after analyzing its social graph, Twitter by far is one of the most active sites that have social links built-in. A tweet (status) is a short message with no more than 140 characters. By following a user, tweets created or retweeted by the user will automatically show up in the follower's home timeline. Being a directed relationship, this follow network has distinguished Twitter from many other social network sites such as Facebook or Google+. Kwak et al. found that most relationships between users of Twitter are asymmetric and public media companies or celebrities use Twitter as a convenient channel to disseminate news to their followers [7]. A tweet can contain hashtags like #fifa to indicate user-defined topics. Authors of a tweet can mention other accounts by using the mention sign such as @worldbank.

2.1   Tag Popularity Prediction

Like other data mining tasks, predicting the popularity of a hashtag needs to choose a suitable set of predictors in the first step. Unlike traditional customer data where most records are retrieved from a relational database system, features of various forms can be extracted from a hashtag, e.g. content based, network based and time series based features, just to name a few.

The inherent content of a hashtag is considered an important factor for its popularity. If a tag can evoke viewers' emotion, it is more likely to be imitated and spreads into a large population. Tsur and Rappoport used content based features such as the number of words contained in the tag, lexical items and emotional characteristics to study the spread of memes in Twitter [8]. In addition to the hashtag itself, contextual features from the tweets have also been used to predict the popularity of newly emerging hashtags. For example, Ma et al. used fraction of tweets containing URL, fraction of tweets containing mention @, and sentiment of tweets as content based features in their study [3]. Suh et al. found that the numbers of URLs and hashtags in a tweet are strongly correlated to the retweetability of the tweet [9]. Thus, hyperlinks and opinion orientation of tweets containing a hashtag may impact the popularity of the tag. Yet, other research with randomized experimental design seems to indicate that internal quality of a meme may only play a small role of popularity prediction due to the strong effect of social influence [10].

Research on social influence has received a lot of attention in the marketing field, because word-of-mouth marketing is thought to be more effective than the conventional mass marketing. Through the follow network, Twitter provides a convenient channel to disseminate information quickly. A basic assumption in social influence theory is influential nodes are more likely to successfully spread messages, though there are many definitions of the influence capability. The in-degree count of a follow network is the easiest indicator to measure the influence capability. Twitter users with a large population of followers enjoy the advantage of exposing their ideas to a large audience immediately. Ma et al. used 11 features derived from the social graph of Twitter to predict the popularity of a hashtag [3]. Weng et al. considered the community structure of Twitter social networks in their prediction task [4]. We ignore most network based features in this study because of two reasons. First, due to the Twitter API rate limit policy, it is difficult to collect a large amount of relationship data in a reasonable time. Second, the Twitter API does not report when a follow relationship is created, thus we cannot be sure whether or not a hashtag is diffused through the follow network unless it is contained in a retweeted status.

Research has shown that early popularity of a meme is closely related to its later popularity, and viral memes are expected to spread more quickly than others [11][12]. Weng et al. used the early adoption time series to predict the popularity of a meme [5]. However, they only considered the simple statistics (mean and standard deviation) of the differential series derived from the adoption series. It is well known that these simple statistics cannot describe the oscillation properties of a time series, e.g. the two series in Fig. 1 have the same mean and standard deviation, but they have quite different oscillation properties. Therefore, in this study, we introduce the HHT to analyze the oscillation properties of the differential time series.

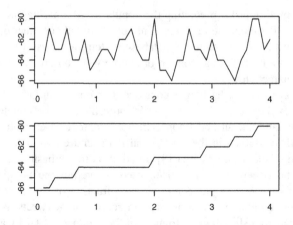

Fig. 1. A time series in the top chart can be rearranged to the bottom chart without changing its mean and standard deviation. However, the oscillation properties of the two series are totally different.

2.2   Hilbert-Huang Transform and Spectrum

Hilbert-Huang Transform is a powerful two-step procedure to analyze complex time series. The first step is called Empirical Model Decomposition (EMD) which is invented by Huang et al. [13], and the second step is the classical Hilbert transform.

Being a data adaptive method, EMD is suitable for the analysis of nonlinear and nonstationary time series. By adaptive, it is meant that EMD uses no preset expansion bases like sinusoids in Fourier analysis or mother wavelets in wavelet analysis.

The expansion bases called IMF in HHT are determined intrinsically from the given data. In the following, we describe the procedure of EMD.

EMD is based on the assumption that any time series consists of several simple intrinsic modes of oscillations. Each of these oscillatory modes is an IMF which must satisfy the following two conditions:

(1) In the whole time domain, the number of extremes and the number of zero-crossings are the same or differ at most by one;

(2) At any point of the time domain, the mean value of the envelope defined by local maxima and the envelope defined by local minima is zero.

IMFs are counterparts to the simple harmonic functions like $\sin(kt)$ in Fourier analysis, except that they are adaptive to the data and come with variable amplitudes and frequencies. IMFs are obtained by a sifting procedure as follows.

Sifting Procedure. Let $x(t), a \le t \le b$ denote a time series. Most likely, only finitely many values are observed for this time series; a and b are respectively the lower and upper bound of the observed data. To obtain IMFs from $x(t)$, conduct the following steps.

(1) Connect all local maxima of $x(t)$ by a cubic spline function. This is designated as the upper envelope.

(2) Connect all local minima of $x(t)$ by a cubic spline function. This is designated as the lower envelope.

(3) The mean of the upper envelope and the lower envelope is designated as m_1. Define $h_1(t) = x(t) - m_1(t)$.

(4) If m_1 satisfies the two conditions for an IMF, then we have obtained the first IMF of the series. Otherwise, repeat the following sifting procedure for h_1:

$$h_{11}(t) = h_1(t) - m_{11}(t)$$
$$h_{1j}(t) = h_{1(j-1)}(t) - m_{1j}(t), j = 2,...,k \tag{1}$$

After a sufficiently large number of siftings, h_{1k} becomes an IMF or a stopping criterion kicks in to stop the sifting procedure. Two stopping criteria are commonly used in the literature. The first one is to stop a sifting procedure when h_{1k} satisfies a Cauchy type of convergence test, and the second one is to stop a procedure when the number of zero-crossings and extremes stays the same for several consecutive steps.

When the sifting procedure stops, we obtain the first IMF $c_1(t) = h_{1k}(t)$. This IMF represents the most vibrant component of $x(t)$. Oscillatory modes of the time series with lower frequencies may still exist. Thus, we consider the residue $r_1(t) = x(t) - c_1(t)$ and apply the sifting procedure to r_1 to find the second IMF c_2. In general, the process is repeated as follows.

$$r_j(t) = r_{j-1}(t) - c_j(t), j = 2,...,n \tag{2}$$

The process can be stopped by any of the following criteria: when c_n or r_n is too small or when the residue r_n becomes a monotonic function from which no more IMF can be extracted. The original time series can be expanded by these empirically determined IMFs as

$$x(t) = \sum_{j=1}^{n} c_j(t) + r_n(t) \tag{3}$$

Since the residue r_n in equation (3) has a very simple oscillation property, we can ignore this part when doing frequency domain analysis. For each IMF, the following Hilbert transform can be applied.

$$d_j(t) = \frac{1}{\pi} p.v. \int_R \frac{c_j(s)}{t-s} ds \tag{4}$$

By combining the IMF and its Hilbert transform, we get an analytic function $c_j(t) + id_j(t) = a_j(t)\exp(i\phi_j(t))$ where $i = \sqrt{-1}$, $a_j(t)$ is the instantaneous amplitude function and $\phi_j(t)$ is the instantaneous phase function. By denoting $\omega_j(t) = \phi_j'(t)$, called the instantaneous frequency function, the original series is expressed as

$$x(t) = \sum_{j=1}^{n} a_j(t)\cos\left(\int^t w(s)ds\right) + r_n(t) \tag{5}$$

We can compute Hilbert-Huang spectrum as follows.

$$H(t,\omega) = \sum_{j=1}^{n} H_j(t,\omega) \tag{6}$$

where $H_j(t,\omega) = a_j(t)$ is defined with the triplet $(t, \omega = \omega_j(t), a_j(t))$ from the Hilbert transform. Finally, the marginal spectrum is defined as

$$h(\omega) = \int_0^T H(t,\omega)dt \tag{7}$$

In practical use, we divide the frequency domain (ω) into a few non-overlapping regions to compute accumulated marginal spectrum.

3   Methodology

In this section, we describe our data collection method and basic statistics of the collected data. Feature selection is an important task for any prediction problem. We explain the features used in our experiments. Finally, the experimental procedure including prediction algorithms is explained.

3.1   Data Collection

Twitter has released two types of APIs (REST and Streaming) allowing authenticated users to collect or manipulate tweet data. The REST API provides programming interfaces to read and write Twitter data, author a new tweet, read author profile and follower data. The Streaming API gives developers low latency access to Twitter's global stream of public tweets. In order to avoid a heavy burden on the service, Twitter implements a rate limit policy on most REST APIs. For example, the GET friendships/show API allows an authenticated user account to retrieve detailed information about the relationship between two arbitrary users. The relationship can be following (friends) or followed_by (followers) between any two queried users. This REST API allows a maximum of 180 calls in a 15-minute window.

In contrast to the REST API, most Streaming APIs do not implement a rate limit policy to restrict the access to public streams. The GET statuses/sample API allows a program to retrieve a small random sample of all public tweets. Tweets returned by the default access level are the same, i.e., two different users connecting to this endpoint will see the same tweets. According to Twitter, this API can return up to a maximum of 1% public tweets that are currently being created [6]. On the other hand, the GET statuses/firehose API returns all public tweets and requires a special permission to access.

We use the Streaming API GET statuses/sample to collect a small sample of public tweets between May 13, 2005 and May 31, 20015. After excluding non-English based tweets, we ended up with more than 16 million tweets. Each tweet is printed in one line starting with the screen name followed by user id, timestamp, status id and the tweet. The tweet may contain RT (indicating a retweet), mentions, hyperlinks or hashtags. The following is an annotated sample.

@MAUREEN_WHITE_ (screen name), 3225914647 (user id), 1433062773000 (time stamp), 604935234283372546 (status id) - RT @Cynthiapoet: Sunset Leopardess by Mark Dumbleton #WeAreAlive #Animals #Photography http://t.co/mu3q6UBk3t (tweet)

3.2   Feature Extraction

After collecting the tweets, we wrote a program to extract user id, time stamp, status id and hashtags. In addition, we also counted how many mentions and hyperlinks are contained in a tweet. The processed raw data have 6 fields (user id, timestamp, status id, mention count, hyperlink count, tag) and 5.67 million records. All these data are stored in a MySQL database for further processing. Table 1 shows two sample records.

Table 1. Two sample hashtags and their metadata.

User id	Timestamp	Status id	m	h	Tag
266043261 6	1431514847000	598442762049040384	1	1	Directioners4Music
274913069	1431514848000	598442766243332096	0	0	alcoharm15

m: count of mentions; h: count of http and https

After grouping records by using the tag field and ignoring case, we found 690917 different hashtags in our collected tweets. In order to conduct the experiments, we delete tags supported by 499 or fewer tweets and tags with one single character. This left us with 1239 tags with a tweet support count ranging from 500 to 244465 (associated with #TeenChoice). Since viral tags are rare, the tweet support count distribution of these remaining tags is seriously skewed towards the lower end. Table 2 lists the count distribution of the tags.

Table 2. Count distribution of experimental tags.

Tweet support count (cnt)	Floor(Log10(cnt))	Number of tags	Popularity level
500~999	2	690	Low (1)
1000~9999	3	512	Medium (2)
10000~99999	4	36	High (3)
100000 and up	5	1	High (3)

Since the last category contains only one tag (#TeenChoice), we combine this category with the previous one to form a high popularity class (3) of 37 tags. The medium popularity class (2) has 512 tags, and the low popularity class (1) has 690 tags. This popularity level (1-3) will be our target variable in the task of popularity prediction.

We use early adoption properties of a tag to prepare the predictor variables. By using the timestamp field in Table 1, we can extract the earliest n (=50 or 100 in our later experiments) tweets that contain the hashtag. Let A denote the author set of these n tweets. Since a user may use the same tag in different tweets, the cardinality of A (denoted as na) is less than or equal to n. If na is relatively small compared to n, then those early tweets are authored by a small population and this may hinder the later diffusion of the tag unless some early adopters have a large size of followers. On the other hand, if na is close to n, then the diversity of early adopters may promote spread of the tag to a wider neighborhood. The number of early adopters na is one of the predictor variables.

The next two predictors are cm and ch which respectively represents the total count of mentions and hyperlinks (http or https) in all early tweets containing the tag. These two variables represent the tweet context of a hashtag. Previous research has indicated that tweets containing mentions and/or links may increase the attention of readers, and thus enhance the spread rate of the tag [3][9].

The next set of variables comes from the timestamps of early tweets. Let $t_1, t_2, ..., t_n$ represent the adoption time of these n tweets. A differential series may be derived from this time series by considering differences between two successive adoptions: $\delta_i = t_i - t_{i-1}, i = 2,3,...,n$. The differential series is nonnegative since each element is nonnegative. Different oscillation patterns of the series may indicate different diffusion patterns of the tag. For example, an increasing differential series indicates that it takes more and more time to spread a tag into the next tweet. Thus, the popularity of the tag may be diminishing slowly. One the other hand, a decreasing differential series indicates the tag is receiving tweet supports faster and faster and the popularity of the tag may be rising. In normal cases, the series is not totally increasing or decreasing, thus diffusion speed of the tag may gear up or slow down from time to time. Due to this reason, we should consider the oscillation characteristics of the differential series in the tag popularity prediction problem.

Previous research using the differential series considers the mean and standard deviation statistics only [5]. We can see in Fig. 1 that a time series can be rearranged without changing the mean and standard deviation, but the new series has a totally different oscillation property. Therefore we need to handle the differential series more carefully.

Let *mu* and *sd* represent the mean and standard deviation of the differential series respectively. Also, let $s_1, s_2, ..., s_{10}$ be the marginal Hilbert-Huang spectrum accumulated in 10 non-overlapping regions. After conducting the HHT and calculating the spectrum, we found the instantaneous frequency spreads between 0 and .5; thus we divide this frequency domain into 10 even regions (0~.05, .05~.1, ..., .45~.5) and accumulate marginal spectrum over those 10 regions. Thus, s_1 represent the total marginal spectrum for frequencies between 0 and .05, and the other *s* variables are interpreted likewise. The various variables (predictor and dependent) used in the prediction are summarized in Table 3.

Table 3. Variables for the tag popularity prediction problem.

Variable	Role	Meaning
na	Input	Number of early adopters
cm	Input	Total count of mentions
ch	Input	Total count of hyperlinks
mu	Input	Mean of differential series
sd	Input	Standard deviation of differential series
$s_1 \sim s_{10}$	Input	Hilbert-Huang marginal spectrum
cla	Output	Popularity level from Table 2

3.3   Experimental Procedure

After extracting features from the collected tweets, we have a table of 1239 records with 16 fields in each record. The *cla* variable is the output variable, while the other 15 variables are the predictor variables. We conduct experiments in two perspectives: using hierarchical linear regressions to examine the model fit from different sets of predictor variables, and using random forest algorithm [14] to measure the prediction accuracy of different sets of predictor variables.

Linear regressions are often used as a preliminary tool to measure the fit between data and linear models. We compare R^2, a common model fit indicator in linear regressions, resulting from different sets of predictors. We also use hierarchical regressions to examine the improvement level of model fit resulted from added predictors.

Random forest (RF) is an ensemble classification algorithm that has been used in many data mining problems. Being a bagging algorithm, RF creates multiple decision trees in the training stage and aggregates decisions from these trees to make a final prediction in the operational stage [14]. Each decision tree is trained with cases sampled with replacements from the original training set. At a decision node, RF chooses a subset of predictors randomly and picks the best predictor from this subset for the node. By using multiple trees in the operational stage, the problem of over-fitted trees can be alleviated.

Since our data set is imbalanced with high concentrations in low and medium popularity tags, the accuracy indicator cannot fully represent the prediction result. For each popularity level, we would like to measure the performance of RF with different feature sets in three perspectives: precision (p), recall (r) and F_1 score. In classification problems, precision is the percentage of predicted samples that are actually relevant, while recall is the percentage of relevant samples that are predicted by the classification algorithm. The F_1 score combines both precision and recall in a simple formula in equation (8), and it is between 0 and 1 with a higher score indicating a better prediction result.

$$F_1 = 2pr/(p+r) \tag{8}$$

4   Results and Discussions

We follow the steps in section 3 to prepare the data set and extract features. This section presents the experimental results in two perspectives: model fitness from regressions and prediction results from the RF algorithm. We also set $n = 50$ and 100 in experiments to extract the earliest n tweets of a hashtag for features preparation.

4.1   Regressions

A linear regression is run with na, cm, ch, $s_1 \sim s_{10}$ predictors (model 1) and na, cm, ch, mu, sd (model 2) predictors separately. The goal is to see which set of predictors fits the data better with a higher adjusted R^2 value. Then, model 2 is expanded in a hierarchical regression by adding the Hilbert-Huang spectrum. The goal of this hierarchical regression is to check the significance level of improvement in model fitness. Tables 4 and 5 show the result for $n=50$ and 100 respectively. Model 1 (with Hilbert-Huang spectrum) beats model 2 when $n=50$, while the result is reversed for $n=100$. In both cases, adding the Hilbert-Huang spectrum improves model fitness significantly ($p < .01$ for $n=50$, $p < .1$ for $n=100$) with a better improvement result for $n=50$.

Table 4.   Regression results for $n=50$.

Model	Predictors	R^2	Adjusted R^2	R^2 change	Significance
1	$na, cm, ch, s_1 \sim s_{10}$	.086	.076	.086	.000
2	na, cm, ch, mu, sd	.078	.074	.078	.000
3	$na, cm, ch, mu, sd, s_1 \sim s_{10}$	.096	.085	.018	.007

Table 5. Regression results for $n=100$.

Model	Predictors	R^2	Adjusted R^2	R^2 change	Significance
1	$na, cm, ch, s_1 \sim s_{10}$	.131	.122	.131	.000
2	na, cm, ch, mu, sd	.144	.141	.144	.000
3	$na, cm, ch, mu, sd, s_1 \sim s_{10}$	.156	.146	.012	.082

4.2   RF Predictions

When using RF to learn classification rules, we need to decide how many decision trees are learned and how many random features are selected. Like the regression experiments, we intended to compare the prediction results from different predictor sets: $\{na, cm, ch, s_1 \sim s_{10}\}$ vs. $\{na, cm, ch, mu, sd\}$ vs. $\{na, cm, ch, mu, sd, s_1 \sim s_{10}\}$. In the following experiments, we created 300 trees in the RF algorithm. When $\{na, cm, ch, mu, sd\}$ is used as the predictor set, randomly selected 3 features were used to determine the decision rule at an internal node. When Hilbert-Huang spectrum was used, 6 features were randomly selected because more predictors were available. We conducted a 10-fold cross validation to yield the final results. Tables 6 and 7 list the predictions results for $n=50$ and 100 respectively.

4.3   Discussions

It seems that results from regressions and RF predictions are quite consistent for both experimental cases of $n=50$ and $n=100$. For example, with $n=50$, regressions show that model 1 (with Hilbert-Huang spectrum) fits the data better than model 2 (with mu and sd). RF predictions also confirm this with a higher accuracy and a higher weighted average F_1 score for model 1. The performance boost of Hilbert-Huang spectrum is reflected in the significant R^2 change from model 2 to model 3 in Table 4 as well as the substantial accuracy and F_1 score improvement from model 2 to model 3 in Table 6. The situation is reversed with $n=100$, where model 2 seems to provide a better fit and prediction results than model 1. By all means, adding Hilbert-Huang spectrum to the predictor set improves model fit and prediction results.

When we examine the RF prediction results more carefully, we see that Hilbert-Huang spectrum provides a better result for low and medium popularity classes, while the simple statistics (mu and sd) tend to provide a better result for the high popularity class.

Table 6. RF prediction results for $n=50$.

Model	Predictors	Accuracy	Class	Precision	Recall	F_1 score
1	na, cm, ch, s_1~ s_{10}	936 (75.54%)	Low	.769	.859	.812
			Medium	.741	.648	.692
			High	.550	.297	.386
			Wghted Av	.751	.755	.749
2	na, cm, ch, mu, sd	920 (74.25%)	Low	.764	.830	.796
			Medium	.715	.652	.682
			High	.591	.351	.441
			Wghted Av	.739	.743	.738
3	na, cm, ch, mu, sd, s_1~ s_{10}	942 (76.03%)	Low	.778	.848	.811
			Medium	.737	.674	.704
			High	.632	.324	.429
			Wghted Av	.757	.760	.756

Table 7. RF prediction results for $n=100$.

Model	Predictors	Accuracy	Class	Precision	Recall	F_1 score
1	na, cm, ch, s_1~ s_{10}	953 (76.92%)	Low	.781	.849	.814
			Medium	.749	.689	.718
			High	.778	.378	.509
			Wghted Av	.772	.774	.772
2	na, cm, ch, mu, sd	959 (77.40)	Low	.797	.835	.815
			Medium	.746	.717	.731
			High	.667	.432	.525
			Wghted Av	.768	.769	.765
3	na, cm, ch, mu, sd, s_1~ s_{10}	963 (77.72)	Low	.790	.849	.818
			Medium	.758	.705	.731
			High	.762	.432	.552
			Wghted Av	.776	.777	.774

5   Conclusions

Online social networks have provided communication channels to spread an idea, behavior, style or usage throughout the world village. Twitter is a special online service that provides both social network and microblog functions. Posting tweets through convenient devices is the main activity of the microblog function while following and retweeting offer the social network function. Users post tweets by encoding topics in the form of hashtags, which are summarized by Twitter to make a list of current trending tags. Like other Internet memes, some plain looking hashtags can go viral unexpectedly, while some high hopes tags just go south. Detecting the later popularity level of hashtags at their early stage has practical applications for marketers because they can use predicted viral tags to run a successful campaign of their products or services.

Even though hashtags are just a single word made by the tweet authors, many features can be extracted from a hashtag. For example, we can count how many lexical words are in the hashtag, are there any digits or special characters? In addition, the tweet content also provides a context for the hashtag. Are there any hyperlinks or mentions in the tweet? What is the sentiment of the tweet? All these properties may affect whether the hashtag will go viral. Social influence resulting from the follow network in Twitter may impact the spread coverage of a hashtag as well.

In this study, we use early adoption properties of a hashtag to predict its later popularity level. Features from the context of early tweets containing the tags are part of the predictor variables. The other predictors include the size of early adopters and the early adoption time series. By early, we mean the first few tweets that have adopted a hashtag.

Our main contribution in this study is to introduce the HHT to analyze the differential adoption time series. A quick reasoning shows that the mean and standard deviation statistics cannot catch the oscillation properties of the differential series. Thus, some frequency domain analysis must be used to accommodate the requirement. The HHT can analyze nonlinear and nonstationary time series. Hence, it is used to calculate the Hilbert-Huang spectrum of the differential series.

By using the Twitter Streaming API, we have collected more than 16 million tweets containing 690 thousand hashtags. Experiments with this data set show that the Hilbert-Huang spectrum can help predict the popularity of hashtags. This confirms our belief that adding the oscillation properties of a time series is beneficial to the popularity prediction task. In addition, computing Hilbert-Huang spectrum is fast with today's machines and algorithms.

In this study, properties from the follow network of Twitter have been largely ignored due to the difficulties of collecting relationship data with Twitter's REST API. In the future, if social graphs of Twitter can be easily obtained, it is possible to consider the impact of network properties in hashtag popularity prediction problems.

The other restriction that should be noted in this study is the possible bias problem resulting from sampled tweets obtained through the Streaming API. Morstatter et al. have found some statistical differences between sampled tweets and all tweets from the Firehose [15]. Popular hashtags mined through sampled tweets may be different from those observed in the full Firehose.

Acknowledgments. This work is supported in part by grants from the ministry of science and technology (Taiwan) under the contract number MOST-103-2410-H-366-001. The author appreciates valuable comments from two anonymous reviewers to improve quality of the paper.

References

1. Huang, N.E., Shen, Z.: The Hilbert-Huang Transform and its Applications. World Scientific Publishing Company, Singapore (2005)
2. Wikipedia: Twitter (accessed June 4, 2015). http://en.wikipedia.org/wiki/Twitter
3. Ma, Z., Sun, A., Cong, G.: On Predicting the Popularity of Newly Emerging Hashtags in Twitter. Journal of the American Society for Information Science and Technology **64**(7), 1399–1410 (2013)

4. Weng, L., Menczer, F., Ahn, Y.Y.: Virality Prediction and Community Structure in Social Networks. Scientific Reports **3**(2522) (2013)
5. Weng, L., Menczer, F., Ahn, Y.Y.: Predicting successful memes using network and community structure. In: Proceedings of the 8th International AAAI Conference on Weblog and Social Media, pp. 535–544. AAAI (2014)
6. Twitter: API Overview (accessed June 4, 2015). https://dev.twitter.com/overview/api
7. Kwak, H., Lee, C., Park, H., Moon, S.: What is twitter, a social network or a news media? In: Proceedings of the 19th International Conference on World Wide Web (WWW), pp. 591–600. ACM (2010)
8. Tsur, O., Rappoport, A.: What's in a hashtag? content based prediction of the spread of ideas in microblogging communities. In: Proceedings of the 5th ACM International Conference on Web Search and Data Mining (WSDM), pp. 643–652 (2012)
9. Suh, B., Hong, L., Pirolli, P., Chi, E.H.: Want to be retweeted? large scale analytics on factors impacting retweet in twitter network. In: Proceedings of IEEE International Conference on Social Computing, pp. 177–184. IEEE (2010)
10. Salganik, M., Dodds, P., Watts, D.: Experimental Study of Inequality and Unpredictability in an Artificial Cultural Market. Science **311**(5762), 754–756 (2006)
11. Yang, J., Leskovec, J.: Patterns of temporal variation in online media. In: Proceedings of ACM International Conference on Web Search and Data Mining (WSDM), pp. 177–186. ACM (2011)
12. Szabo, G., Huberman, B.A.: Predicting the Popularity of Online Content. Communication of the ACM **53**(8), 80–88 (2010)
13. Huang, N.E., Shen, Z., Long, S.R., Wu, M.C., Shih, H.H., Zheng, Q., Liu, H.H.: The Empirical Mode Decomposition And The Hilbert Spectrum For Nonlinear And Nonstationary Time Series Analysis. Proceedings of the Royal Society of London A: Mathematical, Physical and Engineering Sciences **454**(1971), 903–995 (1998)
14. Breiman, L.: Random Forests. Machine Leaning **45**(1), 5–32 (2001)
15. Morstatter, F., Pfeffer, J., Liu, H., Carley, K.M.: Is the sample good enough? comparing data from twitter's streaming api with twitter's firehose. In: Proceedings of the 7th AAAI International Conference on Weblogs and Social Media. AAAI (2013)

Cost/Benefit Analysis of an e-Ambulance Project in Kochi Prefecture, Japan

Yoshihisa Matsumoto[1], Masaru Ogawa[2], Taisuke Matsuzaki[3],
and Masatsugu Tsuji[3(✉)]

[1] Graduate University for Advanced Studies, Tokyo, Japan
matsumoto-yo@itochu.co.jp
[2] Faculty of Business Management, Kobe Gakuin University, Kobe, Japan
CZK07133@nifty.ne.jp
[3] Graduate School of Applied Informatics, University of Hyogo, Kobe, Japan
sumataisuke@gmail.com, tsuji@ai.u-hyogo.ac.jp

Abstract. This study aims at evaluating the economic effect of an e-ambulance project, or emergency telemedicine in the rural areas in Kochi Prefecture in Japan. Ambulances equipped with ICT devices which transmit images of acute patients to remote hospitals are focused on. Kochi Prefecture started an e-ambulance project in Aki and Muroto Cities in 2012. From two cities, it takes approximately one hour to reach emergency hospitals located in Kochi City, the prefectural capital. e-ambulance enhances wellness of residents, since they perceive more secure, and thus the CVM (Contingent valuation method) is applied and WTP (willingness to pay) is used as benefit and estimated based on surveys to residents, which amounts to 1,747 yen per resident per year. Total cost calculated is 381,792,228 yen over three years and accordingly, B/C ratio amounts 0.459. This study calculated another B/C ratio from the viewpoint of two local governments.

Keywords: e-ambulance · WTP · Cost-benefit analysis · CVM

1 Introduction

Telemedicine or e-Health has been implementing all over the world and in some countries e-Health has already passed the experimental stage, and is entering the diffusion stage. In order for the system to be diffused further, there are still lots of obstacles such as the legal framework, economic foundations of implementations, and other regulations. All medical systems were established in the age of face-to face medicine far prior to e-Health. In order to overcome these obstacles, one important effort is to demonstrate its effectiveness, that is, e-Health contributes to efficiency of medical services and enhances wellness of people. One strong measure is to prove its cost-effectiveness by comparing its benefits and costs. The latter consists of equipment such as servers and peripheral devices, salaries and wages of doctors and nurses, and maintenance fees such as communication charges and other miscellaneous operating costs. On the other hand, to indicate its concrete benefits in monetary terms is analytically difficult, since the benefits mainly come from users' subjective satisfaction which is

© Springer-Verlag Berlin Heidelberg 2015
L. Wang et al. (Eds): MISNC 2015, CCIS 540, pp. 509–520, 2015.
DOI: 10.1007/978-3-662-48319-0_42

difficult to measure. Without a firm basis of its cost-effectiveness, the future sustainability of the e-Health cannot be guaranteed.

In order to measure the benefits of public services such as e-ambulance which are not traded in the market, the following methods are utilized in this field: (a) travel cost method: (b) replacement costs method; (c) hedonic approach; and (d) CVM (Contingent valuation method). In what follows, this study employs CVM, which has been recently widely adopted in the fields of health economics and environmental economics. CVM measures the benefits to users in terms of WTP (Willingness to pay), which is the monetary amount that users want to pay for receiving the service. By asking the WTP of each user, we can then construct the surrogate demand function for e-Health. It should be noted that this paper utilizes the results of economic evaluation of e-Health for the assessment of a project which aims to introduce the system. If the system is not introduced yet, it is impossible theoretically to evaluate a project by asking residents about WTP. CVM, however, can be applied in this case, since it asks imaginary questions to people relating to benefits. Although CVM and WTP have a strong theoretical basis, CVM tends to have a bias because it asks for concrete valuation and choice under fictitious circumstances. Care should be taken to clarify what kind of bias it possesses and to remove them.

This paper aims to analyze the Cost Benefit Analysis of the e-ambulance project in two cities in the rural and depopulated areas in Japan: Aki City and Muroto City, Kochi Prefecture. Ambulances are equipped with ICT devices which transmit images of patient to remote hospitals. In the depopulated areas, the number of clinics and medical specialist is small and a patient with acute disease or wounded by accident must be transported to hospitals with full facilities. Traditional ambulances are equipped with the mobile communication system only for voice or facsimile, and accordingly information transmitted from the ambulance to hospitals is limited. Kochi Prefecture started the e-ambulance project in Aki and Muroto Cities in 2012. From these two cities, it takes about approximately one hour to reach emergency hospitals located in Kochi City, the prefectural capital. One of the merits of e-ambulance with the image transmitting system is that doctors in the accepting hospital can monitor real time situation of a patient and prepare for necessary treatment when patient arrives. They thus save time and effort.

The analysis in this study provides not only the firm theoretical basis of evaluating an e-Health project but also practical guideline for regions which plan to implement e-Health. In particular, surveys to residents were conducted not only above tow cities but also Ino Town, Kochi Prefecture, which plans to implement e-ambulance. In this context, we can compare how WTP is different between two types regions which are implementing or not, that is we can compare ex-post and ex-ante WTP. This study therefore leads to useful information to local governments which plan to implement e-Health.

2   Surveys Conducted

2.1   Aki Nad Kuroto City

In this section, let us briefly describe Aki and Muroto city. These neighboring cities are located in the east of Kochi Prefecture. They are mountainous and face the Pacific Ocean. Their primary industries are agriculture and fishery. Their population is declining, while the percentage of the elderly is increasing. Aki City with the area of 317.37km² has 18,657 residents and 8,055 households, and the elderly ratio is 29.14%, while Muroto with the area of 248.25 km² has population 17,490 and 7,598 households, and its elderly ratio is 32.94%. There are 42 in Aki City and 23 clinics in Muroto City, but no tertiary emergency medical facility in two cities and therefore patients who emergency services need have to be transferred to Kochi City. This is the background of the e-ambulance project.

2.2   Surveys to Residents

The surveys were conducted to residents in Ino Town on November 5, Aki City on November 18 and Muroto City on November 19, 2013. We interviewed 62 in Ino Town, 55 Aki City, and 47 Muroto City, totaling 164, and asked questions pertaining to the following: (a) WTP; (b) effectiveness; (c) frequency of usage; and (d) user properties such as age, gender, income, education, and health condition. These are supposed to affect WTP of residents.

Let us examine characteristics of residents who replied to our questionnaires. 76 are males, while 85 are females (Table 1). The average age is 42.7 (Table 2) and regarding education, high school, junior college, and university and higher are equally distributed (Table 3). 153 are still working, while 8 not working (Table 4). The average family size is 2.9 (Table 5), and people living alone and living with wife or husband share more than half. The number of children and grandchildren living in the same city is shown in Table 6. Regarding health condition, more than two-thirds replied either good, fair or all right (Table 7). Accordingly, the average frequency of visiting medical institutions per month is 1.6, and more than 90% reported no necessity for outside medical help, and these are due to their average ages (Table 8). Even though tow cities located depopulated areas, they are enough neighboring clinics, and the most of them live close to clinics (Table 9).

Table 1. Gender

	Freq.	%
1 Male	76	47.2
2 Female	85	52.8
total	161	100.0

Table 2. Age distribution

Age	Freq.	%
20-24	12	7.5
25-29	13	8.1
30-34	6	3.8
35-39	16	10.0
40-44	17	10.6
45-49	24	15.0
50-54	29	18.1
55-59	22	13.8
60-64	10	6.3
65-69	6	3.8
70-74	4	2.5
75-79	1	0.6
Total	160	100.0

Table 3. Education

	Freq.	%
1. Junior high school	5	3.1
2 High school	56	34.8
3 Junior collage	51	31.7
4 University and higher	49	30.4
Total	161	100.0

Table 4. Employment

	Freq.	%
1 Working	153	95.0
2 Not working	8	5.0
Total	161	100.0

Table 5. Number of family

	Freq.	%
0	15	11.1
1	40	29.6
2	27	20.0
3	38	28.1
4	0	0.0
5	12	8.9
6	3	2.2
Total	135	100.0

Table 6. Number of children and grandchildren living in the same city

	Freq.	%
0	123	76.4
1	14	8.7
2	9	5.6
3	9	5.6
4	3	1.9
5	2	1.2
6	1	0.6
Total	161	100.0

Table 7. Diseases treated

Diseases	Freq.	%
1 High blood pressure, Atherosclerosis	11	27.5
2 Heart diseases	1	2.5
3 Diabetes	2	5.0
4 Stroke	1	2.5
5 Chronic Gastritis, Gastric ulcer	3	7.5
6 Assume	1	2.5
7 Backache, arthritis, rheumatism	6	15.0
8 Gglaucoma, cataract	1	2.5
9 Rrenal disease, kidney failure	3	7.5
10 Hemorrhoid	0	0.0
11 Others	11	27.5
Total	40	100.0

Table 8. Frequency of visiting clinic

Number	Freq.	%
1 0	90	57.3
2 1-2	66	42.0
3 3-5	1	0.6
4 6-9	0	0.0
5 more than 10	0	0.0
total	157	100.0

Table 9. Minutes to the nearest clinic

Time	Freq.	%
1 less than 10 minutes	79	50.6
2 10-less than 30	52	33.3
3 30-less than 60	16	10.3
4 60-less than 2 hours	9	5.8
total	156	100.0

3   Estimation of WTP

3.1   Contingent Valuation Method (CVM)

In order to measure the benefits of services which are not traded in the market, the following methods are utilized: (a) travel cost method: (b) replacement costs method; (c) hedonic approach; and (d) CVM. In what follows, we use CVM, which has been recently widely adopted in the fields of health economics and environmental economics. In CVM, the benefits to residents or users of e-Health are measured in terms of WTP, which is the monetary amount which users are willing to pay for receiving the service. By asking the WTP of each user, we can then construct the surrogate demand function for the e-ambulance system. Although CVM and WTP have a strong theoretical basis, CVM tends to have a bias, because it asks for concrete valuation and choice under fictitious circumstances. Care should be taken to clarify what kind of bias it possesses and to remove them.

3.2   Questionnaire

We conducted the surveys to residents in Ino Twon on 5, Aki City on 18 and Muroto City on 19, November 5, 2013, and asked questions pertaining to the following: (a) WTP; (b) whether they know the e-ambulance project, (c) desire to continue the project; and (d) user properties such as age, gender, income, education, and health condition. The questionnaire related to WTP is based on the three-stage double bound method which is expressed in Figure 1 as a tree structure: We begin by asking whether they would be willing to pay monthly charges of 1,500 yen (US$15). This initial value in CVM method is important, since WTP tends to depend on the initial value. If their answer is "yes," we then ask whether they would be willing to pay 2,500 yen (US$25). If they reply "yes" again to 2,500 yen, their WTP is 2,500 yen. If "no", then we lower the amount to 2,000 yen (US$20). If they reply "yes" to 2,000 yen, then that is their WTP. If again their answer is "no," we lower the amount further to 1,500 yen. In the first question of 1,500, if the reply is "no" to 1,500 yen, then we lower the amount to 500 yen. If the reply to 500 yen is "yes," then we ask whether 1,000 yen is acceptable. If the reply to 1,000 yen is "yes," then his/her WTP is 1,000 yen. If not, it becomes 500 yen. On the other hand, the reply to 500 yen is "no," then we ask how about 250 yen. If the reply is "yes," then WTP becomes 250 yen. If it is "no," then we ask how much he/she wants to pay. They reply their acceptable amounts. These series of questions are standard in the evaluation of issues in public services, environments, and so on.

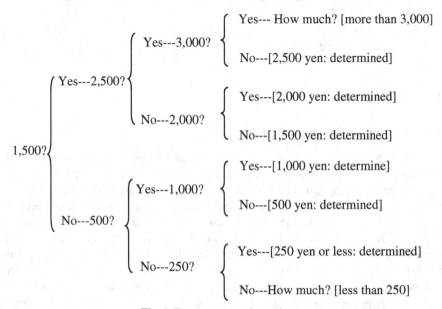

Fig. 1. Tree structure of questions

The distribution of WTP from the survey shown in Table 10 is as follows: more than 3,000 yen (5), 2,500-2,999 yen (5), 2,000-2,499 yen (5), 1,500-1,999 yen (10), 1,000-1,499 yen (18), 500-999 yen (30), 250-499 yen (36), and 1-249 yen (11). The distribution of replies is shown in Table 11. It should be noted that there are 33 residents responded that their WTPs is zero. After checking their reasons, those are considered as "nonresponse," and their replies are omitted from the analysis.

Table 10. Distribution of replied WTP

WTP (JPY)	3,000>	2,500-2,999	2,000-2,499	1,500- 1,999
Number	5	5	5	10
WTP (JPY)	1,000-1,499	500-999	250～499	1～ 249
Number	18	30	36	11

3.3   Estimation of Demand Function and WTP

Based on the above WTP of each respondent, the demand function of the e-ambulance service is estimated; more precisely, the probability of acceptance to amounts questioned is estimated and the number of residents who will agree to pay. The functional form of demand to be estimated is assumed to be logistic, namely

$$\text{Probability of acceptance} = 1 - 1/(1 + exp(-\alpha - \beta log WTP)).$$

The probability of acceptance is the ratio of the residents who reply that they are willing to use the device at the amount of charges provided in the questions. The estimated coefficients α and β are summarized in Table 11. The estimated demand function for e-ambulance is shown in Figure 1. The average WTP is calculated as the area under this demand function, which results in being 1,747 yen (approximately US$17.50) per resident per year. The mean value for WTP, which is the amount that the probability of acceptance is 50%, is estimated at 1,379.45 yen. This paper uses the average value as WTP in the analysis in what follows.

Table 11. Results of estimation

	Coefficients	S.D.	t-value	p-value	
α	18.765	1.276	14.711	0.000	***
β	2.596	0.175	14.854	0.000	***

Note: Log likelihood function: -475.7578

4   Cost Benefit Analysis

4.1   Total Benefits

In a Cost Benefit Analysis, total benefits and costs are compared over the period of several years. In this paper, two kinds of the time span are used: one is three years, which is the same as the project period, while that of five year is also considered, which is usual period of public projects like e-ambulance. WTP obtained above is for per resident per year, and it is multiplied by total number of residents, since all residents have a chance to use ambulance. The population of each city is 18,657 in Aki City and 17,490 in Muroto City as of January 1, 2014, and thus total population is 36,147. Since all residents have possibility of using ambulance, even if they are babies or 100 years old, the numbers of residents who receive benefit are total of two cities, we multiply WTP 1,747 yen by population of 36,147, which ends up with 63,148,089 (US$631,000). That is, one-year benefits of e-ambulance services total approximately 63,148,809 yen (US$631,000). In order to obtain three and five years' worth of benefits, the present values of three or five years' benefits are calculated with a 4% discount rate, and we assume that population of two cities remains at the level for six years. This results in three (five) years' benefits totaling 175,243,694 yen (281,127,278 yen).

4.2   Total Costs

The total costs of the system consist of two major categories; initial fixed and annual operating costs. The former is the items which have to pay at the first year of the project and covers that (i) ICT hardware equipment of the systems of transmitting and receiving images and related equipment, (ii) ambulance, (iii) costs related to software development and the purchase of software, (iv) installment, and (v) initial training cost. The latter, on the other hand, the latter is required annually and contains the followings: (vi) salary of ambulance crew; (vii) maintenance fees which consist of those related to hardware and software; (viii) gasoline mainly for ambulance; and (ix) communications charges for the wireless and wired devices. These cost items are summarized in Table 12 for the annual basis.

In order to obtain total costs of three (five) year period, operational annual costs must be discounted at a 4% discount rate. As a result, the costs of salary in three (five) years are 140,342,126 yen (225,137,915 yen), respectively, while those of miscellaneous expense including maintenance, gasoline and communications are 9,990,328 yen (16,026,560 yen). Thus, total operating costs for three (five) years are 150,332,453 yen (241,164,475 yen), respectively.

4.3   B/C Ratio

From the above calculation, benefits total to 175,243,694 yen (US$1,752,436) over the period of three years and 281,127,278 yen (US$2,811,272) over the period of five years, whereas costs total to 381,792,228 yen (US$3,817,922) over the period of three years and 472,624,250 yen (US$4,726,242) over the period of five years. On the other hand, total costs amount to 381,792,228 yen (US$3,817,922) for three years and 472,624,250 (US$4,726,242) yen for five years. Therefore, the B/C ratio over the period of three years is 0.459, while 0.595 over the period of five years, that is, benefits are about half of costs for three year project, and about 60% for five year project. It can be concluded that benefits are far smaller than costs. Accordingly, these ambulance projects are not recommended from the academic view.

In what follows, let us change the viewpoint; we can evaluate the project from the view of local governments which implementing the project, that is, Aki and Muroto City. They only bear the operating costs, since initial costs are borne by subsidies from the central government, and they can bear only costs for operating the projects. They are only interested in operating costs. Table 13 and 14 show the B/C ratios of the project for three and five years in terms of total costs and operating costs, respectively. The usual B/C ratio (B/TC) is 0.459 as before, while benefits over operating costs (B/OC) is 1.166 for the both of periods, indicating that for two local governments, benefits exceed its costs. Thus from the viewpoint of city, this project is favorable, and worthy to implement.

Table 12. Costs of the e-ambulance project

C. Total Cost (single year)		JPY
1. Initial cost		
Hardware (equipment)		
	1-A Equipment	8,144,662
Ambulance		
	1-B Ambulance	36,000,000
Software		
	1-C Software development	19,341,000
	1-D Software	4,114,803
Installment		
	1-E Installment cost	2,706,784
Training		
	1-F Initial training	1,680,000
	subtotal	71,987,249
2. Operational Cost		
Salary		
	2-A Salary of ambulance crew	50,572,080
Maintenance		
	2-B Software maintenance	893,928
	2-C Hardware maintenance	654,360
Fuel		
	2-D Gasoline	2,400,000
Communication		
	2-E Communication fees	1,200,000
	subtotal	55,720,368
	Total	127,707,617

Table 13. Cost/Benefit: 3 years

	JPY
Total benefit B3	175,243,694
Initial cist IC3	231,459,775
Operating cost OC3	150,332,453
Total Cost TC3	381,792,228
B3/TC3	0.459
B3/OC3	1.166

Table 14. Cost/Benefit: 5 years

	JPY
Total benefit B5	281,127,278
Initial cist IC5	231,459,775
Operating cost OC5	241,164,475
Total Cost TC5	472,624,250
B5/TC5	0.595
B5/OC5	1.166

4.4   Comparison with Other e-Health Projects

There is no economic evaluation of e-ambulance thus far, but the results can be compared with other e-Health projects which are implementation of telecare. Telecare transmits health-related data of its users such as blood pressure, ECG, and blood oxygen to a remote medical institution via a telecommunications network. The system is equipped with a simple device which, when used continuously, records the condition of the elderly or a patient's illness in graphs, which are then used for diagnosis and consultation. Reports sent by the medical institution are also helpful for users to enhance their daily health consciousness and make an effort to maintain good health. Such positive effects have been identified through field surveys [12, 13].

Cost Benefit Analysis of the telehealth system in the following four regions in Japan: Kamaishi City, Iwate Prefecture; Nishiaizu Town and Katsurao Village, Fukushima Prefecture; and Sangawa Town, Kagawa Prefecture. Benefits are expressed in terms of WTP based on CVM, whereas the costs are calculated as the sum of equipment, salaries of doctors and nurses, and other operations. Then, the benefits and costs are compared in terms of the B/C ratio, and the results obtained are shown in Table 15. Since the users receive and perceive benefit from telecare by using every day, while benefits of e-ambulance are less noticed by residents, WTP of e-ambulance tends to be larger than e-ambulance.

Thus the B/C ratios obtain for e-ambulance are similar to those of telecare, which were operated by local governments and received subsides for initial equipment from the central government. As a project, they are less than 1, that is, benefits are smaller than costs, while for local governments it is worthy to implement since benefits to the users are larger than costs which were borne by local governments.

Table 15. Costs and benefits of Telecare by CVM

Unit: JPY

	Kamaishi	Nishiaizu	Katsurao	Sangawa
Number of devices	211	400	325	225
WTP	4,519 yen	3,177 yen	1,640 yen	2,955 yen
Equipment	39.9*	136.7*	111.4*	133.5*
Salaries	8.6*	3.7*	3.36*	4.5*
Others	1.9*	1.9*	10.4*	3.0*
Total costs (6 years)	95.5*	184.5*	184.2*	174.3*
B/C	1.07	0.58	0.54	0.61
(B/C)**	1.87	2.31	1.42	2.60

Note 1: * indicates million yen

Note 2: (B/C)**indicates Benefit/Operating cost

Source: Tsuji, Suzuki, and Taoka (2003a), (2003b)

5   Conclusions

In this paper, WTP in Aki, Kuroto and Ino Town is estimated by CVM and WTP obtained is 1,747 yen. According to our rigorous analysis, we found that this value is not different from the ex-post WTPs estimated in our previous research. The effects of the e-ambulance in Aki and Muroto Cities are also similar to realized ones in the other regions. These results indicate that WTP can be an indicator of potential effectiveness of regional health policy implemented by local governments.

So far, we have conducted surveys of four local governments as shown in Table 15. Except for Kamaishi City, their B/C ratios are approximately 0.5, that is, benefits cover only half of the costs. In addition, regarding the frequency of usage of the device, Kamaishi City also has a much higher ratio than the other local governments. This is due to charges, not free like other region, their efforts to promote usage such as a users' association which organizes events to enhance consciousness towards health, and the participation by medical doctors in this system, which increases the users' reliance on the system. It is clear from our previous studies that telecare is useful for consultation and maintaining the good health of the elderly and patients suffering from chronic diseases who are in stable condition, but it is not for curing disease. It therefore has a psychological effect such as providing a sense of relief to its users by the knowledge of being monitored by a medical institution 24 hours a day. This makes it difficult to estimate its benefits in concrete terms.

On the other hand, benefits of e-ambulance are hardly perceived by residents, except transported by ambulance. But the residents feel wellness because of e-ambulance, since in case of acute diseases they would be treated better than the situation without it. These benefits are less perceived and it is difficult to measure. CVM is an only suitable method to measure benefits.

References

1. Akematsu, Y., Tsuji, M.: An Empirical Approach to Estimating the Effect of eHealth on Medical Expenditure. Journal of Telemedicine and Telecare 16(4), 169–171 (2010)
2. Akematsu, Y., Tsuji, M.: Measuring the Effect of Telecare on Medical Expenditures without Bias Using the Propensity Score Matching Method. Telemedicine and e-Health 18(10), 743–747 (2012)
3. Akematsu, Y., Tsuji, M.: Relation between telecare implementation and number of treatment days in a Japanese town. Journal of Telemedicine and Telecare 19(1), 36–39 (2013)
4. Akematsu, Y., Nitta, S., Morita, K., Tsuji, M.: Empirical analysis of the long-term effects of telecare use in Nishi-aizu Town, Fukushima Prefecture. Japan. Technology and HealthCare 21(2), 173–182 (2013)
5. Garrod, G., Willis, K.G.: Economic Valuation of the Environment: Method and Case Studies. Edward Elgar, London (1999)
6. Loomis, J.B., Walsh, R.G.: Recreation Economic Decisions: Comparing Benefit and Costs, 2nd edn. Venture Publishing Inc., State College, Pennsylvania (1997)
7. Minetaki, K., Akematsu, Y., Tsuji, M.: Effect of e-Health on Medical Expenditures of Outpatients with Lifestyle-related Diseases. Telemedicine and e-Health 17(8), 591–595 (2011)
8. Minetaki, K., Akematsu, Y., Tsuji, M.: Empirical Study of Emergency Medical Service. Journal of eHealth Technology and Application 18(2), 167–176 (2010)
9. Tsuji, M.: The Telehomecare/telehealth System in Japan. Business Briefing: Global HealthCare, London, pp 72–76 (2002)
10. Tsuji, M., Iizuka, C., Taaoka, F.: On the ex-ante and ex-post evaluation of economic benefits of the japanese e-health system. In: Proceedings of 3rd APT Telemedicine Workshop 2005. Multimedia University, Kuala Lumpur, Malaysia, pp. 68–72 (2005)

11. Tsuji, M., Suzuki, W.: An economic assessment of tele-health: the WTP approach. In: Proceedings of APT Conference on Mobile Communication Technology for Telemedicine and Triage, Jakarta, Indonesia (2002)
12. Tsuji, M., Suzuki, W.: The application of CVM for assessing the tele-health system: an analysis of the discrepancy between WTP and WTA based on survey data. In: Aliprantis, C.D., Arrow, K., Hammond, P., Kubler, F., Wu, H.-M., Yannelis, N. (eds.) Assets, Beliefs, and Equilibria in Economic Dynamics: Essays in Honor of Mordecai Kurz, pp. 493–506. Springer (2003)
13. Tsuji, M., Suzuki, W., Taoka, F.: An Empirical Analysis of a Telehealth System in term of Cost-sharing. Journal of Telemedicine and Telecare 9(Suppl. 1), 41–43 (2003)
14. Tsuji, M., Suzuki, W., Taoka, F.: An empirical analysis of the economic assessment of the tele-health system by CVM. In: Proceedings of MODSIM 2003, Townsvill, Australia, pp. 1950–1995 (2003b)

International Workshop on Ethical Issues Related to SNS

The Trap of Quantification of Advertising Effectiveness: A Preliminary Study of the Ethical Challenges of Native Advertising

Hiroshi Koga[⊠]

Faculty of Informatics, Kansai University, 2-1 Ryozenji-cho, Takatsuki,
Osaka 560-1095, Japan
koga@res.kutc.kansai-u.ac.jp

Abstract. As a new advertising form that uses the SNS, the native advertising has been attracting attention. It presents the ad as a natural reading, such as the article. One of the factors that allow such advertising is a big data and life log. However, analysis of big data and life log contains the ethical issues. In this paper, in order to consider the ethical issues, two cultures (that is, Quantification and Relationship Paradigm) that marketing science are focused. In particular, the problems of the trap of quantification, that is, myopia will cause some ethical problems.

Keywords: Social media · Native advertising · Customer relationship management

1 Introduction

After the WWII, Management Science and Operations Research (MS/OR) were applied to corporation management, introducing management decision-making theory based on scientific grounds but not on institution or gut [1]. Marketing science was born as one field of the management decision making theory research. Marketing science is characterized in that the MS/OR techniques such as the Dynamic Programming, the Markov Chain, Regression Analysis, and so on are adopted in the field of marketing decision-making, utilizing ICT (Information and Communication Technology) as a computing function to solve mathematical models - in other words, computer used as the calculating machine.

As the ICT has been further developed, it has become possible to gather, accumulate and analyze huge data of action histories of customers. In the results, the advent of marketing science based on purchase history data has developed. In other words, the marketing science has shifted its focus from studying mathematical models and their solution to analyzing customer behaviors more closely and finding how to respond to them.

The marketing approach based on MS/OR has been modeled on the premises that the customer behaviors are known. For example, the brand transition model, which is

© Springer-Verlag Berlin Heidelberg 2015
L. Wang et al. (Eds): MISNC 2015, CCIS 540, pp. 523–533, 2015.
DOI: 10.1007/978-3-662-48319-0_43

analyzed utilizing the Markov Chain, assumes that the brand switching probability is at constant, ignoring actual bandwagon effects. In such marketing approach, "being scientific" and adopting a mathematical technique are same meaning.

In this approach, it had focused applying a mathematical model to customers and economic and market trends. In such an approach, customer would be understood as collection (market is considered as the collection of each customer). Therefore, maximization of market share has been the purpose.

On the other hand, the recent marketing science is regarded as "being scientific" because it is quantitative marketing based on purchase and/or action history data such as website browsing history, website visit duration history, contract rate after browsing ads. In other words, marketing science of the web era, there is a focus on individual trends rather than the mass or market. Therefore, maximization of customer share (if a different representation, share of wallet or share of stomach) is intended.

Thus, the marketing challenges are to become an only one of customer, not the number one of market. In order to aim the only one, it is necessary to approach the history of the individual's behavior or purchasing. Specifically, when, where, what, how much, what and who bought, personal attitudes, browsing history and so on. Such a new approach of marketing science is based on "Panoptic Sorting [2]. " This is why establishment of information ethics has been called for.

However, the new marketing science has a drawback. It is what the researchers may be trapped with due to the over-demonstrated and quantified advertisement effect. That is, in order to be able to quantitatively measure the advertising effectiveness, it is a strong tendency of the absolute of that number. In particular, the case to be bound by short-term indicators, the author called the "myopia." For example, the absolute view or myopia of the click-through rate of the ads on the user screen of SNS, if such satisfaction of goods are neglected is equivalent to myopia. As will be described later, because it is too focused on small changes in the metrics of Web advertisement, there is a risk that company loses sight of the customer share. As that proverb is saying, "not seeing the wood for the trees" is important.

In this paper, the author demonstrates that excessive pursuance of marketing science is associated with a risk of leading to a direction not towards customer satisfaction (that is, myopia). As examples of myopia problems, Pay-Per-Post[1] in on-line advertisement, stealth marketing, and native advertisement, especially ethical problems regarding native advertisement will be discussed. And the author points out the need of professionalism in marketing in order to overcome the myopia evaluation scale.

2   Two Streams of Marketing as Science

Marketing science is becoming a centerpiece in the marketing research. However, there have been two streams in the arguments regarding "science and marketing". One is the aforementioned MS/OR-based marketing science, which is recent marketing

---

[1] Pay-Per-Post means a service to request a blogger to post an introduction of a product, promising to pay for the posting.

based on customer information. The other stream is a scientific argument based on scientific philosophy [3], [4], [5], [6], [7]. In other word, it is the controversy, that is, to argue "whether marketing is science or art."

The argument based on scientific philosophy has developed to a methodologic debate, noting the effectiveness of not only so-called positive method but also critical method and interpretive method. This is because it has been recognized that various methodologies are necessary in order to grasp the customer's needs.

But the advanced ICT such as database has changed the landscape of marketing science: It has made it possible to analyze and grasp the buying history and characteristics of consumers individually. In the end, the MS/OR-based marketing research made it possible to quantitatively "find out the real needs of customers", which has been the goal of the scientific debate.

This is not the end of story. The MS/OR-based marketing science has eagerly adopted the interpretive method. For example, the customer behavior analysis on persona and context has won the attention as an eclectic method between the quantitative method and the qualitative method. The MS/OR-based marketing science is now the main stream of the marketing as science.

Furthermore, the marketing science has discovered such knowledge that "it is more beneficial for corporations to build a long-term relationship with customers." The conventional mass marketing has taught that it is important to make a wide-ranged approach, aiming to obtain new customers. Therefore, it was a revolutionary finding that it is important to maintain a long-term relationship with existing customers. This leads to the proposal of a marketing concept called CRM (Customer Relationship Management), which emphasizes the importance of retaining leaving customers over obtaining new customers and respects regular customers over first-time customers. Thus, the point of the concept of CRM is in the long-term perspective. In other words, in this concept, the key factor of profit would be considered to build a long-term relationship with customers. If we analogy in operations research terminology, the essence of CRM would be refferd as the approach to "SA: simulated annealing", rather than the branch and bound method.

The purchase history analysis has demonstrated that "good customers high in total purchasing amount and purchase frequency are not only rather insensitive about price (for exampe, members of star-alliance do not take LLC?), but also can be excellent buzz marketing media encouraging the other persons to buy the goods. On this account, it has been recognized as an agenda to extract good customers and perform tailor-made marketing suitable for tastes and interests of the individual good customers. It is an advent of new marketing approach (relational paradigm), which is different from the conventional one by 180 degrees [8], [9].

As explained above, it can be said that the marketing science not only has indicated that "knowing customers", which has been argued in the methodological debate, can be realized by the quantitative method, which have been criticized in the methodological debase, but also have led the trend of the marketing research, for example, by eagerly adopting the interpretive method.

On the other hand, the MS/OR-based marketing science (relational paradigm) is not a "silver bullet", as a matter of course. We can see some problems thereof in and out of shadow.

Firstly, there are legal and ethical problems regarding acquisition and handling of personal data called life logs, which constitute the core element of the MS/OR-based marketing science [10]. Life logs are the "digital records of personal lives, including not only posting at SNS and E-mails, which are output by the living persons at their wills, but also positional information, browsing and buying history data acquired from membership of data SNS and devices [11]. There are concerns about personal information leakage and violation of privacy associated with the life logs, as well as debates on the right of self-control for deletion of the life logs and the right of being forgotten [12].

Secondly, there are problems called a trap of orthogenesis in the scientific approach and positivism. Fundamentally, the concept of CRM is based on a long-term perspective. In addtion, the essence of CRM is intended to separate from the appoarch of the mass marketing, by narrowing to customers who bring long-term benefits.

However, the marketing behavior, especially, the on-line marketing practice has a risk that the observers may become blind to anything but the numerals of the measurement scales (later described), because the effect of the practice can be clearly calculated out. According to some personal attitudes of each customer and their behavior history can be realized, a variety of metrics have been proposed. As a result, if clearing the reference value of the individual metrics, the misconception that could prolong the relationships with all customers occurred. In this paper the excessive emphasis attitude metrics of intermediate stage is referred to as "myopia."

If repeatedly emphasized, the author will discuss such problems as "mistaking means for the end" or "to make a method more important than its purpose" resulted from establishing the objective outcome indexes.

3   Objectiveness and Drawbacks of Marketing Science

3.1   Overreaching of Marketing Science

With the remarkable advancement of ICT, it has become possible to grasp the information regarding behavior history such as on-line browsing history and time-length, membership registration, and goods purchase[2].

---

[2] For example, PV (Page View: the number of Web page displayed in the Web browser of visitors), UU (Unique User: the number of people who visited a specific page within the web site or web site.), Bounce Rate (the percentage of people who landed on a page and immediately left), Exit Rate (the percentage of people who left your site from that page), CTR (Click Through Rate: the percentage of the advertizment is clicked), CVR(ConVersion Rate: the proportion of visitors to a website who take action to go beyond a casual content view or website visit, as a result of subtle or direct requests from marketers, advertisers, and content creators), and so on.

More specifically, it has become possible to grasps the web-site browsing history by using cookie and web beacon. It is possible to numerically find out such information as to whether an individual sees the advertisement to the end, or whether the viewing of the advertisement leads to membership registration, requests for more information, or purchasing. As a result, it has become possible to clearly gauge the cost effectiveness of individual advertisement with ICT, which was not possible with the existing media (such as TV, radio, newspapers, and magazines).

But, it should be kept in mind that overemphasizing the numerical data would be "mistaking means for the end" or "means (that is, metrics) have become ends." Nevertheless, many promotions emphasizing the benefit of the numerical data can be founded. We called this phenomenon as "overreaching" in Sumo's terminology[3]. Please image a case where the audience rating as a measurement scale is weighed unreasonably.

Needless to say, increasing the advertisement metrics (that is, evaluation indexes) is not the solution. The real solution is the construction of a long-term relationship with customers. Then, described with reference to the words of mathematics, individual metrics is the differentially-oriented and long-term relationship is integral-oriented. Since obsessed with differential thinking, we call its atitude as "myopia."

Then, the author would like to explain the myopia as an example the advertisement that is posted on social media such as SNS. Nowaday, the reliability of corporate advertising is shaken. Instead, word-of-mouth (buss or customer reviews) will come to be trusted [13], [14]. In other words, for consumers, advertisement will be regarded as a noise on the SNS [15]. On the other hand, as described above, the metorics of the advertisement effect are many proposed. And, in order to enhance the level of individual evaluation indicators such as PV, UU and so on, improper advertising methods have been born. There is, specifically, Pay-Per-Post, stealth marketing[4], and native advertisement (See below for further details).

Thus, could the use of Pay-Per-Post for promoting browsing of an advertisement be a tool to increase customer's royalty or encourage long-term patronage?

The answer is No!

It is said that the payment for the bloggers participating in Pay-Per-Post is just several dollars per posting. As the information posted for such petty profits is accordingly poor in quality (reliability and accuracy), it is said many of them are no use for readers. Thus, such a risk cannot be denied that a situation "searching for a trash will find only a trash" may become reality when such "spam blogs" overfill the network.

---

[3] The meaning of the word changed to mean "the action of failing to achieve something because one's overconfidence." Or, to borrow the theory on evolution, it would be called "orthogenesis." By the way, the metrics that can be measured quantitatively is, essentially, merely an indicater of the process leading to customer satisfaction. Nevertheless, that would misunderstand the metrics of the intermediate stage as the final object. Then, we will referr to as "overreaching" in this paper. Or, the trend to progress excessively in the direction of the quantification is not an exaggeration as being similar to the "orthogenesis."

[4] Stealth marketing means an advertising so that it is not aware of and promotion to consumers. It also refers as "undercover marketing."

Maybe because of this, the search engine giant Google has announced that they do not consider Pay-Per-Post in their evaluation [16].

Even though the problems of Pay-Per-Post have been pointed out, it is rare that the contents as posted for Pay-Per-Post become a legal issue (provided that Japan has some restrictions set forth by the Pharmaceutical Affairs Act). But it has been pointed out that it would be ethically problematic to post an advertisement disguising as an opinion of an individual [17].

While the ethical problems have been pointed out, the marketing practices disguising buzz marketing have been flourished. One example is something called stealth marketing, which is posting experiences of goods on a user evaluation field on a website or a word-of-mouth communication site by someone pretending a neutral provider for promoting purchasing of the goods. In Japan, it has become a news story that an entertainment personality, who has participated in stealth marketing by introducing "favorite site" on her blog, was forced to refrain from activities in the show business.

Such mistrustful marketing practices are supported by logic that it has an excellent capability as a luring device for guiding to a site to show. Because the direct advertisement cost for posting information in a user evaluation field or word-of-mouth communication site is negligible, it can be a very cost-effective advertising tool once the advertisement succeeded in luring. But such a practice has been pointed out as being problematic ethically. The current situation in which these ethically problematic advertising means have been flourished can be said as "overreaching" caused by objectiveness allowing finding out which site the viewer of an advertisement has passed through to reach the advertisement.

3.2   Marketing Science and Native Advertisement

Recently, more skillful means have come out: native advertisement [18]. At this moment, there is no widely-accepted definition on the native advertisement. The number of debaters arguing for this makes it indefinite. The outline of native advertisement may be "advertisement so natural that it can be cited in SNS, content sites, and curation sites as an article, and can be read as an article by readers" [19]. That is, native advertisement is an advertisement, which is fabricated as looking like an article, hiding its nature of advertisement as much as possible.

Advertisement agencies, promoting native advertisement, insist that the native advertisement is greater in browsing rate and response rate than on-line banner advertisement [20]. Unlike the stealth marketing, which is mainly performed on blogs, the native advertisement is slipped into SNS and content sites as an advertisement. It can be easily imagined that it catches eyes of readers so easily.

Here, this paper introduces one specific example, which is a famous case of failure in U.S. [21], [22]. Just after the noon of January 14 in 2013, a religious group called Scientology posted an advertisement on the website of a magazine, Atlantic, in the same fashion as the other articles thereon. It was labelled as "sponsor's contents", but many readers read it as an article, failing to notice it was actually an advertisement. It is said that the claims "why Atlantic writes up a publicity for a religious group" became a hot topic on the Internet. Consequently, the article was deleted in the midnight of the day, but remembered as the case asking the ethic of native advertisement.

The author also introduces herein one successful example in Japan. Wacoal Corporation, a women's underwear manufacturer posted a PR article on a content site called MY LOHAS [23]. The article called the day of March 12 as "the day of size" (because 3•12 can be pronounced as Sa I Zu in Japanese), introducing how important it is to wear an underwear of right size. This advertisement article was clearly labelled as a sponsor's article, and received no criticism but rather made a large contribution to luring customers to the website of Wacoal [24].

As native advertisement had drawn attentions, IAB (Interactive Advertising Bureau), an American on-line advertisement industry group, introduced "the Native Advertising Playbook" in December 2013, classifying the native advertisement into 6 types for the efforts to dissolve the confusion in concept: that is, In-Feed Units, Paid Search Units, Recommendation Widgets, Promoted Listings, and In-Ad with Native Element Units [25]. Not a few persons in the industry say that it is difficult to classify them so clearly since in reality there are some native advertisements classifiable into two or more of the types

The guideline proposes evaluation axes for native advertisement. Important ones are 1) it can be "naturally" read on the screen of Web sites and SNS, and 2) it is explicitly labelled as an advertisement.

If it was not labelled as an advertisement explicitly, consumers might have a feeling of distrust on the native advertisement as in the stealth marketing. In addition, posting on content site of major media but not on blogs has a large impact. Some criticize that Source Laundering becomes possible for the native advertisement by being posted on a trustful site. Moreover, if native advertisement as bad as stealth marketing has become flourished, the believability of articles on the Internet will be significantly deteriorated, just like bad money drives out good money.

Therefore, disclosure is the most important factor in the native advertisement, as it is necessary to make the readers understand that the article is actually an advertisement. The industry groups are now discussing about self-imposed regulation in order to realize disclosure. The aforementioned guideline is also referred in Japan as well as in U.S.[5]

3.3   Ethical Problems Associated with Native Advertisement

In this paper, it is considered that the problems of information ethics associated with native advertisement can be divided into three.

---

[5] Furthermore, in Japan, JIAA (Japan Interactive Advertising Association) has formulated the "Recommended provisions in native advertising" in March 2015. In there, various representation criterion native advertisement is clarified. For example, the media have to specify clearly that an article is advertisement. This documentation is available at the following URL. http://www.jiaa.org/download/JIAA_rinrikoryo_keisaikijyun2015_03.pdf However it is Japanese.

The first problem is on perspectives of a corporation behind the native advertisement. As discussed above, the perspective on which the market science is supported is the establishment of long-term relationship with customers rather than the acquisition of new customers (maintenance of existing customers and avoidance of their leaving). However, it cannot be denied that there are some promotions, which, at glance, look like trying to acquire new clients. I believe corporations should provide message about what kind of relationship the corporations want to build with customers via the native advertisement. It is a business ethic and an information ethic regarding the content of the advertisement and the method to provide it. I consider that not only quality requirements as to the accuracy and reliability but also the information ethics in letting the readers know what kind of relationship (and value arising therefrom) will be provided.

Secondly, there is a problem regarding association with life logs in performing native advertising. There is a possibility that an ethical problem regarding personal information would arise when the native advertisement is associated with positional information using the GPS function of devices or membership information of SNS. But if the native advertisement can clearly inform which kind of relationship the corporation wants to build with customers as described above, the relationship between the corporation and the customers may become different one. That is, it may change from the equivalent exchange relationship in which a customer provide his/her personal information in exchange of privileges such as discount coupons as a trade-off (or consideration), to a donation-like exchange relationship in which a customer provides their information voluntarily because he/she wants the corporation to know him/her, hoping the corporation will provide detailed services to him/her later on.

The third problem is one regarding disclosure. To emphasize this problem repeatedly, it is not an ethical problem only for the poster of the advertisement but also for the provider of the advertisement media. That is, the professional ethics for the sites providing curation and contents are questioned.

In the stealth marketing, it is rare that bloggers posting information are accused, except that the person posting the information is a notable public figure. But, the poster of the native advertisement is a major content provider such as Forbes, Yahoo, Facebook, Twitter, Google, Bing, Ask, Amazon, Foursquare, and so on. Therefore, it is highly probable that the corporation providing the place for advertising will be questioned for information ethics.

Moreover, Search engines such as Google have announced that they will not reflect websites suspected as stealth marketing to their page ranking, so that the search results by their search engines will not be influenced by blogs for stealth marketing.

However, it is difficult to exclude native advertisements from the search results. Because even the major search engine Google also uses native advertisement, they should not be permitted to make a decision not to reflect a native advertisement of insufficient disclosure to their search results.

4   Conclusion

Mass marketing is not worth the expense of approach to all customers. In other words, mass marketing is not possible to get customers satisfaction, despite it requires a huge cost. From purchase history data, this point has been proved. In addition, the following facts became clear. It is that good customers bring great profits than customer otherwise. As a result, new marketing science has created two cultures, namely quantification and the respect for customer relationship.

However, as the marketing science shifted its weight to the on-line marketing, the quantification has been excessively highly regarded. Marketing science, that tried to adopt a long-term perspective based on the accumulation of individual purchase history data, had come to microscopically analyze every action history.

In this paper, the tendency to focus on every behavior pattern rather than the long-term accumulation was called "Myopia." This results in disrespecting the culture of respecting the customer relationship, and consequently creating techniques for Myopia advertising effect such as Pay-Per-Post and stealth marketing.

Some may point out that the immaturity and confusion of the on-line marketing has allowed creation of such unsophisticated or vice tactics. But it is not accidental that such shadow part has been born. We should accept that the concept of the stealth marketing is originally included in the marketing science, because it is not possible to deny the possibility of admitting the stealth marketing factor when the objectivity, which is one of two factors highly regarded in the marketing science, is excessively emphasized.

Therefore, it can be said that it is an object of corporation ethics or information ethics to consider which kind of marketing message they should provide in view of the customer relationship (maintaining existing customers or preventing leaving thereof).

By the way, currently, the online marketing of the media can be broadly classified into three: 1) owned media, 2) earnd (social) media, and 3) payed media[6]. Nowaday, "three-way of advertising tactics" that take advantage of earned and payed media in order to induce the company's owned site has attracted attention. Such a tacics are called "omni channel."

However, customers do not trust the advertisment (that is, payed media). Rather, it is thought to be noise. However, on the other hand, customers have wanted information. Customer is considered a reliable advertising is less. Therefore, search engine, such as through reviews site, we want to know the experience of using the product. They are seeking the truth that has not been written in the corporate advertising. Thus, the social media has becoming a source of goods for customers.

---

[6] The company'sits, such as a company blog, web site, and other media that are owned and managed in-house are called "owned media." In order to "get trust and recognition of name form users, rather to sell the goods, are called "earnd media." Many of such media is social media. Therefore, it's social media and synonymous. Lastly, media that get me information pays advertising expenses such as advertising frame of web site (or newspaper and television) is called "payed media."

Therefore, stealth marketing and native advertisment can be regarded as bait that was scattered on the water (Web space) to attract fish (attention on the social media). However, would the customer allow the cunning of stealth marketing and native advertisment? Indeed, it may sometimes attract attention by flames. But it is difficult to obtain a long-term interest. If customers are not responsive to conventional advertisements, they should consider what kind of information to provide and how to provide it, in order to provide natural advertisement in the form as it should be.

Furthermore, the business model of the SNS, which rely on advertisment, might have to be reconsidered. However, the introduction of the fee system is there will be resistance for many users. The ethics of established beyond the interests between the user, advertisers (companies), advertising agencies, media that rely on advertising, has been demanded.

Acknowledgments. This study was supported by JSPS KAKENHI Grant Numbers 26380550, and by Kansai University's Domestic-Research-Program (April-September/2014) and Kansai University's Overseas-Research-Program (April-September /2015).

References

1. Charchman, C.W., Ackoff, R.L., Arnoff, J.: Introduction to Operations Research. Wiley and Sons, New York (1957)
2. Gandy, O.H.: The Panoptic Sort: A Political Economy of Personal Information. Westview Press, Boulder (1993)
3. Converse, P.D.: The Development of the science of marketing: An Exploratory survey. Journal of Marketing **10**, 14–23 (1945)
4. Buzzell, R.D.: Is marketing a science? Harv. Bus. Rev. **41**, 32–40 (1963)
5. Hunt, S.D.: Marketing theory: Conceptual foundations of research in marketing. Grid, Columbus (1967)
6. Hunt, S.D.: Positivism and paradigm dominance in consumer research: Toward critical pluralism and rapprochement. J. Consumer Res. **19**, 32–44 (1991)
7. Anderson, P.F.: Marketing, scientific progress, and scientific method. J. Mark. **47**, 18–31 (1983)
8. Payne, A., Frow, P.: A strategic framework for customer relationship management. J. Mark. **69**, 167–176 (2005)
9. Peppers, D., Rogers, M., Dorf, B.: Is your company ready for one-to-one marketing? Harv. Bus. Rev. **77**, 3–12 (1999)
10. Sinpo, F.: The definition of the term 'life-log' and related legal accountabilities: The appropriateness of using personal records for commercial purposes. Johokanri **53**(6), 295–310. doi:10.1241/johokanri.53.295 (in Japanese)
11. O'Hara, K., et al.: Memories for life: a review of the science and technology. J. Royal Soci. Interface **3**, 351–365 (2006)
12. Werro, F.: The right to inform v. the right to be forgotten: A transatlantic clash, Georgetown University Law Center, Public Law and Legal Theory Research Paper Series, Paper No. 2 (2009)
13. Rosen, E.: The anatomy of Buzz: creating word-of-mouth marketing. HarperCollins (2000)

14. Aaker, J., Smith, A.: The dragonfly effect: Quick, effective, and powerful ways to use social media to drive social change. John Wiley & Sons (2010)
15. Gladwell, M.: The tipping point: How little things can make a big difference. Little, Brown (2006)
16. Selling links that pass PageRank, Posted December 1, 2007 in Google/SEO. https://www.mattcutts.com/blog/selling-links-that-pass-pagerank/
17. Summary web site of the Pay-per-post problem. http://marketingis.jp/wiki/%E3%83%9A%E3%82%A4%E3%83%91%E3%83%BC%E3%83%9D%E3%82%B9%E3%83%88%E5%95%8F%E9%A1%8C (in Japanese)
18. DVorkin, L.: Inside Forbes: What's next for native ads? Forbes
19. Japan Interactive Advertising Association, Archives of Native ad Study Group. http://www.jiaa.org/native_ad/index.html (in Japanese)
20. http://www.sharethrough.com/2013/05/infographic-native-advertising-effectiveness-study-by-ipg-media-labs/
21. Erik Wemple, The Atlantic's Scientology problem, start to finish. http://www.washingtonpost.com/blogs/erik-wemple/wp/2013/01/15/the-atlantics-scientology-problem-start-to-finish/
22. http://poynter.org/extra/AtlanticScientology.pdf
23. http://www.mylohas.net/2014/03/036588sizeday.html
24. IAB (Interactive Advertising Bureau), Native Advertising Playbook. http://www.iab.net/media/file/IAB-Native-Advertising-Playbook2.pdf

Regulations of Using SNS for Minors

Joji Nakaya[✉]

Faculty of Business Administration, Kindai University, 3-4-1 Kowakae,
Higashi-Osaka 577-8502, Japan
nakaya@kindai.ac.jp

Abstract. It is very difficult to build regulations on social media for adults. Because adults have the right to do what is wrong. All that we can do for adults is to publicize risks of social media associated with the services that affect users. However, we can impose regulation against minors to use social media, for they have no right to do what is wrong. In order to protect children from the problems of social media, various approaches have to be employed by the adults. And regulation of social media is inefficient for protecting children. Adults should think seriously about what type of life is good for their children, and it is important that adults help the children about this issue.

Keywords: Social Media · Protection of minors · Regulation

1   Self-responsibility Even if Social Media Causes Harms?

In recent years, various problems have arisen as the young generations use social media. Troubles involving social media users which happened in Japan include the following cases.

--- Users confess their own bad conduct such as driving under the influence or cheating on an exam via Twitter; make defamatory remarks about celebrities who are customers of the store they are working part-time for on social media; send a photo of their dealing with foodstuffs in an unsanitary manner at the store they are working part-time for to their friends through social media. As a result, their personal address, name, school or work place is identified and a large number of copies are posted on the internet and exposed to the scrutiny of many others.

--- Users believe in vicious rumors. Chain mail.

--- Users meet unknown others through social media and get involved in trouble.

According to Hisatake Kato the philosopher, the minimum rule of liberalism is the harm principle, to prevent harm to others. In other words, (1) if it is an adult of sound judgment, (2) regarding his/her own life, body and properties, (3) as long as it does not harm others, (4) even if the decision concerned is against his/her own interests, (5) he or she has a right of self-determination[1].

---

[1] Kato [1993], Chapter11.

© Springer-Verlag Berlin Heidelberg 2015

L. Wang et al. (Eds): MISNC 2015, CCIS 540, pp. 534–540, 2015.

DOI: 10.1007/978-3-662-48319-0_44

In brief, the most fundamental rule in the liberalistic society is the principle requiring us "not to harm others"[2]. To put it the other way around, as long as he or she does not do harm to others, an adult can do whatever he/she wants.

In the society of liberalism, an adult has a right to do silly things, that is, a right to do what is wrong[3]. Deadly sports such as winter mountain climbing or diving, smoking, dirtiness, disregard for health or intemperance are among examples of the right to do what is wrong[4]. If we tolerate the right to do what is wrong under the liberalistic system, troubles related to social media as mentioned above can be considered as consequences of exercising the right to do what is wrong.

Driving under the influence is a violation of the Road Traffic Act and cheating is a wrongful act. However, the question here is the pros and cons of an act of confessing such acts in social media. The confession itself does not harm others and only damages the confessor him/herself when his/her wrongful act was revealed. In that case, such confessions in social media can be considered as a sort of right to do what is wrong.

Blindly accepting information in social media which is full of false rumors or voluntarily participating in services to meet unknown others, which are known for a high incidence of trouble, can be also considered as an exercise of the right to do what is wrong, as long as trusting information and using dating services do not involve the harm principle in itself.

Of course, subjects of this right to do what is wrong do not include minors. Therefore, it is necessary to develop systems or regulations to prevent high school students from getting involved in such trouble. However, under the current situation, it is difficult to regulate the right to do what is wrong that adults exercise in social media.

2   Protection of Minors on Social Media

Efforts to widely inform risks of social media should not target adults only. It should target minors as the most imminent issue.

As previously noted, the harm principle explains that an adult of sound judgment has a right of self-determination as long as it does not harm others and this principle does not cover minors.

According to a paper on "the situation of SNS usage by individual users by sex and age" in "The 2012 infrastructure development in our country information economy and society" by the Ministry of Economy, Trade and Industry, the rate of SNS users amounts to 83.3% for teenage boys and 93.0% for girls[5]. Under such present circumstances, we can say that society tolerates that minors suffer irreversible damages from social media.

---

[2] Ibid.
[3] Ibid.
[4] Ibid.
[5] The infrastructure development in our country information economy and society by the Ministry of Economy: http://www.meti.go.jp/press/2013/09/20130927007/20130927007-4.pdf (Seen at 04.28.2015).

Of course, companies operating social media explain that they are addressing the issue of minor protection by excluding users who commit illegal acts from their services or by providing services for usage restriction by parents. With regards to SNS games, companies also put a cap on the usage fee for minors.

However, can we say that minors are sufficiently protected with such ad hoc regulations? It seems appropriate for the government to take more strict measures, such as a regulation on the usage by minors.

3   Difference Between Social Media and Conventional Products for Minors

Among things for which children feel enthusiasm, there have been many things which became social problems. As many children get glued to TV cartoons, there have been many opinions that the existence of television would inhibit the psychological development of children. Video games have been exposed to a similar criticism.

However, the major difference between these items and social media is that in social media, parents are very hard to check the actual services or products their children have purchased. As a family often puts a TV set in their living room, parents can monitor what children are viewing. As for video games, parents can also check what kinds of games they are at the time of purchase and in the first place, it is difficult for children to access dangerous games such as those for adults.

On the other hand, in social media, as new services are provided one after another, parents cannot know the details of services that children are using. Likewise, in the case of cell phones which are related to the right to privacy, there is a problem that parents can hardly monitor the contents as it would infringe the right to privacy of their children.

If a child started to associate with bad friends in real life, parents can get to know these friends by actually seeing them. On the other hand, parents cannot know what kinds of friendship their children have in social media.

In social media, parents cannot figure out what their children purchase, what kinds of services their children play with, or what kinds of friends their children have. Under such circumstances, it is almost impossible to entrust only parents to forestall problems involving children.

In the United States, under the Children's Online Privacy Protection Act, the use of social media by children of 12 years of age or younger is generally prohibited[6]. Likewise, "the Act on Development of an Environment that Provides Safe and Secure Internet Use for Young People" in Japan requires mobile companies to provide a service to filter harmful information for mobile phones used by young people under 18 years-old[7]. However, under the current situation, these laws do not have any punitive

---

[6] Children's Online Privacy Protection Act: http://www.coppa.org/coppa.htm (Seen at 04.28.2015).

[7] Act on Development of an Environment that Provides Safe and Secure Internet Use for Young People: http://www8.cao.go.jp/youth/youth-harm/law/pdf/ja-en.pdf (Seen at 04.28.2015).

clause and the use of social media by children of 12 years of age or younger in the United States or the access to harmful information by minors in Japan remain uncontrolled.

4   Arguments Against the Monitoring of Internet

Kay Mathiesen, a U.S. ethicist, points out problems which would arise as parents monitor the use of internet by their children.

Mathiesen claims that the risks of internet are overestimated. While it had been claimed that comics, movies, TV programs, and video games would expose children to harm, they have not caused any significant problems. Mathiesen criticizes that parents are overreacting although no one will jump out of a personal computer and attack their children[8].

Likewise, Mathiesen explains that monitoring is ineffective. Reading a racist leaflet does not necessarily mean that the reader is a racist and children may be just posting exaggerated or false stories. Mathiesen claims that it is non-sense to react to such postings[9].

According to Mathiesen, monitoring itself may be harmful to children. If parents saw their child writing "I might be a gay" and try not to make their child a gay, parents are infringing the child's autonomy. Parents may make a mistake in determining the best option for their children and parents' monitoring their children's use of internet will prevent the self-reliance of their children[10].

However, harm caused through the internet is changing daily and it is difficult even for adults to use it safely. Mathiesen does not give a clear answer to the question of whether giving children a free hand will remain safe in the future under such circumstances. It seems to be a valid counterargument to say that it is a role of parents to protect children from harms which would arise in the future. Giving minors a free hand without being monitored by parents seems to be an irresponsible argument. In the first place, the principle of individualism includes a principle that minors are subjected to paternalism and given a limited freedom. It seems that the claim of Mathiesen that monitoring by parents is not necessary is like an argument made by teenagers who do not want their parents to intervene in their lives.

5   Regulations of Using Internet for Minors

In order to address repeated problems related to the use of social media by minors, the educational community has started to restrict minors' use of social media.

---

[8] Mathiesen, 267-268.
[9] Ibid.
[10] Ibid.

For instance, in April 2014, the city of Kariya in Aichi prefecture began a new initiative regarding the use of smart phones by minors. The rule requires that (1) parents do not allow their children to possess any unnecessary mobile phone or Smartphone; (2) at the time of contract, parents make an agreement with their children and they receive a "filtering service" which limits the access to harmful sites; (3) parents keep children's phones after 9:00 p.m.

It is not that the board of education of the city or schools made the rule in a top-down manner but the rule was proposed by "Kariya-shi Jidou Seito Aigo Kai (Association for the Protection of Pupils and Students of Kariya city)" consisting of elementary schools, junior high schools, high schools and police stations in Kariya-city. The group called for cooperation of parents jointly with the PTA and schools. According to a source related to the board of education of the city, prior to this effort, serious events which could have resulted in crimes occurred in junior high schools in the city: Students run away from home counting on adults whom they met through LINE or send their nude photos to each other[11].

Nick Bilton, a technology writer of the New York Times, described about Steve Jobs, Apple's then C.E.O.[12]. "So, your kids must love the iPad?" He asked Steve Jobs. "They haven't used it," Jobs told him. "We limit how much technology our kids use at home." Walter Isaacson, the author of "Steve Jobs," wrote "Every evening Steve made a point of having dinner at the big long table in their kitchen, discussing books and history and a variety of things." "No one ever pulled out an iPad or computer. The kids did not seem addicted at all to devices[13]." Bill Gates, the Microsoft founder, said that his 10-year-old daughter, his oldest child, was not a hard-core Internet and computer user. He said that he and his wife Melinda had decided to set a limit of 45 minutes a day of total screen time for games and an hour a day on weekends, plus the time she needed for homework[14].

Nick Bilton found that technology industry leaders commonly include: "Children under 10 seem to be most susceptible to get addicted, so these parents draw a line for not allowing any gadgets during the week. On weekends, there are limits of 30 minutes to two hours on iPad and smart phone use. And 10- to 14-year-olds are allowed to use computers on school nights, but only for homework[15]."

It is very difficult to build regulations on social media for adults. Because adults have the right to do what is wrong. All that we can do for adults is to publicize risks of social media associated with the services that affect users. However, we can impose regulation against minors to use social media, for they have no right to do what is wrong. The challenge of Kariya-city is one good idea of self-regulation for minors. The lessons of Steve Jobs and Bill Gates suggest that we can set strict limits on the use of social media for our children. We can tell our children that even the Jobs' and

---

[11] Asahi News Paper, March 24, 2014.

[12] Written by Nick Bilton. Aversion of this article appears in print on September 11, 2014, on page E2 of the New York edition with the headline: Steve Jobs Was a Low-Tech Parent.

[13] Ibid.

[14] Article of Reuters. http://www.reuters.com/article/2007/02/21/us-microsoft-gates-daughterid USN2022438420070221 (Seen at 04.28.2015).

[15] Bilton, op. cit.

Gates' families set strict limits on technology use, so why we cannot set strict limits on technology? We should talk more about regulations of social media for our children's future.

In order to protect children from the problems of social media, various approaches have to be employed by the adults. And regulation of social media is inefficient for protecting children.

Children who fall into trouble in the age of social media probably had fallen in other forms of trouble when social media did not exist. For the reason of poverty, family problems, or bad grades, children who lost contact with a school or their home tend to involve in risky activities. Once it was like taking illegal drugs or associating with gangsters. Now these risks are like involving in prostitution or bullying by the means of social media[16].

Adults should think seriously about what type of life is good for their children, and it is important that adults help the children about this issue. Children can be protected from trouble by increasing communication between adults and children. For example, even if children have fallen in some trouble, an intervention by an adult may bring an early solution. If we think from that viewpoint, then we would realize that the problem of children and social media is not concerned with the technology of social media, but it is concerned with the relationship between adults and children.

6   Conclusion

We will now discuss how a new technology interacts with the society when it is newly developed.

In terms of relations between a new technology and society, I would like to discuss the technological determinism. Technological determinism is the idea that technology determines the course of the society[17]. This technological determinism includes three categories, which are (1) determinism which descriptively records changes in the culture forming daily lives; (2) pessimism which explains that technologies distinguish those who can enjoy the fruit of technological progress from those who cannot and eventually have a negative impact on human beings; and (3) optimism which claims that technologies will make the society better[18].

In the idea of (3) optimism, which explains that technologies will make the society better, technology is considered as something good which plays a core role in the development and expansion of civilization. Among theories of optimism, naturalistic optimism accords with the current social trend related to social media. In ethics, naturalism means to "entrust everything to what exists" or to "entrust to the value or norm that existing things have in themselves"[19]. Naturalistic optimism believes that the situation will improve if human beings entrust everything to nature.

---

[16] Boyd, Chapter 3 & 4.
[17] Murata, 114-122.
[18] Ibid.
[19] Kato[2001], 151.

Many of the general ideas surrounding current social media belong to this naturalistic optimism. Many people describe social media positively, as a tool which enables them to build new personal relations and obtain information that they could not get from conventional media (such as TV and newspapers). They use it without any hesitation every time a new social medium is introduced, and believe that the society is getting better by doing so. We can understand this trend from the current situation where governmental restrictions or ethical criticism on social media have been limited while the number of social media users is increasing explosively.

Among philosophical theories of technology, there is an idea of social constructivism[20]. It is a theory that society can determine the future of technology. I will give a concrete example of social constructivism.

Guns arrived in Japan in 1543 but after the Tokugawa Shogunate placed gun smitheries under its control, guns were not widely used[21]. Many gun smitheries became fireworks artisans to earn their living. As a result, it is almost impossible to get firearms in Japan to this day. Likewise, Japanese firework skills have established a reputation as the greatest in the world.

We will be able to find a new direction for ethical issues related to social media by introducing this idea of social constructivism. The society must find a way to live with social media ahead of technology, not allowing the technology of social media to build a society. In doing so, we may need to make a bold change in policy related to social media.

Acknowledgments. This work was supported by JSPS KAKENHI (Grant-in-Aid for Scientific Research C) Grant Number 25380491. This study was also supported by the MEXT (Ministry of Education, Culture, Sports, Science and Technology, Japan) Programme for Strategic Research Bases at Private Universities (2012-16) project "Organizational Information Ethics" S1291006.

References

1. Boyd, D.: It's Complicated: The Social Lives of Networked Teens. Yale University Press (2014)
2. Kato, H.: Rinrigakunokiso (Basic of Ethics), hosodaigakukyouikushinkoukai (1993)
3. Kato, H.: Kachikantokagaku/gijyutu (Values and Science/Thechnology), Iwanamishoten (2001)
4. Mathiesen, K.: The Internet, children, and privacy: the case against parental monitoring. Ethics and Information Technology **15**, 263–274 (2013)
5. Murata, J.: Gijyutunotetugaku (Philosophy of Thecnology), iwanamishoten (2009)

---

[20] Murata, 119-122.

[21] Murata, 111.

Influence of the Social Networking Services-Derived Participatory Surveillance Environment over the Psychiatric State of Individuals

Yohko Orito[1(✉)] and Kiyoshi Murata[2]

[1] Faculty of Law and Letters, Ehime University, 3 Bunkyo-cho, Matsuyama,
Ehime 790-8577, Japan
orito.yohko.mm@ehime-u.ac.jp
[2] Centre for Business Information Ethics, Meiji University, 1-1 Kanda Surugadai,
Chiyoda, Tokyo 101-8301, Japan
kmurata@meiji.ac.jp

Abstract. This study attempts to propose hypotheses concerning the connection between the participatory surveillance environment, which we has been created by the widespread use of social networking services (SNS), and mental disorders such as dissociative disorder youngsters would develop, based on observations of individual and organisational surveillance in the SNS sphere and research findings in the field of psychopathology.

Keywords: Social Networking Services · Participatory surveillance environment · Dissociative disorder

1 Introduction

Recent days, social media including social networking services (SNS) such as Faceboook, Google+, LinkedIn and LINE have pervaded society, and many people, especially youngsters, have enjoyed various and convenient services of them. SNS enable individual users to post information on their personal or private life and to watch and respond to other users' posts. Owing to the advent of the technological environment where individuals can access the Internet using mobile devices such as a smartphone and a tablet personal computer, SNS sites have become a 24/7 communication platform for individual users to keep in touch with their friends and acquaintances.

On the other hand, it is alleged that SNS have a negative influence on individuals and society. In fact, the concern that the use of SNS would cause the invasion of the right to privacy (Lawler and Molluzzo [1]; Orito et al. [2]) and the distortion of digital identity (Orito [3]; Murata and Orito [4]) has been expressed. The connection between social media and depression has also been studied by many researchers. For example, Tandoc et al. [5] find out the possible link of the use of Facebook for surveillance, or to keep track of what others are doing, with depression among college students, mediated by feelings of envy that use can evoke. In contrast, Jelenchick et al. [6] state that

© Springer-Verlag Berlin Heidelberg 2015
L. Wang et al. (Eds): MISNC 2015, CCIS 540, pp. 541–549, 2015.
DOI: 10.1007/978-3-662-48319-0_45

the evidence supporting the existence of a relationship between the use of SNS and clinical depression was not found and counselling patients or parents regarding the risk of "Facebook depression" may be premature.

However, negative influence of social media over the psychiatric state of individuals, especially youngsters, and mental growth of them should be investigated beyond depression from a long-range perspective. With this awareness in mind, the authors have discussed negative social impacts of the participatory surveillance environment which has been created due to the widespread use of various kinds of monitoring technology and services such as CCTV, mobile devices and social media over people's mental health referring to as the "schizophrenic society" (Murata and Orito [7] [8]). Carrying the discussion one step further, this study attempts to propose a hypothesis concerning the potential connection between the participatory surveillance environment and mental diseases centred on dissociative disorder which, in particular, youngsters would develop, based on the research findings in the field of phenomenological psychopathology.

2   Participatory Surveillance Environment

2.1   Participatory Surveillance in the Real and Virtual Worlds

Owing to the development and widespread use of information and communication technology (ICT) and ICT-based systems and services, the living situation that can be characterised as the participatory surveillance environment has emerged. In such an environment, a number of organisations and individuals seemingly voluntarily, but often forcibly in reality, participate in ICT-mediated surveillance (Murata and Orito [8]).

The typical situation of participatory surveillance can be observed by taking a look at how SNS are used. SNS users post various forms of information such as text messages, photos and movies, which can include their personal information or information about their private life. Simultaneously, the users can review and respond to contents posted by other users on SNS sites. The ease and effectiveness of interactive online communication SNS enable are a distinctive characteristic and a source of customer satisfaction of such services. On the other hand, throughout their using SNS, users are watching with each other and their online personal relationships and communications are being monitored by third parties (including both individuals and organisations).

It implies that on a social media site, its user, consciously or unconsciously, monitors others' thought and behaviour through looking at posts made by them, is monitored by others when they look at his/her posts, and provide others with opportunities to monitor his/her friends and acquaintances through posting articles, photos or videos about them. That is to say, the use of social media has generated a new form of surveillance. In contrast to the traditional model of surveillance which is vertical and hierarchical, the new form of surveillance on SNS is regarded as mutual and lateral monitoring among the users (Albrechslund [9]). It seems that SNS users are willing to participate in the surveillance sphere, provide their personal information voluntarily and prefer to be monitored by others as well as to monitor other users.

At the same time, user's behaviour on a social media site are monitored and recorded by the social media platform company, which participate in the surveillance environment as an infrastructure builder and a surveillant. A similar situation happens when users conduct a search using a Web search engine. Recent free-of-charge online services assume such a surveillance scheme in which users participate as a surveillee baring the wisdom of crowds [10] as well as a surveillant.

The participatory surveillance environment could be observed not only in the SNS sphere but also in real space. There are many electronic devices which enable relevant organisations to collect, store, analyse and utilise personal information in real space. For example, many people willingly use mobile devices such as smartphones and smart cards for convenience. Point-of-sales (POS) systems and radio-frequency identification (RFID) are widely utilised by companies to streamline their business operations. These technologies allow organisations to participate in surveillance as a surveillant through collecting behavioural and location information of individuals who are, in many cases, insidiously forced to be involved in the surveillance environments. Personal information collected through the monitoring devices in the real space could be stored, analysed and utilised for providing customised or personalised online services or advertisements on, for example, SNS sites.

It is very difficult for anyone to deny the participation in the current surveillance environment. First of all, the fact of participatory surveillance is elusive. If we determine not to have any of a smartphone, a smart card and a credit card, the convenience of our life would significantly deteriorate. In fact, we cannot get on a bus in London without a smart card (Oyster Card). It may be hard to find a grocery store which does not use POS system. If a youngster stops using SNS, he/she would plunge into social isolation. We have been captives of the participatory surveillance environment.

2.2   Three Types of Participation in Surveillance and Its Potential Risk

The participation of a number of "little brothers" in surveillance using human and/or electronic eyes has realised the man-machine omni-optical surveillance, which allows organisations to automatically collect and store a lot of as well as various kinds of personal data including state and behavioural information that are not necessarily structured. In this situation, there are three types of participation in surveillance that are not mutually exclusive:

(a)   participation as a surveillant, who monitors others,
(b)   participation as a surveillee, whom is monitored by others, and
(c)   participation as a builder of infrastructure for surveillance, who contribute to setting up conditions under which organisations and individuals can have opportunities to monitor others (Murata and Orito [8]).

The modern surveillance environment which has been generated both in real and virtual spaces. Actually, the boundary between real space and cyberspace has already been vanished in terms of surveillance. The users of SNS participate in such monitoring sphere as any of the three types of participants, regardless of their intention or recognition. Almost all of SNS users may not realise that they participate in surveillance when they enjoy online communication with their friends and acquaintances on SNS sites. As Albrechslund [9] suggests, the participatory surveillance of online social networking has potential for empowerment and subjectivity building of individuals. However, there are arguments not only for but also against continuous monitoring of individuals through collecting their personal data.

For example, personal information collected and stored in the current surveillance environment is an integral component of dataveillance systems (Clarke [11]), which are designed to provide individuals with personalised services. Specifically, those systems attempt to suggest his/her "desirable" actions to each individual in a timely fashion based on analyses of personal data stored in databases, as part of outbound personal marketing. This type of personalised service can silently control individuals' ways of thinking and acting, leading to threatening intellectual freedom (Orito [12]). That is, those personalised services potentially deteriorate autonomy of individuals in the development of their identity and self.

3   Psychiatric State of Individuals in the Participatory Surveillance Environment

3.1   "Digital Identity" and Personal Relationships Developed in SNS Sphere

Amongst various kinds of risks caused by the SNS-derived participatory surveillance environment, issues concerning users' "digital identity" are deeply related to sense of self and others and development of personal relationships. According to Rannenberg et al. [13], "digital identity" is defined as the representation of the identity of a person in digital environments, in particular in terms of presentation of the characteristics (values associated to a set of attributes) of the person.

Digital identities of SNS users can be developed jointly by their own and others' postings on the SNS sites. However, those who are hooked on social media may distort their digital identity by themselves through posting indiscreet and/or misleading remarks online. In addition, SNS encourage their users to reveal personal information not only of them but of others. Substantially, many SNS users have experienced the revelations of their personal data in the form of text comments, photos and movies on the SNS sites without their permissions, even when those data include personally identifiable information. Information of a person disclosed on social media sites by him/her or someone else can become a significant component of his/her "digital identity" (Rannenberg et al. [13]). In the situation where anyone can face the difficulty in controlling his/her own digital identity, individuals' developing a sense of jealousy against or of inferiority to others' digital identity potentially leads to his/her falling in depression (Tandoc et al. [5]; Krasnova et al. [14]).

On the other hand, under the participatory surveillance environment, it is no wonder that the user tends to play a character expected by his/her imagined online audience or put on his/her personae he/she suppose his/her online audience identify and therefore his/her digital identity is developed in a heteronomous fashion. Actually, if they desire to collect a lot of others' "like", "share" or "retweet" on social media plat-forms, their postings may tend to be what they consider many of their friends favour, resulting in the mental state in which they feel their digital identity, therefore their self, is controlled by the others (Murata and Orito [8]).

3.2 Schizophrenic Society

This sort of negative influence over the individuals under the participatory surveil-lance environment may be serious among people before and at puberty. When they perceive that their thought and action are controlled by others, they may experience the feeling of strangeness that their mind is controlled by others or their independence is taken over by others, which is a symptom unique to schizophrenia (Kimura [15], p. 307).

Kimura ([15], pp. 295–296) describes that schizophrenia is a state of affairs in which the indisputable fact that one is none other than oneself is called into question most directly and seriously and the mental basis of self-perception which enables one to distinguish oneself as a being different from others is fundamentally in crisis. As Kimura explains, schizophrenia harms the fundamental structure of the self, which can bring about an identity crisis with serious symptoms such as delusion of control and withdrawal of thought as immune responses to dysfunction of the mental proc-esses of generating self. The patients of schizophrenia feel that they are controlled by others or invaded from outside, suffering from auditory hallucinations of others.

The feeling of strangeness similar to one schizophrenic patients have can be expe-rienced by SNS users, whose identity and self-consciousness are developed under the influence of the participatory surveillance environment. Considering the current so-cio-technological situation where portable electronic devices such as a smartphone enable individuals to access the Internet and encourage them to use SNS sites to construct and maintain a personal association from their earliest childhood, the SNS-derived participatory surveillance environment entails a potential risk of SNS users', in particular young users', developing schizophrenia.

4   Potential risks of Dissociative Disorder

4.1   Internalisation of Others' Gase and Disappeared Self

On further reflection, the SNS-derived participatory surveillance environments may pose a serious mental disorder other than depression and schizophrenia. When people strongly feel "I have to behave in a disciplined manner, because someone else is al-ways watching me", they may be forced to attempt to internalise others' moral gaze, to integrate it into their own moral values and, consequently, to create their selves following their supposition of what the moral gaze expects. This type of detachment is associated with specific symptoms of dissociative disorder (Shibayama [16]). For example, the loss of sense of self and multiple personality are characterised as typical symptoms of this mental disorder, where self, which is integrated by good rights, is fragmented, true self is lost and stable self-identity or image is diluted.

Such kinds of situations can be observed in the online personal relationship developed in the SNS sphere. On SNS sites, users are recommended to develop or represent their real personal relationships and to control them online as well as offline. In fact, Facebook encourages its users to find and to connect real friends and acquaintance on the site. However, as aforementioned, youngsters tend to play their character in a coordinated fashion at different times. They seem to behave in synchronisation with each other on a moment-to-moment basis. Then, under the recent technological environment in which SNS can be used at any time or any place, they attempt to take on different identities or characters based on the situation they are located in. As Shibayama points out, personal relationships for youngsters tend to be more partial and selective, and online personal relationships could be more readily-resolved than ones in real space (Shibayama [17], p. 167). Personal relationships that can be developed in the SNS sphere as a participatory surveillance field seem to assume unstable and ever-changing characteristics.

This may lead to the situation in which SNS users' self or identities are fragmented and/or diffused and they become an inconsistent entity. When the self is confused through the process of coming into line with others, the mental function to maintain the unity of self can deteriorate. In that event, the mental function would turn out not to work and the self would be transformed into just an "observer" or an entity which only takes a view on the surrounding situations and others. Such self-transformation is at the core of symptoms of dissociative disorder to let patients survive the crisis of their human existence (Shibayama [17], p. 176).

Under the SNS-derived participatory surveillance environment, SNS users can play various kinds of characters or personalities according to each circumstance of the moment. On the other hand, this would lead to the disappearance of an integrated self and the loss of the sense of self of the users. It may consequently result in the mental status of them that they are watching themselves as a distant spectator. The SNS sphere as a participatory surveillance environment seems to have potential risks for provoking mental status or symptoms of dissociative disorder.

4.2   Loss of Sense of Trust with Others and Secure Belongings

When the SNS users are enjoying online communication, sometimes they get in trouble such as online bullying, flaming and cyber stalking in an unexpected manner. In case that what an individual user posts on SNS websites brings about online and/or offline bullying or verbal attacks made by known and/or unknown people, he/she would have trauma which results in dissociative disorder. Considering that this disease deeply relates to patients' feeling of loss of a place where they can feel a sense of belonging (Shibayama [16]), such trauma can be serious for those who are immersed in the virtual world.

Basically, the development of self-identity in a society has fundamental importance in youngsters' attainment of adulthood. In the process of identity development, youngsters' identities can repeatedly be swayed to a certain extent depending on situational changes they face. In general, youngsters tend to concern about their reputation within their social circle, and this sometimes leads them to pretend a good fellow and to be at odds with themselves because they desire to make their social circle a place where they can feel a sense of belongings. However, traumatic experiences online would generate victims' feeling of absolute distrust with others. As described in the above section, SNS users may be forced to feel a sense of uncertainty concerning personal relationship management in the participatory surveillance environment due to its characteristics and architecture. In this regard, young SNS users may be prone to lose the sense of trust with others, and therefore tend not to be able to develop a mature identity in the secure place or sphere where they can feel trust with others and establish the sense of belongings.

A similar kind of concern has been pointed out by psychopathologists in the context of dissociative disorder. Noma ([18], p. 132) explains that patients with dissociative disorder have an absolute distrust towards the world around them. Hence, they cannot trust people surrounding them and maintain ties with the society, and are forced to look down at the world they live in taking a step back from their situation or merely as an observer. Shibayama ([17], p. 169) suggests that the current situation in which anyone faces the difficulty in realising self-identity, his/her own story, community and whereabouts has been developed by the modern society that is filled with uncertainty and inconsistency.

Speaking of youngsters, whereas their experiences of SNS usage from childhood and of the development of their discipline through using SNS have positive influence on their self-development in the modern information age, those experiences may influence them so that their personalities or identities are transformed as if they suffered dissociative disorder due to their immaturity in terms of identity development and mental instability. Of course, the discussions in this study cannot prove the relevant association between use of SNS which are major components of the participatory surveillance environment and the occurrence of psychiatric state. However, the potential risks caused by the participatory surveillance environment are worthy to be discussed.

5   Conclusions

This study attempts to consider the potential risks caused by the SNS-derived participatory surveillance environment. Under such environment, SNS users may create their selves with the feeling that the imagined moral gaze of others controls their selves and they just observe themselves without exerting any influence over their self-development, as patients with dissociative disorder experience. This type of feeling of strangeness would be experienced by youngsters more seriously, due to their immaturity of identity formation and self-development. Of course, discussions in this study are simply a proposal of hypotheses concerning the connection between the participatory surveillance environment and mental disorders. Further validations and examinations are crucially important and necessary.

However, it is indisputable that the analysis of the use of SNS that has been pervasive in our everyday life and has strong influence on our feelings or mental status, from the psychopathologic perspectives is useful to identify risks associated with the modern technological society and to consider both of their immediate impacts and long-term consequences on individuals and society. Utsumi [19] suggests that psychiatric symptoms are reflecting circumstances of the moment, mentioning the followings.

> "The appearance of mental disease reflects the truth of the times, though this has not been discussed in the field of "scientific" psychiatric studies or psychopathology. The appearance always reflects the zeitgeist and culture, and projects what kinds of difficulties people have and attempt to cover up them, as a negative. This is one of the reasons why mental disease is recused from the society" (Ustumi [19], p. 172; translated by the authors)

Mental symptoms which have been analysed to this date including hysteroepilepsy, dysphoria and schizophrenia have marked a passage of time. The use of SNS makes it possible to transform the whole concept of self-identity and personal relationships. Analysis on the potential risks caused by the use may demonstrate the existence of pathology many people are forced to face in the modern information age.

In contrast to schizophrenic symptoms which can be characterised as conflicts between "self and others", the patients with dissociative disorders seem to make the distinction between "self and others" ambiguous and to merge self with others (Matsumoto [20], pp. 23–24) or the environment surrounding them. In the current world which is symbolised dissociative disorder symptoms, the integrated self may be disappeared and individuals can behave as if they had ubiquitous identities. However, their self-identities may not to be found anywhere.

Acknowledgement. This study was supported by the MEXT (Ministry of Education, Culture, Sports, Science and Technology, Japan) Programme for Strategic Research Bases at Private Universities (2012-16) project "Organisational Information Ethics" S1291006, the JSPS Grant-in-Aid for Scientific Research (B) 25285124, and the JSPS Research Grant-in-Aid for Young Scientists (B) 24730320.

References

1. Lawler, J.P., Molluzzo, J.C.: A Study of the Perceptions of Students on Privacy and Security on Social Networking Sites (SNS) on the Internet. Journal of Information Systems Applied Research 3(12) (2007). http://jisar.org/3/12/
2. Orito, Y., Fukuta, Y., Murata, K.: I Will Continue to Use This Nonetheless: Social Media Survive Users' Privacy Concerns. International Journal of Virtual Worlds and Human Computer Interaction 2, 92–107 (2014). doi:10.11159/vwhci.2014.010
3. Orito, Y.: Real name social networking services and risks of digital identity. In: Uesugi, S. (ed.) IT Enabled Services, pp. 217–227. Springer, Wien (2013)
4. Murata, K., Orito, Y.: The paradox of openness: Is an honest person rewarded? In: Proceedings of CEPE 2013, pp. 221–231 (2013)
5. Tandoc Jr., E.C., Ferrucci, P., Duffy, M.: Facebook Use, Envy, and Depression among College Students: Is Facebooking Depressing? Computers in Human Behavior 43, 139–146 (2015)

6. Jelenchick, L.A., Eickhoff, J.C., Moreno, M.A.: Facebook Depression? Social Networking Site Use and Depression in Older Adolescents. Journal of Adolescent Health **52**(1), 128–130 (2013)
7. Murata, K., Orito, Y.: The schizophrenic society. In: Proceedings of ICT, Society and Human Beings 2012, pp. 112–116 (2012)
8. Murata, K., Orito, Y.: The Schizophrenic Society: The Potential Risk of Individual Identity Crisis in the Participatory Surveillance Environment. In: Palm, E. (ed.) ICT-Ethics: Sweden and Japan (Studies in Applied Ethics 15), The Centre for Applied Ethics, Linköping University, pp. 10–23 (2013)
9. Albrechtslund, A.: Online Social Networking as Participatory Surveillance. First Monday **13**(3-3) (2008). http://firstmonday.org/article/view/2142/1949
10. Surowiecki, J.: The Wisdom of Crowds: Why the Many Are Smarter than the Few and How Collective Wisdom Shapes Business, Economies, Societies and Nations. Doubleday, New York (2004)
11. Clarke, R.A.: Information Technology and Dataveillance. Communications of the ACM **37**(5), 498–512 (1988)
12. Orito, Y.: The Counter-control Revolution: "Silent Control" of Individuals through Dataveillance Systems. Journal of Information, Communication and Ethics in Society **9**(1), 5–19 (2011)
13. Rannenberg, K., Royer, D., Deuker, A. (eds.): The Future of Identity in the Information Society: Challenges and Opportunities. Springer, Heidelberg (2009)
14. Krasnova, H., Wenninger, H., Widjaja, T., Buxmann, P.: Envy on facebook: A hidden threat to users' life satisfaction? In: Proceedings of 11th International Conference on Wirtschaftsinformatik, pp. 1–16 (2013). http://warhol.wiwi.hu-berlin.de/~hkrasnova/Ongoing_Research_files/WI%202013%20Final%20Submission%20Krasnova.pdf
15. Kimura, B.: The Self, Relations and Time: Phenomenological Psychopathology. Chikumashobo, Tokyo (2006). (in Japanese)
16. Shibayama, M.: The Structure of Dissociation: Adaptation of Self and Remedy. Iwasaki Gakujutu Shuppansya, Tokyo (2010). (in Japanese)
17. Shibayama, M.: Pathology of Dissociation: Self, Society. Age. Iwasaki Gakujutu Shuppansya, Tokyo (2012). (in Japanese)
18. Noma, S.: Temporality of Body: Psychopathology Living in the Moment. Chikumashobo, Tokyo (2012). (in Japanese)
19. Utsumi, T.: Wandering Self. Psychopathology in Post Modern Age. ChikumaSensyo, Tokyo (2012). (in Japanese)
20. Matsumoto, M.: Exploring Roots in Dissociative Disorder. In: Shibayama, M. (ed.) Pathology of Dissociation: Self, Society, Age, pp. 3–24. Iwasaki Gakujutu Shuppansya, Tokyo (2012). (in Japanese)

Special Session on Information Technology and Social Networks Mining

Privacy Protection Framework in Social Networked Cars

Eric Ke Wang$^{(\boxtimes)}$, Chun-Wei Lin, Tsu-Yang Wu, Chien-Ming Chen, and Yuming Ye

Shenzhen Key Laboratory of Internet Information Collaboration, School of Computer Science and Technology, Harbin Institute of Technology Shenzhen Graduate School, HIT Campus Shenzhen University Town, Xili, Shenzhen, People's Republic of China
{wk_hit,yym}@hitsz.edu.cn, jerrylin@ieee.org,
{wutsuyang,chienming.taiwan}@gmail.com

Abstract. With the social networked cars develops, security issues begin to be concerned. In this paper, we survey the major security threats and attacks types; then present the security objectives for social networked cars; finally, considering the key treat is the privacy challenges, we propose corresponding privacy protection solutions for social networked cars. The experiment has been executed, and preliminary results have been encouraging.

Keywords: Social networked car · Privacy preserving · Compromised-Key attacks

1   Introduction

In many countries, talking and texting on cell phones while people are driving might be forbidden [1]. Recently several major big car firms have developed a kind of social networked cars which enable social networks and communication applications on screens embedded in vehicles.

In 2012, Researchers in Toyota have created a social network that would allow owners of future Toyota electric and plug-in hybrids connect with each other and obtain information on their vehicles. Called "Toyota Friend" ， the service will allows owners of the vehicles to exchange real time updates, or "tweets," with other Toyota drivers, connect with local dealers and remotely obtain diagnostic information about their hybrid vehicles, such as battery usage, the architecture is as shown in Figure 1. [2]

Recently, researchers in university of Michigan, in conjunction with Ford, Microsoft and Intel, have realized systems which enable multiple cars to connect while on car trips in order to share information about how much gas is left in the fuel tanks, competing fuel economy between the cars, shared routes, and land marks and gas stations on the route ahead. In the other words, they employed in-vehicle social networking to help cut fuel consumption, record the real-time fuel economy of a vehicle and then compares that fuel economy to peers. The in-vehicle dashboard enables the driver to see other drivers that have driven the same, or similar routes, and suggests the best route for the best fuel economy.

© Springer-Verlag Berlin Heidelberg 2015
L. Wang et al. (Eds): MISNC 2015, CCIS 540, pp. 553–561, 2015.
DOI: 10.1007/978-3-662-48319-0_46

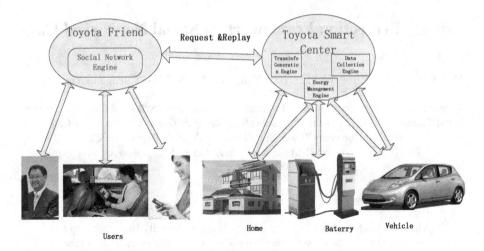

Fig. 1. Toyota Friend

Social networks embedded in cars enrich people's life in the road, however, as the interaction between the physical and cyber systems increases. the physical systems become increasingly more susceptible to the security vulnerabilities in the cyber system. For example, some hackers can find out the cars' location privacy by social networks embedded in cars and execute an kidnapping to the targeted driver.

Hackers can be expected to create problems in this space, whether their motives are financial, to prove that weaknesses exist in the system and/or apps, or simply because they think it is fun. "Most people would rather have malicious software running on their laptop than inside their car braking system. Thus, incorporating strong security solutions will give manufacturers a competitive advantage.

In this paper, we survey the major security threats and attacks types, then present the security objectives for social networked cars, finally, considering the key treat is the privacy challenges, we propose corresponding privacy protection framework for social networked cars.

2   Major Types of Attacks to Social Networked Cars

2.1   Privacy Attacks

Social networks in cars are particularly susceptible to eavesdropping through traffic analysis such as intercepting the monitoring data transferred in networks. Besides, it may suffer the privacy inference attacks.

Actually, many identities are interested in the information that people post or exist on location based services. Privacy issues for social networked cars are increasingly challenging users. Identity thieves, scam artists, debt collectors, stalkers, and corporations looking for a market advantage are using location based applications to gather information about consumers.[3] Companies that operate LBS are themselves collecting a variety of data about their users, both to personalize the services for the users and to sell to advertisers.

Actually, more and more people are willing to use location based services to make queries related locations. As is well known, the values at the core of services – precise, connecting[4] -- However, the more precisely and convenient we become with these services, the more apt we are to share personal details about ourselves and let our guard down as we interact with others. Actually, the facts tell that the majority of LBS users post risky information online, without giving due diligence to privacy and security concerns. Unfortunately the native core value of openness, connecting and sharing are the very aspects which allow cyber criminals to use these sites as a vector for various kinds of bad online behavior. At the same time, cyber criminals are targeting location based services with increasing amounts of malware and online scams, honing in on this growing user base.

Privacy issues for social networks embedded in cars can be divided into 2 categories, one is direct privacy, and the other is indirect privacy. Direct privacy means that users' privacy information is revealed directly from profile or data archived in the system. In other words, direct privacy issues concerns individuals' personal information which should be part of archived or communication data. Some variables often found in social science datasets present problems that could endanger the privacy of research subjects. Commonly these variables point explicitly to particular individuals or units. Examples of direct identifiers include: {Names; Home Addresses, including ZIP codes; Telephone numbers, including area codes; Social Security numbers; other linkable numbers such as driver license numbers, certification numbers, etc.}.

On the other hand, indirect privacy issues for social networks in cars are also becoming challenging. Since service providers want to provide users intelligent active service and create good user experience, it requires that service providers have to record and study drivers' location and behaviors, infer drivers' interests, forecast drivers needs to give precise services. Thus, even if drivers do not register their personal privacy information, it is still possible for service providers to infer sensitive information based on the public information such as drivers' location, drivers' behavior. For example, for the geographical location log cached may be inferred that which is the driver's office location(day time location) and which is the driver's home location(night time location)[5], then the user's name and home telephone number can be found out according to the home address by a yellow page book.

2.2   Identity Theft [6]

Identity Theft is an act of stealing someone's identity or sensitive information, and then pretending to be that person, or using that identity in a malicious way. Social networks are promising targets that attract attackers since they contain a huge number of available user's information. One technique of identity theft is profile cloning. In this technique, attackers take advantage of trust among friends, and that people are not careful when they accept friend requests. Social phishing is another method that can be used to steal social network user's identity.

2.3   Compromise Key Attack

A key is a secret code which is necessary to interpret secure information. Once an attacker obtains a key, then the key is considered a compromised key [7]. An attacker can gain access to a secured communication without the perception of sender or receiver by using the compromised key. The attacker can decrypt or modify data by the compromising key, and try to use the compromised key to compute additional keys, which could allow the attacker access to other secured communications or resources.

Actually, it is possible for an attacker to obtain a key although the process maybe a difficult and resource intensive. For example, the attacker could capture the sensors to execute reverse engineering job in order to figure out the keys inside.

2.4   Man-in-the-Middle Attack

In man-in-the-middle attack [8], attacks makes independent connection with the victims and eavesdrop the communications, then relays messages between them to make them believe that they are talking directly to each other over a private connection when in fact the entire conversation is controlled by the attacker.

2.5   Denial-of-Service Attack

Denial of Service (DoS) attack [9] is one of the network attacks that prevent the legitimate traffics or requests for network resources from being processed or responded by the system. This type of attacks usually transmits a huge amount of data to the network to make busy handling the data so that normal services cannot be provided.

The denial-of-service attack prevents normal work or use of the system. After gaining access to the network, the attacker can always do any of the following:

--Flood the entire social network with traffic until a shutdown occurs due to the overload.

--Send invalid data to networks, which causes abnormal termination or behaviour of the services.

--Block traffic, which results in a loss of access to network resources by authorized elements in the system.

2.6   Physical Attack

Social network drivers may encounter physical threat, which is another issue need to concern. Physical threat is physical harm to a person, or to a person's property such as theft, stalking, blackmailing, or physical harassment. With the characteristics and features provided in the social networks, drivers are at risk of such threats.

The first characteristic is that driver's real identity is not known. Hence, we do not know who we are connecting with. The second is that privacy related information is posted on the social networks. These allow criminals to easily learn about and approach victims. In addition, many of the previous issues mentioned can also lead to physical threats. For example, social network features allow criminals to be able to track victim's behavior and location. It allows social network users to check in and post their current location onto their message board. Also, if social network users use social network application on their smart phones to post something, their rough location will also be posted . Moreover, another feature such as location sharing can also expose user's location, so stalkers or criminal will easily know where the victims are, and can approach them.

3   Security Objectives

Confidentiality. Confidentiality refers to the capability to prevent the disclosure of information to unauthorized individuals or systems. For example, a car on the Internet requires location records to be transmitted from car to the RP. The system attempts to enforce confidentiality by encrypting the record during transmission, by limiting the places where it might appear (in databases, log files, backups, and so on), and by restricting access to the places where it is stored. If an unauthorized party obtains the personal health care in any way, a breach of confidentiality has occurred.

Confidentiality is necessary (but not sufficient) for maintaining the users' privacy in socail networked car. Realizing Confidentiality in car must prevent an adversary from inferring the state and location of car users by eavesdropping on the communication channels between the sensors and the controller, as well as between the controller and the actuator.

Integrity. Integrity refers to data or resources cannot be modified without authorization. Integrity is violated when an adversary accidentally or with malicious intent modifies or deletes important data; and then the receivers receive false data and believe it to be true. Integrity in CPS could be the capability to achieve the physical goals by preventing,detecting, or blocking deception attacks on the informationsent and received by the sensors and the actuators or controllers.

Availability. For any system to serve its purpose, the service must be available when it is needed. It means that the social networked car systems used to store and process the information, the physical controls used to perform physical process, and the communication channels used to access it must be functioning correctly. High availability of the social networked cars aims tom always provide service by preventing computing, controls, communication corruptions due to hardware failures, system upgrades, power outages or denial-of-service attacks.

Authenticity. In computing and communication process it is necessary to ensure that the data, transactions, communications are genuine. It is also important for authenticity to validate that both parties involved are who they claim they are. In social networks embedded in cars, the authenticity aims to realize authentication in all the related process such as sensing, communications, actuations.

4   Privacy Solution

In order to tackle the above privacy problems, we built a privacy protection framework which includes two models (indirect privacy protection model and direct privacy protection model). For indirect privacy protection model, it has three kinds of protection mechanisms which are social relationship based protection, user behavior based protection, and physical location based protection. While, considering that privacy attackers often execute attacks with combination background information involving multiple contexts, we make three protection mechanisms fuse each other by fusion inference method. Besides, we combine indirect privacy protection with direct privacy protection which can adopt access control techniques. Moreover, privacy protections need various policies corresponding to various conditions. To some extent, it is required that these protection policies can be adaptive to environment dynamically. So the framework includes self-adaptive privacy policies mechanism which can dynamically adapt privacy protection policies. Finally, we need an evaluation process to evaluate whether the privacy protection policies are proper. Thus it can give a feedback to the self-adaptive privacy policies module.

In this framework, as shown in figure 2, there are 3 important modules (social relationship based protection, user behavior based protection and physical location based protection).

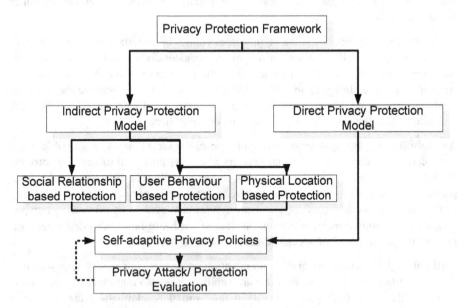

Fig. 2. Privacy Protection Framework

4.1   Social Relationship Based Privacy Protection

In order to be against inference attacks based on social relationships, we need corresponding protection solutions. Our solution is as follows:

(1) Bayesian Inference
We adopt Bayesian Network Classifier to classify location based services. The process is shown in figure 2.

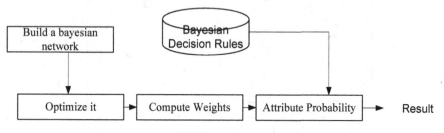

Fig. 3. Bayesian Inference Process

For instance, if we need to infer whether the point X has some attribute A, the steps are as shown in figure 3, firstly, we build a Bayesian network based on point X, then optimize it based on relation types, next, we need to compute the weights of each relations to find out the high impact relations, finally, based on Bayesian Decision Rules we can infer the probability of X has the attribute A.

(2) Obfuscation
Once locating sensitive relationship information, we can employ updated k-anonymity algorithm to obfuscate them. Considering based on traditional k-anonymity algorithm we often do not generate good anonymity sets. From our research result, we found that we can view k-anonymity problem as a clustering problem for k members. Therefore, the k-anonymity set generation problems can be transferred to k members clustering problems.

User Behavior Based Privacy Protection

In privacy inference there are two important factors (user behavior and user identity) which are strongly related to privacy knowledge. We found that the attacker can employ machine learning to execute inference attack based on supervised learning or half-supervised learning method. Besides, we adopt one-way hash function to cover the users' identities. As shown in formula 1, f_k is a one way hash function with k as a key. The user send a value of $H(id)$ which is its identity hash result.

$$H(id) = f_k (ID \oplus Si)$$

Location Privacy Protection

We employ location K-anonymity to protect location information. When user sends a location based service request to service provider, the location information sent out is replace with an area which covers the original location point. For example, as shown in figure 4, the real location of user C is *(x,y)*, however, the area *Rc* is sent out to represent the user *C* location.

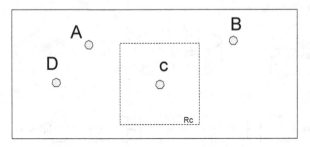

Fig. 4. Location K-anonymity

We built a privacy attack-defense experimental model. As shown in figure 5, we assume that in the attack-defense process, the attackers choose one policy from their policies set to execute an attack. Then defenders select one defense policy to protect. Therefore, each attack-defense round can be assessed in order to get the best defense policy for different attacks.

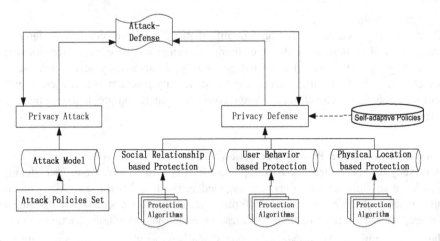

Fig. 5. Privacy attack-defense experimental model

We define our attack-defense model as follows: the triple *G={P,S,U}* is called a experimental set, the attributes are :

P: the set of the parties involved, $P=\{P_a, P_d\}$, P_a represents attack; P_d represents defender;

S: the set of polices with all parties $S=\{S_a, S_d\}$, $S_a=\{S_{a1}, S_{a2},...,S_{am}\}$ is attacker policies set, $S_d=\{S_{d1}, S_{d2},..., S_{dn}\}$ is defense policies set;

U: Utility function $U=\{U_a, U_d\}$ which reflects the preference of attackers and defenders;

U_a is attackers utility function; U_d is defense utility function.

5   Conclusion

In this paper, we investigate the major security threats and attacks types, and point out the security objectives for social networked cars; finally, we propose corresponding privacy protection framework for social networked cars to defend the privacy attacks, and we built a privacy attack-defense experimental model.

Acknowledgment. This research was supported by Shenzhen Technology Innovation (No.CXZZ20140509152928116), National Key Technology R&D Program of MOST China under Grant No. 2012BAK17B08 and National Commonweal Technology R&D Program of AQSIQ China under Grant No.201310087. The authors thank the reviewers for their comments.

References

1. Redelmeier, D.A., Tibshirani, R.J.: Association between Cellular-Telephone Calls and Motor Vehicle Collisions. The New England Journal of Medicine **336**, 453–458 (1997)
2. 'Toyota Friend': Social network for cars, owners, dealers & Toyota, Telematics News (May 23, 2011)
3. Gong, Z., Sun, G.-Z., Xie, X.: Protecting privacy in location-based services using K-anonymity without cloaked region. In: Eleventh International Conference on Mobile Data Management (2010)
4. Zhang, W., Cui, X., Li, D., Yuan, D., Wang, M.: 2010 18th International Conference on Geoinformatics, Beijing, China (2010)
5. Dewri, R., Thurimella, R.: Exploiting Service Similarity for Privacy in Location Based Search Queries. IEEE Transactions on Parallel and Distributed Systems, February 25, 2013
6. Nosko, A., Wood, E., Molema, S.: All about me: Disclosure in online social networking profiles: The case of FACEBOOK. Computers in Human Behavior **26**(3), 406–418 (2010)
7. Chalkias, K., Baldimtsi, F., Hristu-Varsakelis, D., Stephanides, G.: Two Types of Key-Compromise Impersonation Attacks against One-Pass Key Establishment Protocols. Communications in Computer and Information Science **23**(pt. 3), 227–238 (2009)
8. Jian-Zhu, L., Zhou, J.: Preventing delegation-based mobile authentications from man-in-the-middle attacks. Computer Standards & Interfaces **34**(3), 314–326 (2012)
9. Jahid, S., Nilizadeh, S., Mittal, P., Borisov, N., Kapadia, A.: DECENT: A decentralized architecture for enforcing privacy in online social networks. In: 2012 IEEE International Conference on Pervasive Computing and Communications Workshops (2012)

Effects of Monetary Policy on Housing Price Based on ANOVA and VRA Model

Lin Xu[1,2(✉)], Chao-Fan Xie[3], and Lu-Xiong Xu[1,4]

[1] The Institute of Innovative Information Industry, Fuqing Branch of Fujian Normal University,
Fuzhou, China
[2] The School of Ecomomic, Fujian Normal University, Fuzhou, China
xulin@fjnu.edu.cn
[3] Network Center, Fuqing Branch of Fujian Normal University, Fuzhou, China
[4] The School of Mathematics and Computer Science, Fujian Normal University, Fuzhou, China

Abstract. The transmission mechanism of monetary policy is the medium to connect monetary factors and real economy. The impact of the policy can lead to various variables change in the process of economic. In recent years, the research on monetary policy to the transmission mechanism of real estate price has become a hot issue. The monetary policy as a regulation of the Chinese real estate price basic tools. So research and design the mechanism of money supply and interest rate level play an important role in economic. How to make the monetary policy can not only make the consumption and investment stability, but also makes the real estate bubble has decreased. Ultimately achieve the healthy and stable economic growth target .And becomes the focus of policy making. This paper will first use the ANOVA to examine the interaction between money supply and interest rates. Finally, using the VAR model to calculate the weight of each factor.

Keywords: Monetary policy · Interest rates · ANOVA · VAR

1 Introduction

The real estate has dual properties of consumption and investment. Money supply, interest rates, exchange rates and other currency factors are closely related to the real estate changes. In 1998, China carried out the reform of the housing market, over the years the real estate market has been expanding in volume and price. During this period, Easing monetary policy to promote greater market rises, but the tightening monetary policy shift also will bring the real estate market slowdown in volume and price. In all monetary policy tools, particularly the most significant is in the money supply and interest rates which impact on the real estate market. When increasing the amount of money in circulation .In order to make profit, Most Chinese families choose real estate as the safest investment direction, which making real estate prices rise. However, when it is deflationary, most people will take a wait and see attitude, making the real estate prices fall. The nature of the underlying is effect by supply and demand.

© Springer-Verlag Berlin Heidelberg 2015
L. Wang et al. (Eds): MISNC 2015, CCIS 540, pp. 562–571, 2015.
DOI: 10.1007/978-3-662-48319-0_47

When the real estate market supply exceeds demand, the supply and demand mechanism to curb property prices will bring it down. When the real estate market tends to balance supply and demand, and the equilibrium price will form. Monetary policy will be affected supply and demand of the real estate. However, due to the impact of the supply of real estate prices in the short term is almost zero, so it can be considered real estate prices are determined by demand in the short term. When increasing the amount of money in circulation, resulting in increased demand, pushing the equilibrium price rises, while reducing the amount of money in circulation, resulting in reduced demand, so that the equilibrium price drop Bernake and Gertler (1999) pointed out in the study asset prices is to promote the consumption through the wealth effect, and prompting the currency in circulation to enter the real estate market, so that increased demand [1] . Wu Liping (2006) in the study pointed out that real estate prices to consumers than the stock prices lead to higher ROI . As the real estate commodity of great value, indivisible, long life, and therefore it is usually not fully buy its own funds to pay for, but must be carried out by way of mortgage loans and other credit facilities. In this way, consumers only need to pay part of the price, then you can enjoy the full benefits of the entire asset value of the house rises. But the whole family usually invests in stocks with its own funds. Thus comparison, real estate prices arise can offer consumers a higher return on investment, that the real estate price changes greater impact on consumption, real estate wealth has greater wealth effect [2] .

Changes in interest rates play an role in adjusting the supply and demand of funds, from the demand side, if the economic agents believe that the high cost of financing, the financing demand will decline, monetary and credit growth will slow, Interest rates will decline, and thus the formation of a new equilibrium. With the gradual development of China's real estate, the structure of demand Show differentiation, from the objective point of purchase we can divide it into user demand and investment demand; In terms of investment demand, it is the wealth of investment goods with storage function. In the current housing loans become the main way in the background, no matter what the demand is, the loan interest rate changes will directly affect the change of purchase costs, thus affecting the property buyers'   consumption and investment plan, further affecting the real estate prices.

Charles (2007) to study the effect of monetary policy on asset prices, and he finds that not only an increase in interest rates will affect the real estate market, but also affect the rest of the industry in the whole nation [3]. Kenny (1999) found that house prices and interest rates are positively related interest rates, and found that house prices and nominal interest rates are negative correlation [4].Bong Min (2011) used the impulse response function and variance decomposition method to research the housing bubble of South Korea from 1986 to 2003, and found that when interest rates rise, the housing bubble will reduce, at the same time, the impact of interest rates on the housing bubble to family loan and industrial production is relatively small[5].

Domestic researchers also had active research in this field, and achieved some results. Wei Wei (2008) obtained the result in the monetary policy impact on the real estate market that the impact of interest rate policy is obvious and lasting effect, and it is the most effective tool of monetary policy to adjust the real estate market;

The tight monetary credit policy can only suppress the real estate market demand in the short term, and the long term effect is poor;   the impact of money supply to the real estate market is not significant [6]. Hu Ying et al. (2008) in the study of real estate price transmission effect of monetary policy   that exist the obstacles of real estate price transmission of China's monetary policy , and did not show the wealth effect and investment effect, interest rates cannot play a regulatory role in the real estate market [7]. Wang Min et al (2014) pointed out that in the tools of monetary policy impact on real estate prices, interest rates in the short and medium term impact on real estate is negative, but the effect of regulation in the medium term is significant; the impact of money supply on real estate prices in the short and medium term is positive, the long term is negative [8].

2   ANOVA Model

2.1   Data Selection and Variable Declaration

This paper selects from June 2010 to February 2015 in National City real estate data, as well as the corresponding one-year lending interest rate and currency in circulation (M2), studied the impact of monetary policy factors and interactions. Data from the People's Bank of China, China Index Academy Interval division on the level of interest rates is divided into four levels. For the amount of money in   circulation, the amount of money in circulation is divided into 5 levels, the city into three levels according to the division of economic and political influence, which first-tier cities, second tier city, three lines of the city, and select a representative of the city the partition table, as follows.

Table 1. Level of City

First-tier cities	Second-tier city	Third-tier city
Beijing	Ji'nan	Hohhot
Chengdu	Qingdao	Lanzhou
Guangzhou	Dalian	Urumqi
Hangzhou	Ningbo	Guiyang
Nanjing	Xiamen	Changzhou
Shanghai	Wuhan	Nantong
Shenzhen	Harbin	Weifang
Tianjin	Shenyang	Tangshan
	Xi'an	Haikou
	Changchun	Shaoxing
	Changsha	Daqing
	Fuzhou	Zhuhai
	Suzhou	Shantou
	Taiyuan	Weihai
	Hefei	Tai'an
	Nanchang	Jilin

Analysis of variance was used to study interaction strength between the level of interest rates and the amount of currency in circulation, the level of interest rates s_I has 4 levels, $r = 4$ .with the amount of money,it represents the level of circulation, with the interval division can be divided into 5 levels, $s = 5$, Under (s_I^i, s_{M2}^j) The test results in the horizontal $N(u_{ij}, \sigma^2)$ combination of independent distribution, and so

$$u = \frac{1}{rs} \sum_{i=1}^{r} \sum_{j=1}^{s} u_{ij} \tag{1}$$

$$u_{i.} = \frac{1}{s} \sum_{j=1}^{s} u_{ij} \ , i = 1, 2, \cdots, r \tag{2}$$

$$u_{.j} = \frac{1}{r} \sum_{i=1}^{r} u_{ij} \ \ j = 1, 2, \cdots, s \tag{3}$$

$$\alpha_i = u_{i.} - u \ , i = 1, 2, \cdots, r \tag{4}$$

$$\beta_j = u_{.j} - u \ \ j = 1, 2, \cdots, s \tag{5}$$

$$\gamma_{ij} = u_{ij} - u - \alpha_i - \beta_j \tag{6}$$

Effect of α_i for the interest rate factor first i level, β_j is the amount of money in circulation factor j, γ_{ij} for the interaction between monetary factor interest rate factor and i level in the circulation, they satisfy the relation:

$$\sum_{i=1}^{r} \alpha_i = 0 \ , \ \sum_{j=1}^{s} \beta_j = 0 \tag{7}$$

$$\sum_{i=1}^{r} \gamma_{ij} = 0 \ , j = 1, 2, \cdots, s \tag{8}$$

$$\sum_{j=1}^{s} \gamma_{ij} = 0 \ , i = 1, 2, \cdots, r \tag{9}$$

Do the t_{ij} test at each level of combination, the results for the y_{ijk} test model.

$$\begin{cases} y_{ijk} = u + \alpha_i + \beta_j + \gamma_{ij} + \varepsilon_{ijk} \\ \sum_{i=1}^{r} \alpha_i = 0 \ , \sum_{j=1}^{s} \beta_j = 0 \ , \sum_{i=1}^{r} \gamma_{ij} = 0 \ , \sum_{j=1}^{s} \gamma_{ij} = 0 \\ \varepsilon_{ijk} \text{Dependent for each other , obey } N(0, \sigma^2) \\ i = 1, 2, \cdots, r, \ j = 1, 2, \cdots, s, k = 1, 2, \cdots, t_{ij} \end{cases} \tag{10}$$

The model test of the hypothesis with three:

$$H_{01} : \alpha_1 = \alpha_2 = \cdots = \alpha_r = 0$$
$$H_{02} : \beta_1 = \beta_2 = \cdots = \beta_s = 0 \tag{11}$$
$$H_{03} : \gamma_{ij} = 0, \forall i, j$$

2.2 Establishment and Test of Variance Model

Because the data is downloaded from the site of a excel connection, using the Excel data of the ODC technology, the use of SQL language and the PivotTable to integrate data and analysis, due to the large amount of data the following table gives only part of the data.

Table 2. First-tier Cities

First-tier cities	2010.06	2010.07	2010.08	2010.09	2010.10
Shenzhen	23738	23465	23658	24366	24550
Shanghai	24185	23883	23037	23113	23240
Beijing	21911	21867	21734	22161	22317
Hangzhou	20070	19950	19864	20441	20507
Guangzhou	14181	13926	13993	14054	14378
Nanjing	12041	11892	11887	12017	12019
Tianjin	12101	11822	11719	11765	11650
Chengdu	7479	7524	7521	7528	7600

Data import into SPSS, there are first-tier cities as follows :

		Label	N
I	1	Level1	15
	2	Level2	29
	3	Level3	135
	4	Level4	77
m2	1	Level1	56
	2	Level2	48
	3	Level3	29
	4	Level4	67
	5	Level5	56

Fig. 1. Between-Subjects Factors

Source	Type III Sum Of Squares	df	Mean Squares	F	Sig.
Corrected Model	6.041E8	9	6.713E7	1.237	.273
Intercept	5.529E10	1	5.529E10	1018.929	.000
I	5.305E7	3	1.768E7	.326	.807
m2	3.167E7	4	7917389.564	.146	.965
I* m2	2.262E8	2	1.131E8	2.084	.127
Error	1.335E10	246	5.426E7		
Total	1.160E11	256			
Corrected Total	1.395E10	255			

Fig. 2. Tests Of Between-Subjects Effects

As can be seen from the chart, the significant level of factor interest rate is only 0.807,   significant level of factor M2 is only 0.965, One overall adjustment of China's monetary policy alone to the role of real estate is not a particularly effective, this view coincides with Hu Ying and others, but notice the interactions between factors is huge, the significance level is 0.127. The second line, three line cities also obtained similar results. Therefore, can be predicted in the circulation of money supply and interest rates in regulating the real estate price when there is a optimal combination relationship, in other words, the currency regulation exists in the two elements of the optimal reaction function.

3   VAR Model

With interaction between due to factors, here the VAR model must be defined four variables, time t variable and property prices y_t, time t interest rates I_t, time t circulation of money M_t, moment t interaction level $I_t * M_t$. Y_t is time t fac-

tor vector for $\begin{pmatrix} y_t \\ I_t \\ M_t \\ I_t * M_t \end{pmatrix}$,order p VAR model is:

$$Y_t = \phi_1 Y_{t-1} + \phi_2 Y_{t-2} + \cdots + \phi_p Y_{t-p} + \beta + \varepsilon_t \tag{12}$$

the data import Eviews, to the y_t, M_t, $I_t * M_t$ difference after taking logarithm, ADF inspection, get the following results:

Automatic lag length selection based on SIC: 0 to 7
Newey-West automatic bandwidth selection and Bartlett kernel

Method	Statistic	Prob.**	Cross-sections	Obs
Null: Unit root (assumes common unit root process)				
Levin, Lin & Chu t*	-2.65464	0.0040	4	212
Breitung t-stat	0.33028	0.6294	4	208
Null: Unit root (assumes individual unit root process)				
Im, Pesaran and Shin W-stat	-8.56750	0.0000	4	212
ADF - Fisher Chi-square	79.7367	0.0000	4	212
PP - Fisher Chi-square	121.708	0.0000	4	221

** Probabilities for Fisher tests are computed using an asymptotic Chi
   -square distribution. All other tests assume asymptotic normality.

Fig. 3. ADF Inspection

From test results to see the data after the first order difference is smooth, and Johansen method was applied to test the co integration relationship between the indicators, choose lag order number 3, the following figure 4:

Unrestricted Cointegration Rank Test (Trace)

Hypothesized No. of CE(s)	Eigenvalue	Trace Statistic	0.01 Critical Value	Prob.**
None *	0.547871	107.9977	54.68150	0.0000
At most 1 *	0.497044	65.13319	35.45817	0.0000
At most 2 *	0.337808	28.02149	19.93711	0.0004
At most 3	0.101220	5.762739	6.634897	0.0164

Trace test indicates 3 cointegrating eqn(s) at the 0.01 level
* denotes rejection of the hypothesis at the 0.01 level
**MacKinnon-Haug-Michelis (1999) p-values

Fig. 4. Johansen Method

Create VAR model, we through the pulse function to describe the effect of the level of interest rates changes, changes to the amount of money in circulation, and mutual interaction factors to the real estate price, and obtained the following results.

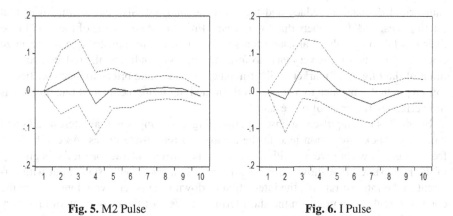

Fig. 5. M2 Pulse   **Fig. 6.** I Pulse

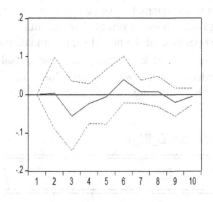

Fig. 7. Interaction Pulse

Figure 5 is the impulse response function diagram of currency changes lead to changes in real estate prices. From figure 5 we can see that, after give a positive impact to the current the amount of money, the real estate price show a short-term growth trend, and then gradually decreased and finally stable after 6 periods. So we can believe that the increase in money supply will lead to real estate prices show short rise, long-term for the real estate prices have little effect. Because, when the money supply increases, will encourage developers to increase the investment in real estate, resulting in an increase in supply, on the other hand, to consumers or investors, money growth will also cause the increase in demand, in the short term the demand increases significantly higher than the supply rate, the increase in demand and stimulate the increase in supply caused prices to fall at last.

Figure 6 is the impulse response function diagram of real interest rate changes lead to the impact of real estate price changes. As can be seen from Figure 6, after give a positive impact to the current real interest rate , through various paths to react to the real estate market gradually , the price of real estate will first appeared a downward trend, but then the trend up, and finally tends to be stable. A rise in interest rates will cause rigid demand to reduce and the short demand will also cut down, lower real estate prices, and developers due to increased financing costs, some of them will take their capital to exit the real estate market. The real estate industry due to the huge demand of funds, so it is very easy to form an oligopoly industry, the reduced competition of the oligopoly market will stimulate the real estate prices to increase, finally form a gradually balance to a stable level. In the long term, the interest rate policy has little effect on the price of real estate.

Figure 7 is the impulse response function diagram changes in real interest rates and monetary shock interaction lead to the changes in real estate prices. As can be seen from Figure 7, when give a positive impact to the current interaction, at the beginning of the real estate price show a downward trend, then rise, finally stable, but with different to the interest rate its, the interaction of down time is relatively long, so in the control of real estate prices in the short term, the effect of the interaction of monetary policy is obvious. But in the long run is not obvious.

From the above analysis, monetary policy in the regulation of the real estate price short-term obviously, and the interaction of currency and interest rate has the greatest impact, but in the long run, real estate prices changed little, the price has always been in accordance with the operation of the law of supply and demand.

At the $\alpha = 10\%$ significant level, using Granger causality test shows that the real estate prices are indeed by the level of interest rates, money supply, and the interaction to determine. The following is the price of real estate inspection table of Granger causality test:

Dependent variable: D_IN_Y

Excluded	Chi-sq	df	Prob.
D_IN_M	3.760827	3	0.2885
I	1.948803	3	0.5831
D_IN_I_M	3.432509	3	0.3296
All	10.15821	9	0.3378

Fig. 8. Granger Causality test

4 Conclusion

This paper use the 2010 June to 2015 February data, select the real estate prices, interest rates, money supply M2 and the interaction between them as variables, constructed the variance analysis model and VAR model, discussed the monetary policy's various factor and the interaction of each factor to the real estate price's effect. Monetary policy Monetary policy in the process of adjusting the real estate prices, all the elements are not isolated, the level of interest rates and the amount of money in circulation has optimal regulation of combination, which Requires government departments in the formulation of policy, should not take piecemeal policy, should also adjust the interest rate and money supply at the same time, in order to make the optimal expected objectives of regulation. Monetary policy in the regulation of the real estate price short-term obviously, and the interaction of currency and interest rate has the greatest impact, but in the long run, real estate prices changed little, the price has always been in accordance with the operation of the law of supply and demand.

Acknowledgement. This research was partially supported by the Department of Fujian Provincial Education,under grant JK2012063, by the the School of Mathematics and Computer Science of Fujian Normal University,by the the Institute of Innovative Information Industry of Fujian Normal University,by the The School of Ecomomic of Fujian Normal University.

References

1. Bemake, B., Gertler, M.: Monetary policy and asset price volatility. In: New Challenges for Monetary Policy: A Symposium Sponsored by the Federal Reserve Bank of Kansas city (1999)
2. Li-ping, W.: Analysis of wealth effect on the real estate prices. Seek knowledge, January 2006
3. Charles, P.: House Price and Monetary Policy. European Economics and Financial Centre **12**, 95–99 (2007)
4. Kenny, G.: Modelling the demand and supply sides of the housing market: evidence from Ireland. Economic Modelling **16**(3), 389–409 (1999)
5. Bong, H.K., Min, H.G.: Household lending, interest rates and housing price bubbles in Korea: Regime switching model and Kalman filter approach. Economic Modelling **28**(3), 1415–1423 (2011)
6. Wei, W.: Dynamic Identification of Monetary Policy Shock on Real Estate Market. Contemporary Finance & Economics **8**, 55–60 (2008)
7. Ying, H., Yao-ming, P., Wei-zhou, Z.: An empirical study on the real estate price transmission effect of monetary policy. Shanghai Financial, November 2008
8. Ming, W., Peng, S., Jin, Y.: Dynamic Study on Effects of Monetary Policy Tools on Housing Prices. Journal of Huazhong Agricultural University, February 2014

A Swarm-Based Approach to Mine High-Utility Itemsets

Jerry Chun-Wei Lin[1]([✉]), Lu Yang[1], Philippe Fournier-Viger[2], Ming-Thai Wu[3],
Tzung-Pei Hong[3,5], and Leon Shyue-Liang Wang[4]

[1] School of Computer Science and Technology, Harbin Institute of Technology
Shenzhen Graduate School, Shenzhen, China
jerrylin@ieee.org, luyang@ikelab.net
[2] Department of Computer Science, University of Moncton, Moncton, Canada
philippe.fournier-viger@umoncton.ca
[3] Department of Computer Science and Information Engineering,
National University of Kaohsiung, Kaohsiung, Taiwan, R.O.C.
wmt@wmt35.idv.tw, tphong@nuk.edu.tw
[4] Department of Information Management, National University of Kaohsiung,
Kaohsiung, Taiwan, R.O.C.
slwang@nuk.edu.tw
[5] Department of Computer Science and Engineering, National Sun Yat-sen
University, Kaohsiung, Taiwan, R.O.C.

Abstract. High-utility itemset mining (HUIM) is a critical issue in recent years since it can reveal the profitable products by considering both the quantity and profit factors instead of frequent itemset mining (FIM) or association-rule mining (ARM). In the past, a GA-based approach was designed to mine HUIs. It suffers, however, the combinational problem to assign the initial chromosomes for later evolution process. Besides, it is a non-trivial task to find the appropriate parameters for GA-based mechanism. In this paper, a binary PSO-based algorithm is thus proposed to efficiently find HUIs. A sigmoid function is adopted in the designed algorithm in the evolution process for discovering HUIs. Substantial experiments on real-life datasets show that the proposed algorithm has better results compared to the state-of-the-art GA-based algorithm of HUIM in terms of execution time and number of discovered HUIs.

Keywords: PSO · Evolutionary computation · Bio-inspired · High-utility itemset mining · Discrete PSO

1 Introduction

Knowledge discovery in database (KDD) is an emerging issue to find potential or implicit information from a very large database. Most of them, frequent itemset mining (FIM) or association-rule mining (ARM) has been extensively developed to mine the set of frequent itemsets in which their occurrence frequencies are no less than minimum support threshold or their confidences are

© Springer-Verlag Berlin Heidelberg 2015
L. Wang et al. (Eds): MISNC 2015, CCIS 540, pp. 572–581, 2015.
DOI: 10.1007/978-3-662-48319-0_48

no less than minimum confidence threshold [1,3]. Since only the occurrence frequency is concerned whether in FIM or ARM, it is insufficient to identify the high profit item/set especially when the item/sets is rarely appeared but has high profit values. To solve the limitation of FIM or ARM, high-utility itemset mining (HUIM) [20,21] was designed to discover the "useful" and "profitable" item/sets from the quantitative databases. Yao et al. concerned the quantity of items as the internal utility and the unit profit of items as the external utility to discover the HUIs [20]. Liu et al. designed the two-phase (TWU) model and developed the transaction-weighted downward closure (TWDC) property for mining HUIs [13]. Lin et al. developed the condensed high-utility pattern (HUP)-tree and related algorithm for discovering HUIs [11]. Tseng et al. then designed the mining algorithms of UP-growth [18] and UP-growth+ [19] to retrieve the HUIs based on the developed UP-tree structure. Other related works of HUIM are still developed in progress [5,12].

The traditional algorithms of HUIM have to handle the "exponential problem" of search space while the number of distinct items or the size of database is very large. Evolutionary computation is an efficient way and able to find the optimal solutions using the principles of natural evolution [2]. The genetic algorithm (GA) [6] is an optimization approach to solve the NP-hard and non-linear problems and used to investigate a very large search spaces to find the optimal solutions based on the designed fitness functions. In the past, Kannimuthu and Premalatha developed the high utility pattern extracting using genetic algorithm with ranked mutation using minimum utility threshold (HUPE$_{umu}$-GRAM) to mine HUIs [7]. In this paper, a binary PSO-based (BPSO) [9] algorithm is adopted and modified for mining the HUIs. Extensive experiments showed that the proposed approach can efficiently identify the set of HUIs from a very condense database with higher minimum utility threshold and has better results compared to the state-of-the-art GA-based algorithm [9] for mining HUIs.

2 Related Work

2.1 Particle Swarm Optimization

In the past, many heuristic algorithms have been facilitated to solve the optimization problems for discovering the necessary information in the evolutionary computation [2]. The simple genetic algorithm (SGA) [6] is a fundamental search technique to find the feasible and optimal solutions in a limit amount of time. Each chromosome in GA is composed by the set of the fixed-length genes as a solution. Many variants of GAs have been extensively studied and applied to a wide range of optimization problems [7,17].

Kennedy and Eberhart first introduced particle swarm optimization (PSO) [8] in 1995, which was inspired by the flocking behavior of birds to solve the optimization problems. Each particle is represented as an optimized solution and updated by the velocities in the evolution process. In the updating process of PSO, the corresponding particle and velocity are described as follows:

$$v^{id}(t+1) = w_1 \times v^{id}(t) + c_1 \times rand() \times (pbest - x^{id}) + c_2 \times rand() \times (gbest - x^{id}). \quad (1)$$

$$x^{id}(t+1) = x^{id}(t) + v^{id}(t). \tag{2}$$

In the equations (1) and (2), w_1 plays a balancing role between global search and local search; v^{id} is represented the id-th particle velocity; x^{id} is represented the id-th particle; $rand()$ is the random number in range of $(0, 1)$; c_1 is the individual factor and c_2 is the social factor, which are usually set as 2. The PSO was originally defined to solve the continues valued spaces. In the past, PSO has been adopted to various real-world applications [10,15]. In real-world situations, many problems are set as the discrete variable spaces such as scheduling and routing problems. Kennedy and Eberhart then designed a discrete (binary) PSO (BPSO) [9] to solve the limitation of continuous PSO. Sarath and Ravi developed a BPSO optimization approach to discover ARs [16]. Other applications adopting PSO to mine the required information are still processed in progress [14]

2.2 High-Utility Itemset Mining

High-utility itemset mining (HUIM) [20,21] is a critical issue and an emerging topic in recent decades, which can be concerned as the extension of frequent itemset mining (FIM) but more factors such as quantity and profit are considered in it. Yao et al. then designed an approach to discover the profitable itemsets by considering both the purchase quantity (also considered as internal utility) and profit (also considered as external utility) of items to reveal HUIs [20,21]. Liu et al. then developed a two-phase (TWU) model and designed the transaction-weighted downward closure (TWDC) property to early prune the unpromising HUIs but still can discover the complete HUIs [13]. Lin et al. designed a high-utility-pattern (HUP)-tree for discovering HUIs [11]. It first finds the high-transaction-weighted utilization 1-itemsets (1-HTWUIs) based on TWU model. The kept 1-HTWUIs are then used to build the HUP-treefor later mining approach. Tseng et al. presented the UP-tree stucture and two mining algorithms of UP-growth [18] and UP-growth+ [19] for discovering HUIs. A novel list-based algorithm called HUI-Miner was also developed to mine HUIs without candidate generation, which requires less memory usage but can efficiently retrieve the HUIs [12]. Fournier-Viger et al. then presented an improved algorithm namely FHM to quickly mine HUIs based on the built Estimated Utility Co-occurrence Structure (EUCS) [5].

Instead of traditional HUIM, Kannimuthua and Premalatha first designed the GA-based algorithm to mine HUIs with the ranked mutation [7]. In their approach, it is a non-trivial task to first find the HUIs as the initial chromosomes for later evolution process. Besides, some specific parameters are required in the evolution process of GAs. In this paper, we first present a BPSO-based algorithm for efficiently mining the HUIs using the sigmoid function.

3   Preliminaries and Problem Statement

3.1   Preliminaries

Let $I = \{i_1, i_2, \ldots, i_m\}$ be a finite set of m distinct items. A quantitative database is a set of transactions $D = \{T_1, T_2, \ldots, T_n\}$, where each transaction

$T_q \subseteq D$ $(1 \le q \le m)$ is a subset of I and has a unique identifier q, called its *TID*. Besides, each item i_j in a transaction T_q has a purchase quantity (internal utility) denoted as $q(i_j, T_q)$. A profit table $ptable = \{pr_1, pr_2, \ldots, pr_m\}$ indicates the profit value pr_j of each item i_j. A set of k distinct items $X = \{i_1, i_2, \ldots, i_k\}$ such that $X \subseteq I$ is said to be a k-itemset, where k is the length of the itemset. An itemset X is said to be contained in a transaction T_q if $X \in T_q$. A minimum utility threshold is set as δ according to users' preferences.

An illustrated example is stated in Table 1 as the running example in this paper. In Table 1, it has 10 transactions and 6 distinct items, denoted from (a) to (f). The profit value (external utility) of each item is shown in Table 2 as the profit table. The minimum utility threshold is set as $(\delta = 30\%)$.

Table 1. A quantitative database.

TID	Transaction (item, quantity)
T_1	b:6, d:1, e:1, f:1
T_2	a:1, c:18, e:1,
T_3	a:2, c:1, e:1
T_4	c:4, e:2
T_5	d:1, e:1
T_6	a:3, c:25, d:3, e:1
T_7	b:1, f:1
T_8	b:10, d:1, e:1
T_9	b:6, c:2, e:2, f:4
T_{10}	a:1, b:1, f:3

Table 2. A profit table.

Item	Profit
a	3
b	9
c	1
d	5
e	6
f	1

Definition 1. The utility of an item i_j in a transaction T_q is denoted as $u(i_j, T_q)$, and is defined as:

$$u(i_j, T_q) = q(i_j, T_q) \times pr(i_j). \tag{3}$$

For example, the utility of items (b), (d), (e), and (f) in transaction T_1 are respectively calculated as $u(b, T_1) = 6 \times 9 \ (= 54)$; $u(d, T_1) = 1 \times 5 \ (= 5)$; $u(e, T_1) = 1 \times 6 \ (= 6)$; and $u(f, T_1) = 1 \times 1 \ (= 1)$.

Definition 2. The utility of an itemset X in transaction T_q is denoted as $u(X, T_q)$, and defined as:

$$u(X, T_q) = \sum_{i_j \subseteq X \wedge X \in T_q} u(i_j, T_q). \tag{4}$$

For example, the utilities of itemsets (bd) and $(bdef)$ in T_1 are respectively calculated as $u(bd, T_1) = u(b, T_1) + u(d, T_1) = 54 + 5 \ (= 59)$ and $u(bdef, T_1) = u(b, T_1) + u(d, T_1) + u(e, T_1) + u(f, T_1) = 54 + 5 + 6 + 1 \ (= 66)$.

Definition 3. The utility of an itemset X in a database D is denoted as $u(X)$, and defined as:

$$u(X) = \sum_{X \subseteq T_q \wedge T_q \in D} u(X, T_q). \tag{5}$$

For example, the utilities of itemsets (a) and (ac) in D are respectively calculated as $u(a) = u(a, T2) + u(a, T_3) + u(a, T_7) + u(a, T_{10}) = 3 + 6 + 9 + 3 \ (= 21)$ and $u(ac) = u(ac, T_2) + u(ac, T_3) + u(ac, T_7) = 21 + 7 + 34 \ (= 62)$

Definition 4. The transaction utility of a transaction T_q is denoted as $tu(T_q)$, and defined as:

$$tu(T_q) = \sum_{X \subseteq T_q} u(X, T_q). \tag{6}$$

For example, $tu(T_1) = u(b, T_1) + u(d, T_1) + u(e, T_1) + u(f, T_1) = 54 + 5 + 6 + 1 \ (= 66)$. The resting transactions from T_2 to T_{10} are respectively calculated as $tu(T_2) \ (= 27)$, $tu(T_3) \ (= 13)$, $tu(T_4) \ (= 16)$, $tu(T_5) \ (= 11)$, $tu(T_6) \ (= 55)$, $tu(T_7) \ (= 10)$, $tu(T_8) \ (= 101)$, $tu(T_9) \ (= 72)$, and $tu(T_{10}) \ (= 15)$.

Definition 5. The total utility of a database D is denoted as TU, and defined as:

$$TU = \sum_{T_q \in D} tu(T_q). \tag{7}$$

For example, the total utility in a database D is calculated as $TU = 66 + 27 + 13 + 16 + 11 + 55 + 10 + 101 + 72 + 15 \ (= 386)$.

Definition 6. An itemset X in a database D is a high-utility itemset (HUI) if and only if its utility is no less than the minimum utility value as:

$$HUI \leftarrow \{X | u(X) \geq TU \times \delta\}. \tag{8}$$

For example, the utility of itemsets (b) and (bc) are respectively calculated as $u(b) \ (= 216)$ and $u(bc) \ (= 56)$. Thus, the itemset (b) is a HUI since $u(b) = 216 > 386 \times 0.3 \ (= 115.8)$. The itemset (bc) is not a HUI since $u(bc) \ (= 56 < 115.8)$.

3.2   Problem Statement

Based on the above definitions, the problem of HUIM is to find the set of high-utility item/sets (HUIs), in which the utility of an item/set X is no less than $(TU \times \delta)$.

4   Proposed BPSO-Based Algorithm of HUIM

For the designed binary PSO (BPSO)-based model [9] for mining HUIs, the high-transaction-weighted utilization 1-itemsets (1-HTWUIs) [13] are first discovered based on TWU model [13]. The particle is thus encoded as the set of binary variables corresponding to the sorted order of 1-HTWUIs. The particles are then evaluated to find the satisfied HUIs. After that, the particles are correspondingly updated by velocities, *pbest*, *gbest*, and the sigmoid function. This iteration is repeated until the termination criteria is achieved. After that, the set of HUIs is discovered. The proposed BSPO-based algorithm is shown as follows.

Algorithm 1. Proposed algorithm

Input: D, a quantitative database; *ptable*, a profit table, δ, the minimum utility
threshold; M, the number of particles of each iteration.

Output: *HUIs*, a set of high-utility itemsets.

1   **for** *each* $T_q \in D$ **do**
2   **for** *each* $i_j \subseteq T_q$ **do**
3         $tu(T_q) = q(i_j, T_q) \times ptable(i_j)$;
4   calculate $twu(i_j) = \sum_{i_j \subseteq T_q} tu(T_q)$;
5   $TU = \sum_{T_q \in D} tu(T_q)$;
6   find 1-HTWUIs [13];
7   set $m = |\text{1-HTWUIs}|$;
8   set $PV \leftarrow normalize(\text{1-HTWUIs})$;
9   initialize $p(t) = $ either 1 or 0;
10 initialize $v(t) = rand()$ in the range of $(0, 1)$;
11 initialize $pbest(t)$ and $gbest(t)$;
12 **while** *termination criteria is not reached* **do**
13   update the $v(t + 1)$ and the $p(t + 1)$;
14   **for** $i \leftarrow 1, M$ *particles* **do**
15   **if** $fitness(p_i(t + 1)) \geq TU \times \delta$ **then**
16           $HUIs \leftarrow GetItem(p_i(t + 1)) \cup HUIs$;
17   find $pbest(t + 1)$ and $gbest(t + 1)$;
18   set $t \leftarrow t + 1$;
19 return HUIs;

---

5   Experimental Results

Substantial experiments were conducted to verify the effectiveness and efficiency of the proposed algorithm compared to the state-of-the-art evolutionary $\text{HUPE}_{\text{umu}}\text{U-GRAM}$ algorithm [7]. The original $\text{HUPE}_{\text{umu}}\text{-GRAM}$ algorithm requires the top-k HUIs as the initial particles for later evolution process, which needs very large computations to find the initial particles. We have thus improved this approach by adopting our designed pre-processing phase to randomly generate the initial particles. The algorithms in the experiments were implemented in C++ language, performing on a PC with an Intel Core2 i3-4160 CPU and 4GB of RAM, running the 64-bit Microsoft Windows 7 operating system. Two real-world datasets called chess [4] and mushroom [4] are used in the experiments. A simulation model [13] was developed to generate the quantities and profit values of items in transactions for all datasets. A log-normal distribution was used to randomly assign quantities in the [1,5] interval, and item profit values in the [1,1000] interval. In the conducted experiments for mining HUIs, the algorithms are all performed for 10,000 iterations and the population size is set as 20. Parameters and characteristics of the datasets used in the experiments are respectively shown in Tables 3 and 4.

Table 3. Parameters of used datasets.

#|**D**|	Total number of transactions
#|**I**|	Number of distinct items
AvgLen	Average transaction length
MaxLen	Maximal length transactions

Table 4. Characteristics of used datasets.

Dataset	#|**D**|	#|**I**|	**AvgLen**	**MaxLen**
chess	3196	76	37	37
mushroom	8124	120	23	23

5.1 Runtime

In the conducted experiments of runtime in two datasets, the state-of-the-art evolutionary $HUPE_{umu}$-GRAM algorithm of HUIM is compared to the designed algorithm. The results are shown in Fig. 1.

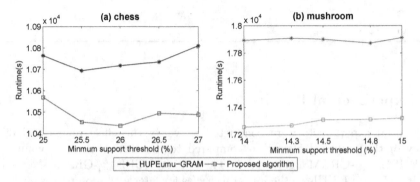

Fig. 1. Runtime w.r.t variants of minimum utility thresholds.

From Figure 1, it can be seen that the proposed PSO-based algorithm has better performance in terms of execution time compared to the improved HUPE$_{umu}$-GRAM algorithm w.r.t. different minimum thresholds in two datasets. For example, the chess dataset shown in Fig. 1(a) with the minimum support threshold 25%, the runtime of improved HUPE$_{umu}$-GRAM algorithm and the proposed algorithms were respectively calculated as 10,764 and 10,568 at 10,000 iteration. The proposed algorithm has always better results than that of the GA-based algorithm.

5.2   Number of HUIs

In this section, the number of HUIs is evaluated to show the performance of the compared algorithms. The state-of-the-art FHM [5] algorithm is used to discover the actual and complete HUIs from the quantitative databases. Experiments are conducted and shown in Figure 2.

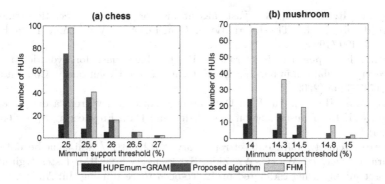

Fig. 2. Number of HUIs w.r.t variants of minimum utility thresholds.

From Figure 2, it can be seen that the proposed PSO-based algorithm can generate nearly the same number of HUIs compared to the state-of-the-art FHM algorithm especially when the minimum utility threshold is set higher in the condense datasets, which were respectively shown in Figure 2(a) and Figure 2(b). The reason is that the size of a particle is associated with the number of 1-HTWUIs, less computations are required when the minimum utility threshold is set higher. For the condense chess and mushroom datasets, the number of 1-HTWUIs is close to the number of discovered HUIs under higher minimum utility threshold; the number of generated HUIs of the designed PSO-based algorithm is close to the traditional way for mining the complete HUIs.

6   Conclusions

High utility itemset mining (HUIM) has emerging as an important topic in recent years since it can reveal the highly profitable products instead of the frequent itemset mining (FIM). Several algorithms have been proposed to efficiently mine the high-utility itemsets (HUIs) from the quantitative databases and most of them applied the statistical analysis to mine HUIs. In the past, a GA-based approach namely $HUPE_{umu}$-GRAM was proposed to mine HUIs based on genetic algorithm. In this paper, a particle swarm optimization (PSO)-based algorithm is proposed to efficiently mine HUIs. The proposed algorithm adopts the TWU model to find the number of high-transaction-weighted utilization 1-itemsets (1-HTWUIs) as the particle size, which can greatly reduce the combinational problem in the evolution process. The sigmoid function of the discrete (binary) PSO (BPSO) is also adopted in the developed algorithm for discovering the HUIs. From the conducted experiments, the proposed PSO-based approach has better results than the GA-based algorithm in terms of execution time and the ability for discovering HUIs.

References

1. Agrawal, R., Srikant, R.: Fast algorithms for mining association rules in large databases. In: The International Conference on Very Large Data Bases, pp. 487–499 (1994)
2. Cattral, R., Oppacher, F., Graham, K.J.L.: Techniques for evolutionary rule discovery in data mining. In: IEEE Congress on Evolutionary Computation, pp. 1737–1744 (2009)
3. Chen, M.S., Han, J., Yu, P.S.: Data mining: An overview from a database perspective. IEEE Transactions on Knowledge and Data Engineering 8(6), 866–883 (1996)
4. Frequent itemset mining dataset repository (2012). http://fimi.ua.ac.be/data/
5. Fournier-Viger, P., Wu, C.-W., Zida, S., Tseng, V.S.: FHM: faster high-utility itemset mining using estimated utility co-occurrence pruning. In: Andreasen, T., Christiansen, H., Cubero, J.-C., Raś, Z.W. (eds.) ISMIS 2014. LNCS, vol. 8502, pp. 83–92. Springer, Heidelberg (2014)
6. Holland, J.: Adaptation in Natural and Artificial Systems. MIT Press, Cambridge (1975)
7. Kannimuthu, S., Premalatha, K.: Discovery of high utility itemsets using genetic algorithm with ranked mutation. Applied Artificial Intelligence 28(4), 337–359 (2014)
8. Kennedy, J., Eberhart, R.: Particle swarm optimization. IEEE International Conference on Neural Networks 4, 1942–1948 (1995)
9. Kennedy, J., Eberhart, R.: A discrete binary version of particle swarm algorithm. In: IEEE International Conference on Systems, Man, and Cybernetics, pp. 4104–4108 (1997)
10. Kuo, R.J., Chao, C.M., Chiu, Y.T.: Application of particle swarm optimization to association rule mining. Applied Soft Computing 11(1), 326–336 (2011)
11. Lin, C.W., Hong, T.P., Lu, W.H.: An effective tree structre for mining high utility itemsets. Expert Systems with Applications 38(6), 7419–7424 (2011)
12. Liu, M., Qu, J.: Mining high utility itemsets without candidate generation. In: ACM International Conference on Information and Knowledge Management, pp. 55–64 (2012)

13. Liu, Y., Liao, W., Choudhary, A.K.: A two-phase algorithm for fast discovery of high utility itemsets. In: Ho, T.-B., Cheung, D., Liu, H. (eds.) PAKDD 2005. LNCS (LNAI), vol. 3518, pp. 689–695. Springer, Heidelberg (2005)
14. Nouaouria, N., Boukadouma, M., Proulx, R.: Particle swarm classification: A survey and positioning. Pattern Recognition **46**(7), 2028–2044 (2013)
15. Pears, R., Koh, Y.S.: Weighted association rule mining using particle swarm optimization. In: Cao, L., Huang, J.Z., Bailey, J., Koh, Y.S., Luo, J. (eds.) PAKDD Workshops 2011. LNCS, vol. 7104, pp. 327–338. Springer, Heidelberg (2012)
16. Sarath, K.N.V.D., Ravi, V.: Association rule mining using binary particle swarm optimization. Engineering Applications of Artificial Intelligence **26**, 1832–1840 (2013)
17. Salleb-Aouissi, A., Vrain, C., Nortet, C.: QuantMiner: a genetic algorithm for mining quantitative association rules. In: International Joint Conference on Artifical Intelligence, pp. 1035–1040 (2007)
18. Tseng, V.S., Wu, C.W., Shie, B.E., Yu, P.S.: UP-growth: an efficient algorithm for high utility itemset mining. In: ACM SIGKDD International Conference on Knowledge Discovery and Data Mining, pp. 253–262 (2010)
19. Tseng, V.S., Shie, B.E., Wu, C.W., Yu, P.S.: Efficient algorithms for mining high utility itemsets from transactional databases. IEEE Transactions on Knowledge and Data Engineering **25**, 1772–1786 (2013)
20. Yao, H., Hamilton, H.J., Butz, C.J.: A foundational approach to mining itemset utilities from databases. In: SIAM International Conference on Data Mining, pp. 211–225 (2004)
21. Yao, H., Hamilton, H.J.: Mining itemset utilities from transaction databases. Data & Knowledge Engineering **59**(3), 603–626 (2006)

Author Index

Printed in the United States
By Bookmasters